Next-Generation Biodegradable and Bioactive Biomaterials

Safaa Najah Saud Al-Humairi
https://orcid.org/0000-0001-5978-7651
Management and Science University, Malaysia

Abdollah Bahador
https://orcid.org/0000-0002-4308-5388
Materials Processing Institute, UK

Fida Majdoub
University of Lyon, France

Nattapon Chantarapanich
Kasetsart University, Thailand

IGI Global
Scientific Publishing
Publishing Tomorrow's Research Today

Vice President of Editorial Melissa Wagner
Director of Acquisitions Mikaela Felty
Director of Book Development Jocelynn Hessler
Production Manager Mike Brehm
Cover Design Jose Rosado

Published in the United States of America by
IGI Global Scientific Publishing
701 East Chocolate Avenue
Hershey, PA, 17033, USA
Tel: 717-533-8845 | Fax: 717-533-7115
Website: https://www.igi-global.com E-mail: cust@igi-global.com

Library of Congress Cataloging-in-Publication Data

LCCN: 2026017024 (CIP Data Pending)
ISBN13: 9798260000601
Isbn13Softcover: 9798260000618
EISBN13: 9798260000625
British Cataloguing in Publication Data
A Cataloguing in Publication record for this book is available from the British Library.

This book is dedicated to patients, caregivers, and clinical communities whose unmet needs continue to motivate the development of safer, smarter, and more sustainable biomedical technologies and interventions.

It is also dedicated to the emerging generation of scientists, engineers, clinicians, and innovators who will advance biodegradable and bioactive biomaterials, transient bioelectronics, and circular bioeconomy solutions in pursuit of a healthier, more equitable, and environmentally responsible future.

Table of Contents

Detailed Table of Contents

Biodegradable and bioactive materials are vital in modern science due to their ability to perform specific functions and degrade safely. Widely used in biomedical fields such as tissue engineering, drug delivery, and implants, their design relies on understanding the relationship between structure, properties, and degradation. Structural features from molecular arrangement to microstructure govern mechanical strength, bioactivity, and degradation behavior across polymers, metals, and ceramics. Polymers degrade via hydrolysis or enzymes, metals through corrosion, and ceramics via dissolution. These mechanisms are influenced by factors like crystallinity, porosity, and composition. The interplay between structure and properties directly affects degradation rates, requiring a holistic design approach. Hybrid materials combine advantages of different classes, enabling improved performance and tunable degradation. Controlled degradation is essential for matching functional lifetimes.

Bioorganometallic polymers represent a new generation of novel biomaterials which combine metal centers and organic polymer backbones to realize novel physicochemical and biological performance. Such materials have become the subject of interest as next generation biodegradable and bioactive materials in biomedical applications by combining polymer versatility with the catalytic, electronic and bioactive properties of organometallic compounds. Addition of biologically relevant metals like iron, zinc, titanium and ruthenium improves antimicrobial, biocompatibility and controlled degradation. Tunable nature enables them to be controlled carefully with respect to mechanical strength, degradation rates and biological interactions. They can therefore be broadly utilized in drug delivery, tissue engineering scaffolds, biosensors, antimicrobial coating and cancer treatment. The developments in synthesis and bioorganometallic chemistry are still allowing designing of multifunctional biomaterials with better performance and safety.

The surging demand for sustainable biocompatible materials from renewable resources has drawn attention towards Tilapia skin waste for tissue repair and generation. Tilapia has abundant source of type I collagen, structurally similar to mammalian collagen with outstanding biocompatibility, low immunogenicity, and no cultural or health restrictions. In order to treat burns wounds, tilapia skin maintains its bioactive qualities for cell adhesion, proliferation, and controlled inflammatory responses. Modern fabrication techniques employ hydrogels, electrospun membranes, and porous scaffolds for skin regeneration. Co-functional bio active molecules improves mechanical stability, moisture retention and improves antimicrobial properties. With the advent of clinical translation, these affordable biomaterials made from tilapia exhibit huge waste utilization potential to be incorporated in circular economy approach for healthcare and environmental purposes.

This chapter advances sustainable transience as a unifying paradigm for next-generation personalised implants that heal, monitor, and disappear without ecological residue. Moving beyond durability-centred biomaterials, it integrates regenerative medicine, biodegradable bioelectronics, and circular green materials science to reconceptualise implants as temporally intelligent socio-technical systems. Drawing on global health, sustainability, and innovation data, it identifies conceptual paradoxes linking longevity, scalability, digital governance, and environmental accountability. The proposed Sustainable Transient Implant (STI) Framework aligns regeneration–degradation kinetics, life-cycle assessment, AI-enabled monitoring, and equity-oriented design. By bridging theory, practice, and regulation, the chapter reframes implantable medicine as clinically effective, digitally integrated, and ecologically responsible.

Transient biomaterials and implants represent a specialized frontier in regenerative medicine, designed to provide temporary structural support or physiological monitoring before degrading within the body. While these materials offer significant clinical benefits and enhance patient privacy by leaving no permanent traces, they present profound challenges for forensic identification and evidence interpretation. These temporal dynamics often result in the rapid deterioration of identification markers, such as manufacturing codes or bioactive coatings, potentially leaving the timeframe between implantation and death indeterminate. This chapter examines the conceptual foundations and various types of transient biomaterials, including biocompatible devices and degradable scaffolds. It further evaluates the limitations of current forensic sampling protocols and analytical techniques, such as infrared spectroscopy, which may fail to distinguish individual traces from surrounding biological residues.

Scorpion venom is known for its richness in bioactive molecules, particularly peptides, which confer a wide range of remarkable biological activities. Notably, it exhibits antimicrobial, antitumor, and immunomodulatory effects. However, its clinical translation remains limited due to its poor stability, dose-dependent toxicity, and potential immunogenicity. In this context, biodegradable biomaterials have emerged as promising platforms to overcome these limitations. They enable precise, targeted, and sustained delivery of scorpion venom peptides. This chapter highlights an innovative approach based on the integration of scorpion venom into biodegradable biopolymer matrices for applications in modern medicine. By leveraging the unique properties of biodegradable biomaterials, this work outlines a roadmap for the development of next-generation clinical applications.

Bone and osteochondral tissue engineering has emerged as a promising strategy to overcome limitations associated with conventional grafting and permanent orthopedic implants. Increasing incidences of trauma, degenerative diseases, and critical-sized bone defects demand advanced regenerative solutions capable of restoring both structural integrity and biological functionality. This chapter discusses recent advances in next-generation biodegradable and bioactive biomaterials focusing on load-bearing scaffolds, graft substitutes, fixation devices, and osteoinductive platforms. Emphasis is placed on biomimetic design principles, material selection, fabrication technologies, biological interactions, and clinical translation challenges. Emerging approaches including additive manufacturing, smart biomaterials, and multifunctional regenerative platforms are highlighted as transformative tools for personalized musculoskeletal regeneration and future orthopedic therapies.

This chapter investigates a new field of biomedical engineering called biocompatible bioelectronics, where bioelectronic devices are designed to serve their therapeutic or diagnostic function and then dissolve naturally in the body without causing any harm. These bioelectronics consist of biodegradable conductors, such as magnesium, zinc, molybdenum, and their alloys, as well as biodegradable substrates, including polylactic acid, polycaprolactone, and silk fibroin, which serve as substrates for silicon nanomembrane semiconductor chips and wireless communications modules that enable precise timing of electronic functionality prior to the complete biological resorption of the device. Because transient implants do not require secondary explanation surgery, they significantly decrease the risk of infection, chronic inflammation and patient morbidity associated with the retention of permanent devices. This chapter will provide an in-depth examination of the materials science, fabrication methods, and clinical pathways for the implementation of biocompatible bioelectronics.

TiO_2 NPs, characterized by their photocatalytic activity, non-toxicity, and high thermal stability, have attracted considerable attention in areas such as environmental remediation, healthcare, energy, and cosmetics. Green synthesis of TiO_2 NPs lessens the use of hazardous chemicals and energy-intensive processes. The process mainly involves harvesting bioactive compounds from the plant materials that help reduce Ti ions to NPs. They exhibit excellent photocatalytic ability to degrade organic pollutants and dyes under UV illumination along with broad-spectrum antibacterial and antifungal abilities. The biocompatibility of green synthesized TiO_2 NPs shows their tremendous opportunities for applications in drug delivery and photothermal therapies in biomedicine. Furthermore, their large surface area promotes the use of these particles in other sectors too. To conclude, the "green" synthesizing of TiO_2 NPs is very promising, sustainable, and innovative by extending their scope of applications while closely following environmental sustainability paradigms.

The growing challenges of food insecurity, climate change, biodiversity decline, and waste mismanagement require practical and regenerative solutions. This paper proposes an integrated bio-cluster model built around three insect systems: Black Soldier Fly (Hermetia illucens), silkworm (Bombyx mori), and bees (Apis species and other pollinators). Each contributes unique ecological and economic value. Black Soldier Fly supports organic waste conversion into protein and compost; silkworms generate high-value silk and biomaterials; bees enhance pollination and agricultural productivity. When combined within a localized bio-cluster, these systems create a circular framework that transforms waste into resources, improves soil fertility, and diversifies rural income. The study presents a holistic approach linking biology, ecology, and economics to demonstrate how insect-based clusters can reduce emissions, lower landfill use, support carbon balance, and strengthen climate-resilient rural development through integrated, multi-stream livelihoods.

In this chapter, the design, formation, and importance of core-shell nanoparticles are emphasized. At the same time, the importance of the production of core-shell nanoparticles from bio-composite polymer materials was also mentioned. Today, nanotechnological studies continue to develop considerably. Nanoparticle research from nanotechnological studies attracts considerable attention. Catalytic effects and performances of nanoparticles have been reported in the studies. However, the controlled release of nanoparticles, especially in drugs, is very important for human health. Direct nanoparticles reaching cells, tissues, or organs can threaten human health. Therefore, it is extremely important to cover the core with a harmless shell, and a harmless controlled drug release is extremely important. While forming core-shell nanoparticles, obtaining from bio nanocomposite polymers will further increase the advantage.

Climate change is an occurrence that has occurred globally and affects the natural world, having overarching effects on human health and well-being, like mental illness. These types of interactions are significant at a time when environmental adjustments are an ugly reality, as is mental illness. Here, it seeks to find studies that document the interaction between gut microbiome and climate change and its impact on mental health. A systematic review of the literature was sought in a number of databases (PubMed, SCOPUS and ProQuest), between 2000 and 2023. Out of 677 studies in the first instance, 10 were finally chosen carefully against pre-decided criteria of inclusion and exclusion. The significant results of this review discovered that there exists an interdependent relationship of such factors that we refer to as brain-gut-climate change vicious cycle, the modification in one compound affects the other, this might be triggered because of climate change. Several mechanisms of modulation of gut microbiota in mental health were also investigated.

Preface

The field of biodegradable and bioactive biomaterials is undergoing a profound transformation, driven by the convergence of materials science, nanotechnology, bioelectronics, regenerative medicine, and sustainability science. As healthcare systems and societies seek solutions that are not only therapeutically effective but also temporally controlled and environmentally responsible, the concept of "smart transience" in biomaterials has moved from a niche research topic to a central design paradigm. *Next-Generation Biodegradable and Bioactive Biomaterials* has been conceived to capture this paradigm shift and to provide an integrated perspective on materials that are engineered to perform complex biological functions and then disappear without leaving a harmful footprint.

Biodegradable and bioactive systems now play a pivotal role in diverse applications: from load-bearing scaffolds and osteochondral grafts to transient implants, biosensors, bioresorbable conductors, and multifunctional nanocomposites. Their capacity to modulate cellular behavior, guide tissue regeneration, deliver therapeutics with spatiotemporal precision, and interface with electronic platforms positions them at the forefront of future medical technologies. At the same time, emerging themes—such as circular economy approaches, green synthesis, venom-inspired bioactivity, AI-enabled rural bio-entrepreneurship, and the intersections between climate, microbiota, and mental health—are expanding the societal and environmental context in which these materials are conceived and deployed.

This volume brings together contributions from an international and multidisciplinary team of researchers and practitioners who collectively address the design, fabrication, characterization, and application of next-generation biodegradable and bioactive platforms. It aims to bridge fundamental structure–property–degradation relationships with translational considerations, including security and forensic challenges, regulatory and clinical implications, and the need for sustainable resource utilization. The overarching goal is to provide readers with a coherent and forward-looking picture of how smart biodegradable and bioactive systems can reshape both healthcare and broader socio-technical landscapes.

Designed for academics, researchers, clinicians, engineers, industry professionals, and graduate students, this book offers both foundational insights and advanced case studies. For researchers and students, it serves as a comprehensive reference on emerging materials concepts and application domains. For clinicians and device developers, it highlights how transient and bioactive systems can be harnessed to improve patient outcomes, reduce long-term complications, and inform new models of care. For those engaged in sustainability, policy, and innovation studies, the book illustrates how biomaterials research can align with circular economy principles, climate resilience, and rural economic empowerment.

ORGANIZATION OF THE BOOK

The book is organized into sixteen chapters that collectively trace the evolution of next-generation biodegradable and bioactive biomaterials from fundamental concepts to specialized applications and broader societal interfaces.

Chapter 1 provides a multimaterial perspective on the intricate structure–property–degradation relationships that underpin the performance of biodegradable and bioactive systems. It discusses how compositional design, microstructure, and processing routes influence mechanical behavior, bioactivity, degradation kinetics, and long-term functionality across polymers, metals, ceramics, and composites.

Chapter 2 focuses on nanotechnology in bioactive biodegradable systems, emphasizing nanoparticles, nanofibers, and nanotopographical cues for cell guidance and immune modulation. The chapter examines how nanoscale architecture can be tuned to regulate cellular adhesion, differentiation, and immune responses, thereby enabling more predictable and personalized regenerative outcomes.

Chapter 3 explores smart bioactive systems driven by nanotechnology, highlighting innovations that extend beyond traditional healthcare settings. Case studies illustrate how nanostructured materials can support advanced diagnostics, theranostics, smart drug delivery, and multifunctional platforms that bridge clinical, environmental, and industrial domains.

Chapter 4 addresses smart biodegradable biomaterials for precision therapeutics and regenerative systems. It reviews materials and device strategies that allow for controlled release, stimuli-responsiveness, and the integration of biological cues, with an emphasis on tailoring degradation profiles and bioactivity to specific therapeutic windows and tissue microenvironments.

Chapter 5 examines smart biodegradable implants and transient biosensors, charting their technological evolution, security challenges, and future directions. Beyond device performance, the chapter considers issues of data integrity, cybersecurity in transient electronics, and the implications of disappearing hardware for regulatory frameworks and clinical practice.

Chapter 6 introduces bioorganometallic polymers as a class of next-generation biodegradable and bioactive biomaterials. It discusses design principles for incorporating organometallic motifs into polymer backbones or side chains, their implications for degradability and bioactivity, and their potential in applications such as antimicrobial surfaces, catalytic therapeutic systems, and responsive implants.

Chapter 7 presents advanced biomaterials for tissue repair and regeneration through a circular economy lens, focusing on the utilization of tilapia skin as a bioresource. The chapter demonstrates how waste-derived biological materials can be valorized into high-value scaffolds and dressings, combining clinical effectiveness with sustainable resource management and local economic benefits.

Chapter 8 explores sustainable transience in personalized implants through regenerative bioelectronics and green materials. It discusses biodegradable electronic components, regenerative interfaces, and material strategies that support personalized therapies while minimizing electronic waste and environmental burden.

Chapter 9 investigates transient biomaterials and implants from a forensic science perspective, highlighting challenges in identification, evidence preservation, and reconstruction when critical components are designed to degrade or disappear. The chapter underscores the need for new forensic methodologies and documentation practices in an era of transient medical and bioelectronic devices.

Chapter 10 examines the emerging intersection of scorpion venom and biodegradable biomaterials, outlining a new class of bioactive systems inspired by venom-derived peptides and proteins. It discusses how these bioactives can be incorporated into degradable matrices for antimicrobial, anticancer, or regenerative applications, while addressing safety, selectivity, and translational hurdles.

Chapter 11 addresses bone and osteochondral tissue engineering, with a focus on load-bearing scaffolds, graft substitutes, fixation devices, and osteoinductive platforms. It analyses how structural design, mechanical performance, and bioactive cues can be integrated into biodegradable constructs that support complex musculoskeletal regeneration.

Chapter 12 centers on bioresorbable and transient bioelectronics, including degradable conductors, sensors, and wireless transient implants for monitoring and therapy. The chapter discusses device architectures, material choices, and system-level integration strategies that facilitate high-performance sensing and actuation during a defined operational lifetime, followed by controlled resorption.

Chapter 13 delves into the green synthesis of titanium dioxide nanoparticles using plant extracts. It highlights how phytochemical routes can reduce toxic precursors and energy inputs, and how such green-synthesized nanomaterials can be engineered for biomedical, environmental, and catalytic applications.

Chapter 14 presents the concept of Trio Insect Bio Clusters as an AI-enabled entrepreneurial blueprint for a circular, climate-resilient rural economy. It demonstrates how insect-based bioresources, advanced materials processing, and artificial intelligence can be integrated to create local value chains, enhance climate resilience, and support sustainable rural development.

Chapter 15 discusses core–shell nanostructure design in polymer bionanocomposites. It elaborates on interfacial engineering, hierarchical structuring, and property optimization strategies that enable enhanced mechanical performance, controlled degradation, and tailored bioactivity in multifunctional composite systems.

Chapter 16 concludes the thematic arc with a systematic review of the climate–microbiota–mental health "trilogy," providing a broader physiological and societal context for biodegradable and bioactive biomaterials. It explores how environmental change, microbiota dynamics, and mental health are interconnected, and how smart biomaterials and devices may contribute to diagnostics, interventions, and preventive strategies within this complex nexus.

Together, these chapters offer a multidimensional journey through the evolving landscape of next-generation biodegradable and bioactive biomaterials, connecting molecular design and device engineering with clinical practice, sustainability, security, forensics, and public health.

CONCLUSION

The chapters collected in *Next-Generation Biodegradable and Bioactive Biomaterials* collectively demonstrate how the field has progressed from the development of isolated biodegradable constructs to the conception of fully integrated, smart, and context-aware systems. Across a broad range of materials classes—polymers, metals, ceramics, composites, and bioorganometallic frameworks—the volume highlights how structure–property–degradation relationships can be deliberately engineered to deliver precise therapeutic functions, control biological interactions, and ensure safe, predictable resorption or disappearance at the end of service life.

A central theme running through the book is the enabling role of nanotechnology in achieving sophisticated bioactivity and transience. Nanoscale design of particles, fibers, and core–shell architectures supports targeted cell guidance, immune modulation, controlled drug delivery, and advanced sensing, while green synthesis routes and bio-derived feedstocks help align these innovations with environmental and circular economy principles. The chapters on transient bioelectronics and smart biodegradable implants further illustrate how electronic functionality, wireless communication, and regenerative interfaces can be embedded into materials that are ultimately resorbed, thereby reducing device burden, long-term complications, and electronic waste.

Another distinctive contribution of this volume lies in extending the scope of biodegradable and bioactive biomaterials beyond conventional clinical boundaries. The use of tilapia skin and other bioresources for tissue repair, venom-inspired bioactive systems, and AI-enabled insect bio clusters exemplify how biomaterials research can intersect with rural development, food and agricultural systems, and climate resilience. At the same time, the discussion of forensic implications, security challenges in transient devices, and the climate–microbiota–mental health trilogy underscores that next-generation biomaterials raise new questions for regulation, ethics, evidence interpretation, and public health.

Taken together, the contributions in this book suggest several clear directions for future work. These include deeper integration of data-driven design and artificial intelligence into materials discovery and performance prediction, more systematic coupling of circular economy metrics with biomaterials development, and closer collaboration between materials scientists, clinicians, forensic experts, social scientists, and policy-makers to anticipate and guide the broader impacts of transient technologies. As the demands on healthcare systems intensify and expectations for sustainability and safety increase, biodegradable and bioactive biomaterials—augmented by smart functionality and informed by systems-level thinking—are poised to become foundational elements of next-generation medical devices, therapies, and bio-integrated technologies.

Ultimately, the hope is that the conceptual frameworks, case studies, and emerging paradigms presented in these chapters will not only inform current practice but also inspire new research trajectories and translational pathways. By consolidating diverse perspectives on smart transience, bioactivity, and sustainability, *Next-Generation Biodegradable and Bioactive Biomaterials* aims to serve as both a reference point and a springboard for continued innovation at the interface of materials, biology, electronics, environment, and society.

Safaa Najah Saud Al-Humairi
Management and Science University, Malaysia

Abdollah Bahador
Materials Processing Institute, UK

Fida Majdoub
University of Lyon, France

Nattapon Chantarapanich
Kasetsart University, Thailand

Acknowledgment

The editors express their sincere gratitude to all authors who contributed chapters to this volume and shared their expertise in biodegradable and bioactive biomaterials, smart transience, and regenerative systems. Their collective efforts have enabled the creation of a multidisciplinary reference that spans materials science, biomedical engineering, nanotechnology, toxicology, bioelectronics, circular economy, and digital innovation.

Special appreciation is extended to the reviewers and pre-publication readers whose constructive comments helped refine the scientific content, structure, and clarity of the chapters. The editors gratefully acknowledge the various funding bodies, universities, and research centers that supported the underlying research reported in this volume, including national and international grant agencies, industrial partners, and philanthropic organizations. Where required, detailed funding acknowledgments are provided within individual chapters. The support of technical staff, laboratory personnel, and students who assisted in experiments, simulations, data analysis, and illustration preparation is also warmly recognized.

Finally, the editors extend heartfelt thanks to their families and close colleagues for their patience, encouragement, and unwavering support throughout the extended period of conceptualizing, coordinating, and completing this book. Their understanding made possible the sustained focus needed to bring together such a diverse and forward-looking collection of work.

The Editors

Introduction

The emergence of next-generation biodegradable and bioactive biomaterials marks a decisive shift in how materials are designed, deployed, and ultimately returned to the environment. Moving beyond the traditional paradigm of inert permanence, contemporary biomaterials are increasingly engineered to interact dynamically with biological systems, modulate cellular responses, integrate with living tissues, and then safely disappear once their therapeutic function is complete.

This edited volume, "Next-Generation Biodegradable and Bioactive Biomaterials," brings together diverse perspectives on structure–property–degradation relationships, nanotechnology-enabled bioactivity, smart transience, and regenerative functions in medical and allied applications. The chapters span topics such as bioactive nanoparticle systems, nanofibers and nanotopography for immune modulation, smart biodegradable implants and transient biosensors, regenerative bioelectronics, organometallic and polymeric platforms, as well as sustainable approaches rooted in circular economy principles and green chemistry.

By integrating case studies on bioactive polymers, transient bioelectronics, load-bearing scaffolds, bioresorbable conductors, and nature-derived materials such as tilapia skin, the book illustrates how advanced material architectures can be tailored for precision therapeutics, tissue repair, and real-time monitoring while minimizing long-term environmental and clinical burdens. The inclusion of emerging themes—such as green synthesis of nanomaterials, venom-inspired bioactive systems, AI-enabled circular bioeconomy models, and the intersection of transient implants with forensic science and mental health—highlights the rapidly expanding frontiers of this field.

The multidisciplinary scope of the volume is intentional. Progress in biodegradable and bioactive biomaterials depends on close collaboration between materials scientists, chemists, clinicians, toxicologists, device engineers, data scientists, and policy-makers. This book provides a platform where fundamental materials design principles meet translational considerations, regulatory and safety challenges, and societal expectations for sustainability and ethical innovation.

For graduate students and early-career researchers, the chapters offer structured overviews, state-of-the-art examples, and forward-looking research questions that can inspire new projects and collaborations. For experienced investigators and practitioners, the volume serves as a consolidated reference on current trends, technological opportunities, and unresolved challenges in biodegradable and bioactive systems.

Looking ahead, it is anticipated that the concepts presented here—smart degradation, bioactive signaling, circular material lifecycles, and digitally augmented regenerative platforms—will play central roles in the design of future medical devices, implants, and therapeutic systems. As healthcare systems worldwide navigate demographic shifts, resource constraints, and climate-related pressures, materials that are both clinically effective and environmentally responsible will become essential rather than optional.

The editors of this volume, Safaa Al-Humairi, Abdollah Bahador, Fida Majdoub, and Nattapon Chantarapanich, are commended for assembling a truly international and interdisciplinary team of contributors and for curating a coherent narrative across such a diverse set of topics. Their collective effort will help guide the next wave of innovation in biodegradable and bioactive biomaterials and serve as a valuable resource for the global biomaterials community.

Chapter 1

Multimaterial Perspectives on Structure–Property–Degradation Relationships in Biodegradable and Bioactive Systems

J. Mangaiyarkkarasi
http://orcid.org/0000-0003-1431-9584
Nadar Mahajana Sangam S. Vellaichamy Nadar College, India

J. Shanthalakshmi Revathy
http://orcid.org/0000-0003-1724-7117
University College of Engineering, Ramanathapuram, India

ABSTRACT

Biodegradable and bioactive materials are vital in modern science due to their ability to perform specific functions and degrade safely. Widely used in biomedical fields such as tissue engineering, drug delivery, and implants, their design relies on understanding the relationship between structure, properties, and degradation. Structural features from molecular arrangement to microstructure govern mechanical strength, bioactivity, and degradation behavior across polymers, metals, and ceramics. Polymers degrade via hydrolysis or enzymes, metals through corrosion, and ceramics via dissolution. These mechanisms are influenced by factors like crystallinity,

DOI: 10.4018/979-8-2600-0060-1.ch001

porosity, and composition. The interplay between structure and properties directly affects degradation rates, requiring a holistic design approach. Hybrid materials combine advantages of different classes, enabling improved performance and tunable degradation. Controlled degradation is essential for matching functional lifetimes.

1. INTRODUCTION

Biodegradable and bioactive multimaterial systems have become a breakthrough in new biomedical, environmental and engineering applications. The systems are conceived by combining two or more different materials, e.g. polymers, ceramics and metals in order to acquire synergistic qualities that are not possible to reach with the help of a single material. The principle of multimaterial systems of varying length scales allows it to have a tight control on structural, mechanical, chemical, and biological functions and makes multimaterial systems very applicable to future applications in tissue engineering, drug delivery, regenerative medicine, and environmentally sustainable devices. Biodegradability is the term used in relation to the fact that a material can decompose in a biological or natural system in a controlled manner, in most cases, by enzymatic or hydrolytic mechanisms (X. Wang et al., 2024). This property is especially significant in medical implants and scaffolds where temporary support is needed in the process of tissue regeneration, and the material is later degraded to remove the need to surgically remove it. Bioactivity on the other hand refers to the ability of material to react favorably with biological systems, like cell adhesion, proliferation and differentiation. Bioactive materials can potentially be used in healing through active participation, such as bonding with local tissues or releasing therapeutic ions. The concept of biodegradability and bioactivity in a single multimaterial platform is highly beneficial. Such as polymer–ceramic composites can be used to incorporate flexibility and degradability of polymers with the ability of bioactive ceramics to stimulate natural tissue hierarchical structure and function, producing materials that replicate the natural tissue hierarchical structure and functionality (Lalonde et al., 2025). Similarly, hybrid structures with metallic elements may be used to increase mechanical strength, and the degradation rates are also under control. These combinations allow customization of properties like degradation kinetics, mechanical stability, and biological response which are important in certain applications.

Recent advances in material synthesis and fabrication techniques, including additive manufacturing, electrospinning, and layer-by-layer assembly, have further expanded the design possibilities of multimaterial systems. These approaches allow precise spatial distribution of materials and the creation of complex architectures that closely resemble natural biological structures (Mukherjee et al., 2023). Consequently, scholars are able to create materials that possess gradient characteristics and multi-functional behavior. In general, biodegradable and bioactive multimaterial systems are a fast-growing subject that spans materials science, biology, and engineering. Their capability to offer regulated performance, flexibility and environmental friendliness makes them a potential solution to present issues in healthcare and sustainability.

2. LITERATURE SURVEY

The connection of the structure in relation to properties and degradation behavior of biodegradable and bioactive multimaterial systems has been a subject of intense discussion in the recent literature and this is an important topic in the design of functional and reliable materials in biomedical and environmental applications. Initial research mainly centered on a single-component biodegradable material including polymers with the parameters of molecular weight, crystallinity, and crosslinking density found to play a significant role in determining the mechanical strength and degradation rates (Amoke & Kalu, 2026). The emphasis has been placed, over time, on the multimaterial systems, in which the interaction of various phases presents a further complexity and performance optimization prospects. Researchers in polymer-based systems have shown that semi-crystalline structure is more likely to degrade slowly than amorphous ones because of a low water permeability and low chain mobility. In the same vein, the addition of hydrophilic groups to polymer scaffolds has been reported to hasten hydrolytic degradation, and hydrophobic regions may retard the process (Sabet-Bokati & Plucknett, 2024). These findings highlight the importance of microstructural design in tailoring degradation kinetics. In contrast, bioactive glasses and calcium phosphates are examples of ceramic materials that have been found to improve bioactivity and dissolution control and release biologically relevant ions which trigger tissue regeneration.

Recent studies on composite and hybrid systems emphasize the role of interfacial interactions in governing overall performance. Strong interfacial bonding between phases can improve load transfer and mechanical integrity, while weak interfaces may lead to premature failure or uneven degradation. As an example, polymer-ceramic composites have been widely explored to use in bone tissue engineering, in which the ceramic component increases osteoconductivity and the polymer component regulates degradation and flexibility (Brochu et al., 2024). The morphology, size and distribution of the reinforcing phase are also reported to have an effect on the mechanical properties and the degradation routes.

Innovations in nanotechnology have also enhanced this area by introducing nanoscale filler that includes nanoparticles, nanotubes and nanofibers. These nanostructures offer high surface area to volume ratio, which allows enhanced contact with the surrounding matrix and biological environment. Research shows that nanocomposites have a tendency to have stronger mechanical behavior and degradation patterns that can be tuned, however, dispersion and stability issues of nanocomposites remain a challenge over time (Parvez et al., 2025).

Besides, recent literature emphasizes the significance of processing methods in defining the ultimate structure and functioning of multimaterial systems. The electrospinning, 3D printing, and solvent casting methods of fabrication can affect the porosity, pore interconnectivity, and material distribution, which affect degradation behavior and biological response. The use of computational modeling and data-driven methods has also become increasingly popular to predict structure-property-degradation relationships to provide fresh knowledge on material design (H. Wang et al., 2025).

Overall, the literature underscores that degradation behavior is not an isolated property but is intrinsically linked to material structure and functional performance. Understanding these interdependencies is essential for the rational design of next-generation biodegradable and bioactive multimaterial systems.

2.1 Research Gaps in Multimaterial Biodegradable and Bioactive Systems

Despite significant advances in biodegradable and bioactive multimaterial systems, several critical research gaps continue to limit their successful translation from laboratory studies to real-world biomedical applications. One of the most prominent challenges is the disparity between in vitro and in vivo performance. Most material evaluations are conducted under simplified laboratory conditions that fail to replicate the complexity of physiological environments, including dynamic mechanical loading, enzymatic activity, immune interactions, and patient-specific biological variability. Consequently, degradation rates, bioactivity, and mechanical stability observed in vitro often differ substantially from those measured in vivo, making accurate prediction of clinical performance difficult.

Another important yet insufficiently explored area concerns the immunological effects of degradation byproducts. While biodegradable materials are generally designed to degrade into biocompatible compounds, the local accumulation of degradation products can alter tissue pH, induce inflammatory responses, or influence immune-cell behavior. For example, acidic degradation products generated from polyesters such as PLA and PGA may trigger localized inflammation and affect tissue regeneration. Similarly, ionic species released from bioactive ceramics and glasses can influence macrophage polarization and immune signaling pathways. However, systematic investigations linking degradation chemistry to long-term immunological outcomes remain limited, creating uncertainty regarding the safety and efficacy of many emerging multimaterial systems.

A further challenge is the lack of universally applicable design principles for multimaterial architectures. Current material development largely relies on empirical approaches and application-specific optimization rather than predictive design frameworks. The complex interactions among composition, microstructure, interfacial bonding, degradation kinetics, and biological response make it difficult to establish generalized structure–property–degradation relationships. As a result, materials that perform well in one application often cannot be readily adapted to another. The absence of standardized datasets and comprehensive computational models further hinders the development of transferable design rules.

Addressing these gaps will require multidisciplinary efforts that combine advanced characterization techniques, long-term biological studies, computational modeling, and artificial intelligence–driven materials informatics. Establishing robust correlations between laboratory and clinical outcomes, understanding immune responses to degradation products, and developing predictive design frameworks represent essential priorities for the next generation of biodegradable and bioactive multimaterial systems. Such advances will be crucial for accelerating clinical translation and enabling the rational design of safer, more effective biomaterials.

3. FUNDAMENTALS OF MATERIAL STRUCTURE AND INTERFACIAL INTERACTIONS

3.1. Hierarchical Structure in Multimaterial Systems

Multimaterial systems have hierarchical structures over several length scales, between atomic structures and macro-architectural ones. The intrinsic properties of individual components are dictated at the atomic and molecular scale by the arrangement of atoms, bonding properties, and chemical composition. An example of this is that crystalline regions are the source of mechanical strength and stability, and amorphous regions are the source of flexibility and increased diffusion pathways. This difference is important in biodegradable systems where the amorphous domains are more vulnerable to hydrolytic and enzymatic degradation than the crystalline regions which are tightly packed. At a microlevel, structural heterogeneity is brought about by the arrangement of phases of polymer matrices strengthened with ceramic or metallic inclusions (Wawrzyniak & Balawender, 2022). The dimensions, shape, and orientation of these phases play a major role in determining the mechanical performance, porosity and transport properties. As an example, porous structures are interconnected and are required in applications such as tissue scaffolds where diffusion of nutrients and infiltration of cells are important. Also, gradient structures, in which composition or porosity varies continuously through the material, can resemble natural tissues and offer better mechanical compatibility.

At the macroscale, the geometry and architecture of the material identify the overall performance of the material in real-world situations. High-technology fabrication methods allow construction of complex structures with customized properties, so that the material can satisfy a set of requirements of its application (Yu et al., 2025). The hierarchical nature of multimaterial systems allows for the integration of multiple functionalities, such as mechanical strength, controlled degradation, and bioactivity, within a single platform. These structural levels and their interrelationships are the basis of improving the performance and durability of biodegradable and bioactive systems.

3.2. Interfacial Interactions and Bonding Mechanisms

Interfaces are central to the determination of the general behavior of multimaterial systems, since they control the interaction between the different phases of materials. The type of interfacial bonding be it physical, chemical or mechanical directly affects the load transfer, structural integrity and degradation patterns. Physical interactions, such as van der Waals forces and hydrogen bonding, are generally weaker but can contribute to flexibility and reversible interactions. In contrast, chemical bonding, including covalent and ionic bonds, provides stronger adhesion and improved mechanical stability (Park & Lee, 2021). Interfacial compatibility is a major element in the attainment of optimality in the performance of polymer-ceramic and polymer-metal composites. Surface modification is frequently used to improve the bonding of dissimilar materials, e.g., by functionalizing or coating reinforcing phases. Enhanced interfacial bonding results in effective transfer of stress between phases which minimises the possibility of crack initiation and propagation. On the contrary, lack of good interfacial adhesion may lead to delamination, formation of voids, and increased degradation.

Interfacial areas are also active areas of chemical reaction and mass transportation in the process of degradation. An example is that degradation products that accumulate at the interfaces may change local pH levels, which can affect further degradation and a biological response. Interfaces can aid ion exchange and biomineralization in bioactive systems, and can be used to support integration of the interface with the surrounding tissues. Thus, to control the mechanical and biological functionality of multimaterial systems, interfacial chemistry and structure must be controlled.

3.3. Structure–Interface Coupling and Its Impact on Performance

Multimaterial systems are characterized by the coupling between material structure and interfaces, which determines the overall properties of materials and functional behavior. The properties of interfaces are directly correlated with structural parameters (like phase distribution, porosity and morphology) and form a complex relationship that defines performance. An illustration is that when a reinforcement phase is densely dispersed and strongly interfaced with one another, mechanical strength and durability can be highly improved, and uniform degradation can be realized. Porosity and pore architecture also affect structure-interface coupling by determining the exposure of internal interface to external environments. In biodegradable, higher porosity is more penetrative to fluid that degrades faster and releases bioactive agents (Lou et al., 2026). Nevertheless, over porosity can affect the mechanical stability, which is why the design should be balanced. Equally, nanoscale interfaces of nanocomposites offer a high interfacial area, which results in enhanced interaction between phases but necessitates a close control to ensure agglomeration is avoided. There is the additional complexity of the changing nature of interfaces with degradation. Mechanical performance and biological interactions may change with time as the material loses its quality, and structure and interfacial properties change. Such changing behavior has to be kept in mind in the practical use of materials like tissue engineering, where the materials should serve as a temporary support that is gradually replaced by natural tissue (Zhu et al., 2026).

Overall, the synergy between structure and interfacial interactions enables the design of advanced multimaterial systems with tailored properties. It is important to have an in-depth knowledge of this coupling to maximize performance, achieve reliability, and desired functional results in biodegradable and bioactive applications.

4. PROCESSING TECHNIQUES AND MULTIMATERIAL FABRICATION STRATEGIES

4.1. Conventional Fabrication Methods for Multimaterial Systems

The basis of developing biodegradable and bioactive multimaterial systems is traditional processing techniques. Solvents casting, melt blending, and compression molding are some of the most common methods because they are simple to use, cost effective and can be scaled. Solvents In solvent casting, the materials are dissolved in a suitable solvent and cast in molds, where the secondary phases (such as ceramics or nanoparticles) can be dispersed fairly uniformly in a polymer matrix. Nevertheless, the final material performance can be impacted by challenges, including solvent residue, inability to control their thickness, and the inability to ensure good interconnectivity between the pores. Another widely used one is melt blending especially of thermoplastic polymers. It is a process that entails blending materials under high temperatures in order to obtain a homogenous composite. The process does not require solvents and can be used in large-scale production, although high temperatures can cause sensitive bioactive components to be lost or change material properties. The compression molding process also makes the structure more refined through the application of heat as well as pressure and the materials become dense and mechanically strong (Ramos et al., 2025).

Although they have their benefits, traditional techniques do not have specific control over microstructure and spatial distribution of components. This means that they are normally applied to bulk materials where complicated architectures are not an issue. However, the methods are still necessary in the creation of baseline composites as well as in processes that need simple geometries and economical production.

4.2. Advanced Fabrication Techniques for Controlled Architectures

In order to eliminate the shortcomings of traditional approaches, there have been developed sophisticated fabrication technologies aimed at gaining a higher level of control over the structure and functionality of materials. Electrospinning is one of the noticeable methods that generate nanofibrous scaffolds which have high surface area and porosity. A wide range of processing parameters and material composition can be used to produce composite fibers, including core-shell structures, which increase both mechanical properties and bioactivity. These nanostructures have found application especially in the biomedical field where replication of natural extracellular matrices is of relevance. Another method of producing highly porous materials is freeze-drying or lyophilization. It is a process of freezing a material solution and then sublimation of the solution, which creates interlocked pore networks. This is particularly useful with tissue engineering scaffolds, which encourages the infiltration of cells and the transportation of nutrients (Karolewska et al., 2024). Also, layer-by-layer assembly can be used to control the deposition of thin films with carefully selected composition and thickness, and multilayered systems can be fabricated with custom properties. These superior methods allow better control of porosity, morphology and phase distribution to create materials with certain functional properties. Nevertheless, they can be more expensive and more complicated processing conditions, which can restrict large-scale use. Figure 1 explains multimaterial fabrication strategies

Figure 1. Multimaterial fabrication strategies

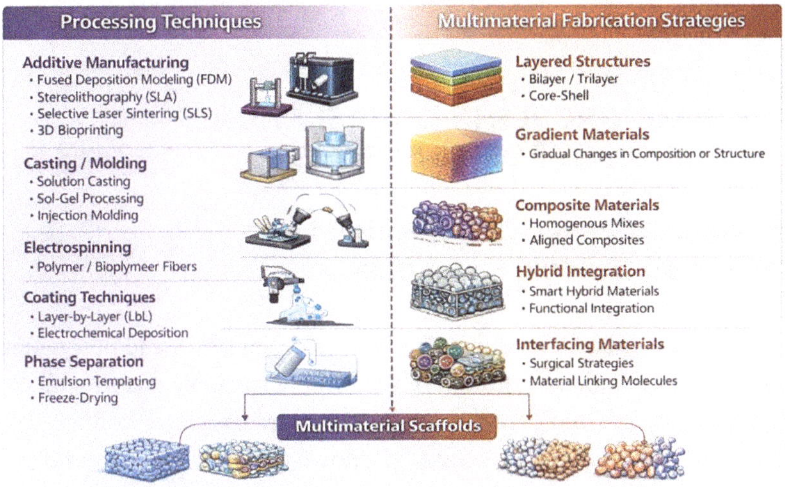

4.3. Additive Manufacturing and Emerging Multimaterial Strategies

Additive manufacturing (also 3D printing) has transformed the manufacturing of multimaterial systems by allowing the spatial control and personalization of the structure. Layers The fused deposition modeling (FDM), stereolithography (SLA) and direct ink writing (DIW) techniques enable building of complex structures through multiple materials, layer by layer. This is especially useful in developing gradient structures and patient-specific implants in the field of biomedicine. Multi-material 3D printing provides the chance to combine materials of different properties into one construct, thus making it possible to design systems with different mechanical stability, degradation, and bioactivity. To illustrate, a scaffold may be designed to have a hard outer coating to provide mechanical support and an inner core that is more degradable to allow regulable tissue regeneration. Also, hybrid fabrication methods that involve 3D printing with other fabrication methods such as electrospinning or incorporation of nanoparticles increase functionality further.

Smart materials and bioinks, which react to environmental factors, are also emerging strategies like temperature, pH, or even biological signals. These inventions provide new opportunities of dynamic systems and adaptive systems. Additive manufacturing is a promising new solution to highly customized and application-specific solutions, despite the challenges associated with material compatibility, resolution, and cost, additive manufacturing is still a great breakthrough in the design and fabrication of biodegradable and bioactive multimaterial systems (Khan & Riccio, 2024).

5. STRUCTURE–PROPERTY CORRELATIONS IN HYBRID AND COMPOSITE SYSTEMS

5.1. Influence of Phase Composition and Distribution on Material Properties

In hybrid and composite systems, the deliberate selection and proportioning of constituent phases enable fine-tuning of material performance across mechanical, thermal, chemical, and biological domains. The matrix phase commonly biodegradable polymers such as poly(lactic acid), polycaprolactone, or natural polymers like chitosan provides structural continuity and processability, while reinforcing phases impart targeted functionalities. Ceramics contribute stiffness, hardness, and osteoconductivity; metals enhance strength and electrical or thermal conductivity; and nanomaterials introduce multifunctionality such as antimicrobial behavior or improved barrier properties. The interaction between these stages is not only additive but in most cases, it leads to emergent properties owing to interfacial interactions. One of the design parameters is the volume fraction of each phase (Nasi et al., 2025). At low filler content, improvements in stiffness and strength may be modest due to limited load transfer. Filler concentration could reach a percolation threshold beyond which a continuous network of reinforcement can create a significant increase in mechanical and functional properties. But above an optimal point, it may be subject to particle clustering, processing difficulties and toughness degradation. Therefore, a determination of the best window in composition is necessary to strike a balance between conflicting needs, e.g. strength vs ductility or bioactivity vs degradation rate.

The phases are spatially distributed, adding more complexity and opportunity. Isotropic behaviour and predictable performance is achieved by uniform dispersion, whereas spatially different properties can be offered by controlled heterogeneity, e.g. layered, core shell or gradient structures. An example is a scaffold to regenerate bone, which could be characterised by a dense outer layer to confer mechanical stability, and an inner layer, which is porous and bioactive to allow penetration of cells and subsequent tissue growth. Equally, stress mismatches in interfaces and long-term strength can be enhanced by functionally graded materials. Such architectures demand a high degree of control over fabrication processes, such as additive manufacturing and phase separation processes. Of special importance is the interdependence between interfacial bonding and phase distribution. Well-spread phases enhance the effective interfacial area, which enhances stress transfer, and crack propagation resistance. Coupling agents or functional groups can be used to surface modify fillers to enhance compatibility with the matrix, which minimizes interfacial energy and agglomeration. This is critical in particular with nanoscale reinforcements, where surface effects prevail over bulk behavior.

Performance is also determined by particle size, shape and orientation. Isotropic reinforcement can be improved by the use of spherical particles whereas fibrous or tubular reinforcement can be better along one particular direction e.g. nanofibers and nanotubes. Plate-like structures have the ability to enhance the properties of barriers and lower permeability. Anisotropic materials can be designed with orientation control, which can be done by methods such as electrospinning or directional solidification. Also, phase composition and distribution have a very strong influence on degradation pathways and kinetics (Promakhov et al., 2022). Hydrophilic phases promote water uptake and accelerate hydrolysis, whereas hydrophobic components slow degradation. The connectivity of phases can create preferential pathways for fluid diffusion, influencing both degradation rate and release of bioactive agents. In bioactive systems, the distribution of ceramic phases governs ion release profiles, which in turn regulate cellular responses and tissue integration.

Ultimately, the ability to engineer phase composition and distribution with high precision allows for the creation of multifunctional materials tailored to specific applications. Whether for load-bearing biomedical implants, drug delivery systems, or environmentally responsive materials, optimizing these parameters is central to achieving desired performance and reliability.

5.2. Insights on Distribution–Property Interplay and Design Optimization

In addition to the basic elements of composition and dispersion, the interaction between phase distribution and the overall system performance can be also explained under the framework of more advanced design concepts and new methods. The transitional regions between different material phases are one of these concepts, which is known as optimization of interphase regions. These areas tend to have characteristics that are distinct to the matrix and reinforcement and can have a considerable effect on the overall behavior. The interphase can be engineered by chemical functionalization or graded bonding to enhance stress transfer, interfacial defect reduction, as well as increase durability during mechanical and environmental loading. The other important factor is the contribution of multiscale architecture in the control of material performance. Hierarchical reinforcement and multifunctionality can be attained through the incorporation of structural features at different nano-, micro-, and macro-scales. As an example, nanoscale fillers improve interfacial strength, microscale porosity facilitates biological interactions and macroscale geometry defines the distribution of loads. The combination of these scales makes sure that the material works effectively in the complicated conditions of real-life (Czapczyk et al., 2023).

The effect of phase distribution is also spread to time-dependent properties, including viscoelasticity, fatigue resistance and long-term degradation. The composite materials, in which reinforcements are well distributed, are likely to have a better resistance to cyclic loading and crack propagation takes a long time. Conversely, when the phases are not well distributed, localized damage can be accumulated and the service life will be shortened. It is especially crucial in biomedical implants and structural parts that are put under repetitive stress. New computational tools and simulation methods are being employed more often to simulate phase distribution and material behavior. Finite element and machine learning methods enable researchers to study a broad set of compositions and architectures in the virtual world and determine the best design before it is experimentally validated. The tools assist in the interpretation of complicated interactions of phases, such as the distribution of stress, diffusion pathways, or degradation patterns. Design strategies are also being determined by sustainability factors. The selection of eco-friendly materials and the optimization of phase distribution for efficient degradation and recyclability are becoming important research directions. For example, incorporating natural fibers or bio-based fillers not only enhances sustainability but also introduces unique structural features that can improve performance.

The impact of the phase composition and distribution is not limited to simple improvement of properties but rather to advanced design concepts, multiscale engineering and predictive modelling. An integrated strategy involving material choice, processing methods and structural design is critical to ensuring that the next generation hybrid and composite systems have better and reliable performance.

5.2. Role of Microstructure and Porosity in Functional Performance

The basic structural parameters that determine the performance, durability and functionality of hybrid and composite systems are microstructure and porosity. Microstructure refers to grain size, phase morphology, crystallinity, and nature of phase boundaries which all determine the response of a material to mechanical loading, environmental conditions and biological interactions. As an example, fine grain microstructures tend to increase strength and toughness by means of grain boundary strengthening, as well as resistance to crack initiation and propagation (Aljabali et al., 2025). Otherwise, finely grained structures can have less strength and high ductility based on the material system. Another important microstructural criterion is crystallinity, especially in biodegradable systems of polymers. Areas with high crystallinity are compressed and less permeable to water and enzymes, which makes them degrade slower and improve mechanical stability. Amorphous regions on the other hand facilitate easy diffusion of fluids, which makes degradation faster, and interaction with the surrounding environment easier. The crystalline and amorphous balance is thus fundamental in tuning the mechanical performance as well as the degradation kinetics. Moreover, the presence and distribution of phase boundaries in composites affect the transfer of stress and may serve as resistant and conduit crack propagation.

The porosity is also important in an equally important way, particularly where mass transport and biological integration are to be performed. The dimensions, form, distribution, and interconnection of the pores are what define the ability of the material to allow fluids, ions, and cellular movement. Macropores (usually >100 μm) are essential in cell penetration and tissue growth, whereas micropores (<10 μm) increase surface area and exchange of nutrients. Nanopores also add to protein adsorption and molecular interactions which play a role in bioactivity (Hernandez & Woodrow, 2022). The desirable is specifically interconnected networks of pore structures, where one can have a continuous pathway to transport, which enhances the overall functionality of pore structures in applications like tissue engineering scaffolds and drug delivery systems. The cost of porosity, however, is usually the cost of mechanical strength and structural integrity. Materials with high porosity are also likely to be less stiff and have lower load bearing capacity because of the existence of voids. This forms a basic trade off between mechanical performance and biological or transport functionality. In order to overcome this difficulty, the design strategies have advanced toward the optimization of pore architecture, as opposed to the mere increase in porosity. Parameters including pore geometry, orientation and distribution are well regulated to make a balance between strength and permeability.

Modern fabrication techniques offer effective instruments to produce microstructure and porosity with great precision. Freeze-drying, electrospinning, and additive manufacturing enable the formation of complex structures, including gradient and hierarchical ones. The gradient porosity allows spatial differentiation in properties, including a dense outer region to support the mechanical and a porous inner region to support the biology. Hierarchical porosity, combining macro-, micro-, and nanoscale pores, that increases multifunctionality by increasing mechanical stability, fluid transport, and surface reactivity simultaneously. Moreover, microstructure and porosity play a great role in degradation behavior. Porous structures enhance the surface area and accessibility of fluids, which enhances the degradation rate, and dense areas degrade slower. The degradation rates across space can be used to fabricate materials that have a specific resorption profile. Also, pore surfaces are frequently active locations of bioactivity, such as in cell attachment and mineralization.

The interaction of microstructure and porosity is the key to the design of high-performance hybrid and composite systems. With a close control of such parameters, the researchers can produce the materials that have the best balance of mechanical strength, degradation rate, and biological functionality, hence addressing the needs of the advanced engineering and biomedical uses.

5.3. Interfacial Effects on Mechanical, Thermal, and Degradation Properties

The boundary between different phases in hybrid and composite systems is one area of immense significance since it controls the interaction, compatibility, and general synergy of constituent materials. In contrast to bulk phases, interfaces can have distinct structural and chemical properties, so they play an important role in defining the macroscopic behavior of the system (Mukasheva et al., 2024). Good interfacial interactions will make sure that the benefits of every phase are maximized whereas bad interfacial design will nullify possible benefits and cause premature failure. Mechanically, the interface is the one that assists in transfer of stress between the matrix and the reinforcement. When strong interfacial bonding exists through covalent bonding, ionic interactions, or mechanical interlocking applied loads are efficiently distributed across phases, resulting in improved strength, stiffness, and fatigue resistance. This is particularly important in load-bearing applications, where repeated mechanical stress can lead to crack initiation at weak interfaces. In contrast, weak interfacial adhesion leads to debonding, fiber pull-out, or particle detachment, all of which reduce the material's load-carrying capacity. Additionally, well-designed interfaces can enhance toughness by promoting mechanisms such as crack deflection, bridging, and energy dissipation, thereby delaying catastrophic failure.

The interfacial properties are also important in defining thermal properties. In composite systems, heat transfer across phases depends on the continuity and compatibility of interfaces. Strong, well-integrated interfaces reduce thermal resistance and facilitate efficient heat conduction, which is essential in applications requiring thermal management. On the other hand, poorly bonded or discontinuous interfaces act as thermal barriers, increasing interfacial resistance and reducing overall thermal conductivity. The dissimilarity between various phases in terms of thermal expansion may also add internal stresses at interfaces on changes in temperature that may cause microcracking and extended deterioration (Zhang et al., 2025).

Interfaces are dynamic in biodegradable and bioactive systems, and they change with time. They can be used as favorable entry points to fluid, allowing water, ions and enzymes to permeate the material. This increases localized degradation and may trigger changes in interfacial chemistry. The rates of degradation vary by phase that generates internal stress, resulting in the generation of voids, microcracks, and interfacial delamination. Such changes do not only affect mechanical integrity but also transport properties and biological interactions. Moreover, interfacial areas may become depositing degradation by-products, which can be able to affect micro-chemical conditions, including changes in pH. Such localized variations in biomedical applications may influence cell viability, tissue integration and general biocompatibility. On the contrary, in bioactive systems interfaces may serve as active sites, where useful processes like ion exchange, protein adsorption, and biomineralization can be promoted, to improve tissue bonding and tissue regeneration.

In order to overcome these challenges, several surface modification and interfacial engineering strategies are used. Chemical functionalization, incorporation of coupling agents, plasma treatment, and coating of reinforcement phases are rather popular techniques to enhance compatibility and adhesion. Stress transfer, degradation behavior, and biological response can be controlled better by tailoring interfacial properties on a molecular or nanoscale level. Alongside, the idea of an interphase a transitional space with its own properties, neither that of the matrix nor the reinforcement, is becoming increasingly significant. Optimization of gradients in mechanical and chemical properties can be achieved by engineering this interphase to reduce stress concentrations and improve the overall performance of a system (Al-Amiery et al., 2024).

Interfacial effects are the key to performance of hybrid and composite systems, which affect mechanical performance, thermal performance, and degradation performance. The accurate control and full understanding of interfacial interactions will allow the creation of materials that are more durable, functional and reliable and suitable in advanced application to biomedical engineering and beyond.

5.4 Quantitative Perspectives on Structure–Property–Degradation Relationships in Biodegradable and Bioactive Multimaterial Systems

A comprehensive understanding of structure–property–degradation relationships is essential for the rational design of biodegradable and bioactive multimaterial systems. While qualitative descriptions provide valuable conceptual insights, quantitative parameters such as degradation rates, elastic modulus, porosity levels, and bioactivity indices are increasingly required to establish predictive design frameworks. These measurable characteristics enable researchers to tailor materials for specific biomedical applications, including tissue engineering, drug delivery, orthopedic implants, and regenerative medicine. Mechanical properties represent one of the most important factors governing the performance of biodegradable materials. Natural soft tissues typically exhibit elastic moduli ranging from approximately 0.01 to 10 MPa, whereas trabecular bone exhibits values between 50 and 500 MPa, and cortical bone ranges from 7 to 30 GPa. Consequently, biodegradable polymers such as poly(lactic acid) (PLA), with modulus values of 2–4 GPa, and polycaprolactone (PCL), with values of 0.2–0.4 GPa, are often combined with bioactive ceramic phases to achieve mechanical properties closer to those of native tissues. Hydroxyapatite (HA), for example, possesses a modulus of approximately 80–120 GPa, while β-tricalcium phosphate (β-TCP) exhibits values between 60 and 90 GPa. Incorporating 10–40 wt.% ceramic reinforcement into polymer matrices can increase scaffold stiffness by 30–300%, depending on composition and microstructural organization.

Porosity is another critical structural parameter that strongly influences both biological performance and degradation behavior. Studies have demonstrated that interconnected porosity levels between 60% and 90% are generally optimal for tissue-engineering scaffolds, promoting nutrient transport, cell migration, and vascularization. Pore sizes between 100 and 400 μm are considered favorable for bone tissue ingrowth, while pores exceeding 300 μm facilitate enhanced vascular formation. However, increasing porosity beyond approximately 90% often results in substantial reductions in compressive strength, highlighting the need for careful optimization. Quantitatively, every 10% increase in porosity may reduce scaffold stiffness by approximately 15–30%, depending on material composition and architecture. Degradation kinetics are equally important because scaffold resorption must be synchronized with tissue regeneration. Biodegradable polymers exhibit widely varying degradation rates. PCL degrades relatively slowly, often requiring 24–48 months for substantial mass loss under physiological conditions. In contrast, PLA typically degrades within 6–24 months, while poly(glycolic acid) (PGA) may undergo significant degradation within 3–6 months. Bioactive ceramics display different dissolution behaviors, with β-TCP degrading approximately three to twelve times faster than HA. For many orthopedic applications, degradation rates of 1–5% mass loss per month are considered desirable to maintain structural support while allowing progressive tissue replacement.

The incorporation of bioactive phases further influences degradation and biological response. Hydroxyapatite-containing composites can increase osteoblast attachment by 20–50% relative to pure polymer scaffolds, while bioactive glass additions often accelerate apatite formation within 24–72 hours of immersion in simulated body fluid. Furthermore, controlled degradation can enhance ion release, including calcium, phosphate, silicon, and magnesium ions, which contribute directly to cellular signaling and tissue regeneration. Future multimaterial design strategies increasingly rely on quantitative structure–property–degradation models capable of predicting performance across multiple scales. By integrating measurable parameters such as modulus ranges, degradation constants, porosity thresholds, and bioactivity metrics into computational and machine-learning frameworks, researchers can move from empirical optimization toward predictive material engineering. Such data-driven approaches will be essential for developing next-generation biodegradable and bioactive systems with precisely tailored functional lifetimes and biological outcomes. Table 1 explains Structure-property relationship.Top of Form

Table 1. Structure-property relationship

Aspect	Structural Features	Key Properties Affected	Impact on Performance	Design Considerations
Phase Composition & Distribution	Ratio of matrix and reinforcement; dispersion; particle size and morphology	Mechanical strength, stiffness, flexibility, degradation rate	Uniform distribution enhances load transfer and durability; agglomeration leads to weak points and failure	Optimize filler content; ensure homogeneous dispersion; control nanoscale interactions
Microstructure & Porosity	Grain size, crystallinity, pore size, pore interconnectivity, hierarchical structure	Mechanical strength, diffusion, bioactivity, degradation kinetics	High porosity improves cell growth and transport but reduces strength; crystallinity slows degradation	Balance porosity and strength; design gradient and hierarchical pore structures
Interfacial Interactions	Bonding type (physical/ chemical), interface area, compatibility between phases	Mechanical integrity, thermal conductivity, degradation behavior	Strong interfaces improve stress transfer and stability; weak interfaces cause delamination and faster degradation	Use surface modification; enhance adhesion; match material compatibility
Degradation Behavior	Phase-dependent degradation rates, interface exposure, porosity-driven fluid penetration	Structural stability over time, mass loss, biological response	Uneven degradation may create voids and cracks; controlled degradation ensures functional lifespan	Tailor degradation rates of components; monitor interfacial stability
Thermal Properties	Interface continuity, phase distribution, material compatibility	Thermal conductivity, thermal stability	Efficient interfaces enhance heat transfer; poor interfaces act as barriers	Improve interfacial bonding; select compatible materials
Nanostructural Effects	Nanoparticles, nanotubes, nanofibers with high surface area	Mechanical reinforcement, degradation tuning, bioactivity	Enhanced strength and functionality; risk of agglomeration	Ensure proper dispersion; stabilize nanofillers

6. MECHANISMS OF DEGRADATION AND BIOACTIVITY IN MULTIMATERIALS

6.1. Hydrolytic and Enzymatic Degradation Mechanisms

Hydrolytic and enzymatic degradation of biodegradable multimaterial systems is mainly controlled by the presence of both hydrolytic and enzymatic degradation, which are highly dependent on material composition and structure. Hydrolytic degradation is a process in which water penetrates the material and breaks vulnerable chemical bonds, such as ester bonds which are present in biodegradable polymers. The result of this process is chain scission, decrease in molecular weight, and eventual loss of mass. Hydrolysis rate is determined by polymer chemistry, crystallinity and porosity. The amorphous regions are more likely to be degraded because they are easily penetrated by water, but crystalline domains cannot be easily penetrated and are, therefore, more resistant to degradation. On the other hand, enzymatic degradation entails the use of biological catalysts, which are selective in degrading certain parts of the material. This effect is especially important in bioactive systems that are used in the in vivo environment, wherein bodily enzymes found in bodily fluids are used to catalyze material degradation. The targeting of enzymes to degrade a particular target can be beneficial in regulated drug delivery or tissue regeneration since it is specific.

Multimaterial systems become more complicated due to the existence of numerous phases. As an example, ceramic component might not hydrolytically degrade but can have an effect on the general degradation behaviour through changing local pH or diffusion of water. Likewise, interfaces of two materials can be preferential sites of degradation as the free energy and structural discontinuities are higher. The combination of hydrolytic and enzymatic processing allows the creation of materials with customized degradation characteristics so that they can be used effectively over a specific time, after which they become resorbed or eliminated safely. Figure 2. depicts Mechanisms and Degradation and bioactivity of the materials.

Figure 2. Mechanisms and degradation and bioactivity of the materials

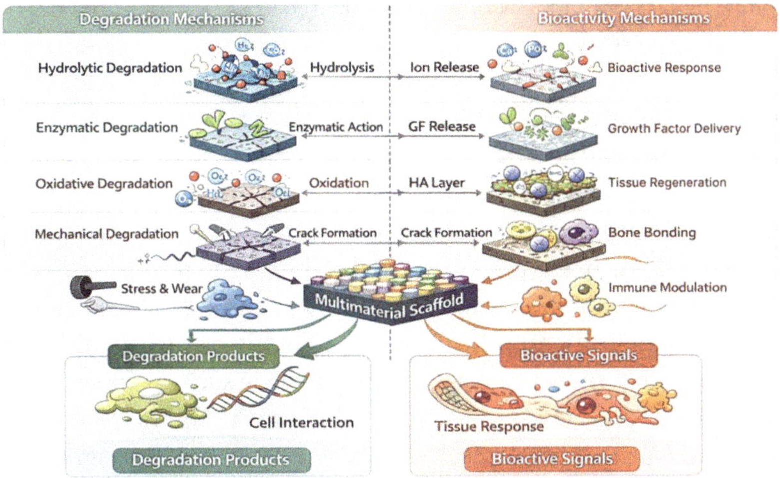

6.2. Surface Erosion, Bulk Degradation, and Interface-Driven Processes

The degradation of multimaterial systems can take place in various forms, which are broadly divided into surface and bulk degradation modes. Surface erosion is the continuous deposition of the layers and the consequence is a uniform and predictable decrease in the thickness of the layers with time. This process is regularly seen in materials with low permeability to water, in which degradation is localized to the surface (Ferreira Rocha et al., 2024). It is especially applicable in processes where there is need to have controlled and constant release of bioactive agents. The opposite process is known as bulk degradation where the water infiltrates the entire material causing degradation of the entire material simultaneously. The resulting process may lead to the rapid loss of mechanical integrity as internal weakening occurs prior to massive loss of mass being experienced. Bulk-degrading materials are usually employed in systems that need quick resorption; however, these materials must be properly designed to prevent premature failures.

The interface-based degradation further complicates the situation in multimaterial systems. The variations in the chemical composition, mechanical behavior, and degradation rate of the phases may result in the building of stress and alterations in the microstructure at the interfaces. The areas tend to serve as crack starters, cavity source, and localized erosion. Also, degradation products can be deposited at the interfaces, changing the local chemical environment and increasing further degradation. The combined effect of these mechanisms determines the overall degradation behavior of the material. Through managing porosity, phase distribution and interfacial bonding, systems can be designed to adopt desired degradation modes and rates thereby guaranteeing optimum performance of a given application.

6.3. Bioactivity and Biological Interactions in Multimaterial Systems

Bioactivity of Multimaterial Systems Multimaterial systems are materials that have the capability of being used to trigger certain biological reactions that improve the functionality of the material especially in biomedical applications. This encompasses cell interactions, cell adhesion, cell proliferation, cell differentiation, and tissue integration. Bioactive materials may include such components as bioactive ceramics or functionalized polymers which have the ability to liberate ions or molecules which facilitate biological activity. Ion exchange between the material and the surrounding environment is one of the most important mechanisms of bioactivity. As an example, calcium phosphate-based components have the ability to release calcium and phosphate ions which aid in the mineralized tissue formation and bone regeneration. Likewise, bioactive glasses are capable of developing a layer of hydroxyapatite-like surface when subjected to physiological fluids, which allows them to be bonded to natural tissue.

Roughness, chemistry and wettability are surface properties, which have a major role in the regulation of the interactions of biology. Multimaterial systems Multimaterial systems contain more than one phase and can be used to design surfaces with modified properties that can improve cellular responses. Also, the products of degradation can affect the bioactivity by changing local pH or releasing bioactive compounds that induce the healing mechanisms (Abdulsalam et al., 2025). Bioactivity in combination with controlled degradation is a key to creating useful materials that facilitate but also take part in tissue regeneration or other biological functions. A balance between degradation and biological performance can be maintained by closely structuring and designing multimaterial systems such that the material is able to perform its intended purpose effectively.

Sterilization Effects

Sterilization is essential for biomedical materials but can significantly alter the structural, mechanical, and degradation properties of biodegradable systems. Methods such as gamma irradiation, ethylene oxide treatment, and steam sterilization may induce polymer chain scission, crosslinking, or changes in crystallinity. Understanding sterilization-induced property modifications is critical for ensuring consistent clinical performance.

Inflammatory Responses to Degradation Products

The degradation of biodegradable materials generates byproducts that may interact with surrounding tissues and immune cells. Acidic degradation products, ionic release, and particulate debris can influence local pH and inflammatory pathways. Comprehensive evaluation of host immune responses is necessary to optimize material biocompatibility and minimize adverse biological reactions.

In-Service Failure Modes

Biodegradable multimaterial systems may experience failure during service due to mechanical fatigue, premature degradation, delamination at material interfaces, or loss of structural integrity. These failure mechanisms are often influenced by physiological loading conditions and environmental factors. Identifying and predicting such failure modes remain essential for improving long-term reliability.

Regulatory Considerations

The clinical translation of multimaterial biomedical systems faces significant regulatory challenges due to their compositional complexity and dynamic degradation behavior. Regulatory agencies require extensive evidence regarding safety, efficacy, degradation products, and manufacturing consistency. Developing standardized testing protocols and regulatory frameworks will be crucial for accelerating approval and commercialization. Table 2. explains the comparison of different Mechanisms and Degradation and bioactivity of the materials

Table 2. Mechanisms and degradation and bioactivity of the materials – A comparison

Mechanism	Key Processes	Controlling Factors	Impact on Material Behavior	Design Strategies
Hydrolytic Degradation	Water-induced bond cleavage (e.g., ester bonds), chain scission, mass loss	Polymer chemistry, crystallinity, porosity, water diffusion	Gradual reduction in molecular weight and strength; faster in amorphous regions	Use copolymers; control crystallinity; tailor porosity for desired degradation rate
Enzymatic Degradation	Enzyme-specific breakdown of material components	Enzyme presence, substrate specificity, surface accessibility	Targeted and accelerated degradation in biological environments	Incorporate enzyme-sensitive linkages; design for controlled bioresorption
Surface Erosion	Layer-by-layer degradation from outer surface	Low water permeability, material density	Predictable thickness reduction; sustained release behavior	Use dense structures; design for controlled drug delivery systems
Bulk Degradation	Uniform degradation throughout material volume	High penetration, amorphous structure	Rapid loss of mechanical integrity before mass loss	Optimize crosslinking; balance diffusion and stability
Interface-Driven Degradation	Degradation at phase boundaries, microcrack formation	Phase compatibility, interfacial bonding, mismatch in degradation rates	Localized failure, void formation, accelerated breakdown	Improve interfacial adhesion; match degradation profiles of phases
Bioactivity (Ion Release)	Release of Ca^{2+}, PO_4^{3-}, or other therapeutic ions	Composition, surface area, dissolution rate	Promotes tissue regeneration and mineralization	Incorporate bioactive ceramics/glasses; tune ion release rates
Surface-Mediated Bioactivity	Cell adhesion, proliferation, and differentiation	Surface roughness, wettability, chemistry	Enhanced biological integration and healing	Functionalize surfaces; design micro/nanostructures
Degradation–Bioactivity Coupling	Interaction between degradation products and biological response	pH changes, by-product release, material composition	Can enhance or hinder biological performance	Balance degradation rate with biological compatibility

7. CHARACTERIZATION TECHNIQUES FOR EVALUATING DEGRADATION AND PERFORMANCE

In order to gain a thorough insight into the degradation behavior and performance of biodegradable and bioactive multimaterial systems, both traditional and advanced characterization methods are used (Liana et al., 2024). Along with the SEM, FTIR, XRD, TGA, and mechanical testing, such sophisticated approaches like DSC, AFM and in situ methods can help to gain further information about the thermal transitions, nanoscale structure and real-time processes of degradation.

1. Scanning Electron Microscopy (SEM)

Principle: High-resolution surface images are produced by the use of an electron beam that is focused.

Explanation: SEM reveals surface morphology, pore structure, cracks, and interfacial changes during degradation. It is essential for visualizing physical deterioration and structural evolution.

2. Fourier Transform Infrared Spectroscopy (FTIR)

Principle: Measures absorption of infrared radiation by chemical bonds.

Explanation: FTIR can determine the presence of functional groups and trace the breaking or formation of chemical bonds, and thus, can be used to understand the degradation process at the molecular level.

3. X-ray Diffraction (XRD)

Principle: Analyzes diffraction patterns of X-rays interacting with crystalline materials.

Explanation: XRD determines crystallinity and phase composition, revealing structural changes such as increased crystallinity during polymer degradation or phase stability in composites.

4. Thermogravimetric Analysis (TGA)

Principle: Measures weight changes with temperature.

Explanation: TGA evaluates thermal stability, decomposition behavior, and composition changes, providing insights into degradation and residual material content.

5. Mechanical Testing (Tensile/Compression)

Principle: Measures response to applied mechanical forces.

Explanation: Assesses strength, stiffness, and durability as degradation progresses, which is critical for load-bearing applications.

6. Differential Scanning Calorimetry (DSC)

Principle: Measures heat flow associated with thermal transitions in materials.
Explanation: DSC is used to study glass transition temperature (Tg), melting temperature (Tm), and crystallization behavior. The variations in these parameters on degradation show how these parameters change with degradation, aiding in correlating thermal properties with degradation kinetics.

7. Atomic Force Microscopy (AFM)

Principle: Uses a nanoscale probe to scan surface topography and measure forces.
Explanation: AFM offers a high-resolution surface image at a nanoscale that is suitable in the analysis of surface roughness, phase distribution, and nanoscale degradation structures. It finds good application in the investigation of interfacial areas and the initial stages of degradation.

8. In Situ Characterization Techniques

Principle: Real-time monitoring of material behavior under operational or environmental conditions.
Explanation: Techniques like in situ SEM, XRD or spectroscopy allow one to observe the degradation processes in real time. The approaches can give dynamic information on the dynamics of structural development, phase, and environment interaction, and can provide a more realistic view of material behavior. Table 3. explain various characterization techniques, principles used and their role in degradation study.

Table 3. Characterization techniques vs role in degradation study

Technique	Principle	Measures	Role in Degradation Study	Key Advantages
SEM	Electron beam interaction	Surface morphology, cracks, porosity	Tracks physical degradation	High-resolution imaging
FTIR	IR absorption	Functional groups, bond changes	Identifies chemical degradation	Non-destructive
XRD	X-ray diffraction	Crystallinity, phases	Monitors structural changes	Accurate phase analysis
TGA	Mass vs temperature	Thermal stability, composition	Evaluates decomposition	Quantitative results

continued on following page

Table 3. Continued

Technique	**Principle**	**Measures**	**Role in Degradation Study**	**Key Advantages**
Mechanical Testing	Applied force response	Strength, modulus	Assesses performance loss	Direct evaluation
DSC	Heat flow measurement	Tg, Tm, crystallinity	Tracks thermal transitions	Sensitive to structural changes
AFM	Probe-based scanning	Surface roughness, nanoscale features	Analyzes early degradation	Nanoscale resolution
In Situ Techniques	Real-time monitoring	Dynamic structural/ chemical changes	Observes degradation in real time	Realistic insights

Figure 3. explains various Characterization techniques applicable for degradation analysis.

Figure 3. Characterization techniques applicable for degradation analysis

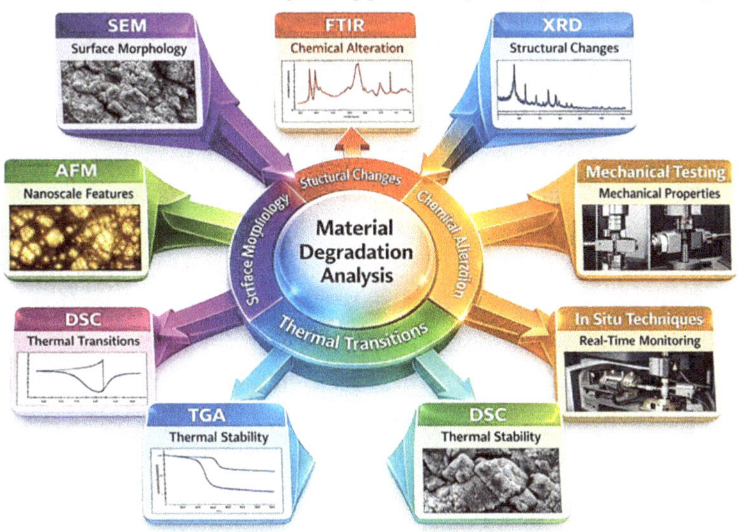

8. FUTURE PERSPECTIVES AND EMERGING TRENDS IN MULTIMATERIAL DESIGN

8.1. Smart, Stimuli-Responsive, and AI-Driven Multimaterial Systems

The future of multimaterial design is increasingly centered on the development of intelligent systems capable of dynamically adapting to their surrounding environments. While conventional materials have traditionally been designed to perform static functions, emerging multimaterial architectures are expected to exhibit active responses to external stimuli such as temperature, pH, light, magnetic fields, mechanical stress, and biological signals. This transition from passive to adaptive material behavior represents one of the most significant priorities in contemporary materials research. In particular, stimuli-responsive systems offer the possibility of real-time modulation of structural, mechanical, and biological properties, enabling materials to perform complex functions that were previously unattainable.

Among the most promising applications are advanced biomedical technologies, where environmental responsiveness can directly improve therapeutic outcomes. For example, pH-sensitive biodegradable materials capable of accelerating degradation in acidic microenvironments may enable precise drug release at diseased sites such as tumors or inflamed tissues. Similarly, shape-memory materials that recover predetermined configurations under specific physiological conditions hold substantial potential for minimally invasive implants and regenerative medicine. These developments suggest that future research efforts should prioritize the design of materials capable of integrating multiple adaptive responses while maintaining biocompatibility and long-term stability.

Equally transformative is the growing incorporation of artificial intelligence (AI) and machine learning (ML) into multimaterial design strategies. The complexity of interactions among composition, microstructure, processing parameters, and degradation behavior often exceeds the predictive capabilities of traditional experimental approaches. AI-driven methodologies address this challenge by extracting meaningful structure–property–performance relationships from large datasets, thereby enabling accelerated material discovery and optimization. Rather than relying solely on trial-and-error experimentation, researchers can increasingly use predictive models to identify optimal material combinations, processing conditions, and performance characteristics before fabrication. This capability is likely to reduce development costs and significantly shorten innovation cycles.

The convergence of AI with advanced manufacturing technologies further amplifies these opportunities. In particular, 4D printing introduces a new paradigm in which fabricated structures are programmed to evolve over time in response to environmental cues. The combination of data-driven design, stimuli-responsive materials, and programmable fabrication creates a pathway toward highly adaptive systems capable of self-adjustment and autonomous functionality. Looking ahead, the most influential advances are likely to emerge at the intersection of these technologies, enabling next-generation solutions in regenerative medicine, soft robotics, wearable electronics, and personalized healthcare. Consequently, the field's primary priority should not merely be the development of smarter materials but the creation of integrated material ecosystems that can sense, process, and respond to environmental information in a coordinated manner (Shangguan et al., 2025).

8.2. Sustainable, Bioinspired, and Multifunctional Material Architectures

As global concerns regarding resource consumption and environmental sustainability continue to intensify, the future direction of multimaterial design is increasingly shaped by the dual objectives of ecological responsibility and functional performance. The emphasis on sustainable materials is no longer a supplementary consideration but a central research priority. Future multimaterial systems must not only achieve superior technical performance but also minimize environmental impact throughout their life cycle. Consequently, increasing attention is being directed toward renewable, biodegradable, and recyclable materials, including biopolymers, natural fibers, and materials synthesized through environmentally benign processes. These approaches are particularly relevant in sectors such as packaging, environmental remediation, and biomedical engineering, where material disposal and degradation present significant challenges.

Beyond sustainability, bioinspiration has emerged as a powerful design philosophy capable of overcoming limitations associated with conventional engineering approaches. Natural systems have evolved over millions of years to achieve remarkable combinations of strength, toughness, flexibility, and functionality using relatively simple constituents arranged in hierarchical architectures. Materials inspired by bone, skin, nacre, and plant structures demonstrate how structural organization across multiple length scales can produce exceptional performance. Rather than focusing solely on material composition, future research is expected to prioritize the replication of natural gradients, interfaces, and hierarchical arrangements that govern biological functionality. Such strategies have already demonstrated the ability to improve fracture resistance, mechanical reliability, and biological compatibility, suggesting that nature-inspired architectures will remain a major source of innovation.

Another defining trend is the pursuit of multifunctionality within a single material platform. Historically, engineering systems have relied on separate components to perform sensing, structural support, healing, and biological functions. However, future multimaterial architectures are expected to integrate these capabilities into unified systems, thereby reducing complexity, weight, and manufacturing requirements. The ability to simultaneously provide mechanical integrity, bioactivity, sensing capability, and self-healing functionality represents a significant advancement in material efficiency and system-level performance. Achieving such integration requires careful control of material interfaces and interactions, an area where advances in nanotechnology and hybrid material engineering are proving particularly valuable.

From a broader perspective, the future success of multimaterial systems will depend on balancing three interconnected priorities: sustainability, bioinspiration, and multifunctionality. While each objective offers distinct advantages, their combined implementation has the potential to produce transformative solutions that address pressing challenges in healthcare, engineering, and environmental protection. The next generation of multimaterial architectures is therefore likely to move beyond merely combining different materials and instead focus on creating integrated, intelligent systems that emulate natural complexity while adhering to principles of sustainable development. Such a holistic approach represents the most promising pathway toward achieving both technological innovation and long-term societal benefit.

CONCLUSION

Biodegradable and bioactive multimaterial systems are a major step forward in the materials science field as they have provided an opportunity that is flexible to solve complex issues in biomedical, environmental, and engineering applications. This chapter has brought to light the importance of structure property degradation relationships in influencing the overall performance and functionality of such systems. By combining several phases of materials, a compromise can be made between mechanical strength, controlled degradation, and biological activity properties that are hard to attain in single-component materials. The fundamental aspects were discussed as to how material behavior is determined at various length scales by hierarchical structures and interfacial interactions. Processing methods, including both traditional fabrication methods and more advanced additive manufacturing, were demonstrated to be very important in the customization of microstructure, porosity, and spatial distribution of components. All these are what directly affect the relationship between structure and properties, and it becomes possible to design materials with performance-specific to the application. An in-depth analysis of the degradation mechanisms such as hydrolytic, enzymatic, and interface-based processes revealed the multicomponent nature of multimaterial systems breakdown. The fact that degradation is coupled to bioactivity further promotes the significance of designing materials that do not only degrade in a controlled way, but also actively take part in biological processes like tissue regeneration and healing. The use of characterization methods was also presented as the key to assessing structural, chemical, thermal and mechanical alterations that give the understanding required to optimize the work of material design. In the future, the future of this field is likely to be defined by such trends as smart and stimuli-responsive materials, AI-driven design solutions, and sustainable and bioinspired architectures. The innovations will make it possible to create adaptive multi-functional systems that are more efficient as well as less harmful to the environment. To sum up, rational design of next-generation biodegradable and bioactive multimaterial systems needs a thorough understanding of the interaction between structure, properties, and degradation. Further interdisciplinary studies involving materials science, biology, and computational tools will be important in realizing their potential and increasing their uses in new advanced technologies.

REFERENCES

Abdulsalam, L., Abubakar, S., Permatasari, I., Lawal, A. A., Uddin, S., Ullah, S., & Ahmad, I. (2025). Advanced Biocompatible and Biodegradable Polymers: A Review of Functionalization, Smart Systems, and Sustainable Applications. *Polymers*, *17*(21), 2901. DOI: 10.3390/polym17212901 PMID: 41228661

Al-Amiery, A. A., Fayad, M. A., Abdul Wahhab, H. A., Al-Azzawi, W. K., Mohammed, J. K., & Majdi, H. S. (2024). Interfacial Engineering for Advanced Functional Materials: Surfaces, Interfaces, and Applications. *Results in Engineering*, *22*, 102125. DOI: 10.1016/j.rineng.2024.102125

Aljabali, A. A. A., Alkaraki, A., Gammoh, O., Qnais, E., Alqudah, A., Mishra, V., Mishra, Y., & El-Tanani, M. (2025). Design, structure, and application of conductive polymer hybrid materials: A comprehensive review of classification, fabrication, and multifunctionality. *RSC Advances*, *15*(34), 27493–27523. DOI: 10.1039/D5RA04634C PMID: 40761904

Amoke, A., & Kalu, I. E. (2026). Advances in biodegradable polymeric materials: Focus on controlling factors and sustainable applications. *Next Research*, *7*, 101501. DOI: 10.1016/j.nexres.2026.101501

Brochu, B. M., Sturm, S. R., Kawase De Queiroz Goncalves, J. A., Mirsky, N. A., Sandino, A. I., Panthaki, K. Z., Panthaki, K. Z., Nayak, V. V., Daunert, S., Witek, L., & Coelho, P. G. (2024). Advances in Bioceramics for Bone Regeneration: A Narrative Review. *Biomimetics*, *9*(11), 690. DOI: 10.3390/biomimetics9110690 PMID: 39590262

Czapczyk, K., Zawadzki, P., & Wierzbicka, N. (2023). Influence of Dispersed Phase Content on the Mechanical Properties of Electroless Nanocomposite Ni-P/Si3N4 and Hybrid Ni-P/Si3N4/Graphite Layers Deposited on the AW-7075 Alloy. *Materials (Basel)*, *16*(18), 6100. DOI: 10.3390/ma16186100 PMID: 37763378

Ferreira Rocha, P. R., Fonseca Gonçalves, G., dos Reis, G., & Guedes, R. M. (2024). Mechanisms of Component Degradation and Multi-Scale Strategies for Predicting Composite Durability: Present and Future Perspectives. *Journal of Composites Science*, *8*(6), 204. DOI: 10.3390/jcs8060204

Hernandez, J. L., & Woodrow, K. A. (2022). Medical Applications of Porous Biomaterials: Features of Porosity and Tissue-Specific Implications for Biocompatibility. *Advanced Healthcare Materials*, *11*(9), 2102087. Advance online publication. DOI: 10.1002/adhm.202102087 PMID: 35137550

Karolewska, K., Szala, G., Trepczyńska-Łent, M., & Ligaj, B. (2024). Improving the mechanical properties of structural elements made of titanium, aluminum alloys, and steel through the additive manufacturing application. *Advances in Science and Technology Research Journal, 19*(1), 48–71. DOI: 10.12913/22998624/193764

Khan, N., & Riccio, A. (2024). A systematic review of design for additive manufacturing of aerospace lattice structures: Current trends and future directions. *Progress in Aerospace Sciences, 149*, 101021. DOI: 10.1016/j.paerosci.2024.101021

Lalonde, J. N., Pilania, G., & Marrone, B. L. (2025). Materials designed to degrade: Structure, properties, processing, and performance relationships in polyhydroxyalkanoate biopolymers. *Polymer Chemistry, 16*(3), 235–265. DOI: 10.1039/D4PY00623B PMID: 39464417

Liana, A., Hałuszczuk, A., Gawor, A., & Bulska, E. (2024). Identification and Structural Characterization of Degradation Products of Linagliptin by Mass Spectrometry Techniques. *International Journal of Molecular Sciences, 25*(5), 2591. DOI: 10.3390/ijms25052591 PMID: 38473837

Lou, J., Li, B., Liao, S., Vishkayi, S. I., Zhang, L., Tian, Y., Yin, L.-J., Tagani, M. B., Zhang, L., & Qin, Z. (2026). Coupling-dependent electronic structure in InSe-based lateral heterostructures. *Physical Review. B, 113*(3), 035302. DOI: 10.1103/439z-t2fh

Mukasheva, F., Adilova, L., Dyussenbinov, A., Yernaimanova, B., Abilev, M., & Akilbekova, D. (2024). Optimizing scaffold pore size for tissue engineering: Insights across various tissue types. *Frontiers in Bioengineering and Biotechnology, 12*, 1444986. Advance online publication. DOI: 10.3389/fbioe.2024.1444986 PMID: 39600888

Mukherjee, C., Varghese, D., Krishna, J. S., Boominathan, T., Rakeshkumar, R., Dineshkumar, S., Brahmananda Rao, C. V. S., & Sivaramakrishna, A. (2023). Recent advances in biodegradable polymers – Properties, applications and future prospects. *European Polymer Journal, 192*, 112068. DOI: 10.1016/j.eurpolymj.2023.112068

Nasi, A., Ushe, K., Dhoska, K., & Pramono, A. (2025). Correlations between Physical and Mechanical Properties of Composite Materials for Civil Structures: A Data-Supported Review. *Mesopotamian Journal of Civil Engineering, 2025*, 68–78. DOI: 10.58496/MJCE/2025/005

Park, H., & Lee, S. H. (2021). Review on Interfacial Bonding Mechanism of Functional Polymer Coating on Glass in Atomistic Modeling Perspective. *Polymers, 13*(14), 2244. DOI: 10.3390/polym13142244 PMID: 34301000

Parvez, M. M. H., Rahman, M. M., Ferdush, J., Al Mohotadi, M. A., Mondal, J., & Uddin, M. N. (2025). State-of-the-art nanocomposites: Tailoring material properties for next-generation applications. *Next Research, 2*(4), 100865. DOI: 10.1016/j.nexres.2025.100865

Promakhov, V., Matveev, A., Schulz, N., Dronov, P., Zhukov, A., & Vorozhtsov, A. (2022). Structure, Properties and Phase Composition of Composite Materials Based on the System NiTi-TiB2. *Materials (Basel), 15*(15), 5327. DOI: 10.3390/ma15155327 PMID: 35955262

Ramos, A., Angel, V. G., Siqueiros, M., Sahagun, T., Gonzalez, L., & Ballesteros, R. (2025). Reviewing Additive Manufacturing Techniques: Material Trends and Weight Optimization Possibilities Through Innovative Printing Patterns. *Materials (Basel), 18*(6), 1377. DOI: 10.3390/ma18061377 PMID: 40141660

Sabet-Bokati, K., & Plucknett, K. (2024). Water-induced failure in polymer coatings: Mechanisms, impacts and mitigation strategies—A comprehensive review. *Polymer Degradation & Stability, 230*, 111058. DOI: 10.1016/j.polymdegradstab.2024.111058

Shangguan, P., Zhou, H., Huang, X., Su, J., Yeong, W. Y., & Sing, S. L. (2025). Artificial intelligence-driven material development for additive manufacturing: A critical review. *International Journal of AI for Materials and Design, 2*(2), 1. DOI: 10.36922/IJAMD025100007

Wang, H., Kumar, R., Pattanaik, A., Kumar, R., Khawaf Aljaberi, A. S. O., & Abass, M. A. (2025). Computational methods and artificial intelligence-based modeling of magnesium alloys: A systematic review of machine learning, deep learning, and data-driven design and optimization approaches. *Frontiers in Materials, 12*, 1645227. Advance online publication. DOI: 10.3389/fmats.2025.1645227

Wang, X., Wang, C., Chu, C., Xue, F., Li, J., & Bai, J. (2024). Structure-function integrated biodegradable Mg/polymer composites: Design, manufacturing, properties, and biomedical applications. *Bioactive Materials, 39*, 74–105. DOI: 10.1016/j.bioactmat.2024.05.024 PMID: 38783927

Wawrzyniak, A., & Balawender, K. (2022). Structural and Metabolic Changes in Bone. *Animals (Basel), 12*(15), 1946. DOI: 10.3390/ani12151946 PMID: 35953935

Yu, P.-J., Lin, Y.-C., & Chen, W.-C. (2025). Review of bioderived and biodegradable polymers/block-copolymers and their biomedical and electronic applications. *Polymer Journal, 57*(3), 233–247. DOI: 10.1038/s41428-024-00980-z

Zhang, A., Wong, J. K. U., Xia, Y., Bilek, M., Yeo, G., & Akhavan, B. (2025). Surface biofunctionalised porous materials: Advances, challenges, and future prospects. *Progress in Materials Science*, *154*, 101518. DOI: 10.1016/j.pmatsci.2025.101518

Zhu, S., Fu, Q., & Yang, L. (2026). Interface coupling effects on Rayleigh–Taylor and Faraday instabilities in granular suspensions under oscillatory forcing. *International Journal of Multiphase Flow*, *194*, 105448. DOI: 10.1016/j.ijmultiphaseflow.2025.105448

Chapter 2
Nanotechnology in Bioactive Biodegradable Systems:
Nanoparticles, Nanofibers, and Nanotopography for Cell Guidance and Immune Modulation

S. Sandhiya
http://orcid.org/0009-0004-3320-9617
Dr. MGR Educational and Research Institute, Chennai, India

S. Janarthanan
http://orcid.org/0009-0006-0488-0342
Dr. MGR Educational and Research Institute, Chennai, India

Aman Suresh Tharayil
http://orcid.org/0009-0000-5377-7428
Dr. MGR Educational and Research Institute, Chennai, India

R. Praveen Javali
http://orcid.org/0009-0001-3737-5813
Sri Raghavendra College of Pharmacy, Chitradurga, India

Adinarayana Andy
http://orcid.org/0000-0002-5163-1094
Weatherwax Family Pharmacies Inc., USA

DOI: 10.4018/979-8-2600-0060-1.ch002

ABSTRACT

Biodegradable bioactive biomaterials have gained significant attention in regenerative medicine due to their ability to interact with biological tissues and gradually degrade after performing their therapeutic function. Unlike traditional inert materials, these advanced biomaterials actively promote cell adhesion, proliferation, and tissue regeneration. Recent progress in material science has enabled the development of systems based on natural polymers, synthetic biodegradable polymers, bioactive ceramics, and composite materials. In addition, smart and stimuli-responsive biomaterials capable of responding to pH, temperature, enzymes, and redox conditions have improved targeted therapeutic delivery. Emerging technologies such as nanostructured scaffolds, biofunctionalization strategies, and 3D bioprinting are further expanding the potential applications of these materials in regenerative medicine and tissue engineering.

INTRODUCTION

Biodegradable bioactive biomaterials are at the forefront of modern regenerative medicine and biomedical engineering, with a class of materials developed not only to function within a biological system, but to do so in a way that is interactive with the living tissues, as well as degrading after the intended therapeutic purpose is accomplished. Current generations of biomaterials were used for structural or inert purposes, such as replacing joints or prosthetics. The newer generations of biomaterials are developed to have an active biological role, which can guide the behaviour of cells, promote regeneration, and revolutionize treatment outcomes for a variety of clinical specialties (Yang, Hu, & Jiang, 2025).

One of the major driving forces for the fast development of these materials is the necessity for the development of innovative biomaterials for tissue repair and healing. Biodegradability is a feature that ensures that the biomaterial used has served the intended purpose by allowing tissue development or the delivery of the necessary therapy; it then biodegrades into harmless products that are easily metabolized by the body, thus eliminating the need for surgical removal of the material.

In the last few years, material scientists have used a wide variety of natural and synthetic polymers for creating scaffolds that closely resemble the natural architecture of the tissue's ECM. Natural biopolymers, including chitosan, collagen, and alginate, possess intrinsic biocompatibility and biodegradability. Conversely, synthetic polymers such as PLA and PCL have demonstrated potential in tissue engineering scaffolds, attributable to their mechanical properties and degradation profiles (Shergujri et al., 2025; Park & De, 2025).

The bioactive agents used, like bioactive glass, hydroxyapatite, and nanostructured agents, improve the effectiveness of these materials for regeneration. For instance, bioactive glass has shown that it can interact with bones, stimulate bone formation, and release bioactive ions that stimulate the development of new blood vessels and signalling (Zhu et al., 2025). Nanotechnology has also improved the effectiveness of biodegradable biomaterials. For instance, biodegradable polymer graphene nanocomposites have shown promising results in terms of increased surface area, mechanical strength, and material-cell interactions for drug delivery systems (Mohiuddin et al., 2025). On the other hand, the development of biomaterials has shown great promise in the development of 3D bioprinting and electrospinning techniques that allow us to customize scaffolds for tissue regeneration. This is important for tissue regeneration since these three factors are essential for tissue regeneration.

1. FUNDAMENTAL PRINCIPLES OF BIODEGRADABLE BIOACTIVE MATERIALS

1.1 Definition and classification

Bioinert materials are non-toxic and bio-inert. The body reacts to these by encapsulating the implant in a fibrous layer. The thickness of the layer varies. Highly bioinert materials such as alumina, zirconia, silicon nitride, and gold induce an extremely thin encapsulation layer. The thickness of the layer is even less if the implant fits snugly. On the other hand, if the implant fits loosely and can move, the layer can be several hundred microns thick. Less bioinert materials such as stainless steel and nitinol induce a thicker layer of fibrous tissue. (Li & Zreiqat, 2019).

Bioactive materials are non-toxic, biologically active materials that create a direct chemical link between the implanted device and the tissue with which it is in contact by initiating a biological response at the tissue-implant interface. Bioactive materials are biodegradable, playing a major role in tissue engineering and regenerative medicine. The body naturally accepts bioactive materials, which can aid the repair and regeneration of the site of the implanted device. Bioactive materials include calcium phosphates, which create a direct bond with hard tissues, including bone, as well as bioactive glasses, which can bond directly with both hard and soft tissues, including muscle and ligament (Li & Zreiqat, 2019).

Biodegradable materials are non-toxic biological materials that are capable of degradation within a specific time frame. The materials are soluble or disintegrate via processes like erosion and enzymatic degradation. The materials are then degraded by the body. The materials can also induce the formation of a fibrous capsule depending on their bioactivity. Dissolvable materials like collagen and β-TCP have positive biological properties but disintegrate within weeks or months. This limits their ability to be used for tissue reconstruction. The disintegration of materials like PCL and HA is slow, making them bioinert. The materials can be retained for years to aid tissue reconstruction. (Li & Zreiqat, 2019).

Figure 1. Classification of biomaterials

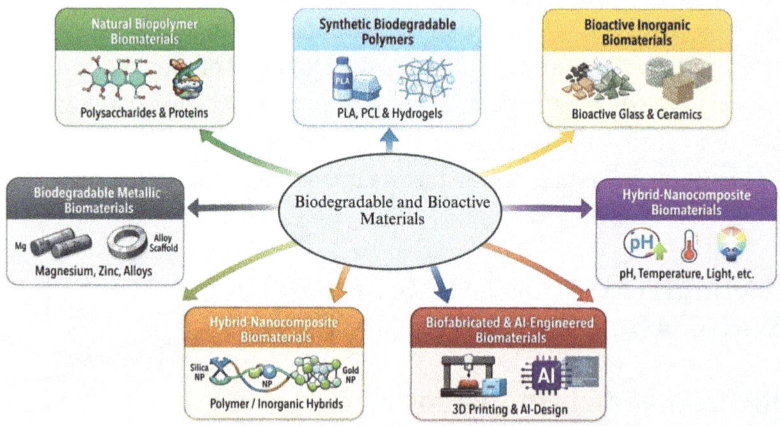

Figure 1 represents the full hierarchy of biodegradable and bioactive materials that are used at the core of next-generation biomedical systems The radial diagram represents six main types of materials diverging out of the center hub -Biodegradable and Bioactive Materials- such as natural biopolymers synthetic biodegradables bioactive inorganic ceramics, degradable metals, hybrid nanocomposites, and state-of-the-art 4D-printed engineered constructs in response to pH/temperature/light triggers. Synergy between platforms and fabrication technologies are visualized by interconnecting arrows and offer a roadmap of nanotechnology and integrating it into tissue engineering, drug delivery, and both regenerative medicine practices.

1.2. Biodegradation Mechanism Hydrolytic degradation

Hydrolytic degradation is the most common type of degradation observed in synthetic biodegradable polymers used in biomedical applications. In this degradation process, water molecules, which are hydrophilic, penetrate the polymer matrix. This penetration breaks the hydrolyzable chemical bonds, such as ester linkages. As a result, the polymer chains break down into smaller oligomers and monomers (D'Amario et al., 2025). Biodegradable polymers, like PLA, PGA, PCL, and PLGA, usually undergo hydrolytic degradation. This is because of the hydrolyzable chemical bonds, specifically ester linkages, in the polymer chains (Yao et al., 2024).

Oxidation

Oxidative degradation is yet another important factor in biomaterial degradation in a biological environment. Oxidative degradation happens when reactive oxygen species and oxidative enzymes degrade the polymer chains. As a result, chemical changes occur in the biomaterials (Rovaletti et al., 2023). Inflammatory cells such as macrophages and neutrophils release reactive oxygen species such as superoxide radicals, hydrogen peroxide, hydroxyl radicals, and hypochlorous acid. All these reactive oxygen species have the potential to produce free radicals in biomaterials. As a result, oxidative degradation happens in biomaterials (Rovaletti et al., 2023). Biomaterials such as polyurethane, polyethylene, and some polymer scaffolds undergo oxidative degradation. In some biomedical implants, oxidative degradation happens along with hydrolytic or enzymatic degradation.

Enzymatic Degradation

Enzymatic degradation is defined as the biological degradation of biomaterials by means of enzyme-catalyzed reactions. In this type of degradation, enzymes from microorganisms or tissues of the host react with the biomaterial surface, leading to the breakdown of specific chemical bonds in the biomaterial, which then results in the fragmentation of the biomaterial and the release of degradation products (Chen et al., 2020). The mechanism of degradation in this case includes the diffusion of enzymes to the biomaterial surface, adsorption of the enzymes to the biomaterial, breakdown of the biomaterial by the enzymes, and finally, the release of degradation products from the biomaterial. The enzymes involved in this type of degradation include lipases, esterases, proteases, and depolymerases. Natural biomaterials such as collagen, gelatin, chitosan, and silk fibroin are prone to this type of degradation since they are similar to components of the extracellular matrix, making it an essential factor in tissue engineering scaffolds, which degrade over time to produce new tissues.(Roohi et al., 2017)

1.3 Bioactivity at the Tissue–Material Interface

Bioactivity at tissue-material interfaces is the ability of materials to interact with biological tissues and elicit specific biological responses that facilitate tissue integration and regeneration. When a material is implanted into the human body, it instantly comes into contact with biological fluids and cells, which triggers processes such as protein adsorption, cell adhesion, immune response, and tissue remodelling. The interaction between materials and biological tissues at the tissue-material interface is crucial for successful tissue integration. This interaction also helps prevent negative biological responses, which could lead to tissue damage and inflammation. The interaction between materials and biological tissues is intended to elicit positive biochemical and biophysical reactions at tissue-material interfaces, which facilitate the development of a strong interfacial bond between materials and biological tissues, thereby improving tissue-engineered devices and biomedical devices. (Hench & Polak, 2002).

1.4 Host–Immune–Material Interactions

The host immune-material interactions are the biological responses elicited when the biomaterial comes into contact with the host immune system. As soon as the biomaterial is implanted, the proteins present in the bodily fluids start to adsorb on the surface of the biomaterial, which then attracts the host immune cells such as macrophages, neutrophils, and dendritic cells. These immune cells then assess the biomaterial and start to produce either a pro-inflammatory or regenerative response, depending on the physiochemical properties of the biomaterial. If the immune response against the biomaterial is high, then the biomaterial can lead to chronic inflammation and fibrous capsule formation, while if the immune response is controlled, then the biomaterial can lead to healing and integration. Thus, the current trend in biomaterial development aims to produce immunomodulatory biomaterials, which can control the macrophage function and immune response for the healing and integration of the biomaterial (Tripathi et al., 2023).

1.5 Surface Energy, Wettability, and Protein Adsorption Dynamics

Surface energy and wettability are two physicochemical properties of great importance in biomaterials science. When biomaterials are subjected to a biological medium, proteins are known to adsorb onto their surface almost instantaneously. The adsorption of proteins is largely determined by thermodynamic interactions between the biomaterial surface, water molecules, and proteins. Biomaterials of varying surface energies and wettability (hydrophobic and hydrophilic properties) are known to influence enthalpic and entropic interactions at the surface, which in turn can affect protein adsorption. Hydrophobic biomaterials are known to enhance irreversible protein adsorption, whereas hydrophilic biomaterials may inhibit nonspecific protein adsorption by the formation of a hydration layer around proteins. Hence, biomaterial surface energy and wettability are of great importance in biomaterials science in controlling protein adsorption dynamics and developing biomaterials for medical applications.

2. MATERIAL PLATFORMS FOR NEXT-GENERATION SYSTEMS

2.1 Natural Polymers

Natural polymers have been extensively studied as part of next-generation biomaterials systems owing to their inherent biocompatibility and biodegradability. Natural polymers like collagen, gelatin, chitosan, alginate, and hyaluronic acid have been found to provide beneficial biological signals that facilitate cell adhesion, proliferation, and tissue engineering. Natural polymers have been found to be processed into scaffolds, hydrogels, and bioactive matrices through advanced manufacturing technologies like electrospinning and 3D bioprinting. Natural polymers have been found to be biomimetic and interact with various cellular components, thereby providing an opportunity to use natural polymers for tissue engineering and regenerative medicine applications.

2.2 Synthetic Biodegradable Polymers

Synthetic biodegradable polymers, such as poly(lactic acid) (PLA), polycaprolactone (PCL), and poly(lactic-co-glycolic acid) (PLGA), are frequently utilized in biomedical contexts. This is attributable to their capacity for customization, allowing for the manipulation of mechanical characteristics and degradation profiles. These synthetic biodegradable polymers can be processed into scaffolds and matrices. These constructs facilitate cellular adhesion and tissue regeneration, concurrently maintaining structural integrity throughout the healing process. PLA, a synthetic polymer made from aliphatic polyester, is often used because it's biocompatible and environmentally friendly. However, it's hydrophobic and breaks down slowly, which means it often needs to be modified or mixed with other materials.

2.3 Bioactive Ceramics

Bioactive ceramics are significant biomaterials due to their capacity to establish direct bonding with biological tissues. Hydroxyapatite, tricalcium phosphate, and bioactive glasses are frequently employed in orthopedic and dental implant applications. These materials, sharing compositional similarities with bone minerals, promote osteogenesis. Upon exposure to physiological fluids, they facilitate the formation of an apatite layer on the implant's exterior. This bioactivity enables the development of strong bonding between the implants and the bone, thereby improving the implants' mechanical properties.

2.4 Composite and Hybrid Biomaterials

Composite and hybrid biomaterials have been developed to combine organic polymers with various inorganic bioactive materials, with the goal of improving mechanical and biological properties. For example, polymer-based composite materials containing bioactive glass or hydroxyapatite nanoparticles have shown good mechanical stability, bioactivity, and cell response. Composite materials containing nanoparticles of various ceramics have been found to improve cell adhesion, proliferation, and differentiation on polymer surfaces while controlling degradation rates. Hybrid composite materials have been increasingly used to study various biological applications, including bone tissue engineering, drug delivery systems, and regenerative medicine.

3. SMART AND STIMULI-RESPONSIVE BIODEGRADABLE BIOMATERIALS

The term "smart" refers to biomaterial qualities that can stimulate or inductively affect tissues in response to external or internal stimuli. Smart on-demand stimuli-responsive biomaterials can maximize therapeutic efficacy and minimize undesirable side effects by rapidly detecting and responding to the disease environment and exerting therapeutic effects while preserving physiologically healthy cells and tissues, thus improving patient quality of life. The next evolutionary step in improving patient-centered care is the use of smart biomaterials, which may offer previously unattainable treatment options. The customized stimuli-responsiveness of smart biomaterials aims to address numerous shortcomings of existing drug delivery systems. Highly controlled smart systems have been produced by incorporating various internal stimuli from the physiological microenvironment, such as pH, redox, temperature, enzyme, and mechanical force, or some easily accessible and relatively easily controlled external stimuli, such as light, ultrasound, electric, and magnetic . Many smart systems have spatiotemporal control, which reduces potential side effects and off-target toxicity while also increasing efficacy.

3.1 pH-Responsive Systems

This type of system is beneficial because many pathological processes, including tumors, inflamed tissues, and infectious processes, have abnormal pH levels compared to normal tissues. Generally, the pH of tumor tissues is slightly acidic, which can cause the degradation of pH responsive systems, thereby controlling the release of a drug.pH responsive systems are generally made from polymers that are capable of responding to pH changes due to the presence of functional groups, which can either donate or accept protons. The main functional groups used for pH-responsive systems are carboxylic acid, amines, or sulfonates. The main difference in the structure of these systems is the type of functional groups used, which can either gain or lose a proton, thereby causing the system to change its structure, affecting its solubility, swelling, and the rate of drug release (Binauld & Stenzel, 2013; Li & Zreiqat, 2019).

3.2 Temperature-Responsive Polymers

Smart biomaterials like temperature responsive polymers are capable of showing reversibility in their properties when exposed to temperature changes. The commonly observed mechanism is based on a polymer that possesses a lower critical solution temperature (LCST). This type of polymer is capable of showing a phase transition from a soluble state to an insoluble state when exposed to temperature changes beyond a particular limit (Roy et al., 2010). Poly(N-isopropylacrylamide) (PNIPAM) is one of the commonly studied temperature responsive polymers owing to its LCST close to physiological temperature. The polymer is hydrophilic when exposed to temperatures lower than its LCST, whereas it becomes hydrophobic when exposed to higher temperatures than its LCST. This type of temperature responsive polymer is of considerable interest for injectable hydrogels, drug delivery systems, and tissue engineering scaffolds that can be responsive to physiological temperature (Roy et al., 2010). Temperature-responsive biodegradable polymers are of particular interest for minimally invasive therapies owing to their ability to be injected in a liquid state, which can later be converted into a stable state when exposed to physiological temperature.

3.3 Enzyme-Responsive Materials

These materials degrade or change their composition in the presence of specific enzymes found in the body. Since the expression of enzymes can vary significantly among tissues and disease states, the response to such materials can be highly specific. This makes them highly valuable as therapeutic agents (Ulijn & Smith, 2008). For instance, some materials have peptide or polymeric bonds that can be cleaved by enzymes such as matrix metalloproteinases (MMPs), proteases, or esterase. These types of materials have been extensively used as tissue engineering scaffolds since MMPs are actively involved in tissue regeneration (Ulijn & Smith, 2008). By incorporating such sensitive bonds into the materials, they can be designed to degrade selectively in response to the enzymes. Such materials have been studied for the treatment of cancer, wound healing, and tissue engineering.

3.4 Redox and ROS-Responsive Systems

The redox-sensitive biomaterials are designed based on the ability of these biomaterials to sense changes in redox potential of cells, primarily based on the differences between intracellular and extracellular glutathione concentrations. In various pathological states, including cancer and inflammation, the levels of ROS are increased, which can be utilized for controlled release (Cheng et al., 2011). Disulfide bonds are generally utilized for designing redox-sensitive polymeric biomaterials. These bonds are stable under normal biological conditions, whereas they are cleaved under reducing environments, such as the intracellular cytoplasm, which contains high concentrations of glutathione. This cleavage of the bonds results in the degradation of the polymeric biomaterial, causing the release of the encapsulated drug (Cheng et al., 2011). ROS-sensitive polymeric biomaterials generally consist of oxidation- sensitive functional groups, including thioketal bonds, which degrade upon exposure to high concentrations of ROS. Such systems are highly useful for the treatment of inflammatory diseases and cancer, as oxidative stress is a major contributing factor for the disease processes.

3.5 Magnetically and Electrically Responsive Biomaterials

Magnetically responsive biomaterials are an advanced kind of smart material that can be remotely controlled. Magnetic responsive biomaterials contain iron oxide particles that can be controlled by an external magnetic field (Reddy et al., 2012). They are typicallly used for the guidance of material placement, drug delivery, or even for hyperthermia therapy. In a similar way, it is possible to utilize electrically responsive materials for regenerating tissues that are electrically active, like neural tissue (Guo & Ma, 2018). Conductive polymers like polypyrrole, polyaniline, or even graphene have shown promise for creating electrically responsive materials for tissue regeneration. This material has the potential for improving tissue regeneration by increasing cell proliferation or differentiation. This material has the potential for improving tissue regeneration by increasing cell proliferation or differentiation. Magnetic responsive materials are also being explored for their use in neural regeneration, cardiac tissue engineering, or even biosensing technology.

4. NANOENGINEERED AND MICROSTRUCTURE BIOMATERIALS

Nano-engineered and micro-structured biomaterials are a new class of advanced biomaterials that mimic the hierarchical structure found in tissues. Biomaterials with nanoscale and microscale structural features are being developed to improve cell-material interaction. Such biomaterials are designed to improve cell-biomaterial interaction by offering a high surface area for the adhesion of proteins and cells. Nanotechnology-based biomaterials are becoming increasingly important in tissue engineering, drug delivery, and regenerative medicine. Nanotechnology-based biomaterials are gaining traction because they can replicate the extracellular matrix and influence cellular activities (Huang et al., 2024; Hasan et al., 2018).

4.1 Nanofibrous Scaffolds (Electrospinning)

Electrospinning is a widely employed technique for fabricating nanofibrous scaffolds, which are engineered to closely resemble the extracellular matrix's architecture. This technique employs a high-voltage electric field, which is applied to a polymer solution or melt, resulting in the formation of very fine fibers. These fibers have diameters that range from nanometers to micrometers. These nanofibers are characterized by significant porosity, a large surface area, and interconnected networks. These properties facilitate cell adhesion and proliferation. Because of these inherent structural characteristics, electrospun scaffolds are well-suited for both tissue regeneration and the targeted delivery of biomolecules (Huang et al.,2024) The structural parallels between electrospun nanofibers and the extracellular matrix facilitate efficient cell adhesion, migration, and subsequent differentiation. Moreover, the porous structure facilitates the efficient transport of nutrients and the removal of metabolic waste. In addition, electrospun scaffolds can be modified with growth factors, drugs, or nanoparticles, which allows for sustained therapeutic effects during tissue regeneration (Bhardwaj & Kundu, 2010).Coaxial and multi-layer electrospinning, advanced electrospinning techniques, enable the creation of multifunctional scaffolds. These scaffolds are characterized by complex structures and compositions that change gradually. These advancements have broadened the scope of electrospun nanofibers in diverse applications, including bone tissue engineering, wound healing, neural regeneration, and vascular grafts (Huang et al., 2024; Bhardwaj & Kundu, 2010).

4.2 3D Bioprinting and Additive Manufacturing

The advent of three-dimensional bioprinting and additive manufacturing has fundamentally altered how biomaterials are created, allowing for the layer-by-layer construction of complex biological structures.These methodologies employ computer-aided design models to produce scaffolds, allowing for meticulous control over geometric configurations, pore dimensions, and mechanical properties. Unlike traditional manufacturing methods, additive manufacturing allows for the creation of patient-specific implants, which are known for their exceptional structural precision (Murphy & Atala, 2014; Matai et al., 2020).In three-dimensional bioprinting, bioinks, which are made from biomaterials, are used to encapsulate living cells and bioactive molecules.These bio inks generally comprise hydro gels or polymer composites, which serve to replicate the extracellular matrix and facilitate cell proliferation. Through the precise placement of cells and bio molecules within the scaffold, bio printing facilitates the creation of functional tissues characterized by intricate structural organization (Murphy & Atala, 2014; Matai et al., 2020).

Recent advancements in bioprinting have facilitated the fabrication of engineered tissues, encompassing bone, cartilage, and blood vessels. Furthermore, bio printed constructs are finding expanded applications in disease modelling, drug evaluation, and personalized medicine (Matai et al., 2020).

4.3 Nanoparticle-Integrated Bioactive Systems

Nanoparticle-integrated biomaterials have become multifunctional systems, improving the biological performance of scaffolds and therapeutic materials. Nanoparticles have unique physical and chemical properties, such as a large surface area, the ability to change their surface chemistry, and increased reactivity. When added to biomaterials, these nanoparticles can offer antimicrobial properties, controlled drug delivery, and improved mechanical properties (Hasan et al., 2018;).Polymeric, metallic, and inorganic nanoparticles are extensively employed in biomaterials for the delivery of therapeutic agents, including growth factors, nucleic acids, and pharmaceuticals. These nanoscale carriers facilitate both targeted and prolonged delivery within the tissue microenvironment, consequently enhancing therapeutic efficacy. As an illustration, silver nanoparticles integrated into scaffolds confer antimicrobial properties, thereby mitigating infection during the wound healing process (Hasan et al., 2018). Nanoparticle-based biomaterials are currently being studied for various biomedical uses, including cancer treatment, regenerative medicine, and controlled drug delivery systems.

4.4 Surface Nanotopography and Cell Signaling

Surface nanotopography encompasses nanoscale structural characteristics present on biomaterial surfaces, which subsequently affect cellular activities. These characteristics, such as nanopillars, nanogrooves, and nanopits, replicate the nanoscale architecture of the extracellular matrix. Consequently, surface nanotopography can modulate cell adhesion, proliferation, and differentiation by influencing cell–material interactions (Dalby et al., 2007).Cells interact with surfaces that have nanoscale features through integrin receptors, which then form focal adhesion complexes.

These interactions activate mechanotransduction pathways, which convert mechanical signals into biochemical responses within the cell.Therefore, these signaling pathways influence gene expression, which then affects how cells develop (Dalby et al., 2007).Nanotopographical alterations to surfaces have been extensively studied within the domains of orthopedic and dental implants, with the aim of augmenting osseointegration and tissue integration. Such modifications possess the capacity to facilitate stem cell differentiation and expedite the process of tissue regeneration.

4.5 Microfluidic Fabrication Techniques

This techniques have become significant methods for creating biomaterials, offering precise control over size, shape, and composition. These techniques use microfluidic devices to manage small amounts of fluid in tiny channels, allowing for precise control of chemical reactions and material creation. As a result, this technology enables the production of uniform particles, fibers, and microgels, which are essential for creating advanced biomaterials Microfluidic systems allow for precise control over flow rates, mixing, and the conditions of chemical reactions. Therefore, these materials show a high degree of consistency and uniform properties. Because of these characteristics, microfluidic methods are especially useful for creating drug delivery systems and cell-based biomaterials. Furthermore, microfluidic platforms can replicate the small environments found in the body, which helps create organ-on-chip systems.

6. BIOFUNCTIONALIZATION STRATEGIES

The process of biofunctionalization of the biomaterial surfaces or structures is the modification of the surface or the structure with biological molecules with the aim of enhancing the interaction of the biomaterial with biological cells or tissues. The current biomaterials offer the biological structures or scaffolds that are needed, but they do not offer the biological signals that are required for the regulation of various cellular activities such as adhesion, proliferation, differentiation, and regeneration. The biofunctionalization strategies aim at offering the biological signals in the biomaterials so that they become active participants in the biological activities rather than passive participants as seen in the current biomaterials. The biofunctionalization strategies include biofunctionalization of thebiomaterialswith growth factors, peptides, antimicrobialpeptides,gene-activated matrices, and extracellular vesicles such as exosomes. Biofunctionalization of the biomaterials enables them to mimic the natural extracellular matrix and offer the optimum microenvironment for regeneration and therapeutic activities (Hersel et al., 2003; Lutolf & Hubbell, 2005).

6.1 Growth Factor Immobilization

Growth factors are defined as signaling molecules that regulate cellular activities like growth, migration, differentiation, and tissue repair. The immobilization of growth factors onto the surface of biomaterials can be an important strategy to enhance the activity of growth factors, which can prevent degradation and diffusion. Growth factor immobilization facilitates the presentation of bioactive molecules to cells in a sustained manner.This enhances the regenerative potential of a biomaterial scaffold (Lutolf & Hubbell, 2005).Different immobilization techniques for growth factors on a biomaterial surface have been reported in scientific literature. The immobilization techniques include physical adsorption, covalent binding, and affinity-based immobilization.Covalent binding is a highly efficient immobilization technique for growth factors. However, in an affinity-based immobilization system, heparin and other growth factor-interacting molecules are used to specifically interact with growth factors in a controlled manner (Lee et al., 2011). The growth factors that are immobilized on the surface of the biomaterials, such as vascular endothelial growth factor (VEGF), bone morphogenetic growth factor (BMP), and fibroblast growth factor (FGF), are used in regenerative medicine. These growth factors promote angiogenesis, osteogenesis, and tissue remodeling. As a result, the biomaterial scaffold's ability to regenerate is improved (Lee et al., 2011; Lutolf & Hubbell, 2005).

6.2 Peptide Functionalization (RGD Motifs)

Another method for the functionalization of biomaterials is peptide functionalization.The peptide sequence consists of a short sequence of amino acids derived from the extracellular matrix proteins.The peptide sequence is incorporated into the biomaterials to improve the interaction with the cells.Among the peptide sequence, the most commonly used peptide sequence is the arginine-glycine-aspartic acid (RGD) peptide motif, which plays an important role in the interaction with the integrin receptor present on the surface of the cells (Hersel et al., 2003).The peptide sequence consists of a short sequence of amino acids derived from the extracellular matrix proteins such as fibronectin, vitronectin, and laminin.The peptide sequence is incorporated into the biomaterials to improve the interaction with the cells.The inclusion of peptide sequences within biomaterials promotes cell adhesion.This adhesion arises from the interaction between the peptide sequence and the integrin receptors located on the cell surface.Consequently, this interaction triggers intracellular signaling pathways, which subsequently govern cytoskeletal migration, differentiation, and organization (Hersel et al., 2003). The incorporation of peptide sequences into biomaterials serves to functionalize them for diverse tissue engineering applications, including bone regeneration, vascular grafts, and neural tissue engineering (Hersel et al., 2003).

6.3 Antimicrobial Functionalization

One of the biggest challenges for biomaterials and scaffolds implanted in the body is the infections they might lead to. Antimicrobial functionalization tries to prevent bacteria from lingering and forming biofilms on the biomaterials. This is done by incorporating antimicrobial materials, for instance, antibiotics, antimicrobial peptides, metal nanoparticles, and polymer materials, into the biomaterials (Campoccia et al., 2013). Antimicrobial peptides, for instance, have shown great hope for the future as they have the ability to target a broad range of bacteria, fungi, and viruses. They have the ability to pierce the bacterial membrane and prevent the formation of biofilms on the biomaterials (Campoccia et al., 2013). Metal nanoparticles, for instance, silver and zinc oxide, have also been used to give biomaterials antimicrobial properties. This works by releasing metal ions that disrupt the bacterial membrane and metabolic processes. Antimicrobial-functionalized biomaterials have been extensively researched for the use of orthopedic implants, wounds, and other medical devices to reduce infections (Campoccia et al., 2013).

6.4 Gene-Activated Matrices

Gene-activated matrices (GAMs) are a novel idea for the biofunctionalization of biomaterials. This concept of biofunctionalization involves the interaction of biomaterials and the genetic engineering of genes.In other words, the biomaterial acts as a scaffold and transfers genetic material, for example, plasmid DNA, mRNA, or virus, to the target cells. After the uptake of the genetic material by the target cells, the expressed genes produce therapeutic proteins, which aid the healing process (Bonadio et al., 1999). GAMs enable the expression of GFs, for example, plasmid DNA, mRNA, or virus, to produce therapeutic proteins.For instance, the use of a plasmid DNA expressing BMP genes for the production of a scaffold biomaterial to induce bone formation in orthopedic materials (Bonadio et al., 1999).The use of the genetic engineering of genes, for instance, the use of a scaffold biomaterial expressing BMP, has the following advantages over the use of GFs: the expressed GFs last longer and the production of therapeutic proteins is cheaper.Gene-activated biomaterials have shown promising results for tissue engineering of bone, wound healing, and vascular tissue engineering.

6.5 Exosome-Loaded Biomaterials

Exosomes are small vesicles secreted by cells that contain proteins, lipids, and nucleic acids to communicate with each other. Exosomes play important roles in tissue repair, immune response, and many other cellular signaling pathways. Catching exosomes and loading them onto biomaterials appears to be very promising for advancing tissue repair (El Andaloussi et al., 2013). When exosomes are loaded onto biomaterials, they can be considered carriers that protect exosomes from degradation and release exosomes slowly at the wound site. Exosomes loaded onto biomaterials can improve cellular communication with each other to promote tissue repair by providing bioactive substances such as microRNAs and growth factors (El Andaloussi et al., 2013). Recently, it has been demonstrated that exosomes loaded onto biomaterials can greatly improve wound repair, bone formation, and cartilage repair. Exosomes loaded onto biomaterials mimic the natural cellular signaling process to open new avenues for tissue repair and tissue engineering (El Andaloussi et al., 2013).

7. IMMUNOMODULATORY AND OSTEOIMMUNOMODULATORY BIOMATERIALS

The immune system significantly influences the outcomes of biomaterial implantation.Historically, biomaterials were engineered to be biologically inert, thereby minimizing immune responses.Nevertheless, contemporary investigations have revealed that controlled immune reactions are critical for both tissue regeneration and the successful integration of implants.Consequently, immunomodulatory biomaterials have been developed to actively engage with immune cells, thereby regulating inflammation and facilitating healing processes.These biomaterials are engineered to precisely modulate immune signaling pathways, thus directing immune cell activity toward regenerative outcomes (Chen et al., 2016; Sadtler et al., 2016).Osteoimmunomodulatory biomaterials are formulated to specifically impact the interplay between the immune system and bone.The functions of immune cells, particularly macrophages and T cells, are crucial for bone regeneration.Subsequently, these cells regulate the functions of osteoblasts and osteoclasts.By altering immune responses, these biomaterials may facilitate bone formation, mitigate inflammation, and ultimately enhance the efficacy of orthopedic implants and bone tissue engineering scaffolds (Chen et al., 2016).

7.1 Macrophage Polarization (M1/M2 Modulation)

Macrophages, which are crucial components of the immune system, manage both inflammatory responses and tissue repair after biomaterials are implanted. These cells are highly adaptable, capable of polarizing into various functional forms depending on the surrounding environment. The two primary macrophage phenotypes are the pro-inflammatory M1 and the anti- inflammatory M2.M1 macrophages are known for producing inflammatory cytokines, such as tumor necrosis factor-α (TNF-α), interleukin-1β (IL-1β), and reactive oxygen species, which start immune responses against foreign substances.M2 macrophages, on the other hand, help repair and grow tissues by releasing growth factors and anti-inflammatory cytokines.

7.2 Immune-Instructive Materials

Immune-instructive biomaterials are different from strategies that try to avoid the immune system.These materials are engineered to actively modulate immune responses.They possess biochemical or physical cues that direct immune cells to undertake actions that facilitate healing and mitigate inflammation.Immune-instructive materials can shift the balance between inflammation and tissue repair through their interactions with immune cells, such as macrophages, dendritic cells, and T lymphocytes (Sadtler et al., 2016).Several design strategies have been employed to create biomaterials capable of altering immune system function.These strategies encompass surface chemistry modifications, the incorporation of immune-regulatory molecules, and the deliberate design of scaffold architectures to govern immune cell recruitment.For example, biomaterials modified to incorporate antiinflammatory cytokines or immune-modulating peptides can promote tissue repair by dampening excessive inflammation (Sadtler et al., 2016).Immune-instructive materials have demonstrated potential applications in regenerative medicine, vaccine delivery, and immunotherapy.These biomaterials offer the possibility of improving tissue regeneration and the lasting effectiveness of implanted devices by carefully controlling immune responses (Chen et al., 2016).

7.3 Cytokine-Responsive Scaffolds

Cytokine-responsive scaffolds represent a new type of biomaterial. These materials are designed to change in response to inflammatory signals in the tissue environment. Cytokines, which are signaling molecules secreted by immune cells, play a critical role in this process. These molecules meticulously regulate inflammation and orchestrate the immune system's activities. For example, during inflammation, pro-inflammatory cytokines like TNF-α, IL-6, and interferon-γ can increase. These molecules can then affect how tissues regenerate and how biomaterials work (Veiseh et al., 2015). Therefore, cytokine-responsive biomaterials are designed to recognize inflammatory signals and then release therapeutic agents. For example, scaffolds can be made to release anti-inflammatory drugs or growth factors when they come into contact with pro-inflammatory cytokines. This ability to respond allows biomaterials to adapt to changes in the body, which helps control therapeutic effects (Veiseh et al., 2015). These advanced biomaterials have been studied for their potential to treat chronic wounds, inflammatory conditions, and tissue regeneration. By responding to cytokine concentrations, these scaffolds can modulate the immune response and promote tissue repair, thereby mitigating excessive inflammation.

7.4 Role in Chronic Inflammation and Autoimmune Conditions

Chronic inflammation and autoimmune diseases significantly hinder the successful integration of biomaterials and the subsequent process of tissue regeneration. Within the context of sustained inflammation, persistent immune activation triggers an overproduction of cytokines, reactive oxygen species, and proteolytic enzymes. Consequently, these elements contribute to the degradation of adjacent tissues.As a result, this harmful environment can hinder the integration of biomaterials and slow down the healing process (Veiseh et al., 2015).To mitigate these difficulties, immunomodulatory biomaterials have been engineered to modulate immune responses and re-establish immune equilibrium.These substances can deliver anti- inflammatory drugs, alter how immune cells work, and help tissues heal during inflammation. For example, biomaterials designed to boost regulatory immune responses could potentially reduce inflammation.This could improve treatments for autoimmune diseases like rheumatoid arthritis and inflammatory bowel disease (Sadtler et al., 2016).Furthermore, biomaterials that can change both the bone and immune systems have shown promise in treating inflammatory bone diseases.

8. ADVANCED DEGRADATION ENGINEERING

Degradation is one of the big deal-makers in how biodegradable biomaterials work in tissue engineering and regenerative medicine. Ideally, the rate of degradation should match the rate of new tissue formation, so that the scaffold is replaced by the tissue over time.Advanced degradation engineering is all about how biomaterials can be engineered to degrade in a way that can be controlled by varying the polymer composition, crosslinking, and their response to their environment.Using these tricks, the scaffold is strong enough in the early stages of repair and then gradually degrades to biocompatible by-products that can be metabolized and/or eliminated.Significant advancements have been achieved in the field of material science, and as a result, there have been smart biodegradable materials developed to sense and respond to their environment.These materials respond to biological signals, for instance, changes in pH, enzymes, and oxidative stress.This evolution of materials has made advanced degradation engineering a key component of the next generation of biomaterials.

8.1 Controlled and Tunable Degradation Kinetics

The rate at which the biomaterials scaffold degrades is another key factor that should be controlled. If the degradation is too fast, the scaffold may not be able to support the regenerating tissue. If the degradation is too slow, then the regeneration of the tissue may be slow. Thus, the degradation rate is important to optimize the performance of the material, as Middleton and Tipton pointed out in 2000.So, what controls the rate at which biodegradable polymers degrade. Several factors influence degradation speed. These include the molecular weight, the crystallinity of the polymers, hydrophilicity, and the density of the polymers. A polymer that is too crystalline in nature tends to degrade slowly. This is due to the inability of water to penetrate the polymer matrix. On the other hand, polymers that are too densely cross-linked have a slower

degradation speed due to hydrolysis. Advances in polymer engineering have provided the opportunity for the creation of biomaterials that have programmable degradation speeds. Such materials can degrade at a particular speed that can be programmed for tissue engineering applications such as bone, cartilage, and soft tissue engineering, as explained by Middleton and Tipton (2000).

8.2 Self-Healing Biodegradable Systems

One of the exciting frontiers in smart biomaterials is self-healing biodegradable materials.These are materials that are capable of healing themselves after being damaged without anyone doing anything to them.They are able to achieve this due to the presence of dynamic covalent bonds and crosslinks that are able to reconnect whenever the material is subjected to stress.There are several approaches that have been investigated in the development of selfhealing materials, and they include the use of dynamic covalent bonds, supramolecular interactions, and hydrogen bonding. When the materials are damaged, the damaged covalent bonds are able to heal themselves, thus extending the period of use of the scaffold.In tissue engineering, self-healing scaffolds play an important role in maintaining the integrity of the structure in the harsh environment.Self-healing hydrogels, for example, have been designed for cartilage and heart tissue engineering, where mechanical forces are always at play.These materials are not only beneficial in providing strength, but they also provide an environment that is favorable for the growth of cells.

8.3 Spatiotemporal Degradation Control

Controlling degradation in terms of time and space allows researchers to orchestrate degradation in a way that matches tissue healing.This is an advanced strategy in biomaterial degradation control.The degradation strategy allows materials to degrade in a certain area or at a certain time to allow sequential release of bioactive cues for tissue development (Lutolf & Hubbell, 2005).A strategy in this degradation control is to use a polymer network with enzyme- cleavable crosslinks. The crosslinks are degraded by enzymes in a tissue environment.This ensures degradation in an area with high cell activity.Another strategy is to use light or pH- sensitive materials.The materials degrade in response to certain external cues (Lutolf & Hubbell, 2005). Degradation control is important in vascular tissue engineering and bone regeneration because certain tissues require certain conditions to develop in distinct stages of development.The degradation pattern of a biomaterial can dynamically support tissue development (Lutolf & Hubbell, 2005).

8.4 Degradation-Driven Drug Release Systems

Biodegradable materials are used in degradation-driven drug delivery systems. In this approach, the drug is incorporated into a polymer matrix, often as a scaffold, and is released in a controlled way as the material breaks down.This method allows for the sustained delivery of drugs, which can reduce side effects throughout the body (Siepmann & Siepmann, 2008).The rate at which the drug is released from the polymer matrix is affected by several factors.These include how quickly the polymer degrades, how the drug moves through the material, and the matrix's porosity. By manipulating these variables, the duration of drug release can be regulated, ranging from several days to several months (Siepmann & Siepmann, 2008).

The utility of degradation-driven drug delivery systems is extensive, encompassing cancer treatment, wound healing, and tissue engineering. For example, scaffolds that break down and include therapeutic agents, such as growth factors or anti-inflammatory drugs, allow for targeted drug delivery. This methodology represents a promising avenue for the application of biomaterials in drug delivery.

9. APPLICATIONS IN REGENERATIVE MEDICINE

Regenerative medicine seeks to restore the structure and function of damaged tissues and organs, utilizing biomaterials, cellular elements, and bioactive molecules. The development of advanced bioactive and biodegradable biomaterials is very important in this area.These materials furnish structural integrity and facilitate the conveyance of biochemical signals, thereby contributing to tissue restoration.Contemporary biomaterials are engineered to emulate the extracellular matrix, modulate cellular activities, and establish an environment conducive to tissue regeneration. Engineered materials are capable of being tailored to possess particular mechanical, chemical, and biological properties, thereby promoting the regeneration of a wide array of tissues, including bone, cartilage, skin, nerve, and cardiovascular tissues (O'Brien, 2011; Langer & Vacanti, 1993).

9.1 Bone Tissue Engineering

Bone tissue engineering represents a significant area of investigation within the domain of regenerative biomaterials.Therapeutic strategies are frequently indispensable for managing bone defects arising from trauma, tumors, congenital defects, or degenerative conditions, with the objective of reinstating both the structural soundness and functional performance of the compromised bone.Conventional treatment approaches, such as autografts and allografts, are constrained by certain drawbacks, including complications at the donor site, limited availability, and the potential for immune rejection.In contrast, scaffolds constructed from biomaterials present a potentially beneficial option, furnishing a three-dimensional framework that promotes the attachment of osteogenic cells, thus facilitating the process of bone regeneration (O'Brien, 2011). Biomaterials employed in bone tissue engineering frequently aim to replicate the mineralized structure inherent in natural bone. Hydroxyapatite, bioactive glass, calcium phosphate ceramics, and biodegradable polymers are commonly selected due to their improved biocompatibility and capacity to stimulate bone regeneration.These materials facilitate the differentiation of mesenchymal stem cells into osteoblasts, thereby fostering the formation of new bone (Hench & Polak, 2002). Furthermore, adding growth factors,as bone morphogenetic proteins (BMPs) and vascular endothelial growth factor (VEGF), to bioactive biomaterials helps with both bone formation and the creation of blood vessels.Including these biochemical signals clearly improves the regeneration of bone tissue, which speedsup the healing process (O'Brien, 2011).

9.2 Cartilage Regeneration

Cartilage's ability to heal itself is limited by its lack of blood vessels and its sparse cellular structure.Consequently, damage to cartilage, especially within weight-bearing joints like the kneeand hip, can precipitate degenerative conditions such as osteoarthritis.Consequently, regenerative biomaterials have become increasingly important in cartilage repair, offering scaffolds that facilitate chondrocyte proliferation and extracellular matrix synthesis (Makris et al., 2015). Hydrogels and polymeric scaffolds are widely utilized in cartilage tissue engineering, owing to their capacity to mimic the hydrated environment inherent to cartilage. Moreover, materials such as collagen, chitosan, hyaluronic acid, and polyethylene glycol can stimulate the differentiation of stem cells into chondrocytes, thus promoting the development of the cartilage matrix (Makris et al., 2015). At the same time, advanced biomaterials are being developed to deliver chondrogenic growth factors, such as transforming growth factor-β (TGF-β) and insulin-like growth factor (IGF).These molecules facilitate cartilage regeneration through the stimulation of extracellular matrix synthesis and the enhancement of the mechanical characteristics of the newly formed cartilage (Makris et al., 2015).

9.3 Skin and Wound Healing.Biomaterials

play a crucial role in regenerative medicine, particularly in the domains of skin regeneration and wound healing.In cases involving chronic wounds, burns, and traumatic injuries, complex treatments are often necessary to re-establish skin integrity and prevent infection.Boateng and his research team showed in 2008 that wound dressings and scaffolds made from biomaterials can act as protective barriers, while also promoting cell growth and helping with tissue repair.Modern wound healing biomaterials are designed to create a moist environment, allow for oxygen exchange, and provide substances that speed up the healing process.Materials like collagen, alginate, chitosan, and hydrogel dressings are often used because they are compatible with the body and support cell movement and the growth of new blood vessels (Boateng et al., 2008).Recent advancements in biomaterials have facilitated the inclusion of antimicrobial agents, growth factors, and nanoparticles in wound dressings.These multifunctional systems have the potential to hinder bacterial infections, reduce inflammation, and encourage the regeneration of skin tissues (Boateng et al., 2008).

9.4 Neural Tissue Engineering

The main goal of neural tissue engineering is to repair or regenerate damaged nervous system tissues, including the brain, spinal cord, and peripheral nerves. Treating injuries to the nervous system is particularly difficult because neurons have a limited ability to regenerate. Biomaterials have been developed to create scaffolds that support neuron growth, guide axonal regeneration, and promote functional recovery (Struzyna et al., 2014). Biodegradable polymers such as polycaprolactone, polylactic acid, and collagen are commonly used in neural scaffolds because they provide structural support and gradually degrade as new tissue forms.In addition, scaffolds can be designed with aligned fibers or microchannels to guide axonal growth, which helps with nerve regeneration (Struzyna et al., 2014).Moreover, neural biomaterials can be improved by adding neurotrophic factors, such as nerve growth factor (NGF) and brain-derived neurotrophic factor (BDNF).

These substances support neuronal survival, axonal outgrowth, and synapse formation, thus contributing to the regeneration of neural tissues (Struzyna et al., 2014).

9.5 Cardiovascular Regeneration

Cardiovascular diseases continue to be a primary cause of death globally.Myocardial infarction frequently leads to the loss of cardiomyocytes and diminished cardiac function due to damage to cardiac tissue.To address this, researchers are studying and creating regenerative biomaterials to repair or regenerate heart tissues, with the goal of supporting cell growth and improving heart function .These biomaterials, used in cardiovascular regeneration, include biodegradable polymers, hydrogels, and decellularized extracellular matrices.These substances provide the necessary structural support for cardiac cells, thus facilitating the formation of functional cardiac tissue.Hydrogels are particularly significant due to their capacity to mimic the mechanical characteristics of heart tissue and promote cell survival .Moreover, biomaterials can be engineered to deliver stem cells, growth factors, or genetic material, potentially stimulating the creation of new blood vessels and the regeneration of cardiac muscle cells.These strategies aim to restore cardiac function and improve patient outcomes subsequent to cardiac injury.

10. NEXT-GENERATION EMERGING TECHNOLOGIES

In recent years, next generation technologies including advances in materials science, biotechnology and computational biology are dramatically changing the biomaterials and regenerative medicine landscape.Traditional biomaterials were mainly developed to offer structural support, but advanced technologies have made possible the fabrication of dynamic, responsive and smart biomaterials that actively engage with biologics systems.Such developments harness ideas from diverse areas including but not limited to AI, gene editing, synthetic biology, and advanced manufacturing to produce biomaterials that possess superior functionality as well as responsiveness (Langer & Tirrell, 2004; Place et al., 2009). Next- generation biomaterials are being designed to replicate the complexity of native tissues and respond to environmental stimuli.4D biomaterials, AI-assisted material design, CRISPRbased genetic engineering, living biomaterials and metamaterials are some of these major breakthroughs that will give you an exciting glimpse into the future of regenerative medicine.Such platforms provide novel opportunities for targeted tissue repair, customized medical therapies and bioactive devices with unique therapeutic capabilities (Langer & Tirrell, 2004).

10.1 4D Biomaterials and Shape-Memory Systems

4D biomaterials are a new evolution of 3D biomaterials, which exhibit time-dependent variations in the material structure or materials properties as the fourth dimension.Such materials can respond to external stimuli like temperature, pH, moisture, light, or magnetic domains and convert from one form to another in a controlled manner.One of the most critical types of smart materials, shape-memory polymers have been extensively studied due to their ability to revert back to their original form once they are triggered by particular environmental factors (Miao et al., 2017).4D biomaterials can help in tissue engineering applications by

changing their properties to suit physiological environments over time.As an example, shape- memory scaffolds can be delivered in a collapsed state and expand to their functional form in vivo.These structures are of great assistance in less invasive surgical procedures and Complex tissue regeneration applications have been explored (Miao et al., 2017).These adaptable materials have been studied for their potential in cardiovascular implants, bone regeneration scaffolds, and smart drug delivery systems.The ability of 4D biomaterials tochange in response to environmental conditions is a key feature.stimuli positions them as promising candidates for advanced regenerative therapies.

10.2 AIGuided Biomaterial Design

Technologies like artificial intelligence (AI) and machine learning are being employed with increasing frequency to expedite biomaterial discovery and design. Traditional biomaterial design has frequently relied on laborious methods, including trial and error.However, computational models based onAI can analyze large datasets and identify patterns to predict the effects of certain material composition and structure on biological responses (Butler et al., 2018).Machine learning algorithms can be trained to predict critical properties of biomaterials such as mechanical properties, degradation, and interaction with cells.These properties can beutilized to design biomaterials with optimized properties for tissue engineering based on the properties required for such engineering (Butler et al., 2018). AI design approaches are being utilized to optimize scaffolds and improve additive manufacturing. This integration of computational modeling and experimental validation can accelerate the development of advanced biomaterials.

10.3 CRISPR-Integrated Bioactive Scaffolds

CRISPR is a type of Clustered Regularly Interspaced Short Palindromic Repeats technology that has revolutionized the field of genetic engineering.CRISPR has enabled the precise and powerful engineering of DNA.CRISPR-integrated biomaterials have opened the door to the possibility of regulating the process of gene expression in the field of tissue engineering (Doudna & Charpentier, 2014).CRISPR-integrated biomaterials may be used as delivery tools to target the process of gene expression with the help of the CRISPR-Cas9 complex or the guide RNA.This may be used to locally target the process of gene expression in target cells, which may then be used to induce the process of regeneration.For example, CRISPR-based biomaterials may be used to induce the process of bone regeneration by activating the osteogenic genes in the target stem cell population that is used for the purpose of implantation (Doudna & Charpentier, 2014).

10.4 Living Biomaterials and Engineered Cells

Living biomaterials constitute a new class of hybrid materials, where biomaterials and living cells, which have the ability to execute biological functions, coexist. These materials have the ability to maintain the viability of the cells and enable the cells to perceive the signals of the environment, produce therapeutic agents, and dynamically interact with the tissues surrounding the biomaterials (Rosales & Anseth, 2016).Engineered cells have the ability to be integrated into biomaterials to produce factors for growth, cytokines, and extracellular matrices, which are useful for tissue engineering.This includes the use of stem cells, which have been

genetically engineered and integrated into biomaterials to produce regenerative agents for the improvement of tissue repair (Rosales & Anseth, 2016). Living biomaterials have the ability to revolutionize regenerative medicine, as they offer a dynamic and adaptive system, which responds to changes occurring in the biological environment. These materials have the ability to find application in the healing of wounds, the immune system, and the regeneration of organs.

10.5 Metamaterials in Tissue Engineering

These are materials that have unusual physical properties not seen in natural materials. These unusual properties are based on the internal architecture of the material rather than the chemistry of the material. For tissue engineering, metamaterials can be designed with certain mechanical properties, such as a negative Poisson ratio, stiffness, or patterns of deformation (Zheng et al., 2014). The ability of the material architecture to be controlled allows for the construction of scaffolds that can mimic the mechanical properties of natural tissues more accurately than those of conventional materials.For example, scaffolds can be designed with graded mechanical properties that can mimic complex tissue structures such as the boundary of bone and cartilage (Zheng et al., 2014).Besides the mechanical advantages, metamaterials can also be used to affect cell behavior based on the architecture of the material.Therefore, the advanced materials provide an exciting prospect for the construction of biomimetic scaffolds for tissue regeneration and improvement of implants.

CONCLUSION AND FUTURE PERSPECTIVES

Biodegradable bioactive biomaterials represent a paradigm shift in which now, biomaterials are not only inert structures but rather dynamic, nanoengineered platforms that actively direct cellular responses and regulate host immune reactions. These platforms combine nanofibrous scaffolds, nanoparticle delivery vectors and biomimetic surface nanotopographies to precisely mimic the signaling microenvironment of the native extracellular matrix to promote multi-tissue regeneration. Controllable spatiotemporal degradation kinetics mimicked to the tissue neo-formation phenomena are obtained using smart, stimuli-responsive material networks that harness spatiotemporal tissue-specific pathophysiological cues (e.g. pH, temperature, oxidative stress) to trigger therapeutic delivery. Importantly, these biomaterials can overcome chronic inflammation by polarizing macrophages from the pro-inflammatory M1 to a pro-healing M2 state in an immune-instructive manner rather than classic manner of immune evasion. In the future, multi-disciplinary technologies will set the limits of biomedical engineering. The integration of 4D shape-memory systems and mechanically tailored metamaterials with computational design, guided by artificial intelligence (AI), and CRISPR-based genome editing, along with living engineered cell matrices, will enable researchers to overcome empirical trial-and-error. In all, these novel convergence platforms are poised to provide adaptive, personalized and translation-ready clinical solutions for complex tissue repair and targeted therapeutics.

REFERENCES

Bhardwaj, N., & Kundu, S. C. (2010). Electrospinning: A fascinating fiber fabrication technique. *Biotechnology Advances*, *28*(3), 325–347. DOI: 10.1016/j.biotechadv.2010.01.004

Binauld, S., & Stenzel, M. H. (2013). Acid-degradable polymers for drug delivery: A decade of innovation. *Chemical Communications (Cambridge)*, *49*(21), 2082–2102. DOI: 10.1039/c2cc36589h

Boateng, J. S., Matthews, K. H., Stevens, H. N. E., & Eccleston, G. M. (2008). Wound healing dressings and drug delivery systems: A review. *Journal of Pharmaceutical Sciences*, *97*(8), 2892–2923. DOI: 10.1002/jps.21210

Bonadio, J., Smiley, E., Patil, P., & Goldstein, S. (1999). Localized, direct plasmid gene delivery in vivo: Prolonged therapy results in reproducible tissue regeneration. *Nature Medicine*, *5*(7), 753–759. DOI: 10.1038/10473

Butler, K. T., Davies, D. W., Cartwright, H., Isayev, O., & Walsh, A. (2018). Machine learning for molecular and materials science. *Nature*, *559*(7715), 547–555. DOI: 10.1038/s41586-018-0337-2

Campoccia, D., Montanaro, L., & Arciola, C. R. (2013). A review of the biomaterials technologies for infection-resistant surfaces. *Biomaterials*, *34*(34), 8533–8554. DOI: 10.1016/j.biomaterials.2013.07.089

Chen, C. C., Dai, L., Ma, L., & Guo, R. T. (2020). Enzymatic degradation of plant biomass and synthetic polymers. *Nature Reviews. Chemistry*, *4*(3), 114–126. DOI: 10.1038/s41570-020-0163-6

Chen, Z., Klein, T., Murray, R. Z., Crawford, R., Chang, J., Wu, C., & Xiao, Y. (2016). Osteoimmunomodulation for the development of advanced bone biomaterials. *Materials Today*, *19*(6), 304–321. DOI: 10.1016/j.mattod.2015.11.004

Cheng, R., Feng, F., Meng, F., Deng, C., Feijen, J., & Zhong, Z. (2011). Glutathione-responsive nano-vehicles as a promising platform for targeted intracellular drug and gene delivery. *Journal of Controlled Release*, *152*(1), 2–12. DOI: 10.1016/j.jconrel.2011.01.030

D'Amario, J., Limsukon, W., Bher, A., & Auras, R. (2025). Impact of hydrolysis pretreatment on the compostability of biodegradable poly(caprolactone) and poly(-lactic acid) films. *RSC Applied Polymers*, *3*(3), 711–721. DOI: 10.1039/D5LP00041F

Dalby, M. J., Gadegaard, N., Tare, R., Andar, A., Riehle, M. O., Herzyk, P., Wilkinson, C. D. W., & Oreffo, R. O. C. (2007). The control of human mesenchymal cell differentiation using nanoscale symmetry and disorder. *Nature Materials*, *6*(12), 997–1003. DOI: 10.1038/nmat2013

Doudna, J. A., & Charpentier, E. (2014). The new frontier of genome engineering with CRISPR-Cas9. *Science*, *346*(6213), 1258096. DOI: 10.1126/science.1258096

El Andaloussi, S., Mäger, I., Breakefield, X. O., & Wood, M. J. A. (2013). Extracellular vesicles: Biology and emerging therapeutic opportunities. *Nature Reviews. Drug Discovery*, *12*(5), 347–357. DOI: 10.1038/nrd3978

Guo, B., & Ma, P. X. (2018). Conducting polymers for tissue engineering. *Biomacromolecules*, *19*(6), 1764–1782. DOI: 10.1021/acs.biomac.8b00276

Hasan, A., Morshed, M., Memic, A., Hassan, S., Webster, T. J., & Marei, H. E. S. (2018). Nanoparticles in tissue engineering: Applications, challenges and prospects. *International Journal of Nanomedicine*, *13*, 5637–5655. DOI: 10.2147/IJN.S153758

Hench, L. L., & Polak, J. M. (2002). Third-generation biomedical materials. *Science*, *295*(5557), 1014–1017. DOI: 10.1126/science.1067404

Hersel, U., Dahmen, C., & Kessler, H. (2003). RGD-modified polymers: Biomaterials for stimulated cell adhesion and beyond. *Biomaterials*, *24*(24), 4385–4415. DOI: 10.1016/S0142-9612(03)00343-0

Huang, T., Zeng, Y., Li, C., Zhou, Z., Xu, J., Wang, L., Yu, D.-G., & Wang, K. (2024). Application and development of electrospun nanofiber scaffolds for bone tissue engineering. *ACS Biomaterials Science & Engineering*, *10*(7), 4114–4144. DOI: 10.1021/acsbiomaterials.4c00028

Langer, R., & Tirrell, D. A. (2004). Designing materials for biology and medicine. *Nature*, *428*(6982), 487–492. DOI: 10.1038/nature02388

Langer, R., & Vacanti, J. P. (1993). Tissue engineering. *Science*, *260*(5110), 920–926. DOI: 10.1126/science.8493529

Lee, K., Silva, E. A., & Mooney, D. J. (2011). Growth factor delivery-based tissue engineering: General approaches and a review of recent developments. *Journal of the Royal Society, Interface*, *8*(55), 153–170. DOI: 10.1098/rsif.2010.0223

Li, J. J., & Zreiqat, H. (2019). Tissue response to biomaterials. In Narayan, R. (Ed.), *Encyclopedia of Biomedical Engineering* (pp. xx–xx). Elsevier. DOI: 10.1016/B978-0-12-801238-3.99880-5

Lutolf, M. P., & Hubbell, J. A. (2005). Synthetic biomaterials as instructive extracellular microenvironments for morphogenesis in tissue engineering. *Nature Biotechnology*, *23*(1), 47–55. DOI: 10.1038/nbt1055

Makris, E. A., Gomoll, A. H., Malizos, K. N., Hu, J. C., & Athanasiou, K. A. (2015). Repair and tissue engineering techniques for articular cartilage. *Nature Reviews. Rheumatology*, *11*(1), 21–34. DOI: 10.1038/nrrheum.2014.157

Matai, I., Kaur, G., Seyedsalehi, A., McClinton, A., & Laurencin, C. T. (2020). Progress in 3D bioprinting technology for tissue/organ regenerative engineering. *Biomaterials*, *226*, 119536. DOI: 10.1016/j.biomaterials.2019.119536

Miao, S., Castro, N., Nowicki, M., Xia, L., Cui, H., Zhou, X., Zhu, W., Lee, S. J., Sarkar, K., Vozzi, G., Tabata, Y., Fisher, J., & Zhang, L. G. (2017). 4D printing of polymeric materials for tissue and organ regeneration. *Materials Today*, *20*(10), 577–591. DOI: 10.1016/j.mattod.2017.06.005

Middleton, J. C., & Tipton, A. J. (2000). Synthetic biodegradable polymers as orthopedic devices. *Biomaterials*, *21*(23), 2335–2346. DOI: 10.1016/S0142-9612(00)00101-0

Mohiuddin, M., Rahman, M. M., Uddin, M. N., Hasan, R., & Rahman, I. (2025). Biodegradable graphene nanocomposites as functional biomaterials: A review of their role in controlled drug delivery and tissue engineering. *RSC Advances*, *15*(53), 45387–45416. DOI: 10.1039/D5RA06280B

Murphy, S. V., & Atala, A. (2014). 3D bioprinting of tissues and organs. *Nature Biotechnology*, *32*(8), 773–785. DOI: 10.1038/nbt.2958

O'Brien, F. J. (2011). Biomaterials and scaffolds for tissue engineering. *Materials Today*, *14*(3), 88–95. DOI: 10.1016/S1369-7021(11)70058-X

Park, J., & De, R. (2025). Next-generation wound healing materials: Role of biopolymers and their composites. *Polymers*, *17*(16), 2244. DOI: 10.3390/polym17162244

Place, E. S., George, J. H., Williams, C. K., & Stevens, M. M. (2009). Synthetic polymer scaffolds for tissue engineering. *Chemical Society Reviews*, *38*(4), 1139–1151. DOI: 10.1039/b811392k

Reddy, L. H., Arias, J. L., Nicolas, J., & Couvreur, P. (2012). Magnetic nanoparticles: Design and characterization, toxicity and biocompatibility, pharmaceutical and biomedical applications. *Chemical Reviews*, *112*(11), 5818–5878. DOI: 10.1021/cr300068p

Roohi, B., Bano, K., Kuddus, M., Zaheer, M. R., Zia, Q., Khan, M. F., Ashraf, G. M., Gupta, A., & Aliev, G. (2017). Microbial enzymatic degradation of biodegradable plastics. *Current Pharmaceutical Biotechnology*, *18*(5), 429–440. DOI: 10.2174/1389201018666170523165742

Rosales, A. M., & Anseth, K. S. (2016). The design of reversible hydrogels to capture extracellular matrix dynamics. *Nature Reviews. Materials*, *1*(2), 15012. DOI: 10.1038/natrevmats.2015.12

Rovaletti, A., De Gioia, L., Fantucci, P., Greco, C., Vertemara, J., Zampella, G., Arrigoni, F., & Bertini, L. (2023). Recent theoretical insights into the oxidative degradation of biopolymers and plastics by metalloenzymes. *International Journal of Molecular Sciences*, *24*(7), 6368. DOI: 10.3390/ijms24076368

Roy, D., Cambre, J. N., & Sumerlin, B. S. (2010). Future perspectives and recent advances in stimuli-responsive materials. *Progress in Polymer Science*, *35*(1–2), 278–301. DOI: 10.1016/j.progpolymsci.2009.10.008

Sadtler, K., Estrellas, K., Allen, B. W., Wolf, M. T., Fan, H., Tam, A. J., Patel, C. H., Luber, B. S., Wang, H., Wagner, K. R., Powell, J. D., Housseau, F., Pardoll, D. M., & Elisseeff, J. H. (2016). Developing a pro-regenerative biomaterial scaffold microenvironment requires T helper 2 cells. *Science*, *352*(6283), 366–370. DOI: 10.1126/science.aad9272

Shergujri, D. A., Khanday, M. A., Noor, A., Adnan, M., Arif, I., Raza, S. N., Mir, R. H., & Khan, N. A. (2025). Next-generation biopolymer gels: Innovations in drug delivery and theranostics. *Journal of Materials Chemistry. B*, *13*(10), 3222–3244. DOI: 10.1039/D4TB02068E

Siepmann, J., & Siepmann, F. (2008). Mathematical modeling of drug delivery. *International Journal of Pharmaceutics*, *364*(2), 328–343. DOI: 10.1016/j.ijpharm.2008.09.004

Struzyna, L. A., Katiyar, K., & Cullen, D. K. (2014). Living scaffolds for neuroregeneration. *Current Opinion in Solid State and Materials Science*, *18*(6), 308–318. DOI: 10.1016/j.cossms.2014.07.004

Tripathi, A. S., Zaki, M. E., Al-Hussain, S. A., Dubey, B. K., Singh, P., Rind, L., & Yadav, R. K. (2023). Material matters: Exploring the interplay between natural biomaterials and host immune system. *Frontiers in Immunology*, *14*, 1269960. DOI: 10.3389/fimmu.2023.1269960

Ulijn, R. V., & Smith, A. M. (2008). Designing peptide-based nanomaterials. *Chemical Society Reviews*, *37*(4), 664–675. DOI: 10.1039/b609047h

Veiseh, O., Doloff, J. C., Ma, M., Vegas, A. J., Tam, H. H., Bader, A. R., Li, J., Langan, E., Wyckoff, J., Loo, W. S., Jhunjhunwala, S., Chiu, A., Siebert, S., Tang, K., Hollister-Lock, J., Aresta-Dasilva, S., Bochenek, M., Mendoza-Elias, J., Wang, Y., & Anderson, D. G. (2015). Size-and shape-dependent foreign body immune response to materials implanted in rodents and non-human primates. *Nature Materials, 14*(6), 643–651. DOI: 10.1038/nmat4290

Yang, L., Hu, J., & Jiang, S. (2025). Editorial: Biodegradable polymers for biomedical applications—Volume III. *Frontiers in Materials, 12*, 1551590. DOI: 10.3389/fmats.2025.1551590

Yao, X., Yang, X., Lu, Y., Qiu, Y., & Zeng, Q. (2024). Review of the synthesis and degradation mechanisms of some biodegradable polymers in natural environments. *Polymers, 17*(1), 66. DOI: 10.3390/polym17010066

Zheng, X., Lee, H., Weisgraber, T. H., Shusteff, M., DeOtte, J., Duoss, E. B., Kuntz, J. D., Biener, M. M., Ge, Q., Jackson, J. A., Kucheyev, S. O., Fang, N. X., & Spadaccini, C. M. (2014). Ultralight, ultrastiff mechanical metamaterials. *Science, 344*(6190), 1373–1377. DOI: 10.1126/science.1252291

Zhu, Y., Zhang, X., Chang, G., Deng, S., & Chan, H. F. (2025). Bioactive glass in tissue regeneration: Unveiling recent advances in regenerative strategies and applications. *Advanced Materials, 37*(2), 2312964. DOI: 10.1002/adma.202312964

Chapter 3
Smart Bioactive Systems Driven by Nanotechnology:
Innovations in Healthcare and Beyond

Atanu Nag
http://orcid.org/0000-0003-0561-1946
Department of Physics, School of Sciences, IFTM University, Moradabad, India

Debika Chaudhuri
http://orcid.org/0000-0002-9026-3191
Department of Electronics and Communication Engineering, School of Engineering and Technology, IFTM University, Moradabad, India

ABSTRACT

Bioactive intelligent systems through nanotechnology stand out as the most revolutionary achievements in contemporary scientific endeavors, facilitating accurate and effective treatments. In this chapter, the fundamentals, materials, and operations of such systems are described, with an extensive focus on their adaptive response to environmental stimuli. Nanomaterials that include nanoparticles, nanocomposites, and nanointerfaces are reviewed in terms of their physical and chemical characteristics and their delivery features. Some of the notable biomedical applications are listed, which include targeted drug delivery systems, biosensors, tissue scaffolds, and implantable intelligent devices. Some of the emerging applications in environmental sensing, agriculture, and bio-hybrid systems are discussed. The challenges associated with the development and implementation of these technologies are identified.

DOI: 10.4018/979-8-2600-0060-1.ch003

1. INTRODUCTION

Nanotechnology and bioactive systems have combined to create a new era of science and engineering. The integration of nanotechnology and bioactive systems has significantly advanced the interaction between materials and biological systems (Singh et al., 2025). This has been made possible by the creation of new kinds of bioactive systems that are not only compatible with biological systems but also intelligent in the sense that they can detect and react to changes in their environment (Zhang, J. et al., 2023). In essence, these smart bioactive systems are a far cry from the old passive materials that were always used for a particular purpose only (Guo et al., 2025). The critical characteristic of smart bioactive systems is their stimulus-responsive behavior. While traditional devices depend on outside factors, intelligent devices are capable of self-activation under certain stimuli, such as variations in pH values, temperature, enzymes, light, magnetic field, or biochemistry (Chen et al., 2024; Damiri et al., 2023). The advantage of stimulus responsiveness is that it enables the system to adapt to changes in real time, thus making applications such as drug delivery, self-healing, and biosensors possible. For instance, the clinically validated nanopharmaceutical Doxil based on liposome technology illustrated how nanoparticles can enhance the circulatory longevity and reduce cardiac toxicity of free doxorubicin, thereby becoming one of the early successes in developing intelligent nanocarriers (Nagpal et al., 2024). Moreover, the lipid-based nanoparticle systems employed for mRNA COVID-19 vaccines proved the concept of nano-enabled bioactive delivery systems through practical implementation at global level (Hou et al., 2021). Intelligent drug delivery systems often release drugs only when exposed to certain stimuli, such as the acidic environment inside cancer cells (Torchilin, 2014; Mura et al., 2013).

Nanotechnology has proven vital owing to the ability of manipulating materials on the nanometer scale (~ 1 to 100 nm) (Whitesides, 2005). Nanomaterials exhibit novel properties because of reasons such as high surface area-to-volume ratio, quantum confinement effects, and the capability to functionalize surfaces through ligands/biomolecules. Increased surface areas lead to enhanced reactions with surrounding environments while quantum confinement dictates how materials behave electrically and optically. Surface functionalization helps facilitate better interactions between nanoparticles and biological entities (Rezaei et al., 2024; Shahalaei et al., 2024). One of the most important benefits of nanoscale technologies is the capability to design material properties to fit specific requirements. When nanofillers are added to a matrix material, there will be nanocomposites that display superior physical and chemical properties, like better mechanical and electrical properties (Singh et al., 2026). Another benefit of nanotechnology is the creation of nanointerfaces, which regulate how nanomaterials interact with biological entities. The manipulation of nanointerfaces can result in selectivity in targeting or integration into biological systems (Li & Wang, 2023; Wang, Y. et al., 2024).

Combination of nanotechnology with bioactive systems has brought about some notable breakthroughs in the field of healthcare. For example, smart drug delivery systems have played an instrumental role in changing the face of drug administration through controlled release. The delivery of drugs may also be tailored according to certain physiological signals, allowing for drug transfer to designated areas or drug release at scheduled times (Salehi et al., 2023; Long et al., 2024). On the other hand, the importance of nanosensors lies in their use in medical diagnostics where the sensors detect even very low levels of biomolecules, making them helpful in disease detection (Ma et al., 2025). An illustration of practical use is the glucose nanosensors that are applied in continuous glucose monitors (CGMs) to detect glucose with a high degree of sensitivity close to real-time (Le et al., 2020).

Moreover, the use of nanotechnology in bioactive systems has brought about some great developments in the areas of tissue engineering and regenerative medicine (Del Bakhshayesh et al., 2023). By developing nano-scaffolds similar to the extracellular matrix, it has been possible to create an environment suitable for cellular growth and differentiation (Narayanan, 2025). Additionally, it is now possible to develop medical implants with integrated nanotechnology capable of monitoring the physiological functions of the body (Suryaprabha et al., 2026). Impact of smart bioactive systems is not limited only to healthcare; it spreads across other sectors including but not limited to environmental monitoring, agriculture, and biohybrid systems. The use of nanosensors in environmental technology allows detecting various types of contaminants with great accuracy, thus ensuring the sustainability of the environment (Çaktü Güler et al., 2024). Nanotechnology is used in agriculture for precision farming, allowing an efficient application of fertilizers and pesticides (Gupta et al., 2025). Biohybrid systems include living organisms combined with artificial nanomaterials (Garmroudi, et al., 2025).

Even with these advances, there are some issues that make it difficult to fully adopt and implement this technology in commercial applications. First, the problem of toxicity and biocompatibility arises. Since the materials used in the development of these devices are nanosized, they may exhibit toxicity in biological systems due to their high reactivity. It is necessary to conduct thorough research on how these substances will interact with the body and whether they will pose any risks in terms of cytotoxicity and immunogenicity (Thomas et al., 2025; Kumar, R. et al., 2025b). However, one of the major obstacles in nanotechnology progress lies in its scalability while ensuring the reproducibility of nanomaterials at a consistent level (Hachhach et al., 2025). Regulatory issues also have a significant impact on the implementation of these technologies from the experimental stage to clinical and commercial use. Furthermore, the current regulations might not be entirely ready to deal with all the intricate questions related to the use of nanomaterials in biotechnology (Rodríguez-Gómez et al., 2025; Mangla et al., 2025). Regulatory standards for the tests, analysis, and safety of these materials need to be standardized in order to gain regulatory approval and social acceptance.

Moreover, regulatory authorities nowadays are demanding more information about physicochemical properties, batch-to-batch consistency, impurities, biodistribution analysis over time, and post-marketing surveillance plans prior to product approval. The lack of international standards tends to hinder the translation process, resulting in higher development expenses. Another issue that needs further investigation relates to the stability of the systems over extended periods of time. The same is true for mass production since most of the processes used for synthesis in the laboratory setting are very sensitive to environmental conditions that are not easily achieved during the manufacturing process. Any slight change in temperature, speed, quality of raw materials, or dehydration technique can affect the characteristics of the nanoparticles created. Consequently, the integration between nanotechnology and bioactivity has resulted in the emergence of novel generations of smart and functional systems, which have the capability to transform several industries (Goshisht et al., 2025). Through the use of nanomaterials, these devices can regulate biological processes in ways that are unparalleled and open the door for new possibilities in diagnostics, treatment, eco-friendly solutions, and many other fields (Hwang et al., 2025; Kumar & Bhanjana, 2025). However, even with the great advances made, there are still only a few systems available today that are capable of meeting all these translation needs. The objective of this chapter is to introduce readers to the basics and advanced concepts associated with smart bioactive systems enabled by nanotechnology.

2. DESIGN PRINCIPLES OF SMART BIOACTIVE SYSTEMS

The design process for smart bioactive systems needs to be an integrative one that takes into consideration the relationship between the structural characteristics of the materials used and their biological performance needs (Sharma & Kishen, 2024). In addition, a good system design depends largely on the ability to ensure functionality in terms of structure and stability (Menghani & Avila, 2025). But the present technology is geared to optimize one specific parameter like efficiency or response while sacrificing stability, manufacturing, or long-term safety.

2.1 Biocompatibility and Biofunctionality

One of the first issues that should be considered in the process of designing the system is the interaction of materials with biological structures. The concept of biocompatibility goes far beyond the issue of toxicity because it means the capacity of a material to ensure beneficial biological functions without triggering an immune response (Zhou, Y. et al., 2026). Material composition, structure, and degradation profile must be chosen carefully in order to achieve high biocompatibility (Wang, P. et al., 2024). Simultaneously, biofunctionality refers to the possibility of providing particular biological interactions (Saeb, 2026). In most cases, surface engineering is vital to achieve this goal because nanoscopic changes might influence cell adhesion and interactions with proteins. For example, functionalization of the system with different proteins, antibodies, or synthetic peptides might allow for selective binding, cell recognition, and signal transduction (Li, H. et al., 2025). Furthermore, adjustment of the surface charge, hydrophilicity, and roughness might be necessary in order to increase protein binding and cellular attachment (Illing et al., 2023). For example, studies have revealed enhanced adhesion of osteoblasts on titanium implant surfaces having nanorough texture in comparison to smooth surfaces (Gongadze et al., 2011). Systematic comparisons among various surface modification techniques under same biological conditions are few.

2.2 Stimuli Responsiveness

A key aspect of designing smart bioactive systems is their responsiveness to certain stimuli. Such responsiveness refers to their ability to change their form in a controlled manner upon exposure to stimuli (Lin et al., 2026). This means the use of responsive elements in the architecture of the biomaterials used in order to transduce the stimulus signal to functional output (He et al., 2025). This depends on the type of stimulus that is being considered and its application. For example, a system that is sensitive to pH will be used in situations where there are variations in the levels of acidity in the surroundings (Wang et al., 2025). Regarding mechanism, pH-sensitive materials often consist of ionizable functional groups including but not limited to carboxyl group, amine group, and imidazole group (Zhuo et al., 2020). Protonation/deprotonation of the groups will result in changes in electrostatic forces, osmotic pressure, and conformation of the polymer chain, resulting in swelling, collapsing, pore formation, and cleavage of the acid-sensitive linkage, which will lead to drug release. For instance, pH-sensitive polymers have demonstrated increased release rate of doxorubicin at pH ~6.5 rather than physiological pH 7.4 in vitro (Lin et al., 2014). Temperature responsive materials experience alterations in their structures or characteristics based on specific temperature conditions (Yadav et al., 2026). Many thermoresponsive polymers utilize a lower critical solution temperature (LCST) or upper critical solution temperature (UCST) as the critical control mechanism for their thermoresponsiveness (Kotsuchibashi, 2020). This means that upon crossing this threshold, there will be changes in the interactions between the polymer and water molecules resulting in hydration-dehydration, phase transition, or network shrinkage. Similarly, enzyme responsive materials are created with cleavages that can only be degraded by specific enzymes (Pi et al., 2025). This class of smart materials contains peptide, ester, or glycosidic linkages, which are selectively cleaved by disease-specific enzymes such as matrix metalloproteinases, phospholipases, and glycosidases, thereby limiting the degradation of the matrix to diseased tissues (Pi et al., 2025). In addition, light (Masai et al., 2024) and magnetic fields (Chen, Q. et al., 2025) can also be applied as external stimuli to induce activation of a process or change in behavior of a system. Typically, light-sensitive responsive systems rely on photochromic units, photocleavable linkages, or photothermal nanoparticles capable of converting energy into heat, thereby providing remotely controllable actuation (Li et al., 2019). Magnetism-sensitive systems commonly utilize superparamagnetic particles, which provide the possibility for generating movements or heat under alternating magnetic fields for external control of the system activity or therapeutic delivery (Hergt et al., 2006). Superparamagnetic iron oxide nanoparticles are also explored clinically as magnetic resonance contrast agents and magnetic hyperthermia for cancer treatment (Weinstein et al., 2009). Unlike single-triggered systems, multistimuli-responsive systems may provide higher precision in applications; however, their development is relatively difficult and not easily reproducible. As an example, pH-enzyme sensitive systems can be more specific to act because they respond only when both stimuli are present (Naeem et al., 2014). Temperature-magnetism systems provide the ability to direct the system to its target location through external guidance followed by releasing it from the target site (Liu et al., 2019).

2.3 Targeting and Selectivity

Precise targeting for smart delivery systems is one of the goals of smart delivery technology. This is accomplished through the application of various targeting methodologies that direct delivery systems to specific destinations while limiting any unintended interactions (Song et al., 2025). This is usually achieved using recognition molecules. Active targeting utilizes interactions between ligands and receptors; ligands that are attached to the delivery system interact specifically with the receptors on the target cell's surface (Yan et al., 2024). Antibodies are used in antibody-based targeting, which is characterized by highly selective targeting due to the antibodies' high affinity for biomarkers. Moreover, various physical targeting strategies, like magnetic targeting or size targeting, can be incorporated alongside biochemical approaches (El-Khawaga et al., 2026). In addition to incorporating the above methods, designing a selectivity strategy involves accounting for various biological barriers, including cell membrane, blood vessels, and the immune system (Lei et al., 2026). The design process should take into account the particle's size and shape and its surface modification to increase its circulation period, increase penetration, and evade elimination before reaching the target site (Öztürk et al., 2024).

2.4 Controlled Release Mechanisms

An important aspect of function for various bioactive systems is the regulation of spatial and temporal release of active agents. This is attained by the development of nanoscale structures capable of delivering, protecting, and finally releasing a certain payload when needed (Zhou, H. et al., 2026). The control of the release can occur via various means like diffusion, degradation of the matrix, and also via certain stimuli (Parvin et al., 2025b). Designing nanoscale structures requires considerations like kinetics of release, amount of the active substance, and its stability (Ortiz et al., 2024). In terms of sustained delivery, nanomaterials are usually made of such components that will allow degradation or diffusion processes to continue (Tao et al., 2023). On the contrary, triggered releases work only when some stimuli affect the structure (Wang, T. et al., 2023). Some complicated approaches involve designs that provide both rapid and subsequent release of the material, or even release of several substances in sequence (Ahmadi et al., 2020).

2.5 Integration and Feedback Capability

Next-generation bioactive smart systems have placed more importance on the creation of a platform that has many functional parts within a compact package for autonomous functionality. Such a platform would require the presence of sensors, signal-processing functions, and actuator components (Wan et al., 2026). The sensors would constantly monitor the environment or internal body factors, like biomarker levels or other parameters. Based on the monitoring process, responses can be generated by the system to form a closed-loop system (Zheng et al., 2024; Del Bono et al., 2025). The implantable device could detect a deviation in physiological factors and release an agent or make changes accordingly. The design of such systems would require close coordination among different parts to achieve compatibility and signal transmission in an efficient way while minimizing energy losses (Gao et al., 2025). The use of feedback-based systems would require other factors, such as reliability and calibration, to be taken into consideration. This is because feedback-based systems will be more relevant when dealing with diseases or applications that are intended to be used for a long period of time (Suryaprabha et al., 2026).

Generally, design concepts are developing into very sophisticated and adaptive systems that can function independently within biological surroundings.

3. MATERIAL INNOVATIONS IN NANOTECHNOLOGY-BASED BIOACTIVE SYSTEMS

The development of intelligent and smart bioactive systems is inherently tied to the advancements made in materials technology, whereby the relationship between structural and property attributes can be programmed (Thakur et al., 2026). Emerging research breakthroughs have expanded the scope of formulation from traditional methods to advanced materials with nanoscale engineering capabilities that support multifunctional applications, scalable production, and system integration (Davis Jr et al., 2025).

3.1 Nanoparticles

While metallic nanoparticles continue to play an integral part in the development of advanced bioactive systems, the latest trends are towards precise design and multifunctionality. For example, gold (Au) and silver (Ag) metallic nanoparticles are being designed for their plasmonic or catalytic activities for use in enhancing imaging contrast, modulating photothermal effects, and generating antimicrobial action (Hang et al., 2024). Au nanoparticles have received much attention for their use in photothermal cancer treatment, as near-infrared radiation causes local heating and killing of tumor cells (Huang et al., 2006). Ag nanoparticles are already present in commercial wound dressing materials due to their wide-spectrum antibacterial activity (Wilkinson et al., 2011). Shape control in nanoparticle designs, including core-shell structures (Huang et al., 2025), and stars (Zhang, F. et al., 2025), offers a higher level of tunability to optical and electronic responses. Polymeric nanoparticles have advanced towards more complex nanostructures through the development of block copolymers and dendrites. They exhibit hierarchical structuring capabilities that facilitate more efficient loading of cargo into the system (Li, Y. et al., 2023, Wang et al., 2022). Polymeric nanoparticles can be used for transporting various agents within one nanoparticle. Lipid-based nanoparticles, especially liposomes and solid-lipid nanoparticles, have been enhanced by improvements in membrane stability and better delivery of cargoes (Hou et al., 2021). Hybrid systems consisting of both lipids and polymers are being explored as an alternative design due to the advantages that polymers bring. In contrast, metal nanoparticles usually possess better imaging and antibacterial capabilities, while polymer and lipid nanoparticles usually perform better in terms of biodegradability and drug loading capacity (Liew et al., 2022). Direct comparative studies on these different platforms are lacking.

3.2 Nanocomposites

Nanocomposites are a prominent class of multifunctional materials resulting from the incorporation of two nanoscale components and the synergism between them. The key recent developments include the optimization of interphase bonding and dispersity of nanofillers (Zhang, K. et al., 2024). However, aggregation of fillers, incompatibility at interfaces, and variation during upscaling remain open issues. Biomedical applications of nanocomposites involve the design of ceramic-polymer composites, which simulate the hierarchical architecture of biological tissues, in particular, in load-bearing structures. The properties of such materials include increased resistance to damage and deformations, the desired level of porosity, and increased bioactivity (Chen et al., 2023). The new generation of conductive nanocomposites includes carbon nanotubes (CNTs), graphitic compounds, as well as metal-based fillers. Such composite materials demonstrate high-level conductivity, flexibility, and compatibility with soft biological tissues (Sharma et al., 2026, Li, L. et al., 2023). Advances in aligning the fillers into networks or percolation channels have led to better conductivity at low concentrations of components, avoiding toxic effects without losing functional properties. It should be noted that current trends include nanocomposites with self-optimizing mechanical or electrical characteristics in accordance with operational requirements.

3.3 Nanointerfaces

Nanointerface design has grown more elaborate, involving precise manipulation of interface effects at molecular and nanometer scales. Instead of simply harnessing surface properties, the modern trend emphasizes active manipulation of interfacial activity (Miller & Medina, 2024). On the other hand, various modifications have been introduced into surfaces in order to manipulate the behavior of cells through controlling cell migration and differentiation among other behaviors. Nanopillars, nanogrooves, and surface architecture have been successfully incorporated to make this possible. Dynamic changes can be achieved via chemical surface modification by virtue of which surfaces can exhibit property change under certain triggers. This may include switching from a hydrophilic surface to a hydrophobic one, or revealing active ligands under certain conditions (Misra & Boriek, 2023, Cheng & Pang, 2023). With respect to biosensors, engineering efforts have been made toward improving signal generation at nanostructured interfaces by means of electron transfer processes as well as optical mechanisms (Koya & Li, 2023). Moreover, strategies such as making devices resistant to non-specific adsorption and contamination have been developed through incorporation of zwitterionic and polymeric brush surfaces.

3.4 Smart Polymers and Hydrogels

There have been great developments in the area of smart polymers and hydrogels during the past few years, whereby not only have they evolved from being mere stimuli-responsive but have also become very intricate in terms of functionalization (Tian et al., 2025). Nowadays, there are highly advanced hydrogels with dynamic crosslinks using either reversible covalent bonds or supramolecular chemistry, enabling the formation of self-healing hydrogels (Qi et al., 2025). For instance, self-healing hydrogels that are used in injection therapy and incorporate Schiff base reactions as well as guest-host complexation processes have been able to regenerate their strength after being deformed cyclically and improve their adherence to damaged tissues in animals (Jiang et al., 2022). It is highly advantageous when these hydrogels are used in applications where mechanical stress or strain is present. Despite all these benefits, some hydrogel materials continue to struggle with poor mechanical properties, slow reaction time, and low stability. Furthermore, it is possible to design polymeric systems that will react to several different stimuli, including temperature and ionic strength. In addition, gradient hydrogels or hydrogel layers can be designed, where the response depends on the location within the gel (Zhang & Wu, 2023). Furthermore, the incorporation of nanofillers into polymeric systems has enabled the development of nano-hydrogels that exhibit improved strength, conductivity, and responsiveness properties (Baishya et al., 2024, Lu et al., 2025). These biomaterials have been created to obtain an optimal combination of biocompatibility and effectiveness, and thus their scope can be very broad, from therapeutic use to engineering. In general, further progress in creating bioactive materials based on nanotechnology leads to the emergence of more sophisticated and multifunctional systems.

A comparison of key types of nanomaterials is given in Table 1, which highlights how nanotechnology helps achieve multifunctionality in diverse areas of bioactive systems via nanoscale engineering.

Table 1. Material innovations in bioactive systems using nanotechnology

Material Class	Representative Systems	Key Structural Features	Functional Properties	Recent Innovations	Applications	References
Nanoparticles	Metallic (Au, Ag), Polymeric nanoparticles, Liposomes	Size/shape tunability (rods, shells, stars), core–shell structures	Optical (plasmonic), catalytic, drug delivery, antimicrobial	Shape-controlled nanoparticles, hybrid lipid–polymer systems, multifunctional loading	Imaging, drug delivery, photothermal therapy	Hou et al., 2021; Huang et al., 2025; Zhang, F. et al., 2025
Nanocomposites	Ceramic–polymer, CNT-based, graphene composites	Multi-phase nanoscale architecture, filler–matrix interfaces	Mechanical strength, conductivity, bioactivity	Controlled filler dispersion, percolation networks, self-optimizing composites	Tissue engineering, flexible electronics, biosensors	Li, L. et al., 2023; Sharma et al., 2026; Zhang, K. et al., 2024
Nanointerfaces	Nanopatterned surfaces, functional coatings	Nanopillars, nanogrooves, chemical functionalization	Cell interaction control, enhanced signal transduction	Stimuli-switchable surfaces, anti-fouling coatings, engineered electron transfer	Biosensors, implants, diagnostic devices	Sahoo et al., 2024; Miller & Medina, 2024; Misra & Boriek, 2023; Cheng & Pang, 2023
Smart Polymers & Hydrogels	Stimuli-responsive polymers, nano-hydrogels	Dynamic crosslinking, gradient structures, supramolecular networks	Self-healing, multi-stimuli responsiveness, tunable mechanics	Reversible bonding, nanofiller-enhanced hydrogels, spatially graded gels	Drug delivery, wearable systems, regenerative medicine	Lin & Hsu, 2025; Tian et al., 2025; Lu et al., 2025; Zhang & Wu, 2023

4. FUNCTIONAL MECHANISMS

The effectiveness of smart bioactive systems relies on the synergistic operation of various functional components that collectively facilitate autonomous and adaptive behavior (Chen et al., 2026). In this case, the systems need to be able to process biological data streams and perform specific actions (Suryaprabha et al., 2026). However, unlike stand-alone elements, sensing, signal analysis, reaction, and adaptation are all integrated in a single platform, enabling intelligent operation under heterogeneous conditions (Wang, Z. et al., 2026).

4.1 Sensing Mechanisms

On the primary level, it is the sensing component that enables detection of minor changes in biological or physicochemical properties. Contemporary nanosensors are developed to have ultra-sensitive detection, fast response, and specificity (Sarikhani et al., 2025). On a higher level of sophistication, multiple sensing can be implemented by using arrays of nanosensors capable of identifying several types of target molecules simultaneously. In this way, a broader spectrum of biological characteristics can be analyzed without focusing on a single factor (Zhang, Y. et al., 2025). Bio-inspired sensing is a promising new technique that employs receptor molecules capable of mimicking biological reactions, although having increased stability and modifiability (Kang et al., 2025). Furthermore, in order to enable real-time sensing, the material has to withstand various environmental factors, including changing pH levels and mechanical agitation (Ding et al., 2025).

4.2 Signal Transduction

Once the signal is acquired, there needs to be effective transformation to a usable form of energy that can be effectively interpreted and utilized. As for signal transduction in smart bioactive systems, a lot of work has been done to increase precision and integrate these materials into computational systems. Hybrid systems of transduction involving different methods of conversion, such as electro-optical or mechanical-electrical ones, have attracted much interest because of increased precision and reliability (Zhang, P. et al., 2023). Sometimes, a biological signal may be associated not only with an electrical signal but also with certain optical characteristics, thus allowing for more precise detection and measurement. Miniaturization of such devices has led to the successful integration of bioactive components with electronic circuitry for effective wireless signal transmission (Yao et al., 2026).

4.3 Actuation and Response

The actuation module turns signals received after processing into output responses allowing for effective interactions with the external environment. The key principles in modern developments are the accuracy, reversibility, and energy-saving nature of the actuation process (Narvaez & Newell, 2025). This module may involve chemical, mechanical, or electrical stimuli depending on the type of actuation needed. Thus, chemical or biological signals may trigger the release of nanoscale valves or gates of some encapsulated substances, whereas mechanical stimuli may change the conformation of certain structures. Electrical signals may influence biological systems by affecting their electrical potentials (Dong & Malliaras, 2025). Modern approaches pay key attention to ensuring high accuracy of signal responses by using triggering thresholds for activating the desired reaction only in cases where the signal exceeds the critical value.

4.4 Feedback and Adaptation

Among the properties of advanced smart bioactive systems, the most critical feature is their capability of being employed in a closed-loop setting, which means that they are constantly guided by feedback signals and adjust their operation accordingly. Such feedbacks imply continuous loops of sensing and responding to what is happening in the surroundings of the system (Leng & Sun, 2025). One of the examples would be when a system regulates its work according to time trends, rather than reacting to one-time events. Another characteristic of smart bioactive systems is related to the element of adaptability. It involves programming response patterns of the system through external influences and accumulated experience (Dong & Malliaras, 2025). In summary, feedback and adaptability make smart bioactive systems highly advanced devices that have an opportunity to not only respond but regulate themselves.

Figure 1 shows the cyclic nature of the process of interactivity between the sensors, processing system, and actuators through adaptive feedback loops.

5. HEALTHCARE APPLICATIONS

The adaptation of smart bioactive systems, which benefit from nanotechnology, to healthcare applications has sped up their use for developing innovative treatment solutions that are accurate, minimally invasive, and customized for individual patients (Goshisht et al., 2025). Such technologies revolutionize clinical practices through early disease diagnosis, effective drug delivery, and enhanced healing processes. The integration of smart bioactive systems into medicine is not restricted to single components but encompasses comprehensive approaches that integrate diagnostics, treatment, and monitoring procedures (Kumar et al., 2025).

Figure 1. Functional architecture of intelligent bioactive systems with a closed loop that highlights the synergy between sensing, signal transmission, actuation, and feedback, thus providing for autonomous and intelligent operation

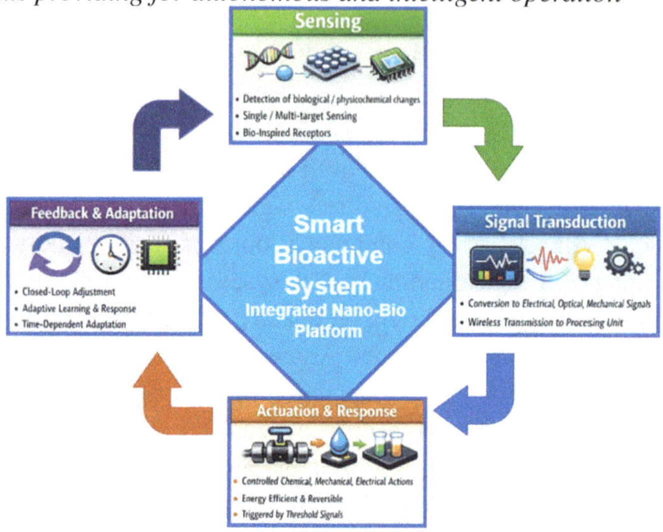

5.1 Smart Drug Delivery Systems

One example in this regard is the design of smart nanocarriers in drug delivery systems to improve pharmacodynamics and pharmacokinetics, which reduces drug loss in the body system. In contrast to traditional means, smart nanocarriers are able to execute complex functions, for example, movement along vessels, passage through cell membranes, and intracellular transport of substances (Mitchell et al., 2021). Although there is no doubt about the efficacy of many nanovehicles that have been tested in laboratories and animals, their clinical applications are still impeded because of poor efficiency, production problems, and other restrictions. One of the most successful cases of the clinical implementation of nanotechnology is the Onpattro drug that delivers siRNA to patients with transthyretin amyloidosis (Adams et al., 2018).

Among recent tendencies in this field, one should mention attempts to overcome physiological barriers to increase effectiveness of treatment. In this respect, there are nanomaterials able to pass through the blood-brain barrier that can be used to treat neurological pathologies. Likewise, materials allowing penetration through the tough matrix of tumors enhance the possibility of delivering drugs into the site of interest (Wilhelm et al., 2016). The delivery of genetic constructs to cells has become a breakthrough in therapy, with various approaches aimed at delivering genes in the form of DNA, RNA, or genetic editing materials into target cells while ensuring the safety of the process (Saraiva et al., 2016). The combination of nanotechnology and gene therapies enables an extremely precise effect on certain genetic abnormalities. Combination therapy is also an important trend, involving several compounds to achieve better results. It is particularly effective in cases when a disease is multifactorial and requires manipulation in different biological pathways at once.

5.2 Biosensors

Diagnostic processes using biosensing systems that rely on nanomaterials have been revolutionized through the ability to detect biomarker indicators of illness in an extremely fast and effective manner (Backiyalakshmi et al., 2024). Biosensors used in diagnostic procedures are becoming increasingly portable and can be worn for extended periods, thus eliminating the need for complex laboratory settings. For instance, devices like Dexcom and FreeStyle Libre, which are CGMs, can be considered exemplary wearables that help conduct biochemical analyses every day (Vashist, 2013). Despite their higher level of portability and capacity to record information non-stop compared to the traditional lab-based instruments, wearable nanosensors are vulnerable to problems like calibration mistakes and biofouling.

Concerning metabolic disorders, biosensing technology allows for continuous detection of biochemical markers, thereby enabling preemptive action (Wang, B. et al., 2026). The same applies to oncology where biosensors that use nanotechnology can detect circulating biomarkers in extremely small amounts, allowing for a quick diagnosis (Rao Bommi et al., 2023). The field of pathogen detection has also made impressive strides by the aid of nanotechnology (Mallick et al., 2025). In addition, nanosensors can rapidly identify pathogens (such as viruses, bacteria, or fungi), often responding immediately (Guliy et al., 2025). When digital health technologies are combined with biosensors, there is an improvement in the efficiency of these technologies and their applications. The information gathered by these sensors is capable of being uploaded to cloud computing servers, which can then analyze and interpret the data (Ren & Cui, 2026). In this way, it is possible to diagnose people remotely and manage their health status in a personalized manner, along with the possibility of making predictions about diseases based on the changes in biomarkers that can lead to new diseases (Xian, 2026; Ren & Cui, 2026).

5.3 Tissue Engineering

Regenerative medicine has been revolutionized by novel scaffolding systems that have been created through advances in nanotechnology. In tissue engineering, there are innovative scaffolds that not only mimic the structure of the tissue but also its biochemistry and mechanics (O'Brien, 2011; Place et al., 2009). Scaffolding systems that can dynamically change have been developed recently as innovations in regenerative medicine. Adaptability in such cases is based on the system's ability to change certain parameters depending on the stage of development of the cell that is growing. Introduction of biological signals into the structure enables spatial regulation of cellular activity.

Cell migration and cell organization can be achieved using gradients of signalling molecules. In addition, efforts are underway to incorporate vascularization techniques that provide an efficient delivery of nutrients and oxygen and overcome another limitation in tissue engineering (Groll et al., 2016). 3D bioprinting and nanomaterials are among the emerging trends in this area which enable the manufacture of complex tissues with increased precision (Murphy & Atala, 2014). For instance, bioprinting patient-specific cartilage and bone grafts with nanohydroxyapatite have proven effective with regards to enhancing cell proliferation and bone mineralization (Huang et al., 2023). This may open up avenues for designing implants such as organs or grafts customized according to the patient's requirement.

5.4 Implantable Devices

Over time, implantable devices have advanced from mechanical components to sophisticated systems able to constantly interact with their surroundings. Nanotechnology improves the functionality of implantable devices by increasing sensitivity and integration of devices with the biological tissues (Lin et al., 2025). Neural interfaces are being created with the help of nanoelectrodes, which make it possible to interact precisely with neurons. An example of how this is achieved in practical settings includes the miniaturization of cochlear implants and better neural interface designs that have greatly aided patients suffering from hearing problems (Vecchi et al., 2024). Using such a method may turn out to be beneficial for the treatment of nervous diseases, recovery of sensations, and creation of highly efficient brain-computer interaction (Yu et al., 2024).

There is cardiac implantation that includes monitors that are able to check the heart rhythm, blood pressure, and other chemical components. This development makes it possible for doctors to detect the risk factors early on (Bauer & Dlaska, 2024). Smart prosthetics incorporating nanosensors and actuators improve their efficiency because they can offer tactile sensations and self-control to the user (Zhuwawu & El-Hussieny, 2025). Efficient energy management plays an essential role in implants; hence, new techniques involving energy harvesters that generate energy from physiological motion or biochemistry have been invented (Yadav et al., 2025).

Table 2 presents a comparison of important uses of medical technologies in the field, which include categories of system used, their features, and nanotechnologies involved in their realization. The table demonstrates the progress made recently and the influence of this progress on medical practice.

Table 2. Applications of nanotechnology-based smart bioactive systems in healthcare

Application Area	System Type	Key Functional Features	Enabling Nanotechnologies	Recent Innovations	Clinical Impact	References
Smart Drug Delivery	Nanocarriers, gene delivery systems	Targeted delivery, controlled release, barrier penetration	Polymeric nanoparticles, lipid nanoparticles, hybrid systems	Blood-brain barrier (BBB)-crossing carriers, tumor-penetrating systems, combination therapy	Precision therapy, reduced side effects	Dilliard & Siegwart, 2023; Mitchell et al., 2021; Saraiva et al., 2016
Biosensors	Wearable/implantable sensors	Real-time monitoring, high sensitivity detection	Nanosensors, plasmonic materials, nanoelectronics	Cloud integration, multiplex sensing, rapid pathogen detection	Early diagnosis, remote healthcare	Wu et al., 2023; Wang, B. et al., 2026; Backiyalakshmi et al., 2024; Ren & Cui, 2026
Tissue Engineering	Scaffolds, 3D bioprinted constructs	Tissue regeneration, cell guidance, biomimicry	Nanocomposite, nano-hydrogels	Dynamic scaffolds, vascularization strategies, gradient signaling	Regenerative medicine, personalized implants	Wang, X. et al., 2024; Murphy & Atala, 2014; Huang et al., 2023; O'Brien, 2011
Implantable Devices	Neural, cardiac, prosthetic systems	Continuous monitoring, bio-integration	Nano-electrodes, nanomaterials, energy harvesters	Smart prosthetics, brain–computer interfaces, self-powered implants	Long-term therapy, improved quality of life	Wang, Y. et al., 2023; Lin et al., 2025; Yu et al., 2024; Vecchi et al., 2024

6. APPLICATIONS OF SMART BIOACTIVE SYSTEMS BEYOND HEALTHCARE

Furthermore, beyond the scope of the medical field, the flexibility of these smart bioactive systems has been harnessed to apply them in other sectors as well. However, most other applications outside the field of medicine are still at an experimental stage.

6.1 Environmental Sensing

Environmental scientists are taking advantage of nanotech sensors in detecting ecological conditions to enable effective monitoring and management of such conditions. They can be used to detect even minute quantities of harmful elements in water, air, or soil, thus facilitating early detection of pollution (Parameswari et al., 2024). For example, sensors developed from nanoparticles such as graphene and Au nanoparticles have been used to identify the presence of ions of lead and mercury in water, even at parts per billion level (Nadumane et al., 2025). Some of the platforms feature continuous in-situ monitoring and are integrated with wireless networking capabilities, making them suitable for use in smart city programs and ecological management (Zeng et al., 2024). Some even have the ability to remediate the detected contamination in an attempt to mitigate its harmful effects (Kumar, R. et al., 2025a).

6.2 Agriculture

Precision farming is being enabled by the use of bioactive nanoparticles in agriculture. They help in making better use of resources by providing inputs such as nutrients and agrochemicals in an efficient manner. Bioactive nanoparticles are useful in developing fertilizers capable of releasing nutrients in a way that conforms to the needs of plants and minimizing nutrient losses through leaching and volatilization (Hamed et al., 2024; Nawadkar et al., 2024). Studies carried out on nanofertilizers like zinc oxide and nano urea have shown better results as compared to traditional fertilizers (Reddy et al., 2024). Moreover, delivery of crop protection materials through nanoparticle technology is becoming more and more common. Using nanosensors helps collect information regarding the state of the soil along with other metrics, which can help make rational choices when it comes to irrigation and harvesting operations (Jali et al., 2024; Zeng et al., 2025). These developments play an important role in enhancing crop yields and managing resources.

6.3 Biohybrid Systems

Biohybrid technologies are the latest technology generation that involves using living organisms along with man-made nanomaterials in the design of superior performing devices compared to artificial systems (Garmroudi et al., 2025). For instance, biohybrid devices include bioactuators that employ the energy produced by living cells to create movement. Biosensors which incorporate biological recognition systems like enzymes into nanostructures are another example that provides highly sensitive detection of diverse substances (Yu et al., 2026). New developments in biohybrids are using microbial systems together with nanomaterials for applications such as biofuels, biomanufacturing, and bioremediation (Li et al., 2025). In order to develop the hybrid bio-systems, a delicate balancing act is required in order for them to function effectively. The development of the systems can be expected to create possibilities in fields as varied as robotics to eco-friendly manufacture (Garmroudi et al., 2025). Together, the various uses of smart bioactive systems based on nanotechnology show that the systems have revolutionary capabilities in many different fields. This will allow them to solve a variety of problems faced by society (Wu et al., 2025).

Figure 2 shows the potential uses of advanced smart bioactive systems using nanotechnology beyond the healthcare sector.

Figure 2. Potential uses of advanced smart bioactive systems using nanotechnology beyond healthcare sector

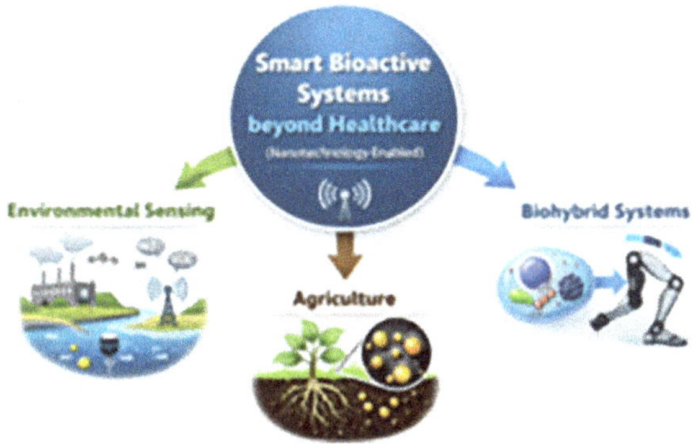

7. CHALLENGES AND LIMITATIONS

Despite the promising possibilities that have been realized through the development of nanotechnology-based intelligent bioactive systems, the move towards practical utilization on a broad scale faces several hurdles (Parvin et al., 2025a). This includes issues pertaining to science and technology, regulation, economy, and society, among others. One common problem with almost all materials is that improving one characteristic usually means degrading another, like improved responsiveness may affect stability, whereas higher function might imply higher costs and complex manufacturing processes.

7.1 Toxicity and Biocompatibility

A major issue during the design and production of nanoscale devices involves the unpredictability concerning their behavior when exposed to biological environments over time (Bose et al., 2026). Although it might be relatively easy to determine the short-term interactions and compatibility of such devices with living tissue, long-term use raises issues of accumulation, biodegradation, and potential unexpected biological reactions (Gao et al., 2026). There are some biological hazards associated with the nanomaterials used for producing nanoscale devices. First, the biological behavior of nanomaterials might depend on their size and geometry; hence, they might cause oxidative stress, inflammation, or other unintended biological responses in organisms (Pathak et al., 2025).

Toxicity levels can also be influenced by the route of exposure (intravenous, oral, inhalation, and dermal), dosage amount, and the patient's age or other predisposing health conditions. In other words, it is crucial to have standardized toxicity testing models that incorporate in vitro, in vivo, and clinical data (Wang et al., 2018). As was shown in a few studies (Carlson et al., 2008; Gliga et al., 2014), smaller sizes of Ag nanoparticles led to more production of reactive oxygen species and higher toxicity than larger sizes under the same mass concentrations. One of the key issues is related to the presence of the protein corona, where biomolecules bind to the surface of nanoparticles, altering their behavior (Soliman et al., 2024). It is essential to consider what happens to the nanomaterials once their designated purpose has been served. If the material lingers inside a living being, the risk of chronic toxicity arises (Sanati et al., 2025). Additionally, the material might degrade into other substances that may also have biological or toxic activity (Mndlovu et al., 2024).

7.2 Scalability and Manufacturing

Scaling up innovations from the nano-level to an industry level involves certain technical challenges. Nanoparticles are usually synthesized in lab conditions which are hard to reproduce in industrial settings (Jundale et al., 2024). Scaling-up leads to alteration in heat transfer, mass transfer, shear stress, residence time, and mixing; all of these can impact nucleation and growth of nanoparticles (Suda et al., 2018). As such, substances that work perfectly in grams can be completely different in kilograms or tons. For instance, variations in temperature, precursor concentration, and kinetics are likely to produce inconsistency in particle size, structure, and functionality (Srinivasan & Rana, 2024).

Reproducibility becomes important especially when the product needs high quality control measures, such as is the case in the biomedical industry. Slight variations in nanoscale characteristics may cause notable disparities in the operation of the system (Rezaei et al., 2024). In addition, incorporation of different functional elements in one platform poses another manufacturing challenge. Assemblies at such a small scale require complex processes that may not be easily scalable at present (Fan et al., 2025). In general, scaling up nanotechnology innovations involves overcoming challenges related to quality control, reproducibility, and complexity in fabrication.

7.3 Regulatory and Ethical Issues

Currently, the regulatory framework applicable to the use of nanotechnologies in the creation of new systems remains in instability. Nevertheless, the current legal framework formulated under existing material regulations is incapable of handling the unique features of nanotechnology-based inventions (Rodríguez-Gómez et al., 2025). As an illustration, there is uncertainty about whether a nanotechnology-enhanced medical product must be classified as a drug, device, biologic, or combination product, and different pathways would involve the generation of evidence and regulatory time frames. This raises concerns among the inventors and developers who are uncertain about their inventions' future applications. Thus, one of the biggest problems faced by researchers and scientists is the absence of standardized testing protocols to assess the safety, efficiency, and environmental impacts of such innovations (Rasmussen et al., 2025). There are other regulatory matters such as life-cycle regulation relating to sources of raw materials, handling of workers during the manufacturing process, transport, disposal of nanomaterials as waste, and the environmental impact resulting from their release after product use.

Furthermore, differences in regulation laws among various regions contribute to hindering commercialization processes (Ma et al., 2024). The ethical aspect cannot be ignored, especially where biohybrids have a direct connection to the biological system. The matter is connected with the data protection of a particular device, and nowadays, there appears to be an increased need in devices for physiological parameters' monitoring (Capulli et al., 2025). With regard to smart implantable or wearable devices, cybersecurity and data privacy have become equally important regulatory considerations. The equitable availability of advanced technologies for diverse populations also represents an ethical consideration that should be taken into account (Vayena et al., 2018).

7.4 Stability and Reliability

Consistent performance over time is vital for the success of the intelligent bioactive devices in any application where there will be continuous use or implantation of such systems. The materials that constitute nanomaterials are highly susceptible to changes brought about by the environment, especially temperature, pH, ionic strength, and physical strain, all of which could cause deterioration or decrease in functionality of the device (Hu & Dong, 2024). Biological environments are known to cause changes to the system through interaction of proteins, cells, or other biological molecules leading to changes in behavior of the system, and hence affecting its performance and reliability (Zhang, L. et al., 2024), like for example, alteration of size or change in the properties of the nanomaterials due to agglomeration (Jiang et al., 2024).

Another factor to be considered is reliability of device performance in relation to its multiple functionalities in a consistent manner (Mengesha & Nagessar, 2026).

7.5 Cost and Accessibility

The economic factor related to smart bioactive systems has a lot of influence on the application and practical importance of this invention. Expensive development and manufacturing processes may prevent the widespread usage of these systems, especially if we are talking about developing countries (Torres, 2025). The cost factors include expensive materials, high-tech equipment, and manufacturing processes. However, apart from the price factors, additional costs related to the infrastructure required for implementing and operating the system need to be considered. For instance, some devices require advanced processing units or control units that may be unaffordable in certain locations due to the lack of technical resources (Sodagar et al., 2025). Therefore, an important issue is finding ways to make this system inexpensive while still ensuring maximum efficiency. The implementation of a modular system architecture and using easily available construction materials would greatly help in this case (Kim et al., 2025). Other potential solutions may include technology transfer for impoverished countries, distributed manufacturing models, and streamlined modular manufacturing systems.

Figure 3 shows five main barriers relating to issues of biocompatibility, scalability, regulation, stability, and cost, which are obstacles in the application of smart bioactive systems through nanotechnology.

Figure 3. Multi-dimensional barriers to the implementation of nanotechnology-driven smart bioactive systems

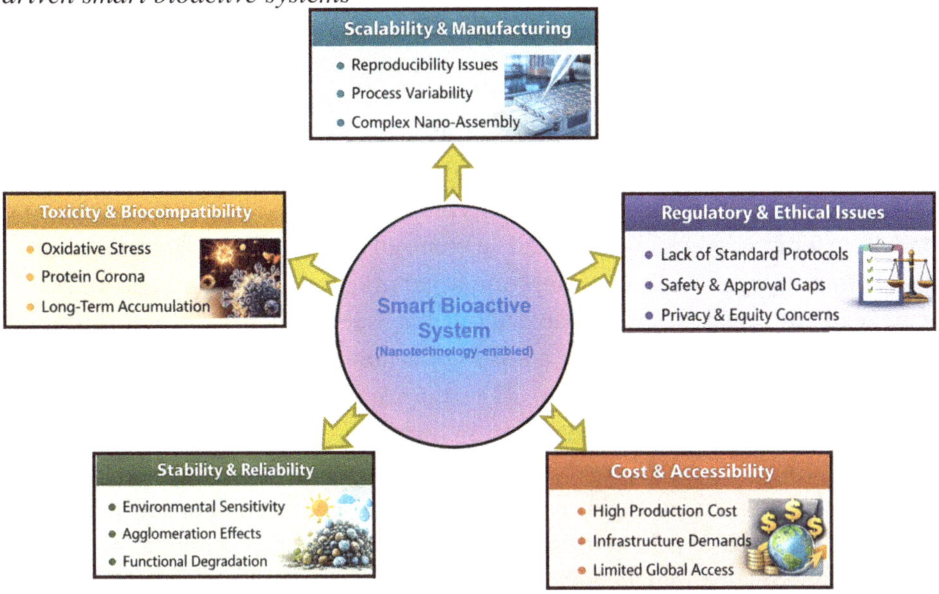

8. FUTURE PERSPECTIVES

The future path of smart bioactive systems will be determined largely by their convergence with other scientific and technological areas of knowledge. Such a convergence will provide an opportunity to create systems with greater levels of functionality, independence, and flexibility that will allow them to function efficiently in different settings (Zhang, W. et al., 2025). Research in the future should focus mainly on the standardization of benchmarking procedures, sustainable green manufacturing techniques, safety assessment for the long term, and adaptive controls facilitated by artificial intelligence (AI). Equal focus should be on harmonizing regulatory structures across countries, establishing toxicological information databases, and implementing scalable roadmaps for transitioning technologies developed in labs to commercialization processes. Self-powered nanosystems (Wang et al., 2012) can be an area for the development of the next generation. Such systems will harness energy from external environmental resources through mechanical vibrations, body movements, temperature differences, or chemical processes (Gołąbek & Strankowski, 2024). Self-powered nanosystems will eliminate the need for an external battery source and help in building implantable systems and remote devices for longer periods. In addition, AI can be the game-changer here (Jiang et al., 2020). By processing large volumes of biological data, machine learning (ML) algorithms might provide real-time responses and adaptations in response to biological data (Topol, 2019). Some recent experiments have already shown how wearables can help in predicting glucose levels through ML (van Doorn et al., 2021). For example, if a system detects particular patterns, it changes its functioning. Technological developments related to fabrication methods, especially in additive manufacturing (Mani & Thepperumal, 2024) and 3D bioprinting (Lee et al., 2024), are creating new prospects for developing complex and customized designs. These novel techniques enable controlling spatially the composition and structure of materials and the development of functional systems capable of fulfilling specific demands. Multi-material constructs will become an effective means of achieving greater flexibility in smart bioactive systems (Jose et al., 2026). The next major trend will be connected to the design of autonomous biohybrid systems combining the advantages of biological intelligence and engineered devices. A particular autonomous biohybrid can behave intelligently and perform sophisticated functions like adaptation or self-regulation (Chen, C. et al., 2025). In addition, with recent progress in synthetic biology, it becomes possible to implement the potential of a particular biohybrid with the help of programmable bio-components (Garmroudi et al., 2025). Further success in this sphere would depend on cooperation within different fields, namely materials science, biology, engineering, and data sciences. Partnerships between academic, industrial, and governmental bodies would be crucial for the successful work in this area (Rabani et al., 2026).

9. CONCLUSION

Nanobiomaterials become a breakthrough in functional materials as far as they provide unique opportunities for solving difficult problems. Such materials can interact with their surroundings within biological environments in an exact manner. Nevertheless, using such materials is associated with numerous challenges, the most critical among which relate to safety, long-term toxicity, scalability, regulatory compliance, cost-effectiveness, and reliability. In particular, their widespread commercial adoption requires overcoming issues of batch consistency, safety standardization, and scalable production systems. Resolving them will demand not only technical progress but also building of adequate infrastructure and cooperation mechanisms. With the continuous development of research in the field, the combination of nanotechnology with other branches of knowledge is seen as a catalyst for more advanced and complex systems. This process offers huge potential in the improvement of human healthcare, the development of sustainable environments, and new opportunities for bioengineering. However, progress in the future needs to extend from proof-of-concept studies to reliable, clinically proven, economically feasible, and industrially scalable technology. At the end of the day, the effectiveness of intelligent biomaterials depends on striking the right balance between creativity and accountability.

REFERENCES

Adams, D., Gonzalez-Duarte, A., O'Riordan, W. D., Yang, C.-C., Ueda, M., Kristen, A. V., Tournev, I., Schmidt, H. H., Coelho, T., Berk, J. L., Lin, K.-P., Vita, G., Attarian, S., Planté-Bordeneuve, V., Mezei, M. M., Campistol, J. M., Buades, J., Brannagan, T. H.III, Kim, B. J., & Suhr, O. B. (2018). Patisiran, an RNAi therapeutic, for hereditary transthyretin amyloidosis. *The New England Journal of Medicine, 379*(1), 11–21. DOI: 10.1056/NEJMoa1716153

Ahmadi, S., Rabiee, N., Bagherzadeh, M., Elmi, F., Fatahi, Y., Farjadian, F., Baheiraei, N., Nasseri, B., Rabiee, M., Dastjerd, N. T., Valibeik, A., Karimi, M., & Hamblin, M. R. (2020). Stimulus-responsive sequential release systems for drug and gene delivery. *Nano Today*, *34*, 100914. DOI: 10.1016/j.nantod.2020.100914

Backiyalakshmi, G., Snekhalatha, U., & Salvador, A. L. (2024). Recent advancements in non-invasive wearable electrochemical biosensors for biomarker analysis: A review. *Analytical Biochemistry*, *692*, 115578. DOI: 10.1016/j.ab.2024.115578

Baishya, G., Parasar, B., Limboo, M., Kumar, R., Dutta, A., Hussain, A., Phukan, M. M., & Saikia, D. (2024). Advancements in nanocomposite hydrogels: A comprehensive review of biomedical applications. *Discover Materials*, *4*(1), 40. DOI: 10.1007/s43939-024-00111-8

Bauer, A., & Dlaska, C. (2024). Implantable cardiac monitors: The digital future of risk prediction? *European Heart Journal. Digital Health*, *5*(4), 397–398. DOI: 10.1093/ehjdh/ztae036

Bose, P., Ray, D., Sah, S. K., & Kaity, S. (2026). Implantable nanoelectronics: Material considerations and biointerface interactions. *Sustainable Materials and Technologies*, *47*, e01837. DOI: 10.1016/j.susmat.2025.e01837

Çaktü Güler, K., Göktürk, I., Yılmaz, F., Araz, A., & Denizli, A. (2024). Plasmonic nanosensors for environmental pollutants sensing: Recent advances and perspectives. *Essential Chemistry*, *1*(1), 1–19. DOI: 10.1080/28378083.2024.2386522

Capulli, E., Druda, Y., Palmese, F., Butt, A. H., Domenicali, M., Macchiarelli, A. G., Silvani, A., Bedogni, G., & Ingravallo, F. (2025). Ethical and legal implications of health monitoring wearable devices: A scoping review. *Social Science & Medicine*, *370*, 117685. DOI: 10.1016/j.socscimed.2025.117685

Carlson, C., Hussain, S. M., Schrand, A. M., Braydich-Stolle, L. K., Hess, K. L., Jones, R. L., & Schlager, J. J. (2008). Unique cellular interaction of silver nanoparticles: Size-dependent generation of reactive oxygen species. *The Journal of Physical Chemistry B*, *112*(43), 13608–13619. DOI: 10.1021/jp712087m

Chen, C., Liu, Z., Shi, P., Zhao, Y., Duan, S., Du, Y., Yan, Y., Si, M., Iwasaki, T., & He, X. (2025). Bio-inspired multimodal soft actuator with environmental self-adaptation. *Nature Communications*, *16*(1), 7630. DOI: 10.1038/s41467-025-62328-2

Chen, D., Zheng, W., Zhang, Z., Yu, S., Hang, X., Wu, H., Xiang, X.-W., Mu, W., Jiao, Y., Dong, Z., & Chang, L. (2026). Nanozymes integrated biochips toward smart detection system. *Advanced Science (Weinheim, Baden-Wurttemberg, Germany)*, *13*(11), e19136. DOI: 10.1002/advs.202519136

Chen, Q., Furrer, R., Jamilpanah, L., Chumakov, A., Bulut, Y., Harder, C., Müller-Buschbaum, P., Roth, S. V., & Braun, A. (2025). Responsive magnetic polymer nanocomposites through thermal-induced structural reorganization. *ACS Nano*, *19*(6), 6165–6179. DOI: 10.1021/acsnano.4c14311

Chen, X., Wu, D., & Chen, Z. (2024). Biomedical applications of stimuli-responsive nanomaterials. *MedComm*, *5*(8), e643. DOI: 10.1002/mco2.643

Chen, Y.-H., Zhao, Y.-H., Tian, S.-H., Wang, P.-F., Qiu, F., & Yi, T.-F. (2023). Recent progress and strategic perspectives of high-voltage $Na_3V_2(PO_4)_2F_3$ cathode: Fundamentals, modifications, and applications in sodium-ion batteries. *Composites. Part B, Engineering*, *266*, 111030. DOI: 10.1016/j.compositesb.2023.111030

Cheng, Y., & Pang, S. W. (2023). Effects of nanopillars and surface coating on dynamic traction force. *Microsystems & Nanoengineering*, *9*(1), 6. DOI: 10.1038/s41378-022-00473-0

Damiri, F., Fatimi, A., Santos, A. C. P., Varma, R. S., & Berrada, M. (2023). Smart stimuli-responsive polysaccharide nanohydrogels for drug delivery: A review. *Journal of Materials Chemistry. B*, *11*(44), 10538–10565. DOI: 10.1039/D3TB01712E

Davis, R.Jr, Duggal, I., Peppas, N. A., & Gaharwar, A. K. (2025). Designing the next generation of biomaterials through nanoengineering. *Advanced Materials*, *37*(39), e2501761. DOI: 10.1002/adma.202501761

Del Bakhshayesh, A. R., Saghebasl, S., Asadi, N., Kashani, E., Mehdipour, A., Nezami Asl, A., & Akbarzadeh, A. (2023). Recent advances in nano-scaffolds for tissue engineering applications: Toward natural therapeutics. *Wiley Interdisciplinary Reviews. Nanomedicine and Nanobiotechnology*, *15*(6), e1882. DOI: 10.1002/wnan.1882

Del Bono, F., Di Trani, N., Demarchi, D., Grattoni, A., & Motto Ros, P. (2025). Active implantable drug delivery systems: Engineering factors, challenges, opportunities. *Lab on a Chip*, *25*(15), 3608–3629. DOI: 10.1039/D5LC00131E

Dilliard, S. A., & Siegwart, D. J. (2023). Passive, active and endogenous organ-targeted lipid and polymer nanoparticles for delivery of genetic drugs. *Nature Reviews. Materials*, *8*(4), 282–300. DOI: 10.1038/s41578-022-00529-7

Ding, M., Xie, P., Wang, J., Guo, W., Li, H., Hu, S., Li, D., Li, B., Wang, N., Wong, C.-Y., Sun, J., & Ho, J. C. (2025). Biomimetic microstructure design for ultrasensitive piezoionic mechanoreceptors in multimodal object recognition. *Nature Communications*, *16*(1), 8129. DOI: 10.1038/s41467-025-63115-9

Dong, C., & Malliaras, G. G. (2025). Recent advances in stimuli-responsive materials and soft robotic actuators for bioelectronic medicine. *Advanced Materials*, *37*(27), 2417325. DOI: 10.1002/adma.202417325

El-Khawaga, A. M., Youssef, O. I., El-Dydamoni, O. A., Ragab, W. M., & Mohammed, R. A. E. (2026). Advances in green-synthesized magnetic nanoparticles for targeted cancer therapy: Mechanisms, applications, and future perspectives. *European Journal of Medical Research*, *31*(1), 347. DOI: 10.1186/s40001-025-03804-9

Fan, P., Liu, Y., Pan, Y., Ying, Y., & Ping, J. (2025). Three-dimensional micro- and nanomanufacturing techniques for high-fidelity wearable bioelectronics. *Nature Reviews Electrical Engineering*, *2*(6), 390–406. DOI: 10.1038/s44287-025-00174-6

Gao, M., Nikolayev, D., Sipus, Z., & Skrivervik, A. K. (2025). Physical insights and design principles for efficient wireless implantable bioelectronics. *Cell Reports. Physical Science*, *6*(7), 102627. DOI: 10.1016/j.xcrp.2025.102627

Gao, S., Lei, T., & Zhang, Z. (2026). Biocompatible conjugated polymers for bioelectronics. *Advanced Functional Materials*, *28239*, e28239. Advance online publication. DOI: 10.1002/adfm.202528239

Garmroudi, A., Tushar, M. A. K., Liu, C., & Li, Z. (2025). Biohybrid living robotics: A comprehensive review of recent advances, technological innovation, and future prospects. npj. *Npj Robotics*, *3*(1), 43. DOI: 10.1038/s44182-025-00056-x

Gliga, A. R., Skoglund, S., Odnevall Wallinder, I., Fadeel, B., & Karlsson, H. L. (2014). Size-dependent cytotoxicity of silver nanoparticles in human lung cells: The role of cellular uptake, agglomeration and Ag release. *Particle and Fibre Toxicology*, *11*(1), 11. DOI: 10.1186/1743-8977-11-11

Gołąbek, J., & Strankowski, M. (2024). A review of recent advances in human-motion energy harvesting nanogenerators, self-powering smart sensors and self-charging electronics. *Sensors (Basel)*, *24*(4), 1069. DOI: 10.3390/s24041069

Gongadze, E., Kabaso, D., Bauer, S., Slivnik, T., Schmuki, P., van Rienen, U., & Iglič, A. (2011). Adhesion of osteoblasts to a nanorough titanium implant surface. *International Journal of Nanomedicine*, *6*, 1801–1816. DOI: 10.2147/IJN.S21755

Goshisht, M. K., Goshisht, A., Bajpai, A., & Bajpai, A. (2025). Recent advances in biomedical applications of smart nanomaterials: A comprehensive review. *RSC Pharmaceutics*, *2*(6), 1227–1267. DOI: 10.1039/D5PM00137D

Groll, J., Boland, T., Blunk, T., Burdick, J. A., Cho, D.-W., Dalton, P. D., Derby, B., Forgacs, G., Li, Q., Mironov, V. A., Moroni, L., Nakamura, M., Shu, W., Takeuchi, S., Vozzi, G., Woodfield, T. B. F., Xu, T., Yoo, J. J., & Malda, J. (2016). Biofabrication: Reappraising the definition of an evolving field. *Biofabrication*, *8*(1), 013001. DOI: 10.1088/1758-5090/8/1/013001

Guliy, O. I., Chumakov, D. S., Evstigneeva, S. S., Liu, Q., & Zavyalova, E. G. (2025). Nanozymes: A dual approach to bacterial colorimetric and luminescent detection and to bacterial pathogen killing. *Biosensors & Bioelectronics*, *288*, 117835. DOI: 10.1016/j.bios.2025.117835

Guo, Z., Xiao, Y., Wu, W., Zhe, M., Yu, P., Shakya, S., Li, Z., & Xing, F. (2025). Metal–organic framework-based smart stimuli-responsive drug delivery systems for cancer therapy: Advances, challenges, and future perspectives. *Journal of Nanobiotechnology*, *23*(1), 157. DOI: 10.1186/s12951-025-03252-x

Gupta, P., Dhar, H., Bagal, Y. S., & Jaglan, S. (2025). Smart nano-fertilizers: A path to sustainable agriculture. *Environmental Geochemistry and Health*, *47*(10), 443. DOI: 10.1007/s10653-025-02749-6

Hachhach, M., Bayou, S., El Kasmi, A., Saidi, M. Z., Akram, H., Hanafi, M., Achak, O., El Moujahid, C., & Chafik, T. (2025). Towards sustainable scaling-up of nanomaterials fabrication: Current situation, challenges, and future perspectives. *Eng*, *6*(7), 149. DOI: 10.3390/eng6070149

Hamed, R., Jodeh, S., & Alkowni, R. (2024). Nano bio fertilizer capsules for sustainable agriculture. *Scientific Reports*, *14*(1), 13646. DOI: 10.1038/s41598-024-62973-5

Hang, Y., Wang, A., & Wu, N. (2024). Plasmonic silver and gold nanoparticles: Shape- and structure-modulated plasmonic functionality for point-of-care sensing, bio-imaging and medical therapy. *Chemical Society Reviews*, *53*(6), 2932–2971. DOI: 10.1039/D3CS00793F

He, X., Han, Z., Ruan, Y., Wang, Z., Liao, B., Li, X., Tan, J., Han, X., Shen, J., & Bai, D. (2025). Smart responsive biomaterials for spatiotemporal modulation of functional tissue repair. *Materials Today. Bio*, *33*, 102063. DOI: 10.1016/j.mtbio.2025.102063

Hergt, R., Dutz, S., Müller, R., & Zeisberger, M. (2006). Magnetic particle hyperthermia: Nanoparticle magnetism and materials development for cancer therapy. *Journal of Physics Condensed Matter*, *18*(38), S2919–S2934. DOI: 10.1088/0953-8984/18/38/S26

Hou, X., Zaks, T., Langer, R., & Dong, Y. (2021). Lipid nanoparticles for mRNA delivery. *Nature Reviews. Materials*, *6*(12), 1078–1094. DOI: 10.1038/s41578-021-00358-0

Hu, J., & Dong, M. (2024). Recent advances in two-dimensional nanomaterials for sustainable wearable electronic devices. *Journal of Nanobiotechnology*, *22*(1), 63. DOI: 10.1186/s12951-023-02274-7

Huang, J., Huang, Z., Xiong, J., Xia, J., Wang, Y., Yang, L., & Liang, Y. (2023). 3D bioprinted scaffolds of polysaccharide hydrogels in osteochondral and cartilage tissue engineering. *Designed Monomers and Polymers*, *26*(1), 258–272. DOI: 10.1080/15685551.2023.2284482

Huang, J., Tao, L., Wei, H., Huang, H., Zhang, Q., & Zhou, B. (2025). Full-color tuning in multi-layer core-shell nanoparticles from single-wavelength excitation. *Nature Communications*, *16*(1), 2378. DOI: 10.1038/s41467-025-57622-y

Huang, X., El-Sayed, I. H., Qian, W., & El-Sayed, M. A. (2006). Cancer cell imaging and photothermal therapy in the near-infrared region by using gold nanorods. *Journal of the American Chemical Society*, *128*(6), 2115–2120. DOI: 10.1021/ja057254a

Hwang, N. C., Lim, D. M., Goh, T. S., Kang, J. M., Kim, J., Kim, S., Kim, Y. H., & Kim, D. (2025). Recent advances in theranostic nanomaterials for overcoming traumatic brain injury. *Journal of Nanobiotechnology*, *23*(1), 692. DOI: 10.1186/s12951-025-03685-4

Illing, B., Mohammadnejad, L., Theurer, A., Schultheiss, J., Kimmerle-Mueller, E., Rupp, F., & Krajewski, S. (2023). Biological performance of titanium surfaces with different hydrophilic and nanotopographical features. *Materials (Basel)*, *16*(23), 7307. DOI: 10.3390/ma16237307

Jali, P., Acharya, S., & Mahalik, G. (2024). Antimicrobial efficacy of nano-particles for crop protection and sustainable agriculture. *Discover Nano*, *19*(1), 117. DOI: 10.1186/s11671-024-04059-9

Jiang, D., Shi, B., Ouyang, H., Fan, Y., Wang, Z. L., & Li, Z. (2020). Emerging implantable energy harvesters and self-powered implantable medical electronics. *ACS Nano*, *14*(6), 6436–6448. DOI: 10.1021/acsnano.9b08268

Jiang, L., Guo, Y., Liu, Z., & Chen, S. (2024). Computational understanding of the coalescence of metallic nanoparticles: A mini review. *Nanoscale*, *16*(11), 5521–5536. DOI: 10.1039/D3NR06133G

Jiang, X., Zeng, F., Yang, X., Jian, C., Zhang, L., Yu, A., & Lu, A. (2022). Injectable self-healing cellulose hydrogel based on host–guest interactions and acylhydrazone bonds for sustained cancer therapy. *Acta Biomaterialia*, *141*, 102–113. DOI: 10.1016/j.actbio.2021.12.036

Jose, S. A., Evtimow, A., & Menezes, P. L. (2026). Advances in 3D bioprinting: Materials, processes, and emerging applications. *Micromachines*, *17*(3), 282. DOI: 10.3390/mi17030282

Jundale, R. B., Sonawane, J. R., Palghadmal, A. V., Jaiswal, H. K., Deore, H. S., & Kulkarni, A. A. (2024). Scaling-up continuous production of mesoporous silica particles at kg scale: Design & operational strategies. *Reaction Chemistry & Engineering*, *9*(7), 1914–1923. DOI: 10.1039/D3RE00707C

Kang, J., Liu, J., Geng, Y., Yuan, Y., Liu, S., Tan, Y., Du, L., & Wu, C. (2025). Recent advances in biomimetic taste-based biosensors and their applications. *Sensors & Diagnostics*, *4*(1), 24–34. DOI: 10.1039/D4SD00311J

Kim, H. J., Koo, J. H., Lee, S., Hyeon, T., & Kim, D.-H. (2025). Materials design and integration strategies for soft bioelectronics in digital healthcare. *Nature Reviews. Materials*, *10*(9), 654–673. DOI: 10.1038/s41578-025-00819-w

Kotsuchibashi, Y. (2020). Recent advances in multi-temperature-responsive polymeric materials. *Polymer Journal*, *52*(7), 681–689. DOI: 10.1038/s41428-020-0330-0

Koya, A. N., & Li, W. (2023). Multifunctional charge transfer plasmon resonance sensors. *Nanophotonics*, *12*(12), 2103–2113. DOI: 10.1515/nanoph-2023-0196

Kumar, M., Sethi, P., Shiekmydeen, J., Rastogi, S., Mahmood, S., Chopra, S., Thomas, S., Kumar, D., & Bhatia, A. (2025). A recent review on smart sensor-integrated wound dressings: Real-time monitoring and on-demand therapeutic delivery. *International Journal of Biological Macromolecules*, *313*, 144251. DOI: 10.1016/j.ijbiomac.2025.144251

Kumar, R., Jaiswal, S., Rai, A., Rai, V. K., & Singh, M. (2025a). Electroactive advanced nanomaterials for wastewater monitoring. *Discover Electrochemistry*, *2*(1), 67. DOI: 10.1007/s44373-025-00081-3

Kumar, R., Kumar, A., Bhardwaj, S., Sikarwar, M., Sriwastaw, S., Sharma, G., & Gupta, M. (2025b). Nanotoxicity unveiled: Evaluating exposure risks and assessing the impact of nanoparticles on human health. *Journal of Trace Elements and Minerals*, *13*, 100252. DOI: 10.1016/j.jtemin.2025.100252

Kumar, S., & Bhanjana, G. (2025). Editorial: Nanomaterials for affordable biomedical devices, environmental and energy applications. *Frontiers in Nanotechnology*, *7*, 1693985. DOI: 10.3389/fnano.2025.1693985

Le, L. V., Chendke, G. S., Gamsey, S., Wisniewski, N., & Desai, T. A. (2020). Near-infrared optical nanosensors for continuous detection of glucose. *Journal of Diabetes Science and Technology*, *14*(2), 204–211. DOI: 10.1177/1932296819886928

Lee, S. J., Jeong, W., & Atala, A. (2024). 3D bioprinting for engineered tissue constructs and patient-specific models: Current progress and prospects in clinical applications. *Advanced Materials*, *36*(49), e2408032. DOI: 10.1002/adma.202408032

Lei, Y., Xiao, T., Hu, X., & Lin, H. (2026). Advanced drug delivery strategies for overcoming biological barriers: Tumor microenvironment and blood–brain barrier. *Drugs in R&D.*, *26*(1), 1–30. Advance online publication. DOI: 10.1007/s40268-026-00538-9

Leng, Y., & Sun, R. (2025). Multifunctional biomedical devices with closed-loop systems for precision therapy. *Advanced Healthcare Materials*, *14*(31), 2500860. DOI: 10.1002/adhm.202500860

Li, H., Zhang, F., Wang, D., Luo, S., Ding, Z., Bao, H., Zhang, S., Fan, C., Ji, W., & Wang, S. (2025). Specific cell adhesion at nano-biointerfaces: Synergistic effect of topographical matching and molecular recognition. *Nano Letters*, *25*(17), 7097–7106. DOI: 10.1021/acs.nanolett.5c01197

Li, J., & Wang, H. (2023). Selective organ targeting nanoparticles: From design to clinical translation. *Nanoscale Horizons*, *8*(9), 1155–1173. DOI: 10.1039/D3NH00145H

Li, L., Han, L., Hu, H., & Zhang, R. (2023). A review on polymers and their composites for flexible electronics. *Materials Advances*, *4*(3), 726–746. Advance online publication. DOI: 10.1039/D2MA00940D

Li, L., Scheiger, J. M., & Levkin, P. A. (2019). Design and applications of photoresponsive hydrogels. *Advanced Materials*, *31*(26), e1807333. DOI: 10.1002/adma.201807333

Li, S., Zuo, X., Carpenter, M. D., Verduzco, R., & Ajo-Franklin, C. M. (2025). Microbial bioelectronic sensors for environmental monitoring. *Nature Reviews Bioengineering*, *3*(1), 30–49. DOI: 10.1038/s44222-024-00233-x

Li, Y., Zhu, H., Liu, R., Zhao, Y., & Sun, L. (2023). Hierarchical microcarriers loaded with peptide dendrimer-grafted methotrexate for rheumatoid arthritis treatment. *Small Science*, *4*(1), 2300097. DOI: 10.1002/smsc.202300097

Liew, K. B., Janakiraman, A. K., Sundarapandian, R., Khalid, S. H., Razzaq, F. A., Ming, L. C., Khan, A., Kalusalingam, A., & Ng, P. W. (2022). A review and revisit of nanoparticles for antimicrobial drug delivery. *Journal of Medicine and Life*, *15*(3), 328–335. DOI: 10.25122/jml-2021-0097

Lin, C., Huang, X., Xue, Y., Jiang, S., Chen, C., Liu, Y., & Chen, K. (2025). Advances in medical devices using nanomaterials and nanotechnology: Innovation and regulatory science. *Bioactive Materials*, *48*, 353–369. DOI: 10.1016/j.bioactmat.2025.02.017

Lin, C., Yang, M., Zhang, F., Liu, Y., & Leng, J. (2026). Stimuli-responsive smart materials for biomedical applications. *Materials Science and Engineering R Reports*, *167*, 101126. DOI: 10.1016/j.mser.2025.101126

Lin, S. H., & Hsu, S. H. (2025). Smart hydrogels for in situ tissue drug delivery. *Journal of Biomedical Science*, *32*(1), 70. DOI: 10.1186/s12929-025-01166-2

Lin, W., Nie, S., Xiong, D., Guo, X., Wang, J., & Zhang, L. (2014). pH-responsive micelles based on $(PCL)_2(PDEA\text{-}b\text{-}PPEGMA)_2$ miktoarm polymer: Controlled synthesis, characterization, and application as anticancer drug carrier. *Nanoscale Research Letters*, *9*(1), 243. DOI: 10.1186/1556-276X-9-243

Liu, J. F., Jang, B., Issadore, D., & Tsourkas, A. (2019). Use of magnetic fields and nanoparticles to trigger drug release and improve tumor targeting. *Wiley Interdisciplinary Reviews. Nanomedicine and Nanobiotechnology*, *11*(6), e1571. DOI: 10.1002/wnan.1571

Long, J., Liang, X., Ao, Z., Tang, X., Li, C., Yan, K., Yu, X., Wan, Y., Li, Y., Li, C., & Zhou, M. (2024). Stimulus-responsive drug delivery nanoplatforms for inflammatory bowel disease therapy. *Acta Biomaterialia*, *188*, 27–47. DOI: 10.1016/j.actbio.2024.09.007

Lu, D., Mo, Y., Sun, S., Wang, Q., Wu, Z., Wang, W., & Zhu, M. (2025). Mechanically reinforced nanocomposite hydrogels and advanced applications in biosensing and bioelectronics. *Chemistry of Materials*, *37*(11), 3871–3902. Advance online publication. DOI: 10.1021/acs.chemmater.5c00498

Ma, T., Sun, J., Xue, N., Hassan, J. N. A., & Abbas, A. (2025). Nano-engineered sensor systems for disease diagnostics: Advances in smart healthcare applications. *Biosensors (Basel)*, *15*(12), 777. DOI: 10.3390/bios15120777

Ma, X., Tian, Y., Yang, R., Wang, H., Allahou, L. W., Chang, J., Williams, G., Knowles, J. C., & Poma, A. (2024). Nanotechnology in healthcare, and its safety and environmental risks. *Journal of Nanobiotechnology*, *22*(1), 715. DOI: 10.1186/s12951-024-02901-x

Mallick, T., Joshi, R. K., & Mishra, R. (2025). Nanoparticle-based biosensors for pathogen detection: Current updates and future prospects. *BioNanoScience*, *15*(3), 523. DOI: 10.1007/s12668-025-02143-7

Mangla, B., Kumar, P., Javed, S., Pathan, T., Ahsan, W., & Aggarwal, G. (2025). Regulating nanomedicines: Challenges, opportunities, and the path forward. *Nanomedicine (London)*, *20*(15), 1911–1927. DOI: 10.1080/17435889.2025.2533107

Mani, G., & Thepperumal, S. K. (2024). Recent trends on additive manufacturing biomaterial composites in tissue regeneration: Future perspectives, challenges, and road maps to clinics for biomedical applications—A review. 3D Printing and Additive Manufacturing. Advance online publication. https://doi.org/DOI: 10.1089/3dp.2024.0123

Masai, H., Nakagawa, T., & Terao, J. (2024). Recent progress in photoreactive crosslinkers in polymer network materials toward advanced photocontrollability. *Polymer Journal*, *56*(4), 297–307. DOI: 10.1038/s41428-023-00875-5

Mengesha, W. G., & Nagessar, K. (2026). A critical review on electronic materials properties and multifunctional applications. *Discover Materials*, *6*(1), 38. DOI: 10.1007/s43939-025-00517-y

Menghani, R. R., & Avila, R. (2025). Functional criteria for substrates in soft and stretchable bioelectronic systems. npj. *Soft Matter*, *1*(1), 11. DOI: 10.1038/s44431-025-00012-7

Miller, M. A., & Medina, S. (2024). Life at the interface: Engineering bio-nanomaterials through interfacial molecular self-assembly. *Wiley Interdisciplinary Reviews. Nanomedicine and Nanobiotechnology*, *16*(3), e1966. DOI: 10.1002/wnan.1966

Misra, R. D. K., & Boriek, A. M. (2023). Mechanistic understanding of the interaction of cells with nanostructured surfaces within the framework of biological functions. *Materials Technology*, *38*(1), 2216529. DOI: 10.1080/10667857.2023.2216529

Mitchell, M. J., Billingsley, M. M., Haley, R. M., Wechsler, M. E., Peppas, N. A., & Langer, R. (2021). Engineering precision nanoparticles for drug delivery. *Nature Reviews. Drug Discovery*, *20*(2), 101–124. DOI: 10.1038/s41573-020-0090-8

Mndlovu, H., Kumar, P., du Toit, L. C., & Choonara, Y. E. (2024). A review of biomaterial degradation assessment approaches employed in the biomedical field. npj Materials Degradation, 8, 66. https://doi.org/DOI: 10.1038/s41529-024-00487-1

Mura, S., Nicolas, J., & Couvreur, P. (2013). Stimuli-responsive nanocarriers for drug delivery. *Nature Materials*, *12*(11), 991–1003. DOI: 10.1038/nmat3776

Murphy, S. V., & Atala, A. (2014). 3D bioprinting of tissues and organs. *Nature Biotechnology*, *32*(8), 773–785. DOI: 10.1038/nbt.2958

Nadumane, S. S., Biswas, R., & Mazumder, N. (2025). Eco-friendly colorimetric detection of lead and mercury using L-cysteine-functionalized gold nanoparticles: A step towards greening the exposome. *RSC Advances*, *15*(15), 11573–11579. DOI: 10.1039/D5RA01445J

Naeem, M., Kim, W., Cao, J., Jung, Y., & Yoo, J.-W. (2014). Enzyme/pH dual sensitive polymeric nanoparticles for targeted drug delivery to the inflamed colon. *Colloids and Surfaces. B, Biointerfaces*, *123*, 271–278. DOI: 10.1016/j.colsurfb.2014.09.026

Nagpal, S., Jie, J. P. Y., Malinovskaya, J., Kovshova, T., Jain, P., Naik, S., Khopade, A., Bhowmick, S., Shahi, P., Chakra, A., Bhokari, A., Shah, V., Gelperina, S., & Wacker, M. G. (2024). A design-conversed strategy establishes the performance safe space for doxorubicin nanosimilars. *ACS Nano*, *18*(8), 6162–6175. DOI: 10.1021/acsnano.3c08290

Narayanan, K. B. (2025). Nanotopographical features of polymeric nanocomposite scaffolds for tissue engineering and regenerative medicine: A review. *Biomimetics (Basel, Switzerland)*, *10*(5), 317. DOI: 10.3390/biomimetics10050317

Narvaez, D., & Newell, B. (2025). A review of electroactive polymers in sensing and actuator applications. *Actuators*, *14*(6), 258. DOI: 10.3390/act14060258

Nawadkar, A., Chaskar, J., Tiwari, A., & Chaskar, A. (2024). Controlled release of nanofertilizers: A promising approach for efficient delivery of nutrients in plants. *Journal of Crop Improvement*, *38*(3), 189–218. DOI: 10.1080/15427528.2024.2336259

O'Brien, F. J. (2011). Biomaterials & scaffolds for tissue engineering. *Materials Today*, *14*(3), 88–95. DOI: 10.1016/S1369-7021(11)70058-X

Ortiz, A. C., Jackson, N., Buelvas, N., Jerez, A., López-Muñoz, R. A., Morales, J., & Arriagada, F. (2024). Release kinetics approach of stimuli-responsive mesoporous silica nanocarriers: pH-sensitive linker versus pH-sensitive framework. *Journal of Drug Delivery Science and Technology*, *91*, 105212. DOI: 10.1016/j.jddst.2023.105212

Öztürk, K., Kaplan, M., & Çalış, S. (2024). Effects of nanoparticle size, shape, and zeta potential on drug delivery. *International Journal of Pharmaceutics*, *666*, 124799. DOI: 10.1016/j.ijpharm.2024.124799

Parameswari, P., Belagalla, N., Singh, B. V., Abhishek, G. J., Rajesh, G. M., Katiyar, D., Hazarika, B., & Paul, S. (2024). Nanotechnology-based sensors for real-time monitoring and assessment of soil health and quality: A review. *Asian Journal of Soil Science and Plant Nutrition*, *10*(2), 157–173. DOI: 10.9734/ajsspn/2024/v10i2272

Parvin, N., Aslam, M., Alam, M. N., & Mandal, T. K. (2025a). Nanotechnology driven innovations in modern pharmaceutics: Therapeutics, imaging, and regeneration. *Nanomaterials (Basel, Switzerland)*, *15*(22), 1733. DOI: 10.3390/nano15221733

Parvin, N., Joo, S. W., & Mandal, T. K. (2025b). Biodegradable and stimuli-responsive nanomaterials for targeted drug delivery in autoimmune diseases. *Journal of Functional Biomaterials*, *16*(1), 24. DOI: 10.3390/jfb16010024

Pathak, G., Mangla, S., Gupta, G. K., Bhan, V., & Kapoor, R. K. (2025). Toxicological assessment and risk management of nanoparticles mediated composite materials—Critical review: State of the art. *Discover Polymer*, *2*(1), 12. DOI: 10.1007/s44347-025-00023-7

Pi, Y., Ganabady, K., & Celiz, A. D. (2025). Enzyme-responsive biomaterials for biomedical applications. *Communications Materials*, *6*(1), 263. DOI: 10.1038/s43246-025-00983-0

Place, E. S., Evans, N. D., & Stevens, M. M. (2009). Complexity in biomaterials for tissue engineering. *Nature Materials*, *8*(6), 457–470. DOI: 10.1038/nmat2441

Qi, Y.-L., Zhou, H.-Y., & Han, G.-Z. (2025). A novel photo-regulated self-healing hydrogel based on hollow SiO_2@g-C_3N_4@TiO_2 and PVA. *RSC Advances*, *15*(27), 21300–21310. DOI: 10.1039/D5RA01507C

Rabani, M. S., Shrivastav, M., Fayaz, M., Pathak, A., Khan, M., Bashir, S. U., Hamid, I., Hussain, H., Sharma, J. K., & Gupta, M. K. (2026). Mechanistic advances in microbial nanobiotechnology and their applications in sustainable agriculture, environment and biomedicine. *Discover Nano*, *21*(1), 81. DOI: 10.1186/s11671-026-04509-6

Rao Bommi, J., Kummari, S., Lakavath, K., Sukumaran, R. A., Panicker, L. R., Marty, J. L., & Yugender Goud, K. (2023). Recent trends in biosensing and diagnostic methods for novel cancer biomarkers. *Biosensors (Basel)*, *13*(3), 398. DOI: 10.3390/bios13030398

Rasmussen, K., Sayre, P., Kobe, A., Gonzalez, M., & Rauscher, H. (2025). 25 years of research and regulation: Is nanotechnology safe to commercialize? *Frontiers in Toxicology*, *7*, 1629813. DOI: 10.3389/ftox.2025.1629813

Reddy, K. S., Shivay, Y. S., Kumar, D., Pooniya, V., Prasanna, R., Shrivastava, M., Mandi, S., Nayak, S., & Baral, K. (2024). Relative performance of urea and nano-urea in conjunction with zinc fertilization on growth, productivity, and nitrogen use efficiency in spring wheat. *Journal of Soil Science and Plant Nutrition*, *24*(2), 3593–3609. DOI: 10.1007/s42729-024-01780-3

Ren, Z., & Cui, Y. (2026). Wearable biosensors for disease diagnostics and health monitoring: Recent progress and emerging technologies. *Lab on a Chip*, *26*(5), 1444–1470. DOI: 10.1039/D5LC00892A

Rezaei, B., Yari, P., Sanders, S. M., Wang, H., Chugh, V. K., Liang, S., Mostufa, S., Xu, K., Wang, J. P., Gómez-Pastora, J., & Wu, K. (2024). Magnetic nanoparticles: A review on synthesis, characterization, functionalization, and biomedical applications. *Small*, *20*(5), e2304848. DOI: 10.1002/smll.202304848

Rodríguez-Gómez, F. D., Monferrer, D., Penon, O., & Rivera-Gil, P. (2025). Regulatory pathways and guidelines for nanotechnology-enabled health products: A comparative review of EU and US frameworks. *Frontiers in Medicine*, *12*, 1544393. DOI: 10.3389/fmed.2025.1544393

Saeb, M. (2026). Biofunctional materials: Fundamentals and classification. Biofunctional Materials, 1, 0004. https://doi.org/DOI: 10.55092/bm20260004

Sahoo, J., Sharma, R., Pachauri, V., & Gandhi, S. (2024). Biomimetic/Bioderived Nanoengineered Interfaces for Biosensor Applications: A Review. *ACS Applied Nano Materials*, *7*(17), 19854–19875. DOI: 10.1021/acsanm.4c04554

Salehi, S., Naghib, S. M., Garshasbi, H. R., Ghorbanzadeh, S., & Zhang, W. (2023). Smart stimuli-responsive injectable gels and hydrogels for drug delivery and tissue engineering applications: A review. *Frontiers in Bioengineering and Biotechnology*, *11*, 1104126. DOI: 10.3389/fbioe.2023.1104126

Sanati, S., Bakhti, A., & Mohammadipanah, F. (2025). Long-term toxic effects of nanoparticles on human microbiota. *Journal of Trace Elements in Medicine and Biology*, *91*, 127723. DOI: 10.1016/j.jtemb.2025.127723

Saraiva, C., Praça, C., Ferreira, R., Santos, T., Ferreira, L., & Bernardino, L. (2016). Nanoparticle-mediated brain drug delivery: Overcoming blood–brain barrier to treat neurodegenerative diseases. *Journal of Controlled Release*, *235*, 34–47. DOI: 10.1016/j.jconrel.2016.05.044

Sarikhani, E., Mahato, K., Casanova, A., Rahmani, K., Wang, J., & Jahed, Z. (2025). Nanosensors for real-time intracellular analytics. *Nature Nanotechnology*, *20*(10), 1374–1387. DOI: 10.1038/s41565-025-02032-w

Shahalaei, M., Azad, A. K., Wan Sulaiman, W. M. A., Derakhshani, A., Banaee Mofakham, E., Mallandrich, M., Kumarasamy, V., & Subramaniyan, V. (2024). A review of metallic nanoparticles: Present issues and prospects focused on the preparation methods, characterization techniques, and their theranostic applications. *Frontiers in Chemistry*, *12*, 1398979. DOI: 10.3389/fchem.2024.1398979

Sharma, S., & Kishen, A. (2024). Bioarchitectural design of bioactive biopolymers: Structure–function paradigm for diabetic wound healing. *Biomimetics (Basel, Switzerland)*, *9*(5), 275. DOI: 10.3390/biomimetics9050275

Sharma, S. K., Gajević, S., Sharma, L. K., Sharma, Y., Sharma, M., Ivanović, L., Milojević, S., & Stojanović, B. (2026). Self-healing polymer nanocomposites: Mechanisms, structure–property relationships, and emerging applications. *Polymers*, *18*(2), 276. DOI: 10.3390/polym18020276

Singh, H., Balusamy, S. R., Sukweenadhi, J., Saravanan, M., Aruchamy, M., Mijakovic, I., & Singh, P. (2025). Smart hybrid nanomaterials for chronic infections: Microbiome-responsive and sustainable therapeutic platforms. *Journal of Nanobiotechnology*, *23*(1), 698. DOI: 10.1186/s12951-025-03712-4

Singh, S. K., Palaniappan, S. K., Arora, G., Rangappa, S. M., & Siengchin, S. (2026). Recent advances in applications of nanocomposites: A brief overview. *Discover Materials*, *6*(1), 58. DOI: 10.1007/s43939-025-00524-z

Sodagar, A. M., Khazaei, Y., Nekoui, M., & Shaeri, M. (2025). Real-time neural signal processing for high-density brain-implantable devices. *Bioelectronic Medicine*, *11*(1), 17. DOI: 10.1186/s42234-025-00177-6

Soliman, M. G., Martinez-Serra, A., Antonello, G., Dobricic, M., Wilkins, T., Serchi, T., Fenoglio, I., & Monopoli, M. P. (2024). Understanding the role of biomolecular coronas in human exposure to nanomaterials. *Environmental Science. Nano*, *11*(11), 4421–4448. DOI: 10.1039/D4EN00488D

Song, B., Wu, M., Qin, L., Liang, W., & Wang, X. (2025). Smart design of targeted drug delivery system for precise drug delivery and visual treatment of brain gliomas. *Advanced Healthcare Materials*, *14*(4), e2402967. DOI: 10.1002/adhm.202402967

Srinivasan, L. V., & Rana, S. S. (2024). A critical review of various synthesis methods of nanoparticles and their applications in biomedical, regenerative medicine, food packaging, and environment. *Discover Applied Sciences*, *6*(7), 371. DOI: 10.1007/s42452-024-06040-8

Suda, A., Kumatani, N., Sato, K., Inagaki, M., Yamamura, K., Hatanaka, M., & Morikawa, A. (2018). Continuous fabrication of monodisperse ceria–zirconia–yttria composite oxide nanoparticles using a novel high-shear agitation reactor. *ACS Omega*, *3*(6), 6560–6565. DOI: 10.1021/acsomega.8b00730

Suryaprabha, T., Choi, C., Wu, Y., Liu, L., & Hwang, B. (2026). Smart wearable and implantable biosensors for continuous health monitoring: Materials, biocompatibility, and AI integration. npj Flexible Electronics, 10, 46. https://doi.org/DOI: 10.1038/s41528-026-00560-6

Tao, R., You, C., Qu, Q., Zhang, X., Deng, Y., Ma, W., & Huang, C. (2023). Recent advances in the design of controlled- and sustained-release micro/nanocarriers of pesticide. *Environmental Science. Nano*, *10*(2), 351–371. DOI: 10.1039/D2EN00446A

Thakur, A., Kim, J., Wyatt, B. C., Gogotsi, Y., & Anasori, B. (2026). Composition–structure–property relationships in MXenes. *Nature Reviews. Materials*. Advance online publication. DOI: 10.1038/s41578-026-00911-9

Thomas, J., Kumar, V., Sharma, N., John, N., Umesh, M., Huligowda, L. K. D., Kaur, K., & Utreja, D. (2025). Recent approaches in nanotoxicity assessment for drug delivery applications: Challenges and prospects. *Medicine in Drug Discovery*, *25*, 100204. DOI: 10.1016/j.medidd.2025.100204

Tian, X., Wen, Y., Zhang, Z., Zhu, J., Song, X., Phan, T. T., & Li, J. (2025). Recent advances in smart hydrogels derived from polysaccharides and their applications for wound dressing and healing. *Biomaterials*, *318*, 123134. DOI: 10.1016/j.biomaterials.2025.123134

Topol, E. J. (2019). High-performance medicine: The convergence of human and artificial intelligence. *Nature Medicine*, *25*(1), 44–56. DOI: 10.1038/s41591-018-0300-7

Torchilin, V. P. (2014). Multifunctional, stimuli-sensitive nanoparticulate systems for drug delivery. *Nature Reviews. Drug Discovery*, *13*(11), 813–827. DOI: 10.1038/nrd4333

Torres, D. (2025). Scaling low-cost biomedical devices: Challenges and successes in global health deployment. *Journal of Bioengineering & Biomedical Science, 15*, 480.

van Doorn, W. P. T. M., Foreman, Y. D., Schaper, N. C., Savelberg, H. H. C. M., Koster, A., van der Kallen, C. J. H., Wesselius, A., Schram, M. T., Henry, R. M. A., Dagnelie, P. C., de Galan, B. E., Bekers, O., Stehouwer, C. D. A., Meex, S. J. R., & Brouwers, M. C. G. J. (2021). Machine learning-based glucose prediction with use of continuous glucose and physical activity monitoring data: The Maastricht Study. *PLoS One, 16*(6), e0253125. DOI: 10.1371/journal.pone.0253125

Vashist, S. K. (2013). Continuous glucose monitoring systems: A review. *Diagnostics (Basel), 3*(4), 385–412. DOI: 10.3390/diagnostics3040385

Vayena, E., Blasimme, A., & Cohen, I. G. (2018). Machine learning in medicine: Addressing ethical challenges. *PLoS Medicine, 15*(11), e1002689. DOI: 10.1371/journal.pmed.1002689

Vecchi, J. T., Claussen, A. D., & Hansen, M. R. (2024). Decreasing the physical gap in the neural-electrode interface and related concepts to improve cochlear implant performance. *Frontiers in Neuroscience, 18*, 1425226. DOI: 10.3389/fnins.2024.1425226

Wan, Z., Fu, Y., Wang, S., Wei, R., & Wang, Y. (2026). Sensor-actuator integration in intelligent devices: From functional synergy to emerging application. *Composites. Part B, Engineering, 312*, 113339. DOI: 10.1016/j.compositesb.2025.113339

Wang, B., Sanli, A., Lai, Z., Fu, J., Adeel, M., Zhu, X., Yu, M., Qu, L., Li, A., Yang, Z., Song, F., Huang, J., Zeng, S., Mu, C., Güder, F., Wan, Y., & Ledesma-Amaro, R. (2026). Next-generation biosensing for in situ monitoring. *Nature Sensors, 1*(2), 111–130. DOI: 10.1038/s44460-025-00016-w

Wang, D., Yao, H., Ye, J., Wang, Z., Cong, H., & Yu, B. (2025). pH-responsive polymers: Classification, response mode, properties and biomedical applications. *Chemical Engineering Journal, 523*, 168589. DOI: 10.1016/j.cej.2025.168589

Wang, J., Li, B., Qiu, L., Xiao, X., & Yang, H. (2022). Dendrimer-based drug delivery systems: History, challenges, and latest developments. *Journal of Biological Engineering, 16*(1), 18. DOI: 10.1186/s13036-022-00298-5

Wang, P., Wang, M., Wei, X., Xie, L., Tian, L., Yang, Z., Zhou, Z., & Chen, H. (2024). In vitro and in vivo degradation profile, biocompatibility of poly-L-lactic acid porous microspheres. *International Journal of Biological Macromolecules, 272*(Pt 2), 132876. DOI: 10.1016/j.ijbiomac.2024.132876

Wang, T., Wu, C., Hu, Y., Zhang, Y., & Ma, J. (2023). Stimuli-responsive nanocarrier delivery systems for Pt-based antitumor complexes: A review. *RSC Advances, 13*(24), 16488–16511. DOI: 10.1039/D3RA00866E

Wang, X., Wei, W., Guo, Z., Liu, X., Liu, J., Bing, T., Yu, Y., Yang, X., & Cai, Q. (2024). Organic–inorganic composite hydrogels: Compositions, properties, and applications in regenerative medicine. *Biomaterials Science, 12*(5), 1079–1114. DOI: 10.1039/D3BM01766D

Wang, Y., Wang, S., Feng, X. N., Yan, L. C., Zheng, S. S., Wang, Y., & Zhao, Y. H. (2018). The impact of exposure route for class-based compounds: A comparative approach of lethal toxicity data in rodent models. *Drug and Chemical Toxicology, 41*(1), 95–104. DOI: 10.1080/01480545.2017.1320405

Wang, Y., Yang, X., Zhang, X., Wang, Y., & Pei, W. (2023). Implantable intracortical microelectrodes: Reviewing the present with a focus on the future. *Microsystems & Nanoengineering, 9*(1), 7. DOI: 10.1038/s41378-022-00451-6

Wang, Y., Zhang, X., & Yue, H. (2024). Two-dimensional nanomaterials induced nano-bio interfacial effects and biomedical applications in cancer treatment. *Journal of Nanobiotechnology, 22*(1), 67. DOI: 10.1186/s12951-024-02319-5

Wang, Z., Lin, J., Zhu, Y., Sun, R., Lin, Q., Hu, X., Sun, J., Jiang, Y., You, K., Fu, H., Guo, X., Wang, Y., Li, D., Lin, L., Li, F., Liu, Y., Zhao, Y., & Huang, W. (2026). Bioinspired flexible sensing-processing-visualizing integrated system towards tactile-visual signal recognition. *Nature Communications, 17*(1), 603. DOI: 10.1038/s41467-025-67316-0

Weinstein, J. S., Varallyay, C. G., Dosa, E., Gahramanov, S., Hamilton, B., Rooney, W. D., Muldoon, L. L., & Neuwelt, E. A. (2009). Superparamagnetic iron oxide nanoparticles: Diagnostic magnetic resonance imaging and potential therapeutic applications in neurooncology and central nervous system inflammatory pathologies, a review. *Journal of Cerebral Blood Flow and Metabolism, 30*(1), 15–35. DOI: 10.1038/jcbfm.2009.192

Whitesides, G. M. (2005). Nanoscience, nanotechnology, and chemistry. *Small, 1*(2), 172–179. DOI: 10.1002/smll.200400130

Wilhelm, S., Tavares, A. J., Dai, Q., Ohta, S., Audet, J., Dvorak, H. F., & Chan, W. C. W. (2016). Analysis of nanoparticle delivery to tumours. *Nature Reviews. Materials, 1*(5), 16014. DOI: 10.1038/natrevmats.2016.14

Wilkinson, L. J., White, R. J., & Chipman, J. K. (2011). Silver and nanoparticles of silver in wound dressings: A review of efficacy and safety. *Journal of Wound Care*, *20*(11), 543–549. DOI: 10.12968/jowc.2011.20.11.543

Wu, J., Liu, H., Chen, W., Ma, B., & Ju, H. (2023). Device integration of electrochemical biosensors. *Nature Reviews Bioengineering*, *1*(5), 346–360. DOI: 10.1038/s44222-023-00032-w

Wu, X., Ye, Y., Sun, M., Mei, Y., Ji, B., Wang, M., & Song, E. (2025). Recent progress of soft and bioactive materials in flexible bioelectronics. Cyborg and Bionic Systems, 6, 0192. https://doi.org/DOI: 10.34133/cbsystems.0192

Xian, X. (2026). Recent Advances in Wearable Biosensors for Human Health Monitoring. *Biosensors (Basel)*, *16*(1), 27. DOI: 10.3390/bios16010027

Yadav, J., Chahal, S., Kumar, P., & Kumar, C. (2026). Thermo-responsive smart hydrogels: Molecular engineering, dynamic cross-linking strategies, and therapeutics applications. *Gels (Basel, Switzerland)*, *12*(1), 12. DOI: 10.3390/gels12010012

Yadav, N., Zhang, Q., Qi, D., Yadav, A., & Zheng, H. (2025). AI-integrated wearable strain sensors: Advances in e-skin, robotics, and personalized health monitoring. *Nanoscale Advances*, *7*(16), 4803–4819. DOI: 10.1039/D5NA00574D

Yan, S., Na, J., Liu, X., & Wu, P. (2024). Different targeting ligands-mediated drug delivery systems for tumor therapy. *Pharmaceutics*, *16*(2), 248. DOI: 10.3390/pharmaceutics16020248

Yao, M., Hsieh, J.-C., & Wang, H. (2026). Hydrogel-integrated multimodal physiological and modulation systems. *Materials Horizons*, *13*(10), 4819–4838. Advance online publication. DOI: 10.1039/D5MH01852H

Yu, X., Yuan, Y., Zhou, J., Chang, M., Zeng, S., & Zhuang, S. (2026). Sensitivity enhancement of surface plasmon resonance biosensors based on versatile nanostructures: Principle, fabrication, and illustrative applications. *Microsystems & Nanoengineering*, *12*(1), 45. DOI: 10.1038/s41378-025-01118-8

Yu, Z., Liu, C., Wang, C., & Zhao, G. (2024). The past and future depicted by neural electrode technology. *The Innovation Medicine*, *2*(3), 100087. DOI: 10.59717/j.xinn-med.2024.100087

Zeng, F., Pang, C., & Tang, H. (2024). Sensors on Internet of Things Systems for the Sustainable Development of Smart Cities: A Systematic Literature Review. *Sensors (Basel)*, *24*(7), 2074. DOI: 10.3390/s24072074

Zeng, S., Shakoor, N., & Rui, Y. (2025). Nanotechnology and agricultural sustainability: A review. *Nanomaterials (Basel, Switzerland)*, *15*(23), 1755. DOI: 10.3390/nano15231755

Zhang, F., Xu, L., Wang, Y., & Wang, P. (2025). Engineering plasmonic Au nanostars: Enhanced plasmonic properties, preparation and biomedical application. *Materials Today. Bio*, *33*, 102022. DOI: 10.1016/j.mtbio.2025.102022

Zhang, J., Tang, W., Zhang, X., Song, Z., & Tong, T. (2023). An overview of stimuli-responsive intelligent antibacterial nanomaterials. *Pharmaceutics*, *15*(8), 2113. DOI: 10.3390/pharmaceutics15082113

Zhang, K., Gao, Q., Jiang, J., Chan, M., Zhai, X., Jin, L., Zhang, J., Li, J., & Liao, W.-H. (2024). High energy dissipation and self-healing auxetic foam by integrating shear thickening gel. *Composites Science and Technology*, *249*, 110475. DOI: 10.1016/j.compscitech.2024.110475

Zhang, L., Liu, G., Xia, Q., & Deng, L. (2024). Research progress on blood compatibility of hemoperfusion adsorbent materials. *Frontiers in Bioengineering and Biotechnology*, *12*, 1456694. DOI: 10.3389/fbioe.2024.1456694

Zhang, P., Zhu, B., Du, P., & Travas-Sejdic, J. (2023). Electrochemical and electrical biosensors for wearable and implantable electronics based on conducting polymers and carbon-based materials. *Chemical Reviews*, *124*(3), 722–767. DOI: 10.1021/acs.chemrev.3c00392

Zhang, W., Zeng, X., Deng, X., Yang, F., Ma, X., & Gao, W. (2025). Smart biomaterials: As active immune modulators to shape pro-regenerative microenvironments. *Frontiers in Cell and Developmental Biology*, *13*, 1669399. DOI: 10.3389/fcell.2025.1669399

Zhang, Y., Gao, S., Li, H., Yang, T., Zheng, K., Guo, Z. M., Shi, J., Huang, X., Zou, X., Picchetti, P., & Biedermann, F. (2025). Design principles of nanosensors for multiplex detection of contaminants in food. *Small*, *21*(26), e2412271. DOI: 10.1002/smll.202412271

Zhang, Y., & Wu, B. M. (2023). Current advances in stimuli-responsive hydrogels as smart drug delivery carriers. *Gels (Basel, Switzerland)*, *9*(10), 838. DOI: 10.3390/gels9100838

Zheng, Z., Zhu, R., Peng, I., Xu, Z., & Jiang, Y. (2024). Wearable and implantable biosensors: Mechanisms and applications in closed-loop therapeutic systems. *Journal of Materials Chemistry. B*, *12*(35), 8577–8604. DOI: 10.1039/D4TB00782D

Zhou, H., Zheng, H., Yao, W., Sun, H., Yang, Y.-G., Li, Z., Song, D., Zhang, Y., & Sun, T. (2026). Spatiotemporally controlled delivery of biomacromolecules via injectable hydrogels for precision modulation of the tumor immune microenvironment. *Journal of Nanobiotechnology*, *24*(1), 82. DOI: 10.1186/s12951-025-03970-2

Zhou, Y., Xu, T., Han, P., Wang, C., Bartold, M., & Ivanovski, S. (2026). Biological responses to biomaterials: Mechanisms and opportunities for enhanced oral tissue regeneration. *Journal of Periodontal Research*, jre.70078. Advance online publication. DOI: 10.1111/jre.70078

Zhuo, S., Zhang, F., Yu, J., Zhang, X., Yang, G., & Liu, X. (2020). pH-sensitive biomaterials for drug delivery. *Molecules (Basel, Switzerland)*, *25*(23), 5649. DOI: 10.3390/molecules25235649

Zhuwawu, S. S., & El-Hussieny, H. (2025). Leveraging LSTM, tactile sensors, and haptic feedback to augment prosthetic control via grasp type prediction and grasp type feedback. *Scientific Reports*, *15*(1), 29781. DOI: 10.1038/s41598-025-92651-z

Chapter 4
Smart Biodegradable Biomaterials for Precision Therapeutics and Regenerative Systems

Kasturi Laturkar
http://orcid.org/0000-0002-4985-909X
Validation Associates LLC, USA

Kaustubh Laturkar
http://orcid.org/0000-0002-2643-3942
Procon Global LLC, USA

ABSTRACT

Smart biodegradable biomaterials with tailored biological interactions, adaptive functionality, and controlled disintegration are revolutionizing precision and regenerative medicine. This chapter covers transport phenomena, degradation kinetics, structure-property-function connections, and biointerface dynamics, which control these systems. Stimuli-responsive biomaterials that react to pH, enzymatic activity, temperature, and mechanical inputs are also discussed. Multifunctional biomaterial systems are possible by advances in fabrication techniques including electrospinning, additive manufacturing, and hybrid scaffold design. The chapter also looks at modern clinical and industrial uses of these biomaterials. Critical evaluations in this chapter include biological variability, industrial scalability, degradation-product safety, predictive design, and regulatory translation. Short-, medium-, and long-term development approaches emphasize sustainability, autonomous treatment systems, clinical integration, and computational design for future perspectives.

DOI: 10.4018/979-8-2600-0060-1.ch004

INTRODUCTION

In modern therapeutic methodologies, biodegradable biomaterials are of paramount importance, particularly in applications necessitating temporary structural or functional support. These include drug delivery systems, scaffolds designed for tissue regeneration, and implanted medical devices that are designed to degrade after fulfilling their intended purpose (Khan et al., 2022; Nii & Katayama, 2021). Their ability to perform a specific function and subsequently undergo safe resorption eliminates the necessity for further surgical interventions, thereby reducing healthcare costs, enhancing overall treatment outcomes, and mitigating patient morbidity (Dreiss, 2020; Teixeira do Nascimento et al., 2025). Additionally, their transient qualities enables development of more flexible therapeutic approaches, allowing device lifetime to be tailored to specific clinical needs (Cedillo-Servin et al., 2025).

Over the past few decades, conventional biodegradable polymers, such as poly(-lactic acid) (PLA), poly(glycolic acid) (PGA), and poly(lactic-co-glycolic acid) (PLGA), have been extensively utilized, owing to their established biocompatibility and predictable hydrolytic degradation characteristics (Alaswad et al., 2022; Giosafatto et al., 2023). Ester bond hydrolysis serves as the principal degradation pathway for these compounds, generating byproducts that can be further metabolized through natural biochemical processes (Sharma et al., 2021). Their extensive application in controlled drug delivery systems, orthopedic fixation devices, and sutures highlights their dependability and regulatory approval. Despite these advantages, their functional range remains largely predetermined during fabrication. Key material properties, such as degradation rate, mechanical strength, and drug release kinetics, exhibit minimal responsiveness to physiological variations (Badeau & Deforest, 2019; Nii & Katayama, 2021).

This limitation becomes significant within the highly dynamic nature of biological systems. Physiological systems vary spatially and temporally, with local conditions changing among tissues and altering as a disease progresses or heals. Regenerating tissues undergo continuous changes in cellular signalling and extracellular matrix composition, while inflamed tissues frequently exhibit increased oxidative stress, elevated enzyme activity, and a lower pH. Tumor microenvironments are similarly characterized by hypoxia, acidity, and abnormal vascularization (Aflori & Poni, 2021; Kumar et al., 2022). These disparities influence both biological response and therapeutic effectiveness.

Consequently, static material systems may fail to maintain optimal performance throughout treatment. Drug delivery systems with a fixed release profile may underdeliver or overdeliver medication during different stages of treatment. Likewise, scaffolds with predetermined degradation rates might degrade too slowly, thereby hindering tissue regeneration and integration, or too rapidly, thus compromising structural integrity (Poomathi et al., 2020; X. Wang et al., 2019). These limitations highlight the demand for materials capable of dynamically adjusting their properties in response to environmental stimuli.

Smart biodegradable biomaterials address these challenges by incorporating stimulus-responsive components into their design. These materials are engineered to detect specific stimuli, including fluctuations in temperature, pH, enzyme concentration, redox states, or mechanical stress, and respond through controlled modifications of their chemical or physical properties. Such responses may include changes in degradation rates, drug release patterns, mechanical stiffness, and surface properties (Alaswad et al., 2022; Chen et al., 2019). This adaptability enhances therapeutic precision and effectiveness.

The development of these systems requires a thorough understanding of relationships between material composition, structure, transport processes and biological function. The chemical composition determines the types of bonds and functional groups available for interaction and breakdown. Microstructural features, such as porosity, pore size distribution, and network architecture, influence transport properties and mechanical behavior (Kumar et al., 2022; Poomathi et al., 2020). Biological interactions regulate cellular responses, immune activity, and tissue integration, whereas transport mechanisms control the movement of nutrients, waste products, and therapeutic agents. Together, these factors operate through interconnected feedback loops (Liu et al., 2026).

The behavior of smart biodegradable biomaterials is further governed by coupled multiphysics processes. Material degradation modifies the internal structure, porosity and mechanical stability, which subsequently affect molecular transport (Deirram et al., 2019). Simultaneously, biological activities such as immune responses and enzyme release can modify degradation kinetics, creating a dynamic interaction between the material and its environment (Nii & Katayama, 2021).

The design of advanced biomaterials therefore requires a systems-level perspective. Increasingly, predictive modelling is being used to integrate degradation kinetics, transport phenomena and biological interactions, enabling identification of critical design parameters and optimization of performance in physiologically relevant conditions (Dreiss, 2020; Giosafatto et al., 2023). At the same time, advancements in manufacturing technologies allow for precise control of material structure, which helps turn theoretical ideas into practical biomedical applications.

The development of new biodegradable biomaterials is being driven by the combination of modern manufacturing methods, the principles of responsive material design, and predictive modelling (Laturkar & Laturkar, 2024b). These systems possess the potential to enhance clinical outcomes, generate tailored therapeutic responses, and adapt to complex biological environments. The integration of functionality, precision, and adaptability is expected to redefine the role of biomaterials in medicine, evolving them from passive support structures into active contributors within therapeutic processes.

DEGRADATION MECHANISMS AND KINETICS

The degradation of smart, biodegradable biomaterials is dictated by interactions between molecular structure, environmental conditions, transport processes, and mechanical stresses. A thorough understanding of these factors is essential for creating materials with predictable performance and tunable functionality for advanced biomedical applications (Kuperkar et al., 2024; Park et al., 2022).

1. Molecular and Physicochemical Basis

Hydrolytic degradation is the primary degradation process for synthetic biodegradable polymers containing ester, anhydride, or carbonate linkages (Giosafatto et al., 2023; Sharma et al., 2021). Water molecules diffuse into the polymer matrix and break down susceptible bonds through nucleophilic attack. The degradation rate depends on water absorption, polymer hydrophilicity, and the availability of reactive sites (Sánchez-Cid et al., 2022). As a result, polymers with increased hydrophilicity undergo accelerated degradation due to enhanced water absorption (Kuperkar et al., 2024; Lou & Mooney, 2022).

Chain mobility also influences degradation behaviour. Flexible polymer chains increase bond cleavage and water diffusion, whereas rigid polymer structures may hinder decomposition and impede water transport. The incorporation of plasticizers or copolymerization with flexible segments can substantially alter degradation patterns by modifying chain dynamics (Gul et al., 2022; Nathanael et al., 2023).

Microstructural attributes, specifically crystallinity, exert a considerable influence on degradation kinetics (Gaharwar et al., 2020). Crystalline regions resist hydrolysis because of their dense packing and reduced free volume. Amorphous regions are more susceptible to degradation due to their elevated permeability and enhanced chain mobility (Atala, 2020). Therefore, degradation frequently initiates within amorphous domains, subsequently progressing toward crystalline regions.

The orientation of polymer chains and crosslink density also affects degradation. Highly oriented polymer chains may reduce access to reactive sites and alter diffusion pathways, while dense crosslinking restricts swelling and water penetration, slowing degradation. In contrast, loosely crosslinked networks absorb more water and degrade more rapidly (Dreiss, 2020; Lou & Mooney, 2022).

Additives, copolymers or fillers can either speed up or slow down degradation depending on their chemical properties (Maurya et al., 2024). Hydrophilic components can increase water absorption, and inorganic fillers may slow down diffusion. These compositional differences allow for more precise control over degradation behavior. As degradation progresses, molecular weight decreases, resulting in reduced mechanical strength and increased brittleness. Once a critical molecular weight is reached, structural integrity is lost and mass loss begins (Giosafatto et al., 2023; Sharma et al., 2021).

2. Enzymatic and Environment-Triggered Degradation

Hydrolytic degradation establishes a fundamental mechanism; however, biological environments introduce additional pathways that facilitate more adaptive and selective degradation processes. Enzymes such as proteases, lipases, and esterases catalyze bond cleavage with high specificity, allowing materials to degrade preferentially in tissues where these enzymes are expressed (Almutary et al., 2025; Teixeira do Nascimento et al., 2025).

In regenerative environments, enzyme levels often increase during specific phases of healing, such as tissue remodeling or inflammation (Dziki & Badylak, 2019). Materials containing enzyme-sensitive linkages can synchronize their degradation with biological healing processes, facilitating matrix deposition and cellular infiltration (Badeau & Deforest, 2019; Teixeira do Nascimento et al., 2025).

Environmental factors like pH and redox potential also play a crucial role in governing degradation. pH-responsive materials use alterations in acidity to enhance hydrolysis or activate acid-sensitive bond, enabling targeted degradation in tumors and inflamed tissues while remaining stable in healthy environments (Majumder & Minko, 2021; Teixeira do Nascimento et al., 2025).

Similarly, redox-responsive systems use chemical linkages that respond to oxidative or reductive environments. Reactive oxygen species (ROS) levels in inflammatory and disease states can accelerate degradation whereas intracellular reducing conditions may trigger therapeutic release through cleavage of redox-sensitive bonds like disulfides (Badeau & Deforest, 2019; Teixeira do Nascimento et al., 2025).

Temperature-responsive polymers offer an additional regulatory mechanism through phase transitions that affect permeability and degradation behavior. (Badeau & Deforest, 2019). Although physiological temperature variations are generally limited, externally applied stimuli can induce localized degradation.

The incorporation of several stimuli-responsive systems into a single material facilitates intricate and highly tailored degradation. For example, a material could be engineered to undergo degradation solely when exposed to both acidic pH and enzymatic activity, thereby offering a high degree of specificity.

3. Reaction-Diffusion Modeling

Chemical reactions and mass transport are fundamentally interconnected in the degradation of biodegradable biomaterials (Guo et al., 2022; Zhang et al., 2019). These processes determine how water, degradation products and therapeutic agents move through the material while reacting with the polymer matrix.

Diffusion controls the entry of water and the elimination of soluble breakdown products in biodegradable polymers. Hydrolytic cleavage of vulnerable bonds takes place as water permeates the substance, gradually altering the molecular structure. Degradation rates vary throughout the material in both space and time due to this linked transport response characteristic.

A key complexity in these systems is that the transport properties are not constant. As degradation progresses, changes in porosity, swelling and polymer chain breakage alter the internal structure of the material. These structural changes modify the rate at which molecules can diffuse through the matrix, creating a feedback mechanism between degradation and transport (Gaharwar et al., 2020).

Hydrolysis of ester bonds is the first step in the degradation of several biodegradable polymers, namely PLA and PLGA. Acidic byproducts are created during this process and may build up inside the polymer matrix. Internal gradients in material structure and spatially heterogeneous deterioration can result from the local buildup of these products, which can speed up hydrolysis (Alaswad et al., 2022).

Because both transport and reaction processes evolve over time, degradation behavior cannot be described as uniform. Instead, it is governed by spatially dependent interactions between material structure, chemical reactivity and environmental conditions. This results in non-uniform erosion patterns, which are commonly observed in bulk-degrading polymer systems.

These linked phenomena are often studied using numerical and computational techniques. Diffusion, degradation progression, and mass loss over time can be simulated by researchers using finite element and associated discretization techniques. In the biomedical field, these methods are frequently employed to forecast drug release behavior, scaffold degradation patterns, and structural evolution (Guo et al., 2022).

In practical systems, boundary conditions at material-environment interface also play an important role. Interactions with surrounding biological fluids, proteins or cellular layers can influence transport resistance and modify degradation rates (Galli et al., 2026). For implantable biomaterials, where surface contacts change over time and influence material behavior, these interfacial effects are especially important.

4. Mechanical Coupling and Degradation

The degrading behavior of biodegradable biomaterials is significantly influenced by mechanical forces, especially in dynamic physiological conditions. Mechanical stress can speed up the breaking of chemical bonds by changing the shape of polymer chains, increasing segmental mobility, and making reactive sites more accessible (Karami et al., 2019; L. Kong et al., 2022). This combination of chemical and mechanical processes adds another important factor to consideration when designing implantable and load-bearing biomaterials.

In physiological systems such as bone or cardiovascular systems, biomaterials are exposed to cyclic and variable stresses. These conditions can induce fatigue-related microstructural damage, which contribute to the formation of microcracks and localized defects within the material. Such structural changes can facilitate water penetration and increase internal stress concentration, thereby accelerating degradation (Atala, 2020).

The degradation behavior under mechanical loading is not uniform and depends strongly on intrinsic material properties. Polymer viscoelasticity plays a key role in determining how mechanical energy is absorbed and dissipated. Viscoelastic materials can relax applied stress over time, which may reduce the extent of mechanically induced degradation. In contrast, more brittle polymer systems are prone to rapid crack formation and accelerated structural failure under similar loading conditions (Giosafatto et al., 2023).

Mechanical deformation also affects the internal microstructure of the material. Changes in pore size, pore connectivity and network architecture can occur under compression or repeated loading. These structural changes influence the transport of water and degradation byproducts, thereby modifying local degradation rates. As a result, mechanical loading indirectly affects chemical and transport processes through its impact of material structure (Gaharwar et al., 2020).

In tissue engineering applications, this coupling between mechanical behavior and degradation can be beneficial. Scaffolds can be designed to degrade preferentially in regions of high mechanical stress, enabling gradual load transfer to regenerating tissue. This dynamic behavior supports tissue maturation and enhances structural integration between the scaffold and host tissue (Nii & Katayama, 2021).

Overall mechanical, chemical and structural factors are strongly dependent in biodegradable biomaterials. A comprehensive understanding of these interactions is essential for designing materials that maintain functionality under physiological loading conditions while degrading in a controlled and predictive manner.

TRANSPORT PHENOMENA AND CONTROLLED RELEASE

Diffusion, degradation, and structural evolution are interconnected processes that govern transport and controlled release in biodegradable biomaterials (Mircioiu et al., 2019). These interactions determine how therapeutic agents, nutrients, and degradation products move through the material while simultaneously being influenced by ongoing structural changes. (Zhao et al., 2020). As a result, transport behavior must be understood as a dynamic, time-dependent property rather than a fixed material characteristic.

1. Diffusion and Permeability

Mass transport in biodegradable biomaterials is primarily governed by diffusion, which is driven by concentration gradients and modulated by the material's internal structure (Benson et al., 2022). Transport occurs through interconnected pores in scaffolds or through the free volume of polymer networks and is strongly influenced by the size of diffusing species relative to the characteristic dimensions of the matrix (Buenzli et al., 2020). Small molecules diffuse easily while bigger macromolecules have limited mobility as a result of this selective transport characteristic (Elkhoury et al., 2020).

The effectiveness of diffusion is determined by structural parameters such as porosity and tortuosity. Tortuosity characterizes the intricacy of diffusion channels, whereas porosity determines the available void space for transport (Graczyk & Matyka, 2020). If pore connection is extremely complex, even highly porous materials may show limited transport (Allen, 2021). The combined effect of these parameters is commonly expressed using an effective diffusivity concept, where structural features reduce transport relative to free solution conditions (Marlena et al., 2021).

In hydrated polymer systems, swelling further modifies transport behavior. Increased network mesh size due to water uptake improves solute mobility and lowers steric hindrance (Lou & Mooney, 2022). This effect depends on crosslink density, polymer-solvent interactions, and environmental conditions such as pH and ionic strength (Sánchez-Cid et al., 2022). While swelling enhances permeability, it may simultaneously reduce mechanical integrity, requiring a balance between transport efficiency and structural stability.

Transport may also include convective contributions in systems exposed to fluid flow, particularly in porous or vascularized environments. Beyond diffusion alone, interstitial flow can boost mass transfer and solute mobility (Dreiss, 2020). Furthermore, non-uniform transport areas are introduced by spatial heterogeneity in polymer composition, crystallinity, or crosslinking, resulting in spatially varied diffusion behavior (X. Li et al., 2019).

Temperature acts as an external regulator of transport by influencing molecular mobility and polymer chain dynamics. Increased temperature generally enhances diffusivity by reducing matrix viscosity and increasing segmental motion (Mohamed Nasser et al., 2024). However, structural and compositional considerations usually take precedence over these effects in physiological systems.

2. Coupled Transport and Degradation

Degradation and transport processes are closely related because diffusion channels are constantly altered by structural evolution (Kliem et al., 2020). A feedback loop between deterioration and diffusion is created as hydrolysis proceeds due to changes in porosity, molecular weight, and network density that modify transport characteristics (Pielichowski et al., 2023).

In polymer matrices, water intrusion causes bond cleavage, which results in chain scission and enhanced permeability. By enabling deeper water penetration and making it easier to remove soluble byproducts, this improved transport speeds up additional degradation. Therefore, in many bulk-eroding systems, degradation is a self-accelerating process.

Byproducts of degradation can also have an impact on local chemical environments, especially pH, which influences reaction kinetics and speeds up hydrolysis in susceptible regions (Mircioiu et al., 2019). This results in spatially heterogeneous degradation, especially in systems where byproducts accumulate internally. In contrast, surface-eroding systems maintain more uniform structure and exhibit more predictable transport behavior (Cheng et al., 2019; Pielichowski et al., 2023).

Diffusion and degradation interact to produce controlled release mechanisms. Diffusion is the rate-limiting step in diffusion-controlled systems because transport across the matrix is slower than degradation. In degradation-controlled systems, release depends primarily on matrix breakdown. Swelling-controlled systems exhibit release governed by dynamic changes in network structure (Pielichowski et al., 2023).

In most practical systems, these mechanisms coexist and evolve over time. A system may change from diffusion-dominated to degradation-dominated release as deterioration advances. This dynamic behavior necessitates models that account for time-dependent changes in transport and structural properties (Alkekhia et al., 2020; Huang & Kumacheva, 2026).

Transport processes also strongly influence biological responses by regulating nutrient delivery, waste removal, and signaling molecule distribution. Cell viability, proliferation, and tissue regeneration are all directly impacted by these mechanisms (Trucillo, 2024a). Therefore, transport behavior indirectly governs biological outcomes through its coupling with degradation and microenvironmental changes (Huang & Kumacheva, 2026).

3. Sequential and Combination Release

Many biomedical applications require controlled delivery of multiple therapeutic agents with precise temporal and spatial regulation. Inflammation, proliferation, and remodeling are examples of the progressive stages of biological healing processes that call for different therapeutic approaches (Mittapelly et al., 2024).

Sequential release systems, which frequently use compartmentalized architectures or multilayered structures, are made to deploy agents in a predetermined order. A particular medicinal chemical is present in each layer, which breaks down at a regulated pace to allow for gradual release (Kargozar et al., 2019). Release timing is governed by material properties such as degradation rate, layer thickness, and structural composition.

Combination release systems enable simultaneous or overlapping delivery of multiple agents to achieve synergistic therapeutic effects. This is especially crucial in applications where numerous bioactive compounds must operate simultaneously to provide both tissue regeneration and antimicrobial protection (Alkekhia et al., 2020).

Gradient-based systems provide additional control by spatially varying material composition or drug concentration. These gradients are particularly helpful in complex tissue regeneration scenarios because they affect local release rates and can direct cellular behavior including migration and differentiation (Trucillo, 2024).

Mathematical modeling plays a crucial role in designing these systems by predicting release profiles based on material properties and degradation behavior. In order to anticipate release kinetics more accurately, these models must take into consideration changing structural alterations and interactions between various transport systems (Suvarnapathaki et al., 2019). The focus of recent developments is on stimulus-responsive systems, which modify release behavior in response to environmental cues including pH, inflammatory signals, or enzyme activity. By matching release patterns to shifting physiological conditions, these adaptive systems enhance therapeutic accuracy (Badeau & Deforest, 2019).

STRUCTURE, PROPERTY AND FUNCTION RELATIONSHIPS

The performance of biodegradable biomaterials arises from the interdependence of material structure, physicochemical properties, and biological function. Mechanical characteristics dictate structural performance, biological interactions impact tissue integration and treatment results, and microstructural characteristics control transport behavior and degradation patterns. Understanding these relationships enables the rational design of biomaterials that achieve controlled functionality within complex physiological environments (Refer to Figure 1).

1. Microstructure and Transport

Microstructure serves as a critical link between material design and functional performance. The flow of chemicals, fluids, and cells within a biomaterial is influenced by properties including porosity, pore size distribution, interconnectivity, and structural anisotropy (Aristotelous & Haider, 2014; Graczyk & Matyka, 2020). These structural factors work together to determine transport efficiency rather than operating separately. Anisotropic designs can guide transport along favored paths, while highly linked pore networks promote cellular infiltration and molecular diffusion (Buenzli et al., 2020). Hierarchical pore structures containing macro-, micro-, and nanopores are particularly advantageous because they accommodate multiple transport requirements simultaneously, including nutrient delivery, waste removal, and tissue ingrowth (Allen, 2021).

Microstructure also evolves throughout the degradation process. Transport behavior can change over time due to modifications in network structure and pore architecture, leading to dynamic rather than static functionality. Consequently, biomaterial design increasingly focuses on controlling both initial microstructure and its subsequent evolution under physiological conditions (Dreiss, 2020; Sánchez-Cid et al., 2022).

Achieving optimal performance requires balancing transport efficiency with other functional requirements. The development of materials suited to particular biomedical applications is made possible by advanced production processes, such as additive manufacturing and templating approaches, which offer fine control over microstructural architecture.

2. Mechanical Properties and Structural Integrity

Mechanical performance is essential for maintaining functionality throughout the lifetime of a biodegradable biomaterial. The mechanical requirements of the target tissue environment must be met by characteristics including stiffness, strength, toughness, and time-dependent deformation behavior (Zhao et al., 2020).

Biomaterials are anticipated to offer short-term structural support in regenerative applications while progressively shifting load to newly created tissue. Therefore, coordinating degradation behavior with tissue development is essential for successful performance. Overly persistent materials can obstruct tissue remodeling and integration, while materials that disintegrate too quickly may lose structural function too soon (Sharma et al., 2021).

Mechanical behavior is significantly influenced by microstructure. Increased porosity and other characteristics that enhance biological accessibility and transport typically decrease load-bearing capacity and increase vulnerability to localized stress concentrations (Buenzli et al., 2020). Consequently, mechanical design requires balancing structural support with biological functionality.

Beyond structural issues, mechanical qualities have biological importance. Through mechanotransduction pathways, cells actively detect and react to the stiffness of their surroundings. Mechanical design is a crucial tool for guiding tissue regeneration since substrate mechanics can control cellular adhesion, proliferation, migration, and differentiation (Gaharwar et al., 2020). As a result, mechanical characteristics support both biological function regulation and structural integrity.

3. Biological Interactions and Biointerface Dynamics

The biointerface, defined as the region where a biomaterial interacts with biological systems, plays a central role in determining therapeutic performance (Karami et al., 2019). Proteins quickly adsorb onto the material's surface after implantation, creating a conditioning layer that controls later cellular reactions. Surface chemistry, wettability, charge, and topography all affect this layer's composition and arrangement (Guo et al., 2022).

Surface properties control receptor interactions and protein adsorption, which in turn controls cell attachment and signaling. Furthermore, through mechanosensitive pathways, micro- and nanoscale surface characteristics offer physical cues that affect cell shape, motility, and differentiation (Galli et al., 2026).

Immune responses are another critical aspect of biointerface dynamics. Inflammation, tissue repair, and long-term integration are all impacted by the constant interactions between macrophages and other immune cells and implanted materials. Through surface modification, the addition of bioactive compounds, or regulated degradation behavior, biomaterials can be designed to support regenerative immune responses (Nathanael et al., 2023).

Surface functionalization strategies further enhance biological performance by enabling selective control of cell adhesion and signaling. Methods that enhance tissue integration and direct desired cellular responses include immobilization of bioactive ligands, extracellular matrix elements, or peptide sequences (Khan et al., 2022).

The biointerface is intrinsically dynamic since biological interactions change throughout time. To determine if the material is successfully integrated or replaced by newly produced tissue, protein adsorption is followed by cellular attachment, matrix deposition, and tissue remodeling (Alkekhia et al., 2020). Designing biomaterials that support each stage of this progression is essential for achieving long-term therapeutic success.

Figure 1. Structure, property and function framework for smart biodegradable biomaterials

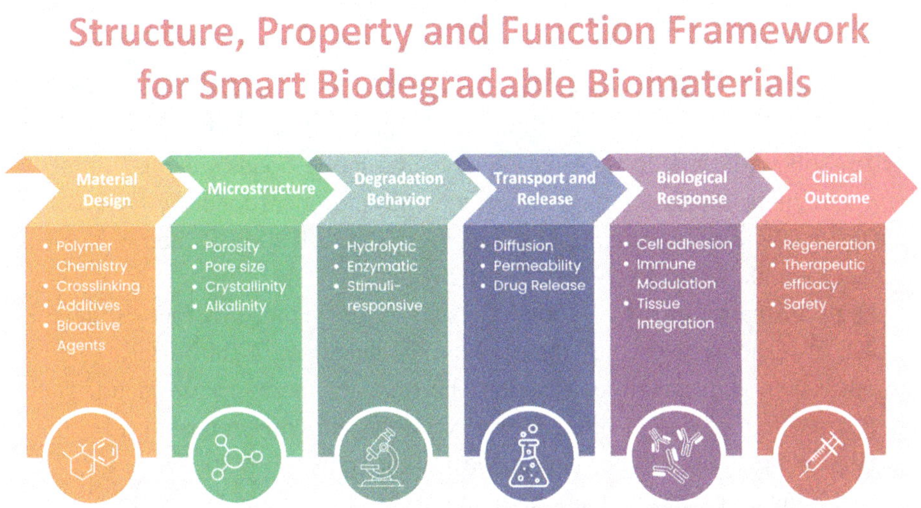

STIMULI-RESPONSIVE SYSTEMS

Stimuli-responsive biodegradable biomaterials represent an advanced class of therapeutic materials capable of adapting their behavior in response to specific environmental cues. These materials are intended to experience controlled physicochemical changes in response to internal or external stimuli, in contrast to traditional biodegradable systems with preset attributes. Such responsiveness enables spatially and temporally regulated therapeutic activity, improving treatment precision while minimizing off-target effects (Badeau & Deforest, 2019; Teixeira do Nascimento et al., 2025).

Stimuli may originate from endogenous biological conditions, such as pH variations, enzyme activity, redox states, and disease-associated biochemical signals, or from externally applied triggers including temperature, mechanical loading, magnetic fields, and light. he integration of responsive functional groups, engineered network architectures, and carefully chosen material compositions that convert environmental information into predictable material reactions are essential to these systems' efficacy.

1. pH-Responsive Systems

pH-responsive biomaterials exploit naturally occurring acidity differences between healthy tissues and pathological environments. Selective activation is possible because tumors, inflammatory tissues, and ischemia areas frequently have lower pH values than the surrounding healthy tissue (Zhuo et al., 2020).

Acid-sensitive chemical bonds or ionizable functional groups are frequently used to produce responsiveness. The charge state of ionizable groups is changed by variations in proton concentration, which modifies network organization and polymer-solvent interactions. Alternatively, acid-labile bonds such as hydrazone, acetal, and imine linkages remain stable under physiological conditions but undergo cleavage in acidic environments, enabling localized activation of therapeutic functions (Z. Li et al., 2022; Pacifici et al., 2020).

The performance of pH-responsive systems is influenced by factors including the pKa of responsive groups, material architecture, and local buffering effects. Responses can be irreversible, allowing one-time activation events such triggered degradation or medication release, or reversible, allowing repeated adaption to changing circumstances, depending on the design. Maintaining sufficient specificity while avoiding premature activation remains a central challenge in the development of pH-responsive biomaterials.

2. Enzyme-Responsive Systems

Enzyme-responsive biomaterials utilize biologically specific catalytic reactions to achieve highly selective activation. These materials can only react in conditions where the target enzyme is present when enzyme-sensitive substrates are incorporated into polymer networks (Pi et al., 2025; M. Wang et al., 2022).

Peptide sequences susceptible to cleavage by matrix metalloproteinases (MMPs) are among the most widely studied examples because of the important role these enzymes play in tissue remodeling and inflammatory processes (Minehan & Del Borgo, 2022). Localized therapeutic responses are made possible by the selective character of enzyme recognition, which closely aligns biomaterial behavior with biological activity.

One of the principal advantages of enzyme-responsive systems is their ability to synchronize material behavior with biological events. These materials can offer a degree of biological specificity that is challenging to attain through traditional degradation mechanisms alone since enzyme expression varies among tissues and changes throughout healing and disease progression (Kargozar et al., 2019).

However, system performance depends strongly on enzyme availability and activity, which may vary among patients and pathological conditions. As a result, preserving consistent behavior in a variety of biological contexts continues to be a crucial design factor.

3. Temperature and Mechanical Responsiveness

Temperature-responsive biomaterials rely on reversible phase transitions that occur near defined critical temperatures. Under regulated temperature settings, these transitions can be exploited to affect material behavior by changing the organization of polymers(Majumder & Minko, 2021).

A common application involves injectable systems that remain fluid during administration and subsequently form stable structures at physiological temperature. After implantation, these materials preserve structural integrity and facilitate minimally invasive delivery (Galli et al., 2026). By using localized heating techniques, temperature responsiveness can also help with externally prompted therapeutic activation.

Mechanical responsiveness represents another approach to adaptive biomaterial design. These systems respond to forces such as compression, tension, or shear by altering their structural organization or activating force-sensitive molecular mechanisms (Rahmati et al., 2020). These reactions are especially important in environments that are mechanically active, such as cardiovascular and musculoskeletal tissues.

To adapt to shifting physiological demands, mechanically responsive materials can be designed to modify their behavior in response to local loading circumstances. This capability is especially attractive for regenerative medicine applications, where material performance must remain coordinated with evolving tissue function (Kliem et al., 2020).

Increasingly, researchers are integrating multiple response mechanisms within a single material platform. Combining thermal and mechanical responsiveness, for example, enables more sophisticated control strategies than can be achieved through a single stimulus alone (Frazar et al., 2020; Rahmati et al., 2020). These methods promote the creation of next-generation adaptive therapeutic systems and increase the functional diversity of biodegradable biomaterials.

The successful implementation of these systems depends on achieving an appropriate balance among sensitivity, stability, and long-term reliability. To ensure consistent performance during the intended duration of use, responsive elements must continue to be stable under normal physiological conditions and sufficiently sensitive to detect target inputs.

FABRICATION STRATEGIES AND DESIGN INTEGRATION

The successful development of advanced biodegradable biomaterials depends on the integration of material design principles with fabrication technologies capable of producing precisely controlled architectures. The translation of desirable structural, mechanical, and biological properties into functional scaffolds suited for particular biomedical applications is made possible by modern manufacturing techniques.

1. Electrospinning and Microstructural Control

Electrospinning is a versatile fabrication technique used to produce nanofibrous scaffolds that emulate key structural features of the extracellular matrix. This method produces ultrafine fibers that are gathered as nonwoven networks by applying a high-voltage electric field to a polymer solution or melt.

A major advantage of electrospinning is the ability to control fiber diameter, morphology, and orientation through adjustment of process parameters such as solution properties, applied voltage, flow rate, and collector configuration (Mosher et al., 2021). Specifically, aligned fiber architectures can be produced by rotating collectors, which makes it possible to create anisotropic structures that are pertinent to tissues with directional organization.

By using sophisticated configurations like coaxial electrospinning, electrospinning also facilitates functional integration. In order to incorporate medicinal medicines or other functional aspects into a protected interior compartment, core-shell fibers can be created to spatially segregate various material components (Teng et al., 2025). Furthermore, scaffold functionality and application-specific performance can be further customized using post-fabrication surface modification techniques.

2. Additive Manufacturing and Spatial Design

Additive manufacturing provides unprecedented control over scaffold geometry through layer-by-layer fabrication of three-dimensional structures (Guzzi & Tibbitt, 2020; Mirzaali et al., 2022). This approach enables the production of highly complex architectures that would be difficult or impossible to achieve using conventional fabrication methods.

The capacity to accurately determine spatial organization across a construct is one of additive manufacturing's most important benefits. Scaffolds can be designed with region-specific features that represent the needs of various functional zones by altering composition, architecture, or printing settings during fabrication. This capability is particularly valuable for applications involving interfaces between distinct tissue types.

By allowing the integration of several polymers, reinforcing phases, or bioactive elements into a single construct, multi-material printing further increases design versatility (Mirzaali et al., 2022). In addition to supporting patient-specific scaffold production, additive manufacturing can be used in conjunction with digital design workflows and medical imaging data to customize geometry and architecture to meet specific therapeutic needs (Laturkar & Laturkar, 2023).

3. Hybrid Scaffold Design

Hybrid scaffold design combines multiple fabrication approaches to create hierarchical structures that integrate complementary features across different length scales (Wasyłeczko et al., 2020). Hybrid systems can attain levels of functionality that are challenging to obtain using a single fabrication approach by utilizing the advantages of different production techniques.

Combining electrospinning and additive manufacturing is a popular tactic. In these systems, electrospun fibers introduce nanoscale architectural elements that more closely match native tissue settings, while a printed framework offers overall structural structure (Oladapo et al., 2019). This combination enables independent optimization of scaffold architecture and surface characteristics.

To increase performance, hybrid systems can additionally include sensing components, nanoparticles, microscale carriers, or other functional features. Separating structural and functional roles among different scaffold components provides greater design flexibility and facilitates the development of multifunctional biomaterials for increasingly complex therapeutic applications. Overall, hybrid fabrication approaches represent an important step toward the creation of integrated biomaterial platforms capable of combining structural support, biological functionality, and application-specific performance within a single engineered system.

INDUSTRIAL AND CLINICAL APPLICATIONS OF BIODEGRADABLE BIOMATERIALS

In recent years, there has been a notable acceleration of the transfer of biodegradable biomaterials from laboratory research to clinical and industrial use. The commercialization of biodegradable systems in several healthcare sectors has been made possible by developments in material design, manufacturing technology, and regulatory approval procedures. This part emphasizes how biodegradable biomaterials are currently being employed in practical medical applications, in contrast to earlier sections that concentrated on material behavior and design concepts.

1. Orthopedic and Musculoskeletal Devices

One of the biggest commercial sectors for biodegradable biomaterials is orthopedic applications. To minimize the need for additional removal procedures, bioabsorbable screws, pins, plates, and fixation devices offer short-term mechanical support during bone healing and progressively resorb after tissue recovery. Biodegradable magnesium, zinc, and composite-based implants that enhance osseointegration and infection resistance are examples of recent advancements that go beyond traditional PLA and PLGA systems. Patient-specific implants made with functionally graded biomaterial composites and additive manufacturing methods are the focus of current industrial efforts (Filip et al., 2022).

Commercial adoption is particularly evident in fracture fixation, spinal fusion, craniofacial reconstruction, and sports medicine applications, where temporary structural support is clinically desirable. Systems made of composite polymers and ceramics are being developed to combine improved bone regeneration with mechanical competency (J.-W. Li et al., 2019).

2. Cardiovascular Applications

The use of biodegradable biomaterials in cardiovascular procedures that need for short-term therapeutic support is growing (Tetali et al., 2023; Y. Wang et al., 2022). Bioresorbable vascular scaffolds and biodegradable stent technologies have been developed to restore vessel patency while avoiding the long-term complications associated with permanent metallic implants. The concept of temporary vascular support followed by gradual resorption continues to drive innovation in next-generation cardiovascular devices (Y. Wang et al., 2022).

Biodegradable materials are being researched for cardiac patches, vascular grafts, and regenerative therapies that enable tissue healing after myocardial infarction, in addition to stents. These applications aim to minimize long-term foreign-body reactions while offering short-term biological and structural support (Theus et al., 2019).

3. Wound Healing and Soft Tissue Regeneration

Biodegradable biomaterials have been used by industry for tissue regeneration matrices, improved dressings, and temporary wound covering systems. In addition to offering physical protection, contemporary biodegradable dressings facilitate tissue regeneration, moisture control, and targeted medicinal administration. Hydrogel-based systems, bioactive matrices, and multifunctional wound-healing platforms are examples of recent advancements intended to enhance results in chronic and challenging-to-heal wounds (Choi et al., 2024; F. Kong et al., 2023).

Biodegradable scaffolds are being employed in soft tissue engineering for ligament regeneration, tendon repair, and skin reconstruction. During the healing process, these materials offer short-term support before being progressively replaced by newly created tissue (Mazzoni et al., 2021).

4. Long-Acting Drug Delivery Systems

Controlled drug distribution is one of the most developed industrial uses of biodegradable biomaterials. Long-acting injectable formulations, microspheres, implants, and localized drug delivery systems all make extensive use of biodegradable polymer platforms based on PLA, PLGA, and related materials. These technologies remove the requirement for device retrieval following treatment and allow for persistent therapeutic release spanning weeks or months (Trucillo, 2024).

Personalized drug delivery systems, implanted depots, injectable gels, and site-specific treatment systems that enhance patient compliance and lower dosage frequency have been the focus of recent industry development. Applications include the treatment of chronic diseases, ophthalmology, cancer, and pain management.

5. Emerging Personalized and Smart Biomaterial Platforms

Multifunctional biodegradable systems that combine improved manufacturing, sensing capabilities, and adaptive therapeutic responses are the focus of recent industrial innovation. The field of customized medicine is growing thanks to innovations like smart hydrogel platforms, injectable biomaterial systems, and 3D-printed patient-specific implants. These systems seek to integrate biological responsiveness, medicinal administration, and structural support onto a single platform (Trucillo, 2024).

The next generation of biomedical products is anticipated to be driven by the confluence of biodegradable materials with nanotechnology, additive manufacturing, and regenerative medicine. Future commercial systems will prioritize integration with patient-specific therapy techniques, less invasive delivery, and customization (Dobrzyńska-Mizera et al., 2024).

6. Representative Experimental and Translational Studies

Recent experimental investigations illustrate how advanced biomaterial technologies are addressing existing clinical limitations. Enzyme-responsive hydrogels have demonstrated improved synchronization between scaffold degradation and tissue modelling, resulting in enhanced cellular infiltration and tissue regeneration (Choi et al., 2024). Similarly, pH-responsive nanoparticle systems have shown increased therapeutic localization within tumor microenvironments while reducing off-target drug exposure.

In regenerative medicine, electrospun nanofibrous scaffolds incorporating growth factors have demonstrated improved tissue integration and vascularization compared with conventional scaffolds (Teng et al., 2025). Additively manufactured patient-specific implants have also shown promising outcomes in craniofacial and orthopedic reconstruction by improving anatomical conformity and reducing surgical complications.

These studies demonstrate that modern biomaterial systems are increasingly designed not only to provide structural support but also to address specific clinical limitations associated with traditional biodegradable materials, including insufficient targeting, limited adaptability and poor integration with host tissues.

CHALLENGES

Despite significant progress, several barriers still prevent smart biodegradable biomaterials from being widely used in clinical and commercial settings. These difficulties go beyond material design and include biological variability, manufacturing repeatability, predictive modeling constraints, and regulatory restrictions. Coordinated efforts from the fields of materials science, engineering, biology, manufacturing, and clinical research will be necessary to address these problems.

1. Predictive Design and System Complexity

One of the primary challenges in developing smart biodegradable biomaterials is the difficulty of accurately predicting long-term performance. These systems frequently display dynamic, time-dependent behavior that changes over the course of their useful lives. It is challenging to establish trustworthy correlations between initial design parameters and final clinical outcomes because material qualities may vary during implantation (Maurya et al., 2024; Nathanael et al., 2023).

Although computational modeling has become increasingly sophisticated, predictive accuracy remains limited by uncertainties in material parameters and biological inputs. Experimental validation is often resource-intensive, and discrepancies between laboratory conditions and physiological environments can reduce the reliability of model predictions (Guo et al., 2022). As a result, it is still very difficult to convert design ideals into consistent clinical results.

2. Manufacturing Scalability and Reproducibility

One major obstacle is still the shift from laboratory-scale manufacturing to industrial output. It may be challenging to reliably replicate manufacturing procedures that yield highly regulated structures under research settings on a commercial scale (Mirzaali et al., 2022).

Batch-to-batch variability is particularly problematic because small deviations in processing conditions can alter material architecture and functional performance. Therefore, strong process control, sophisticated characterization methods, and extensive quality assurance systems are necessary to ensure reproducibility. These specifications raise the cost and complexity of manufacture.

Another issue is material stability throughout production. Processing factors including heat, moisture, sterilizing techniques, or mechanical pressures may affect responsive components. For commercialization to be successful, functionality must be maintained during production, storage, and distribution (Teng et al., 2025).

3. Biological Variability and Patient Heterogeneity

Biological variability presents a major obstacle to achieving consistent therapeutic outcomes. Due to differences in age, genetics, illness status, immunological function, and healing capacity, patients have different physiological conditions (Gomes et al., 2025). As a result, materials that perform effectively in one patient may exhibit different behavior in another.

Preclinical testing and clinical evaluation are further complicated by patient-specific variations. Predicting performance across large patient groups is challenging because experimental models frequently fall short of accurately simulating the variety of human biological responses. Personalized biomaterial techniques provide new production, economic, and regulatory approval issues, even though they might alleviate some of these restrictions.

4. Degradation Product Toxicity and Long-term biocompatibility

Although biodegradable biomaterials are designed to degrade into biologically acceptable products, the safety of degradation by products remains an important consideration for clinical translation. Under some circumstances, the buildup of acidic breakdown products produced by polymers like PLA and PLGA can cause localized pH drops, which may lead to inflammatory reactions and delayed tissue recovery (Sharma et al., 2021; Alaswad et al., 2022).

Beyond initial cytocompatibility, long-term biocompatibility involves assessing immunological responses, fibrous encapsulation, chronic inflammation, and tissue reorganization throughout deterioration. The rate of degradation is crucial because too slow degradation can prolong foreign-body reactions, while too fast degradation may induce a transient build-up of degradation products (Nathanael et al., 2023).

A thorough analysis of degradation kinetics, degradation-product distribution, systemic exposure, and long-term tissue responses is necessary for post-implantation safety assessment. Therefore, toxicity assessment should be incorporated into future biomaterial development at every stage of the material lifespan, from initial implantation to full resorption. Accelerating clinical acceptance and enhancing regulatory compliance will require standardized approaches for long-term safety evaluation.

5. Regulatory Considerations and Clinical Translation

Regulatory approval remains one of the most demanding stages in the development of smart biodegradable biomaterials. These systems need thorough characterization and safety since they frequently integrate structural, biological, and medicinal capabilities (Sikder & Bhaduri, 2019).

Demonstrating long-term safety and consistency over the whole material lifecycle presents a special challenge. Evidence pertaining to material composition, degradation behavior, biocompatibility, and production repeatability is required by regulatory bodies. There may be additional pharmaceutical regulatory requirements for systems that use therapeutic drugs (Suhag, 2024).

There are additional difficulties with clinical evaluation. The adaptable and dynamic behavior of smart biomaterials may not be sufficiently captured by traditional trial designs, necessitating the development of novel performance measurements and evaluation techniques. Furthermore, current legal frameworks are sometimes outpaced by emerging technology, which raises questions about standardization standards and approval processes. Developing evaluation standards that promote innovation and guarantee patient safety will require ongoing cooperation between researchers, producers, doctors, and regulatory bodies.

The usual translational approach for biodegradable biomaterial technology is depicted in Figure 2. Preclinical testing to determine safety, functionality, and biological performance comes after laboratory-scale production and material design for successful translation. Before being commercialized and widely used in clinical settings, promising systems go through regulatory approval and clinical evaluation. Many biomaterial technologies fail during translation because to issues with manufacturing scalability, regulatory permission, reproducibility, cost-effectiveness, and long-term safety evaluation, even though scientific innovation is still crucial. Therefore, the integration of technical, regulatory, manufacturing, and clinical factors throughout the development process is just as important to translational success as material performance (Brassesco et al., 2024).

Figure 2. Translational pathway for smart biodegradable biomaterials

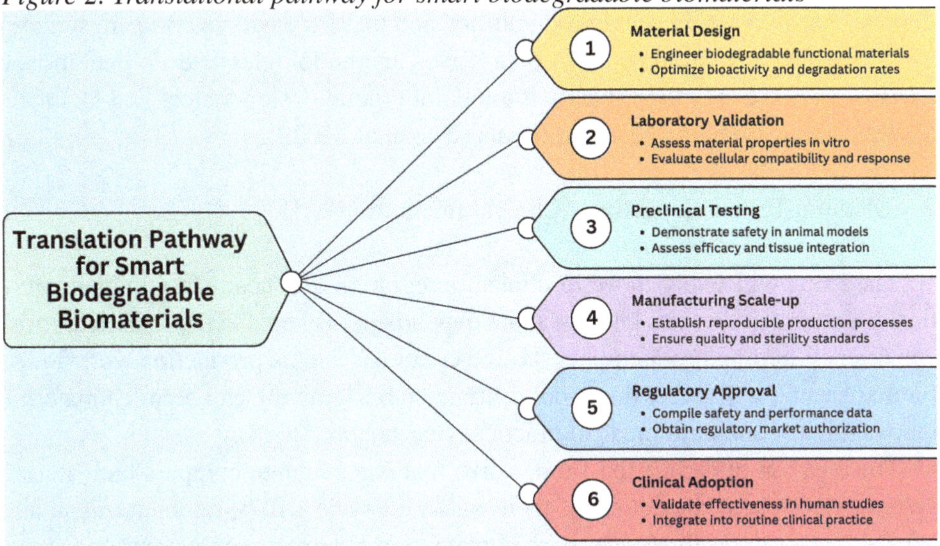

FUTURE DIRECTIONS

Future developments in smart biodegradable biomaterials will rely not only on enhanced material performance but also on the capacity to get around present constraints in implementation, translation, and prediction. It is anticipated that as the field develops, research goals will shift from enhancing design reliability to facilitating clinically integrated biomaterial platforms and, eventually, creating sustainable biomedical ecosystems. There are three ways to look at these developments: short-term, medium-term, and long-term perspectives (Refer to Figure 3).

1. Short-Term Perspectives: Improving Predictability and Standardization

Improving the predictability of biomaterial performance across various biological settings will be a top focus soon. Disparities between laboratory findings and in vivo results often persist despite substantial advancements in material development. More reliable validation techniques, standardized testing procedures, and enhanced techniques for connecting preclinical data with clinical performance are all necessary to close this disparity.

Additionally, efforts are anticipated to concentrate on developing widely recognized characterization techniques for assessing biological safety, long-term functionality, and degradation behavior. Increased uniformity would speed up regulatory evaluation, enhance reproducibility, and make it easier to compare studies. Simultaneously, developments in data-driven methodologies and computational analysis may enhance the capacity to pinpoint crucial design factors and facilitate the creation of evidence-based materials (Guo et al., 2022).

2. Medium-Term Perspectives: Clinical Integration

The focus will likely move to clinical integration and healthcare implementation as material platforms become more dependable. By enhancing manufacturing consistency, cutting production costs, and creating scalable production workflows, future research will increasingly concentrate on bridging the gap between material innovation and standard medical practice (Cecen et al., 2024).

The creation of evaluation frameworks that can evaluate complex biomaterial systems throughout the course of their whole lifecycle will be another crucial avenue. Long-term performance, post-implantation behavior, and patient outcomes are all included in this perspective. To develop translation channels that promote innovation and patient safety, closer cooperation between researchers, physicians, manufacturers, and regulatory bodies will be required (Sikder & Bhaduri, 2019).

It is anticipated that biomaterials will become more integrated into larger healthcare systems. Future platforms may be designed to function alongside diagnostic technologies, clinical monitoring tools, and digital health infrastructures, enabling more comprehensive approaches to patient care.

3. Long-Term Perspectives: Autonomous and Sustainable Biomedical Platforms

It is anticipated that the sector will eventually advance from individual biomaterial products to integrated biomedical systems that can facilitate ongoing treatment management. A greater level of clinical autonomy may be made possible by these platforms' integration of material functionality with real-time monitoring, adaptive decision-making, and automated therapeutic action.

Integrating sustainability concepts into the biomaterial lifecycle will be equally crucial. Waste minimization techniques, resource-efficient manufacturing, and environmentally conscious procurement are anticipated to be given top priority in future development initiatives. In order to guide material selection and manufacturing processes, the use of life-cycle assessment methodologies and circular economy concepts may become more crucial (Laturkar & Laturkar, 2024a; Panahi, 2025).

Achieving a balance between technological sophistication, therapeutic utility, economic viability, and environmental responsibility will ultimately determine the long-term success of smart biodegradable biomaterials. Advances in these fields will contribute to the transformation of biodegradable biomaterials from specialized therapeutic instruments into extensively used elements of healthcare systems of the future (Vinchurkar et al., 2025).

Figure 3. Future directions roadmap for biodegradable biomaterials

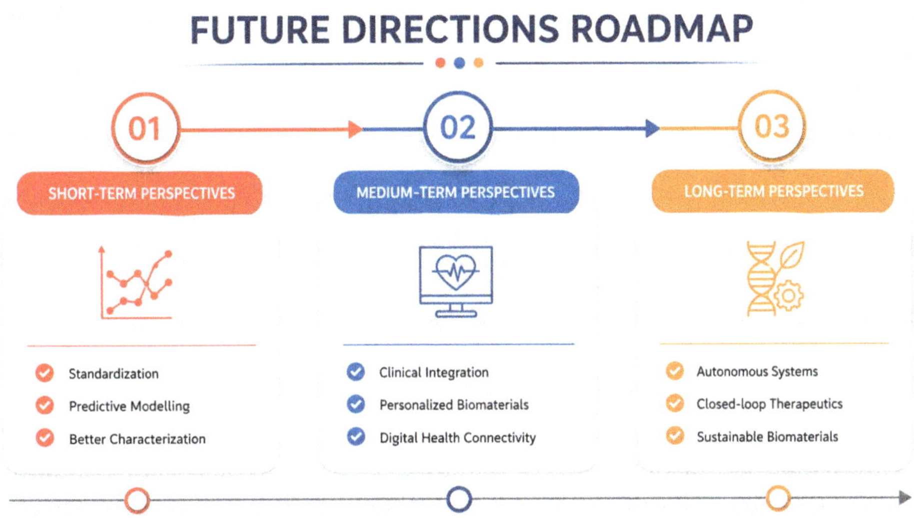

CONCLUSION

By combining controlled degradation, sensitive functionality, and tailored interactions with biological systems, smart biodegradable biomaterials mark a substantial advancement above traditional biodegradable materials. The intricate interactions between degradation kinetics, transport processes, microstructural organization, mechanical behavior, and biointerface dynamics control their performance. These systems are increasingly made to actively react to physiological situations instead of acting as passive materials, allowing for more accurate therapeutic interventions and better clinical results.

This chapter has shown that no single mechanism can fully explain material behavior. Transport controls biological signaling and therapeutic release, degradation processes impact transport pathways, and structural development constantly alters cellular connections and mechanical performance. These interrelated connections emphasize how crucial it is to develop next-generation biomaterials from a systems-level approach. By allowing materials to react selectively to biochemical and physical inputs, developments in stimuli-responsive systems further expand this capability and enhance spatiotemporal control over therapeutic effects.

Advances in fabrication methods, including electrospinning, additive manufacturing, and hybrid scaffold design, have made it possible to customize biomaterial architecture on a variety of length scales. The development of multifunctional platforms that can integrate regulated medicinal delivery, biological activity, and structural support into a single system has accelerated thanks to these manufacturing techniques. The significant therapeutic and commercial effect of these developments is demonstrated by the expanding integration of biodegradable biomaterials into orthopedic devices, cardiovascular applications, wound healing products, regenerative therapies, and long-acting drug delivery systems.

Despite these successes, significant obstacles still exist. Because degradation, transport, mechanical loads, and biological reactions are intertwined, it is still challenging to predict long-term material performance in complicated physiological contexts. Widespread clinical deployment is still hampered by manufacturing scalability, reproducibility, patient-to-patient variability, and changing regulatory constraints. Closer integration of materials science, computer modeling, production engineering, biology, and clinical research will be necessary to address these issues.

It is anticipated that the research would advance toward biomaterial platforms that are more therapeutically integrated, adaptable, and predictive. More dependable material development will be supported by advancements in computational design, data-driven optimization, and sophisticated characterization techniques. Future innovation will be shaped concurrently by a stronger focus on clinical translation, healthcare integration, and sustainability. It is anticipated that these developments will combine to create biomaterials that are not only biocompatible and biodegradable but also intelligent, customizable, and able to interact dynamically with their biological surroundings.

In the end, smart biodegradable biomaterials' future depends on striking a balance between therapeutic usefulness and functional sophistication. Finding this balance will be crucial to converting cutting-edge ideas into safe, scalable, and efficient healthcare solutions that can handle progressively challenging treatment issues.

REFERENCES

Aflori, M., & Poni, P. (2021). Smart Nanomaterials for Biomedical Applications—A Review. *Nanomaterials 2021, Vol. 11, Page 396, 11*(2), 396. DOI: 10.3390/nano11020396

Alaswad, S. O., Mahmoud, A. S., & Arunachalam, P. (2022). Recent Advances in Biodegradable Polymers and Their Biological Applications: A Brief Review. *Polymers 2022, Vol. 14, Page 4924, 14*(22), 4924. DOI: 10.3390/polym14224924

Alkekhia, D., Hammond, P. T., & Shukla, A. (2020). Layer-by-Layer Biomaterials for Drug Delivery. *Annual Review of Biomedical Engineering, 22*(Volume 22, 2020), 1–24. https://doi.org/DOI: 10.1146/ANNUREV-BIOENG-060418-052350/CITE/REFWORKS

Allen, M. B. (2021). *The mathematics of fluid flow through porous media.* 212. https://books.google.com/books/about/The_Mathematics_of_Fluid_Flow_Through_Po.html?id=4G0yEAAAQBAJ

Almutary, A. G., Chauhan, P., Baldaniya, L., Menon, S. V., Kumar, M. R., Chaturvedi, B., Sharma, N., Chauhan, A. S., Abomughaid, M. M., Dhivyadharshni, M., Paiva-Santos, A. C., Lakhanpal, S., & Jha, N. K. (2025). Overcoming challenges in the design of drug delivery systems targeting the central nervous system. *Nanomedicine (London), 20*(1), 5–8. DOI: 10.1080/17435889.2024.2421157 PMID: 39564781

Aristotelous, A. C., & Haider, M. A. (2014). Evaluation of Diffusive Transport and Cellular Uptake of Nutrients in Tissue Engineered Constructs Using a Hybrid Discrete Mathematical Model. *Processes 2014, Vol. 2, Pages 333-344, 2*(2), 333–344. DOI: 10.3390/pr2020333

Atala, A. (2020). Introduction: 3D Printing for Biomaterials. *Chemical Reviews, 120*(19), 10545–10546. DOI: 10.1021/acs.chemrev.0c00139 PMID: 33050702

Badeau, B. A., & Deforest, C. A. (2019). Programming Stimuli-Responsive Behavior into Biomaterials. *Annual Review of Biomedical Engineering, 21*(Volume 21, 2019), 241–265. https://doi.org/DOI: 10.1146/ANNUREV-BIOENG-060418-052324/CITE/REFWORKS

Benson, H. A. E. Roberts, M. S., Williams, A., & Liang, X. (2022). *Fundamentals of drug delivery.* 554. https://books.google.com/books/about/Fundamentals_of_Drug_Delivery.html?id=Mv5FEAAAQBAJ

Brassesco, M. E., Cortez, J., Triacca, V., Payen, J., & Bayon, Y. (2024). Translating Ideas into Commercial Biomaterial Devices: A Path to Improving Healthcare. *ACS Biomaterials Science & Engineering*, *10*(4), 1910–1920. DOI: 10.1021/acsbiomaterials.3c01441 PMID: 38452343

Buenzli, P. R., Lanaro, M., Wong, C. S., McLaughlin, M. P., Allenby, M. C., Woodruff, M. A., & Simpson, M. J. (2020). Cell proliferation and migration explain pore bridging dynamics in 3D printed scaffolds of different pore size. *Acta Biomaterialia*, *114*, 285–295. DOI: 10.1016/j.actbio.2020.07.010 PMID: 32673750

Cecen, B., Hassan, S., Li, X., & Zhang, Y. S. (2024). Smart Biomaterials in Biomedical Applications: Current Advances and Possible Future Directions. *Macromolecular Bioscience*, *24*(3), 2200550. Advance online publication. DOI: 10.1002/mabi.202200550 PMID: 37728061

Cedillo-Servin, G., Al-Jehani, E. A. A., Rossy, T., Teixeira, S. P. B., Sage, F., Domingues, R. M. A., Raman, R., & Castilho, M. (2025). Meta-adaptive biomaterials: Multiscale, spatiotemporal organization and actuation in engineered tissues. *Trends in Biotechnology*, *43*(11), 2709–2723. DOI: 10.1016/j.tibtech.2025.05.004 PMID: 40473558

Chen, Z., Wang, Z., & Gu, Z. (2019). Bioinspired and Biomimetic Nanomedicines. *Accounts of Chemical Research*, *52*(5), 1255–1264. DOI: 10.1021/acs.accounts.9b00079 PMID: 30977635

Cheng, D. B., Wang, D., Gao, Y. J., Wang, L., Qiao, Z. Y., & Wang, H. (2019). Autocatalytic Morphology Transformation Platform for Targeted Drug Accumulation. *Journal of the American Chemical Society*, *141*(10), 4406–4411. DOI: 10.1021/jacs.8b13512 PMID: 30801185

Choi, H., Choi, W.-S., & Jeong, J.-O. (2024). A Review of Advanced Hydrogel Applications for Tissue Engineering and Drug Delivery Systems as Biomaterials. *Gels (Basel, Switzerland)*, *10*(11), 693. DOI: 10.3390/gels10110693 PMID: 39590049

Deirram, N., Zhang, C., Kermaniyan, S. S., Johnston, A. P. R., & Such, G. K. (2019). pH-Responsive Polymer Nanoparticles for Drug Delivery. *Macromolecular Rapid Communications*, *40*(10), 1800917. DOI: 10.1002/marc.201800917 PMID: 30835923

Dobrzyńska-Mizera, M., Dodda, J. M., Liu, X., Knitter, M., Oosterbeek, R. N., Salinas, P., Pozo, E., Ferreira, A. M., & Sadiku, E. R. (2024). Engineering of Bioresorbable Polymers for Tissue Engineering and Drug Delivery Applications. *Advanced Healthcare Materials*, *13*(30), 2401674. Advance online publication. DOI: 10.1002/adhm.202401674 PMID: 39233521

Dreiss, C. A. (2020). Hydrogel design strategies for drug delivery. *Current Opinion in Colloid & Interface Science*, *48*, 1–17. DOI: 10.1016/j.cocis.2020.02.001

Dziki, J. L., & Badylak, S. F. (2019). Acellular Biologic Scaffolds in Regenerative Medicine: Unacceptable Variability with Acceptable Results. *Regenerative Engineering and Translational Medicine 2019 5:4, 5*(4), 414–419. DOI: 10.1007/s40883-019-00106-5

Elkhoury, K., Koçak, P., Kang, A., Arab-Tehrany, E., Ward, J. E., & Shin, S. R. (2020). Engineering Smart Targeting Nanovesicles and Their Combination with Hydrogels for Controlled Drug Delivery. *Pharmaceutics 2020, Vol. 12, Page 849, 12*(9), 849. DOI: 10.3390/pharmaceutics12090849

Filip, N., Radu, I., Veliceasa, B., Filip, C., Pertea, M., Clim, A., Pinzariu, A. C., Drochioi, I. C., Hilitanu, R. L., & Serban, I. L. (2022). Biomaterials in Orthopedic Devices: Current Issues and Future Perspectives. *Coatings*, *12*(10), 1544. DOI: 10.3390/coatings12101544

Frazar, E. M., Shah, R. A., Dziubla, T. D., & Hilt, J. Z. (2020). Multifunctional temperature-responsive polymers as advanced biomaterials and beyond. *Journal of Applied Polymer Science*, *137*(25), 48770. DOI: 10.1002/app.48770 PMID: 34305165

Gaharwar, A. K., Singh, I., & Khademhosseini, A. (2020). Engineered biomaterials for in situ tissue regeneration. *Nature Reviews. Materials*, *5*(9), 686–705. DOI: 10.1038/s41578-020-0209-x

Galli, C., Colangelo, M. T., Meleti, M., & Guizzardi, S. (2026). Morphospace engineering: Morphological computation in scaffold design. *Bio Systems*, *259*, 105676. DOI: 10.1016/j.biosystems.2025.105676 PMID: 41391689

Giosafatto, L., Mariniello, L., Restaino, O. F., Kurowiak, J., Klekiel, T., B̦., R., & Nski, E. (2023). Biodegradable Polymers in Biomedical Applications: A Review—Developments, Perspectives and Future Challenges. *International Journal of Molecular Sciences 2023, Vol. 24, Page 16952, 24*(23), 16952. DOI: 10.3390/ijms242316952

Gomes, Y. V. R., Tavares, A. A., Barbosa, R. C., Tomaz, A. F., Sousa, W. J. B., Oliveira, L. C. C., Silva, S. M. L., & Fook, M. V. L. (2025). Biological responses to biomaterials: A review. *Brazilian Journal of Medical and Biological Research = Revista Brasileira de Pesquisas Médicas e Biológicas*, *58*, e14599. DOI: 10.1590/1414-431x2025e14599 PMID: 40367018

Graczyk, K. M., & Matyka, M. (2020). Predicting porosity, permeability, and tortuosity of porous media from images by deep learning. *Scientific Reports 2020 10:1, 10*(1), 21488-. DOI: 10.1038/s41598-020-78415-x

Gul, K., Gan, R. Y., Sun, C. X., Jiao, G., Wu, D. T., Li, H., Kenaan, A., Corke, H., & Fang, Y.-P. (2022). Recent advances in the structure, synthesis, and applications of natural polymeric hydrogels. *Critical Reviews in Food Science and Nutrition, 62*(14), 3817–3832. DOI: 10.1080/10408398.2020.1870034 PMID: 33406881

Guo, Y., Calve, S., & Tepole, A. B. (2022). Multiscale mechanobiology: Coupling models of adhesion kinetics and nonlinear tissue mechanics. *Biophysical Journal, 121*(4), 525–539. DOI: 10.1016/j.bpj.2022.01.012 PMID: 35074393

Guzzi, E. A., & Tibbitt, M. W. (2020). Additive Manufacturing of Precision Biomaterials. *Advanced Materials, 32*(13), 1901994. DOI: 10.1002/adma.201901994 PMID: 31423679

Huang, Y., & Kumacheva, E. (2026). Hydrogels with multiple characteristic pore dimensions: From transport properties to multifunctional materials. *Progress in Polymer Science, 173*, 102065. DOI: 10.1016/j.progpolymsci.2025.102065

Karami, D., Richbourg, N., & Sikavitsas, V. (2019). Dynamic in vitro models for tumor tissue engineering. *Cancer Letters, 449*, 178–185. DOI: 10.1016/j.canlet.2019.01.043 PMID: 30763717

Kargozar, S., Ramakrishna, S., & Mozafari, M. (2019). Chemistry of biomaterials: Future prospects. *Current Opinion in Biomedical Engineering, 10*, 181–190. DOI: 10.1016/j.cobme.2019.07.003

Khan, H. M., Liao, X., Sheikh, B. A., Wang, Y., Su, Z., Guo, C., Li, Z., Zhou, C., Cen, Y., & Kong, Q. (2022). Smart biomaterials and their potential applications in tissue engineering. *Journal of Materials Chemistry. B, 10*(36), 6859–6895. DOI: 10.1039/D2TB01106A PMID: 36069198

Kliem, S., Kreutzbruck, M., & Bonten, C. (2020). Review on the Biological Degradation of Polymers in Various Environments. *Materials 2020, Vol. 13, Page 4586, 13*(20), 4586. DOI: 10.3390/ma13204586

Kong, F., Mehwish, N., & Lee, B. H. (2023). Emerging albumin hydrogels as personalized biomaterials. *Acta Biomaterialia, 157*, 67–90. DOI: 10.1016/j.actbio.2022.11.058 PMID: 36509399

Kong, L., Gao, X., Qian, Y., Sun, W., You, Z., & Fan, C. (2022). Biomechanical microenvironment in peripheral nerve regeneration: From pathophysiological understanding to tissue engineering development. *Theranostics, 12*(11), 4993–5014. DOI: 10.7150/thno.74571 PMID: 35836812

Kumar, V., Minocha, N., Garg, V., & Dureja, H. (2022). Nanostructured materials used in drug delivery. *Materials Today: Proceedings*, *69*, 174–180. DOI: 10.1016/j.matpr.2022.08.306

Kuperkar, K., Atanase, L. I., Bahadur, A., Crivei, I. C., & Bahadur, P. (2024). Degradable Polymeric Bio(nano)materials and Their Biomedical Applications: A Comprehensive Overview and Recent Updates. *Polymers 2024, Vol. 16, Page 206, 16*(2), 206. DOI: 10.3390/polym16020206

Laturkar, K., & Laturkar, K. (2023). Internet of Things: Architectures, Applications, and Challenges. Https://Services.Igi-Global.Com/Resolvedoi/Resolve.Aspx?Doi=10.4018/978-1-6684-8145-5.Ch023, 456–475. DOI: 10.4018/978-1-6684-8145-5.ch023

Laturkar, K., & Laturkar, K. (2024a). Harmonizing Innovation - Bridging Industry 4.0 with Supply Chain Sustainability. Https://Services.Igi-Global.Com/Resolvedoi/Resolve.Aspx?Doi=10.4018/979-8-3693-1363-3.Ch006, 133–156. DOI: 10.4018/979-8-3693-1363-3.ch006

Laturkar, K., & Laturkar, K. (2024b). *Revolutionizing Pharmaceutical Supply Chains - Harnessing the Power of Blockchain Technology* (pp. 92–112). DOI: 10.4018/979-8-3693-0482-2.ch006

Li, J.-W., Du, C.-F., Yuchi, C.-X., & Zhang, C.-Q. (2019). Application of Biodegradable Materials in Orthopedics. *Journal of Medical and Biological Engineering*, *39*(5), 633–645. DOI: 10.1007/s40846-019-00469-8

Li, X., Liu, Z., Cui, S., Luo, C., Li, C., & Zhuang, Z. (2019). Predicting the effective mechanical property of heterogeneous materials by image based modeling and deep learning. *Computer Methods in Applied Mechanics and Engineering*, *347*, 735–753. DOI: 10.1016/j.cma.2019.01.005

Li, Z., Zhu, Y., & Matson, J. B. (2022). pH-Responsive Self-Assembling Peptide-Based Biomaterials: Designs and Applications. *ACS Applied Bio Materials*, *5*(10), 4635–4651. DOI: 10.1021/acsabm.2c00188 PMID: 35505454

Liu, M., Zhou, Y., Mei, X., Yu, Z., Guan, B., Xiao, Y., Liu, S., Wang, H., & Qin, Y. (2026). AI-driven biomaterial design: An intelligent closed loop from reverse design to biological response. *Frontiers in Cell and Developmental Biology*, *13*, 1755565. DOI: 10.3389/fcell.2025.1755565 PMID: 41561627

Lou, J., & Mooney, D. J. (2022). Chemical strategies to engineer hydrogels for cell culture. *Nature Reviews Chemistry 2022 6:10, 6*(10), 726–744. DOI: 10.1038/s41570-022-00420-7

Majumder, J., & Minko, T. (2021). Multifunctional and stimuli-responsive nanocarriers for targeted therapeutic delivery. *Expert Opinion on Drug Delivery*, *18*(2), 205–227. DOI: 10.1080/17425247.2021.1828339 PMID: 32969740

Marlena, J., Tan, J. K. S., Lin, Z., Li, D. X., Zhao, B., Leo, H. L., Kim, S., & Yap, C. H. (2021). Monolithic polymeric porous superhydrophobic material with pneumatic plastron stabilization for functionally durable drag reduction in blood-contacting biomedical applications. *NPG Asia Materials 2021 13:1, 13*(1), 58-. DOI: 10.1038/s41427-021-00325-9

Maurya, A. K., de Souza, F. M., Dawsey, T., & Gupta, R. K. (2024). Biodegradable polymers and composites: Recent development and challenges. *Polymer Composites*, *45*(4), 2896–2918. DOI: 10.1002/pc.28023

Mazzoni, E., Iaquinta, M. R., Lanzillotti, C., Mazziotta, C., Maritati, M., Montesi, M., Sprio, S., Tampieri, A., Tognon, M., & Martini, F. (2021). Bioactive Materials for Soft Tissue Repair. *Frontiers in Bioengineering and Biotechnology*, *9*, 613787. Advance online publication. DOI: 10.3389/fbioe.2021.613787 PMID: 33681157

Minehan, R. L., & Del Borgo, M. P. (2022). Controlled Release of Therapeutics From Enzyme-Responsive Biomaterials. *Frontiers in Biomaterials Science*, *1*, 916985. DOI: 10.3389/fbiom.2022.916985

Mircioiu, C., Voicu, V., Anuta, V., Tudose, A., Celia, C., Paolino, D., Fresta, M., Sandulovici, R., & Mircioiu, I. (2019). Mathematical Modeling of Release Kinetics from Supramolecular Drug Delivery Systems. *Pharmaceutics 2019, Vol. 11, Page 140, 11*(3), 140. DOI: 10.3390/pharmaceutics11030140

Mirzaali, M. J., Moosabeiki, V., Rajaai, S. M., Zhou, J., & Zadpoor, A. A. (2022). Additive Manufacturing of Biomaterials—Design Principles and Their Implementation. *Materials 2022, Vol. 15, Page 5457, 15*(15), 5457. DOI: 10.3390/ma15155457

Mittapelly, N., Djehizian, A., Telaprolu, K. C., McNally, K., Puttrevu, S. K., Arjmandi-Tash, O., Polak, S., & Bois, F. Y. (2024). Mechanistic Model for Drug Release from PLGA-Based Biodegradable Implants for In Vitro Release Testing: Development and Validation. *ACS Applied Bio Materials*, *7*(11), 7453–7465. DOI: 10.1021/acsabm.4c01054 PMID: 39424342

Mohamed Nasser, S., Abbas, M., & Trari, M. (2024). Understanding the rate-limiting step adsorption kinetics onto biomaterials for mechanism adsorption control. *Progress in Reaction Kinetics and Mechanism*, *49*, 14686783241226858. Advance online publication. DOI: 10.1177/14686783241226858

Mosher, C. Z., Brudnicki, P. A. P., Gong, Z., Childs, H. R., Lee, S. W., Antrobus, R. M., Fang, E. C., Schiros, T. N., & Lu, H. H. (2021). Green electrospinning for biomaterials and biofabrication. *Biofabrication, 13*(3), 035049. DOI: 10.1088/1758-5090/ac0964 PMID: 34102612

Nathanael, A. ; J., Xu, L., Sreena, R., & Nathanael, A. J. (2023). Biodegradable Biopolymeric Nanoparticles for Biomedical Applications-Challenges and Future Outlook. *Materials 2023, Vol. 16, Page 2364, 16*(6), 2364. DOI: 10.3390/ma16062364

Nii, T., & Katayama, Y. (2021). Biomaterial-Assisted Regenerative Medicine. *International Journal of Molecular Sciences 2021, Vol. 22, Page 8657, 22*(16), 8657. DOI: 10.3390/ijms22168657

Oladapo, B. I., Zahedi, S. A., & Adeoye, A. O. M. (2019). 3D printing of bone scaffolds with hybrid biomaterials. *Composites. Part B, Engineering, 158*, 428–436. DOI: 10.1016/j.compositesb.2018.09.065

Pacifici, N., Bolandparvaz, A., & Lewis, J. S. (2020). Stimuli-Responsive Biomaterials for Vaccines and Immunotherapeutic Applications. *Advanced Therapeutics, 3*(11), 2000129. DOI: 10.1002/adtp.202000129 PMID: 32838028

Panahi, O. (2025). Innovative Biomaterials for Sustainable Medical Implants: A Circular Economy Approach. *European Journal of Innovative Studies and Sustainability, 1*(2), 20–29. DOI: 10.59324/ejiss.2025.1(2).03

Park, H. J., Hong, H., Thangam, R., Song, M. G., Kim, J. E., Jo, E. H., Jang, Y. J., Choi, W. H., Lee, M. Y., Kang, H., & Lee, K. B. (2022). Static and Dynamic Biomaterial Engineering for Cell Modulation. *Nanomaterials 2022, Vol. 12, Page 1377, 12*(8), 1377. DOI: 10.3390/nano12081377

Pi, Y., Ganabady, K., & Celiz, A. D. (2025). Enzyme-responsive biomaterials for biomedical applications. *Communications Materials 2025 6:1, 6*(1), 263-. DOI: 10.1038/s43246-025-00983-0

Pielichowski, K. Njuguna, James., & Majka, T. M. (2023). *Thermal degradation of polymeric materials*. 379. https://books.google.com/books/about/Thermal_Degradation_of_Polymeric_Materia.html?id=waRXEAAAQBAJ

Poomathi, N., Singh, S., Prakash, C., Subramanian, A., Sahay, R., Cinappan, A., & Ramakrishna, S. (2020). 3D printing in tissue engineering: A state of the art review of technologies and biomaterials. *Rapid Prototyping Journal, 26*(7), 1313–1334. DOI: 10.1108/RPJ-08-2018-0217

Rahmati, M., Silva, E. A., Reseland, J. E. A., Heyward, C., & Haugen, H. J. (2020). Biological responses to physicochemical properties of biomaterial surface. *Chemical Society Reviews*, *49*(15), 5178–5224. DOI: 10.1039/D0CS00103A PMID: 32642749

Sánchez-Cid, P., Jiménez-Rosado, M., Romero, A., & Pérez-Puyana, V. (2022). Novel Trends in Hydrogel Development for Biomedical Applications: A Review. *Polymers 2022, Vol. 14, Page 3023, 14*(15), 3023. DOI: 10.3390/polym14153023

Sharma, S., Sudhakara, P., Singh, J., Ilyas, R. A., Asyraf, M. R. M., & Razman, M. R. (2021). Critical Review of Biodegradable and Bioactive Polymer Composites for Bone Tissue Engineering and Drug Delivery Applications. *Polymers 2021, Vol. 13, Page 2623, 13*(16), 2623. DOI: 10.3390/polym13162623

Sikder, P., & Bhaduri, S. B. (2019). Regulatory aspects of medical devices and biomaterials. *Biomaterials in Translational Medicine*, 23–36. DOI: 10.1016/B978-0-12-813477-1.00002-5

Suhag, D. (2024). *Regulatory and Ethical Considerations*. 355–372. DOI: 10.1007/978-981-97-5906-4_11

Suvarnapathaki, S., Wu, X., Lantigua, D., Nguyen, M. A., & Camci-Unal, G. (2019). Breathing life into engineered tissues using oxygen-releasing biomaterials. *NPG Asia Materials 2019 11:1, 11*(1), 65-. DOI: 10.1038/s41427-019-0166-2

Teixeira do Nascimento, A., Stoddart, P. R., Goris, T., Kael, M., Manasseh, R., Alt, K., Tashkandi, J., Kim, B. C., & Moulton, S. E. (2025). Stimuli-Responsive Materials for Biomedical Applications. *Advanced Materials*, *37*(36), e07559. DOI: 10.1002/adma.202507559 PMID: 40801175

Teng, Y., Song, L., Shi, J., Lv, Q., Hou, S., & Ramakrishna, S. (2025). Advancing electrospinning towards the future of biomaterials in biomedical engineering. *Regenerative Biomaterials*, *12*, rbaf034. Advance online publication. DOI: 10.1093/rb/rbaf034 PMID: 40416647

Tetali, S. S. V., Fricker, A. T. R., van Domburg, Y. A., & Roy, I. (2023). Intelligent biomaterials for cardiovascular applications. *Current Opinion in Biomedical Engineering*, *28*, 100474. DOI: 10.1016/j.cobme.2023.100474

Theus, A. S., Tomov, M. L., Cetnar, A., Lima, B., Nish, J., McCoy, K., Mahmoudi, M., & Serpooshan, V. (2019). Biomaterial approaches for cardiovascular tissue engineering. *Emergent Materials*, *2*(2), 193–207. DOI: 10.1007/s42247-019-00039-3

Trucillo, P. (2024). Biomaterials for Drug Delivery and Human Applications. *Materials 2024, Vol. 17, Page 456, 17*(2), 456. DOI: 10.3390/ma17020456

Vinchurkar, K., Bukke, S. P. N., Jain, P., Bhadoria, J., Likhariya, M., Mane, S., Suryawanshi, M., Veerabhadrappa, K. V., Eftekhari, Z., & Onohuean, H. (2025). Advances in sustainable biomaterials: characterizations, and applications in medicine. *Discover Polymers 2025 2:1, 2*(1), 2-. DOI: 10.1007/s44347-025-00014-8

Wang, M., Gao, B., Wang, X., Li, W., & Feng, Y. (2022). Enzyme-responsive strategy as a prospective cue to construct intelligent biomaterials for disease diagnosis and therapy. *Biomaterials Science, 10*(8), 1883–1903. DOI: 10.1039/D2BM00067A PMID: 35293402

Wang, X., Rivera-Bolanos, N., Jiang, B., & Ameer, G. A. (2019). Advanced Functional Biomaterials for Stem Cell Delivery in Regenerative Engineering and Medicine. *Advanced Functional Materials, 29*(23), 1809009. DOI: 10.1002/adfm.201809009

Wang, Y., Li, G., Yang, L., Luo, R., & Guo, G. (2022). Development of Innovative Biomaterials and Devices for the Treatment of Cardiovascular Diseases. *Advanced Materials, 34*(46), 2201971. DOI: 10.1002/adma.202201971 PMID: 35654586

Wasyłeczko, M., Sikorska, W., & Chwojnowski, A. (2020). Review of Synthetic and Hybrid Scaffolds in Cartilage Tissue Engineering. *Membranes 2020, Vol. 10, Page 348, 10*(11), 348. DOI: 10.3390/membranes10110348

Zhang, N., Cheng, Y., Hu, X., & Yeo, J. (2019). Toward rational algorithmic design of collagen-based biomaterials through multiscale computational modeling. *Current Opinion in Chemical Engineering, 24*, 79–87. DOI: 10.1016/j.coche.2019.02.011

Zhao, Z., Ukidve, A., Kim, J., & Mitragotri, S. (2020). Targeting Strategies for Tissue-Specific Drug Delivery. *Cell, 181*(1), 151–167. DOI: 10.1016/j.cell.2020.02.001 PMID: 32243788

Zhuo, S., Zhang, F., Yu, J., Zhang, X., Yang, G., & Liu, X. (2020). pH-Sensitive Biomaterials for Drug Delivery. *Molecules 2020, Vol. 25, Page 5649, 25*(23), 5649. DOI: 10.3390/molecules25235649

Chapter 5
Smart Biodegradable Implants and Transient Biosensors Evolution:
Security Challenges and Future Directions

Rachana Yogesh Patil
http://orcid.org/0000-0001-9973-6998
Pimpri Chinchwad College of Engineering, India

Yogesh Patil
http://orcid.org/0000-0001-9968-8287
D.Y. Patil College of Engineering, Akurdi, India

ABSTRACT

Smart biodegradable implants and transient biosensors are emerging as key enablers of next-generation healthcare sys-tems, combining bioactivity, controlled degradation, and real-time monitoring. This chapter presents a bibliometric and conceptual analysis of research on smart implantable systems from 2005 to 2025, with a particular focus on bioresorb-able electronics and biodegradable biomaterials. Using 201 Scopus-indexed publications, the study examines temporal trends, key contributors, and thematic evolution, while identifying critical gaps in security, long-term performance, and clinical translation. The findings highlight a growing disconnect between advances in biodegradable materials and the development of secure, integrated healthcare systems. This chapter emphasizes the need for inter-disciplinary ap-proaches that combine biomaterials science, bioelectronics, and cybersecurity to enable safe, transient, and intelligent implantable devices.

DOI: 10.4018/979-8-2600-0060-1.ch005

INTRODUCTION

The healthcare sector has undergone a major revolution especially due to the integration of smart implants and biosensors. Many of these technologies are revolutionizing healthcare practice and delivery by changing the way healthcare professionals diagnose (Rooney et al., 2019), constantly assess, and manage patients, through a more patient-centered approach driven by the use of information. As the global smart healthcare market expectedly to grow at a CAGR of 19.2% in the next 6 years to reach USD 432.5 billion (Sarwal et al., 2021), these technologies are fast becoming an integral part of any current and future healthcare system. But as these technologies advance, security issues that need to be solved to protect both patient safety and privacy are put at the forefront (Kasoju et al., 2023). This section explores the functions of smart implants and biosensors in today's healthcare, the prospect they hold, and the increasing necessity of security.

Recent advances in biodegradable and transient biomaterials have accelerated the development of smart implantable biosensors and bioresorbable healthcare devices. These systems combine sensing, monitoring, and therapeutic functionalities with controlled degradation characteristics, enabling safer and patient-centric healthcare solutions without the need for secondary surgical removal.

1.1 Smart Implants: Revolutionizing Medical Treatment and Monitoring

Smart implants are sophisticated medical devices that are intended to be placed within the human body and to interface with biological systems in order to bolster or replace, monitor, control or cure multiple pathologies Mohammadi et al., 2024). Smart implants are not like the passive implants that are conventionally used and are equipped with sensors, actuators and wireless communication systems for monitoring and delivering real time data (Raymond et al., 2024). Some of the most common types of smart implants include:

- **Cardiac Implants:** For instance, implantable cardioverter-defibrillators, devices to control the heart rates, such as pacemakers (Bussooa et al., 2018). These implants apart from regulating heart rate can also recognize and manage arrhythmias by giving out shocks that help regulate the heart.
- **Orthopedic Implants**: Implanted prosthesis/ joints are intelligent and can track the healing and movement of the body. Many of these implants contain monitoring features such as sensors that are capable of identifying a patients ability to move around and the data is relayed back to doctors to modify the

patients rehabilitation schedule or monitor the effectiveness of the implant (Ledet et al., 2018).

- **Neurostimulation Implants:** Implements applied in the management of neurological disorders including Parkinson's disease, epilepsy and chronic pain. These implants give electrical signals to the certain brain area or nerve; they are designed to mitigate symptoms and enhance patients' daily experience (Abyzova et al., 2023).

Smart implants are also gradually finding their way in biomedical engineering applications to diagnose chronic diseases, administer drugs and treat patients. These devices can be adjusted to deliver drugs in certain amount according to the real-time physiological parameters, which adds to the therapeutic functions.

Nevertheless, smart implants have a lot of benefits, but their vulnerabilities are numerous. Because they transmit signals to outside equipment or care givers, they are susceptible to hacking, eavesdropping, and signal jamming (Gil et al., 2022). Said security breaches could not only result in loss of patient information but also in disruption of the functions of the device, which poses a threat to patient's lives.

1.2 Biosensors: Enabling Real-Time Health Monitoring

A biosensor is an analytical device that integrates biological or biochemical components with electrochemical systems for the identification of biochemical or biological indicators. These sensors are usually compact, transportable and can supply data in real-time and are thus very useful in the area of Personalized Medicine. It can identify a number of associated biological parameters, such as glucose, oxygen, and cholesterol, and enable patients and care givers to track their health status in real time. Some notable applications of biosensors include:

- **Glucose Monitoring:** CGM systems have brought about a major change in diabetes management through their ability to monitor the patient's glucose levels in real time. These biosensors give constant information to the patient regarding their blood glucose level and hence avoid complications arising from diabetes (Heo & Takeuchi, 2013).
- **Wearable Health Devices**: Biosensors found with fitness trackers, smartwatches, and heath monitoring patches to monitor signs such as blood pressure, respiratory rate, SpO2, temperature, and movement (Sharma et al., 2021). These devices continue to gather information about a patient's health, thus helping to monitor for signs of possible future health problems.
- **Point-of-Care Diagnostics**: They are also used in the rapid diagnostic tests which are tests that give real time results for infections, blood biomarkers or

any other medical condition. This is especially so given that many such areas may not have established health facilities that are required to support the implementation of such solutions (Ghafar-Zadeh, 2015).

Being real time devices that offer information on the overall health status of a patient, biosensors are essential in enhancing the accuracy of diagnosing a disease and health monitoring. But with the growth of their usage, the questions about the data quality and privacy arise. These devices collect and transmit personal health information and can be hacked, breached, or even manipulated so that the data they provide will be incorrect and can lead to detrimental health decisions.

1.3 Security Challenges in Smart Healthcare Systems

The advances in smart implants, biosensors, and other Industries 4.0 tools have provided a basis for highly interdependent healthcare systems. These systems are cloud-based with data analytics and remote monitoring and are very effective but expose a large attack surface (Patil, 2024). The interconnectivity of healthcare devices raises several security concerns:

- **Data Privacy and Confidentiality**: Health data is undoubtedly some of the most personal details that a person can provide. Its exposure can be dangerous since intruders facing it can cause major problems such as identity theft, insurance fraud, blackmailing among others (Nidhya et al., 2022). Since the devices are collecting and transmitting data they come under some laws such as HIPAA in USA and GDPR in Europe which regulates on how personal health information to be managed.
- **Cyber Threats and Attacks**: Recently, more and more often, healthcare organizations become the target of cyber attacks, such as ransomware and malware attacks on medical devices. Such attacks can affect the normal performance of smart implants or biosensors, patient services, and cause delays in important treatments (Patil et al., 2024).
- **Device Interoperability**: Many healthcare facilities have on their inventory devices from different manufacturers thereby resulting in variations in device standards and security measures (Abounassar et al., 2022). This lack of standardization can create vulnerabilities that will be exploited in the way of communication between devices which can lead to havoc among the system.
- **Real-time Data Reliability**: When the data is coming from smart implants and biosensors, the real-time data makes the integrity of that data paramount. An infected device may give wrong results that can result in wrong diagnoses, wrong treatments or failure to alert users of emergent health dangers.

1.4 Evolution of Smart Implants, Biosensors, and Security in Healthcare

The advancement in smart implants, biosensors, and healthcare security started in the 1950s and has advanced to future trends expected in the 2030s and the future. Every age has been characterized by new breakthroughs that show that there is a close link between technology and health care. Early Developments biosensors, and healthcare security has undergone significant advancements over the decades, beginning with early developments in the 1950s and continuing into future trends expected in the 2030s and beyond. Each era has been defined by innovative breakthroughs, reflecting the growing synergy between technology and healthcare as shown in Figure 1.

- **Early Developments (1950s–1970s)**

The journey began in the 1950s to 1970s, an era that laid the foundation for modern healthcare technology. One of the most notable inventions was the first pacemaker in 1958, which transformed cardiac care. During this time, early glucose biosensors were introduced, offering initial solutions for monitoring diabetes (Andreu-Perez et al., 2015). Orthopedic and cochlear implants also emerged, demonstrating the potential of integrating mechanical and electronic devices with the human body. These early innovations provided life-saving and life-enhancing functionality, paving the way for future developments.

- **Advancements in Biosensors (1980s–1990s)**

Biosensors went through a lot of improvements in the 1980s and the 1990s. There were constant blood glucose monitoring through glucose biosensors and new oxygen level measurement through pulse oximeters (Batista et al., 2021). During this time advances in the field of biocompatible materials were made so that implants could be safely and successfully incorporated into a patient's body thus enhancing the patient's quality of life. It was an era of accuracy, safety and utility of diagnostics and other instruments

- **Smart Implants Era (1990s -2000s)**

The 2000s can be referred to as the Smart Implants Era when devices are connected with the help of the Internet of Medical Things (IoMT). When connected orthopedic implants and other related IoMT devices started reporting real-time health status, the health monitoring and management improved in efficiency. Nevertheless, the connection of these devices created security threats because unauthorized parties could use wireless links to interfere or obtain confidential patients' information. This has underlined the fact that healthcare technologies have become more complex due to the two demanding aspects of innovation and security

Figure 1. Timeline of the evolution of smart implants, biosensors, and security in healthcare

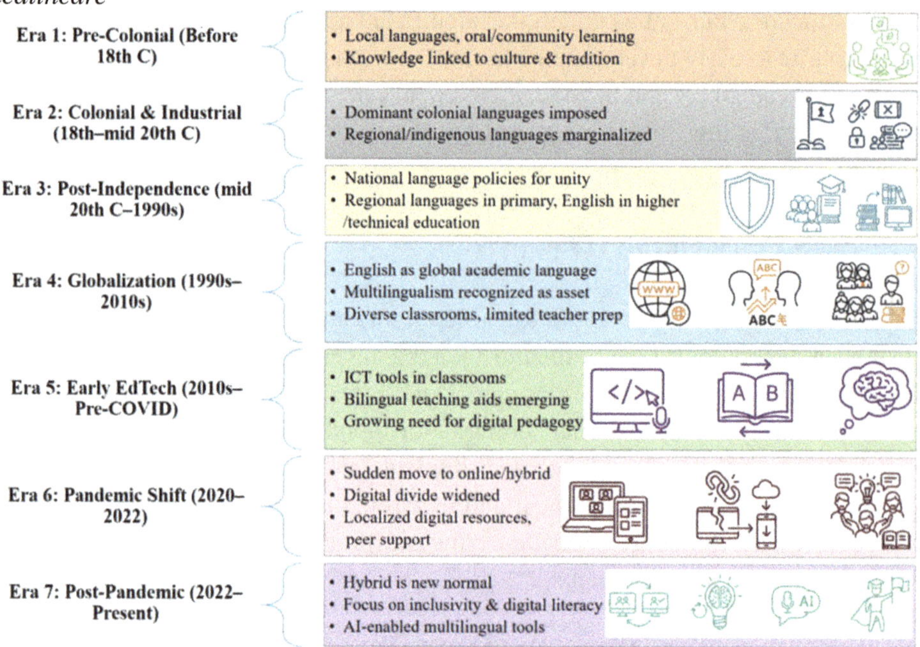

- **IoMT and Biosensor Renaissance (2000s- 2010s))**

The 2010s saw an IoMT and Biosensor Boom where the usage of small wearable devices and neural implants increased remarkably. Advanced systems were developed basing on AI which improved diagnosis, and provided tailor-made treatments. However, with the advancement of IoMT devices several new challenges like data breaches and unauthorized access emerged as the major threats (Patil et al., 2024).

- **AI, Blockchain, and Security (2010s-2020s)**

In the 2020s, the healthcare industry embraced technologies like AI and blockchain to address the growing challenges of innovation and security. AI-powered implants enabled smarter, real-time decision-making for personalized care, while blockchain technology offered secure, decentralized methods for safeguarding patient data and device communications. Governments and organizations implemented global cybersecurity regulations to mitigate the risks associated with the growing interconnectedness of medical systems (Deore et al., 2021; Deore et al., 2022). This era reflects a concerted effort to balance innovation with the need for robust security measures.

- **Future Trends (2030s and Beyond)**

Considering the trends for the further years of the 2030s and even more, the development of smart healthcare will be oriented to sustainability, precision, and even higher levels of security. It is thought that the biodegradable implants will have shorter-term effects on the environment while still being effective. Multiple biosensors will improve diagnosis by providing simultaneous results of various health parameters (Gupta et al., 2019). Meanwhile, quantum-safe cryptographic systems will solve the security issues that quantum computing will bring in the future to health care systems.

Smart implants, biosensors and security in Health care show that innovation is an ever-changing process that is constantly threatened by the need for security. Conventionally, right from the advancement in the pacemakers to the advent of quantum-safe technology all such changes symbolize how far the world has traveled and how much more is there to go in this vital field.

2. LITERATURE REVIEW

Recent research has shown significant advances in the design of smart implantable and biosensors that are transient and biodegradable for monitoring applications in the medical field, electrotherapy, neural interfacing, and biochemical sensing (Alam et al., 2024). As shown in Table 1, the recent research trends show that the integration of biodegradable materials, wireless communication technologies, sensing and therapeutic functionalities are currently the major research focus in the field of transient implantable platforms (Cho et al., 2024). The comparative analysis also reveals that biodegradable polymers, bioresorbable metals, and transient electronic materials are being used more and more to aid in the real-time monitoring and personalisation of therapeutic interventions in next-generation healthcare systems.

Table 1. Comparative analysis of recent research in smart biodegradable implants and transient biosensors

Study	Application	Key Functionality	Major Contribution	Limitation
(Alam et al., 2024)	Gastric sensor	pH monitoring	Early leakage detection	Limited lifetime
(Cho et al., 2024)	Deep-tissue monitoring	Ultrasonic sensing	Real-time tissue monitoring	Complex fabrication
(Choi et al., 2022)	Electrotherapy	Sensing + therapy	Closed-loop healthcare system	Security concerns
(Duttagupta et al., 2023)	Wound healing	Impedance sensing	Monitoring with stimulation	Limited degradation control
(Hu et al., 2024)	Neural implant	Neural recording/ stimulation	Advanced neural interface	Signal instability
(Kaveti et al., 2023)	Glucose biosensor	Continuous glucose sensing	Biodegradable glucose monitoring	Calibration drift
(Kwarteng & Cebe, 2022)	Surgical mesh	Pressure sensing + drug delivery	Multifunctional implant	Limited clinical validation
(Li et al., 2023)	Energy storage	Wireless charging	Implantable power system	Partial biodegradability
(Li et al., 2024)	Transient pacemaker	Leadless pacing	Miniaturized pacemaker	Manufacturing complexity
(Liu et al., 2024)	Chemical biosensors	Biomarker sensing	Review of transient sensing	Selectivity challenges
(Sheng et al., 2023)	Implant security	Threat analysis	Cybersecurity overview	Limited biodegradable focus
(Zhang et al., 2025)	Secure communication	Key establishment	Lightweight security model	Limited transient applications

While much progress has been made in miniaturization, wireless control, and multiple therapeutic functions, there are still challenges in the technology and its ability to cross the translational valley of death (Choi et al., 2022). There are still some challenges with current studies, such as controlling the degradation of materials, maintaining reliable encapsulation, managing the power efficiently, reducing the variations in signals over time, and ensuring the device's performance is stable over time (Duttagupta et al., 2023; Hu et al., 2024; Kaveti et al., 2023). Besides, although wireless healthcare communication is a key aspect of transient implantable systems, security integration is relatively under-researched. The majority of existing studies are devoted to the device functionality and the performance of the biomaterial, while the development of lightweight cryptographic frameworks, secure communication architectures and privacy preserving mechanisms that are appropriate for resource-constrained, biodegradable, implantable devices is somewhat neglected (Kwarteng & Cebe, 2022).

The literature also shows an increasing shift from passive biodegradable implants to more intelligent and bioactive healthcare systems that are able to sense, monitor, communicate and treat (Li et al., 2023; Li et al., 2024).

Recent advances in transient electrotherapy platforms, wireless bioresorbable sensors, and implantable biochemical monitoring systems suggest that future healthcare technologies will continue to be increasingly dependent on the development of biomaterials science, bioelectronics, wireless communication, and secure healthcare informatics (Liu et al., 2024).

The overall findings from these studies indicate that the following areas are worthy of further research: development of programmable biodegradable biomaterials, development of energy efficient wireless architectures, development of integrated sensing-therapy systems, and development of secure transient healthcare platforms that can support safe, adaptive, and patient-centric medical applications (Sheng et al., 2023; Zhang et al., 2025). Moreover, improved inter-disciplinary interaction between biomaterials research, biomedical engineering, clinical practice, and cyber security is crucial to overcome current challenges and clinical translate smart transient biosensors and implants.

Biosensors are increasingly becoming part of the smart implantable devices that are used in the medical field, which further emphasizes the demand for safe, reliable and clinically adaptable healthcare systems. These technologies can practically enable real-time patient monitoring, tailor-made treatment, decrease surgical burden, and enhance post-surgery care. But as more and more wireless communication and implantable sensors become integrated, there are also patient privacy, cybersecurity, interoperability, and regulatory compliance concerns. Thus, future healthcare strategies and policies should emphasize: the standardization of security measures, guidelines for safe deployment, ethical guidelines for patient data handling, and regulations for biodegradable and transient implantable devices. For clinical use of these technologies to become widely available, working together with researchers, healthcare professionals, industry and policy makers will be crucial.

3. RESEARCH QUESTIONS (RQS) AND OBJECTIVES

This section outlines the research questions (RQs) that guide the investigation, focusing on the role of Smart Implants and Biosensors in Secure Healthcare Systems. The following RQs and objectives will serve to explore key aspects of this domain, particularly its temporal evolution, influential contributors, research methodologies, and areas that require further exploration:

- **RQ1: How has the volume of research on smart implants and biosensors in secure healthcare systems evolved over time?**

This research question aims to examine the temporal trends in the volume of publications concerning smart implants and biosensors within the context of secure healthcare systems. Specifically, it will analyze the growth or decline in the number of research papers published over time, identifying periods of increased interest or focus in this field. The objective is to chart how research in this domain has evolved, tracking developments in both the technological and security aspects of smart healthcare solutions.

- **RQ2: Who are the most influential authors, journals, and institutions in the field of smart implants, biosensors, and secure healthcare systems?**

RQ2 focuses on identifying and analyzing the most influential authors, journals, and institutions contributing to the study of smart implants, biosensors, and their integration into secure healthcare systems. By examining citation patterns and publication metrics, the objective is to uncover key thought leaders and leading publications that have shaped the direction of research in this area, highlighting those who have made significant contributions to the advancement of both healthcare technologies and their secure implementation.

- **RQ3: What are the predominant thematic areas and methodologies in the research on smart implants, biosensors, and secure healthcare systems?**

This research question explores the thematic areas and methodologies prevalent in the literature on smart implants and biosensors within the scope of secure healthcare systems. The objective is to identify the core research topics—such as biometric sensing, data security, health monitoring, and integration with IoT systems—and the primary methodologies applied in these studies, such as data encryption, machine learning algorithms, and wireless communication protocols. Understanding the trends in these areas will provide insights into the evolving research focus and the technological approaches being adopted to address security concerns in healthcare systems.

- **RQ4: What gaps exist in the current research on the smart implants, biosensors, and their role in secure healthcare systems?**

The final research question focuses on identifying gaps in the current body of research related to smart implants, biosensors, and secure healthcare systems. By critically reviewing the existing literature, the objective is to pinpoint underexplored areas, such as long-term security challenges, privacy issues, or the real-time application of biosensors. This will guide the identification of research opportunities where further investigation is needed, paving the way for future innovations and developments in the field of secure healthcare technologies.

3.1 Dataset Overview and Key Metrics

Figure 2. Dataset overview and key metrics

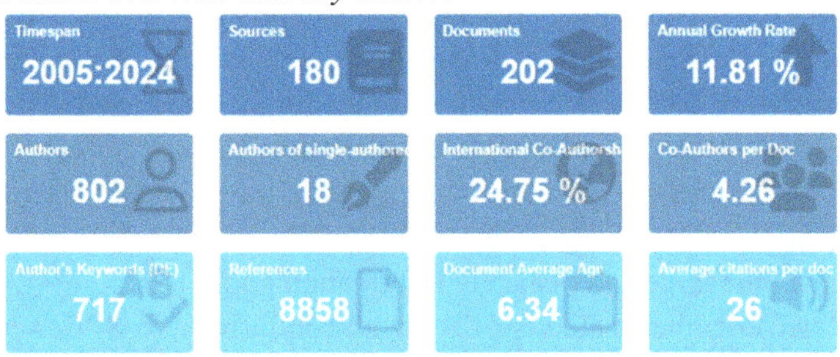

The bibliometric analysis focuses on research trends related to "Evolution of Smart Implants and Biosensors in Secure Healthcare Systems," using data extracted from Scopus for the period 2005–2024. The study encompasses a total of 202 documents, sourced from 180 distinct publications, highlighting the breadth of research in this interdisciplinary field. With an annual growth rate of 11.81%, the field demonstrates a steady increase in scholarly interest and contributions.

Key insights are visualized in Figure 2, which highlights metrics such as the dataset involves the contributions of 802 authors, with only 18 documents authored by single researchers, indicating a strong emphasis on collaborative research. This is further supported by an international co-authorship rate of 24.75%, showcasing the global nature of studies in secure healthcare systems. On average, there are 4.26 co-authors per document, reflecting the complexity and interdisciplinary expertise required in this domain.

Author keywords are central to bibliometric analysis, and the dataset contains 717 unique author keywords that cannot but give a sense of the thematic variety in the field. Also, the dataset includes 8,858 references which proves the richness of the previous literature that researchers have relied on. The average document age of 6.34 years indicates that there is still room for foundational studies, but also comparatively recent developments. The scrutiny also confirms the high concern and applicability of this research by averaging 26 citations per document.

4. METHODOLOGY

The methodology reflected in the Figure 3 is the structured approach to conducting the bibliometric analysis in order to reveal the trends and the gaps in the area of biosensors, smart implants, and data privacy in healthcare (Guo et al., 2020; Jalali et al., 2019). The process starts with a search query entered into Scopus which is designed to pull up the right publications. The query is constructed by using biosensors, smart implants, implantable devices, data privacy and security within the healthcare domain. The analysis is limited to the publications that were published between the years 2004 and 2025 and only in the English language. After applying these filters, documents that are not in English such as Chinese, Czech, French, and German papers are removed, and the remaining the dataset contains 210 documents. They make the selection that guarantees the further analysis of the relevant set of data rather than the large and disparate information pool.

The selected documents are visualized with Biblioshiny, VoS Viewer, and Raw Graphs, which are bibliometric analysis tools. These tools assist in the presentation of the publication trend, collaboration, and thematic map. To that end, the analysis is guided by four over-arching research questions (RQs) that each correspond to a particular aspect of the research space.

RQ1 aims at investigating temporal patterns of publications. It analyses trends in the number of publications, participation of different countries, main journals that focus on this subject, and author and publication profiles. RQ2 explores popular authors, journals, and institutions. First, it defines the main figures and organizations in the field, investigates author collaboration networks, and reveals highly cited journals and institutions.

RQ3 focuses on the thematic areas and methodologies. It provides an insight of the clusters of emerging topics, temporal changes in themes and keyword frequency analysis to determine focus research areas. Finally, RQ4 aims at revealing the gaps in the literature and offering recommendations on how these gaps can be filled, so that it outlines a research agenda.

We have followed this comprehensive methodology in writing our book chapter. By adopting this approach, we ensured a detailed understanding of the research landscape, highlighting significant trends and gaps. This structured analysis forms the foundation of our study, offering valuable insights for advancing knowledge in this domain.

Figure 3. Methodology of the study

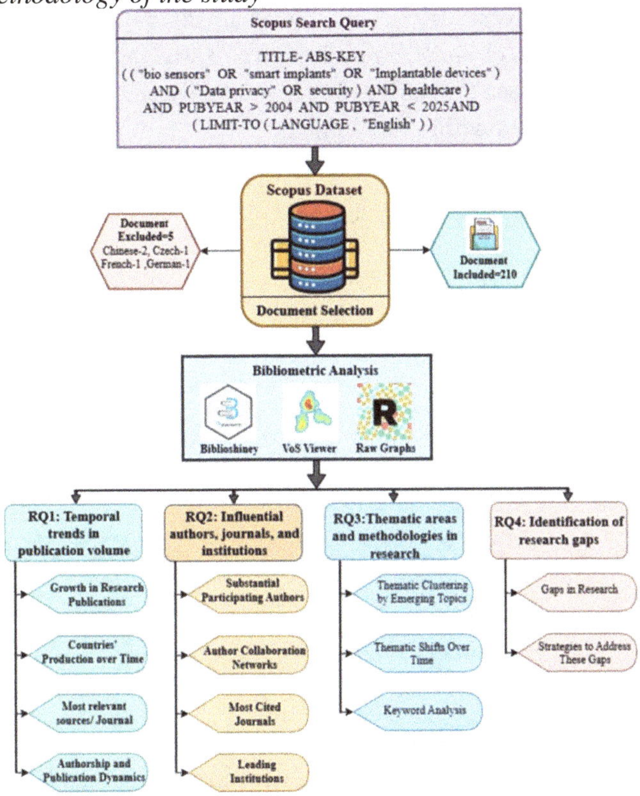

5. TEMPORAL TRENDS IN PUBLICATION VOLUME (RQ1)

5.1 Analysis of Growth in Research Publications

Figure 4 shows the dynamic growth trajectory in the research domain of smart implants and biosensors for secure healthcare system from 2005 to 2024 based on the document frequency. Altogether, the total number of articles published for this investigation has revealed the increasing trend of interest in the academic and industrial sectors.

In the first three years (2005–2007), the field was still quite young, and there was comparatively little published (3, 1, and 5 articles, respectively). These are early exploratory studies which on average have a low citation rate indicating low dissemination and impact. Specifically, 2008 has indicated the growth of interest in the field, with 8 articles published and an average citation per article equal to 2.87.

The publication numbers are moderate and have been fluctuating slightly from 3 to 9 per year from the year 2009-2014. The average citation rates in this period fluctuated like 6.20 in 2011 and 1.86 in 2013, yet it was inconsistent yet effective research activity. This phase probably refers to gradually progressive improvements and various topical studies within the domain.

Figure 4. Analysis of growth in research publications

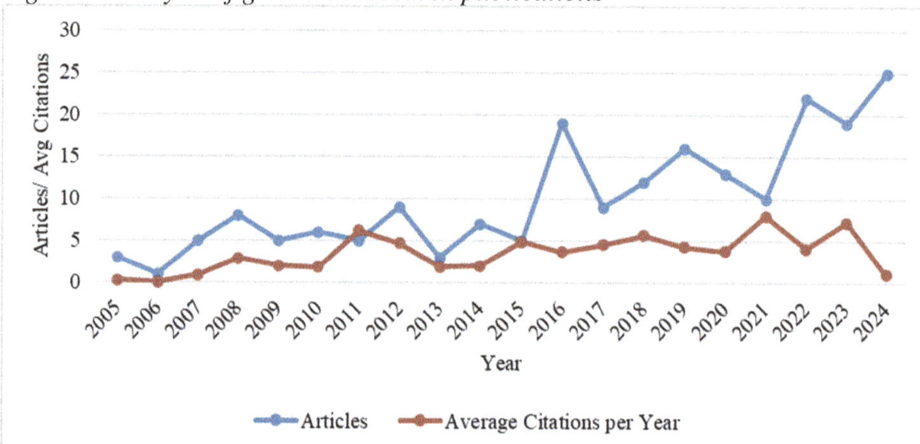

The number of publications rose notably after the year 2015 and 2016 registered the maximum number of articles with 19. This table is also significant for the average citation rates after 2015, such as 4.86 in 2015 and 5.68 in 2018, which indicates scholarly research contribution of high citations. It could be inferred that the year 2021 was dominated by influential papers, as the average citation rate reached 8.00.

Over the last few years, the trend shows that the number of publications has been on the rise, and in 2024, there were 25 articles published, the maximum number. Still, the average citation per year for 2024 (1.08) is a sign that the publications are not well-cited at the moment, probably because they are recent. This observation goes a long way in showing us that we have to be careful when comparing the growth trends of different research outputs with the quantity and quality of the researches.

The analysis of the data presents two trends for the development of the publications: the increase in the number of publications and the increase in the citation rate. As shown in citation averages in a particular year such as 2021, there are probable achievements or a few significant studies to have elicited interest in the specified domain. On the other hand, below par citation rates in productive years such as 2024 show that citations take time to gather for new publications.

In general, the temporal aspect reveals that context of smart implants and biosensors in secure healthcare is progressing at a very fast pace and is shifting from the stage of exploring novel possibilities in the society to becoming a well-defined and actively investigated subject. This evolution is illustrated in Figure 1, which compares the dynamics of publication output and citation rate.

5.2 Countries' Production over Time

This figure 5 shows the steady trend of research activity in area of smart implants and biosensors in secure healthcare systems during 2005 to 2024. Firstly, as shown in Figure 2, the research output was relatively low from 2005-2010 owing to a nascent stage of the research field. This was a time characterized by relatively few technological changes and less emphasis placed on applying technology to health care organizations.

Growth was detected from 2011 through 2015, especially in the USA, UK, and India and it came about as the health care industry started adopting technological advancement such as smart implants and biosensors. This steady increase in these years was attributed to early stage R&D, greater investment and enhanced emphasis on health security.

With the increasing demand for robust and reliable healthcare systems, there has been a steep rise in the number of publications between 2016 and 2020. This rise was also backed up by the rise in investments in health technological advancements and partnerships between countries such as China and India which started to experience growth at this time.

Figure 5. Countries' production over time

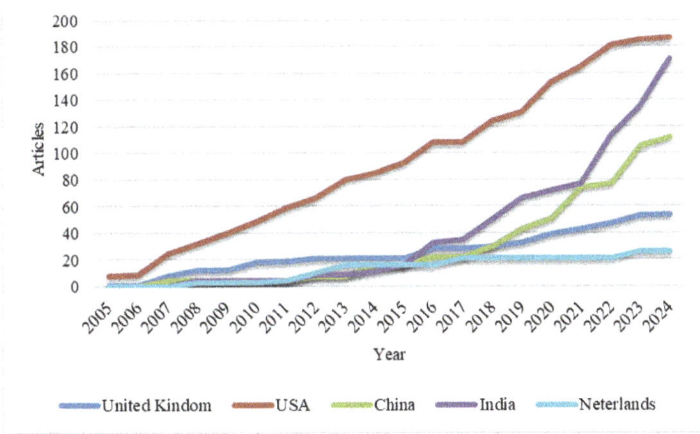

The studies said that the production of research was at its highest for the period 2021 to 2024 especially in China and India. The increase of the output in these years can be explained by the technological development and the need for safe health-care systems, and pandemic. As for authors, researchers from emerging economies like China and India are slowly but surely catching up with the US and UK counterparts.

Figure 6. Country collaboration map

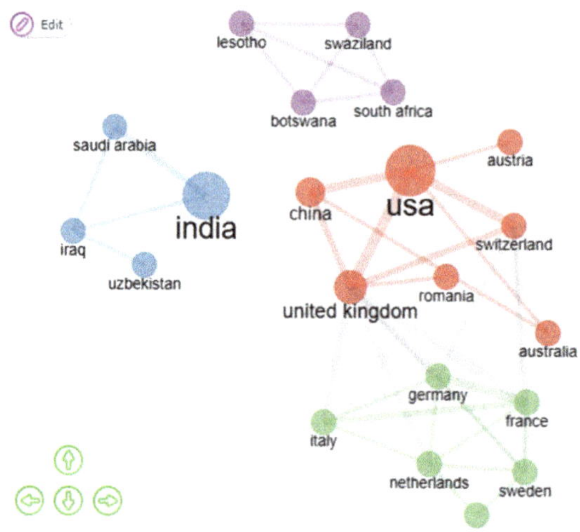

Table 2. Countries and their collaborative index

Node	Cluster	Color	Betweenness	Closeness	PageRank
usa	1	Red	17.9775	0.005319	0.074913
china	1		2.6625	0.005263	0.041588
united kingdom	1		33.14566	0.005464	0.112637
switzerland	1		3.416595	0.005348	0.051828
australia	1		0	0.005051	0.02249
romania	1		0	0.005155	0.013746
austria	1		0	0.005025	0.014634
india	2	Blue	0	0.00277	0.051118
iraq	2		2	0.002778	0.063193
saudi arabia	2		0	0.00277	0.051118
uzbekistan	2		0	0.002762	0.025048

continued on following page

Table 2. Continued

Node	Cluster	Color	Betweenness	Closeness	PageRank
netherlands	3	Green	5.409091	0.005319	0.052554
germany	3		0.601003	0.005319	0.060557
france	3		0.806266	0.005319	0.064018
italy	3		0	0.005263	0.039957
spain	3		0	0.005076	0.020978
sweden	3		4.981385	0.005319	0.049147
south africa	4	Purple	0	0.002778	0.047619
botswana	4		0	0.002778	0.047619
lesotho	4		0	0.002778	0.047619
swaziland	4		0	0.002778	0.047619

Table 2, offer an analytical view of collaborations by partitioning nations into roles and power affiliation. The measures: Betweenness, Closeness, and PageRank provide information about how central or peripheral each nation is in the global network as shown in Figure 6.

Cluster 1 (Red) includes the USA, China and the United Kingdom which are world's super powers. The United Kingdom is the most central country in the cluster with Betweenness of 33,14566 and PageRank of 0.112637. USA comes next with a Betweenness of 17.9775, and PageRank of 0.074913 further placing it in a more strategic mid-position. Switzerland is in the moderately active position with Betweenness value of 3.416595 and China is also moderately active with Betweenness value of 2.6625. On the other hand, some countries with Betweenness equals to zero indicates that they have peripheral roles in collaborations despite being in the cluster, these countries include Australia, Romania and Austria.

Cluster 2 (Blue) has such new entrants as India, Iraq, Saudi Arabia, and Uzbekistan. The Betweenness values in this cluster are also considerably lower; Iraq's value of 2.0 means it has little involvement as an intermediary. India and Saudi Arabia are still in the position of regional power, although their Betweenness is quite low, 0.051118. Uzbekistan is situated on the periphery of the network and has a PageRank of 0.025048 and zero Betweenness. These low Closeness values especially like the value of 0.00277assigned to India show that these nations are more concerned with regional cooperation than international.

Cluster 3 (Green) is the most compact and consists of the European countries such as Netherlands, Germany, France, Italy, Spain, and Sweden. The Netherlands is with the Betweenness centrality of 5.409091 and PageRank of 0.052554 an important in-between point. Sweden again emerged with its Betweenness of 4.981385 and the PageRank of 0.049147 in affirming its position as a bridge. Germany and France are the most influential sites within the cluster having PageRank values of 0.060557 and 0.064018, respectively. On the other hand, Italy and Spain with the PageRank values equal to 0.039957 and 0.020978, respectively, and zero Betweenness, seem to the play a more limited role in the network.

Cluster 4 (Purple) consists of African countries like South Africa, Botswana, Lesotho and Swaziland. All these countries have the same Closeness of 0.002778 and the same PageRank of 0.047619 which underlined that they have the middle rank connectivity and influence. The lack of Betweenness for all the members of this cluster suggests that they do not occupy any intermediary positions and therefore are likely to be involved in only localized or regional cooperation. This cluster therefore highlights the need to enhance their connectivity into other global structures in order to enhance their leverage.

Key findings emerge from this analysis. The United Kingdom, with the highest Betweenness (33.14566) and PageRank (0.112637) across all clusters, is identified as the most central player in the network. The USA (Betweenness: 17.9775, PageRank: 0.074913) and the Netherlands (Betweenness: 5.409091, PageRank: 0.052554) also serve as vital hubs. Countries such as India and Saudi Arabia, despite low Betweenness, maintain moderate PageRank values, indicating potential for regional leadership. Cluster 4 nations, with zero Betweenness and uniform metrics, reflect the need for global engagement initiatives to amplify their roles. European nations in Cluster 3 highlight strong interconnectivity, with countries like Sweden and the Netherlands serving as key intermediaries.

5.3 Top 10 most relevant sources/ Journal Analysis

The analysis of the top 10 most relevant sources described in Table 3, reveals diverse contributions and varying levels of influence in the domain of smart implants and biosensors. Among these, a notable comparison can be made between the Proceedings of SPIE - The International Society for Optical Engineering and Biosensors and Bioelectronics. The Proceedings of SPIE has contributed four publications with a total of eight citations, reflecting a steady but modest impact. Its relatively low citation count indicates that its publications cater to a specialized audience, focusing on niche topics with limited broader visibility.

Table 3. The top 10 most relevant sources

Source	h_ index	g_ index	m_ index	TC	NP	PY_ start
IEEE ACCESS	4	5	0.444	521	5	2016
PROCEEDINGS OF SPIE - THE INTERNATIONAL SOCIETY FOR OPTICAL ENGINEERING	2	2	0.1	8	4	2005
ACM INTERNATIONAL CONFERENCE PROCEEDING SERIES	1	1	0.125	3	4	2017
ADVANCES IN INTELLIGENT SYSTEMS AND COMPUTING	2	3	0.222	18	3	2016
TRAC - TRENDS IN ANALYTICAL CHEMISTRY	2	3	1	20	3	2023
IFMBE PROCEEDINGS	1	2	0.063	7	3	2009
PROCEDIA COMPUTER SCIENCE	2	2	0.2	100	2	2015
SENSORS INTERNATIONAL	2	2	0.667	38	2	2022
BIOSENSORS AND BIOELECTRONICS	1	2	0.111	117	2	2016
IEEE INTERNATIONAL CONFERENCE ON COMMUNICATIONS	1	2	0.05	12	2	2005

In contrast, Biosensors and Bioelectronics, despite having only two publications, has achieved a remarkable 117 citations, highlighting its profound impact in the field. The high citation-per-publication ratio of this journal underscores the transformative nature of its contributions, which resonate strongly with the academic and research community. Each publication demonstrates significant quality and relevance, making it a critical driver of advancements in biosensor technology.

This comparison illustrates the varying dynamics of research influence among the top sources. While the *Proceedings of SPIE* provides a platform for emerging and exploratory studies, *Biosensors and Bioelectronics* focuses on publishing fewer but highly impactful works. This analysis underscores the importance of both specialized and high-impact journals in shaping the research landscape, where influence is determined not only by the volume of publications but also by their quality and contribution to the advancement of the field.

5.4 Authorship and Publication Dynamics Across Journals

The alluvial diagram in Figure 7 shows how authors, years of publications and titles of sources are interconnected and how research in smart implants and biosensors for secure heath care systems is multi-disciplinary and dynamic. Publications include those from 2008 to 2023, with authors Kim H.J, Hirayama K, and Li J making a major contribution.

Figure 7. Authorship and publication dynamics across journals

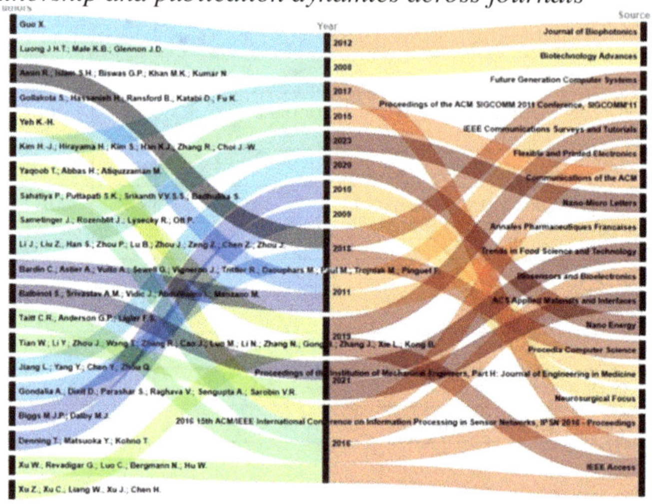

The distribution of publications through the years also shows that there has been an overall rise in the number of research conducted and published in this field especially in 2016, 2018 and 2023. Biosensors & Bioelectronics, Nano-Micro Letters, IEEE Communications Surveys & Tutorials represent a wide range of interdisciplinary context of this research area including sensor technologies, nanoscale, and secure communication systems. It is clearly seen that over time, research interest has shifted from the basics of material sciences and biophotonics, as identified in earlier works such as the Journal of Biophotonics published in 2012, to state-of-the-art flexible and printed electronics of the year 2023. Emerging journals, such as *Neurosurgical Focus* and *Nano Energy*, highlight specialized applications, including neurological monitoring and energy harvesting for implants. Overall, the diagram underscores the sustained growth and evolution of the field, driven by collaborative efforts across disciplines and a shift toward practical and system-level innovations.

6. INFLUENTIAL AUTHORS, JOURNALS, AND INSTITUTIONS (RQ2)

6.1 Substantial Participating Authors

Table 4 reveals critical insights into the contributions of the top 20 influential authors in the domain of Smart Implants and Biosensors in Secure Healthcare Systems. GUO XIAOWEI emerges as the most highly cited author, with 383 total citations (TC), despite having an h-index and g-index of only 1. This indicates that their influence is concentrated in a single highly impactful publication, which underscores the significance of their research but highlights the lack of breadth in their publication portfolio. Similarly, GLENNON JEREMY D., LUONG JOHN H.T., and MALE KEITH B. have garnered 343 citations each, also with an h-index and g-index of 1, emphasizing their niche but highly influential contributions to the field.

Table 4. Top 20 influential authors and impact scores

Author	h_index	g_index	m_index	TC	NP	PY_start
GUO XIAOWEI	1	1	0.077	383	1	2012
GLENNON JEREMY D.	1	1	0.059	343	1	2008
LUONG JOHN H.T.	1	1	0.059	343	1	2008
MALE KEITH B.	1	1	0.059	343	1	2008
FU KEVIN	2	2	0.143	305	2	2011
GOLLAKOTA SHYAMNATH	2	2	0.143	305	2	2011
HASSANIEH HAITHAM	2	2	0.143	305	2	2011
KATABI DINA	2	2	0.143	305	2	2011
RANSFORD BENJAMIN	2	2	0.143	305	2	2011
AMIN RUHUL	1	1	0.143	269	1	2018
BISWAS G.P.	1	1	0.143	269	1	2018
ISLAM SK HAFIZUL	1	1	0.143	269	1	2018
KHAN MUHAMMAD KHURRAM	1	1	0.143	269	1	2018
KUMAR NEERAJ	1	1	0.143	269	1	2018
YEH KUO-HUI	2	2	0.222	221	2	2016
CHOI JI-WOONG	1	1	0.125	205	1	2017
HAN KI JIN	1	1	0.125	205	1	2017
HIRAYAMA HIROSHI	1	1	0.125	205	1	2017
KIM HAN-JOON	1	1	0.125	205	1	2017
KIM SANGHOEK	1	1	0.125	205	1	2017
ZHANG RUI	1	1	0.125	205	1	2017

A group of authors which include FU KEVIN GOLLAKOTA SHYAMNATH HASSANIEH HAITHAM KATABI DINA and RANSFORD BENJAMIN also has higher productivity with h-index and g-index of 2 and total citation of 305 divided in two papers each. This speaks of a longer and perhaps more effective career in the field than those authors who have one highly cited article. These researchers are a valuable group who are extending the depth and width of the research in this field.

As seen from the PY_start of 2018, the below researchers are emerging contributors in the field: AMIN RUHUL, BISWAS G.P., ISLAM SK HAFIZUL, KHAN MUHAMMAD KHURRAM, and KUMAR NEERAJ. I only found one publication each for them and 269 citations; they contribute significantly within a short time and could be promising future scholars. Likewise, the authors like YEH KUO-HUI who has higher m-index of 0.222 and has started publishing from 2016 seem to have a consistent growth slope of highly cited contributions.

Lastly, the derived analysis reveals nine more authors who started writing in 2017: CHOI JI-WOONG, HAN KI JIN, HIRAYAMA HIROSHI, KIM HAN-JOON, KIM SANGHOEK, and ZHANG RUI; their total citations are 205. All three authors have similar bibliometric profiles within this group, which emphasise the interdependent and integrated nature of the research in this area at this level of citation while remaining relatively small in scale at this point.

In this analysis while some authors have shown how they dominate through individual pieces that receive many citations, there are others who have displayed sustained production and impact. The presented works indicate that emerging researchers are already contributing to the development of major breakthroughs in the field and the relatively short time in which they have been active demonstrates the fluctuating and progressing nature of the research in Smart Implants and Biosensors for Secure Healthcare Systems. Such a distribution of contribution highlights the need to both extend the number of topics and deepen the investigation of each in future studies.

6.2 Author Collaboration Networks and Institutional Affiliations

The co-authorship network shown in Figure 8 concerning smart implants and biosensors in secure healthcare systems reveals different clusters and different levels of collaboration among the authors. While on one side of the network, Chattopadhyay K.K., Nag A., Amin R., Islam S.H., Biswas G., and Garg A. are clustered, this shows that the authors work closely together and their research interests may overlap with each other. This cluster seems to be coherent and motivated, which might play a great role in enhancing certain more specialized domains.

Figure 8. The co-authorship network

On the other side of the network, the authors are Kumar M., Hussain S.Z., Chulerttiyawong D.M., and Jamalipoor who are linked through different collaborations. This cluster has moderate denseness of connections within the cluster and with other clusters of the network, depicting interdisciplinary or cross institutional collaboration. The link between the two major clusters represents an overlap or commonality between two otherwise separate research areas, that allows for flow of information and fusion of ideas.

The network structure reveals a balance between tightly-knit research groups and cross-disciplinary partnerships. The presence of bridging authors or clusters suggests opportunities for further collaboration to strengthen the overall network and promote innovation across the field of secure healthcare systems.

Figure 9. Institute collaboration networks

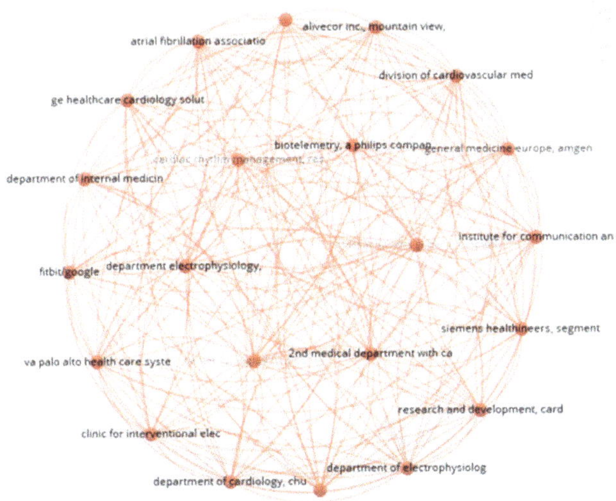

The analysis of institute collaboration networks as shown in Figure 9 for smart implants and biosensors in secure healthcare systems reveals a diverse and hierarchical research ecosystem. Leading institutions such as the School of Electrical Engineering and Automation and the School of Optoelectronic and Information Engineering stand out with the highest citation counts (383 each) and strong total link strengths (249.33), highlighting their central role and extensive collaborative impact. Mid-tier contributors, including the Massachusetts Institute of Technology (MIT) and the University of Massachusetts, Amherst, also show significant contributions with two documents each, garnering 305 citations and a moderate total link strength of 76. These institutions demonstrate active participation in the network with a focus on broader document output. Emerging players, such as the Department of Biomedical Engineering and the Epstein Department of Industrial and Systems Engineering, exhibit notable link strengths (217) despite moderate citation counts, signifying their growing influence in niche areas. Additionally, institutions like the National Dong Hwa University and the Division of Electronics and Electrical Engineering contribute effectively, with citations ranging from 205 to 214 and high link strengths near 200, indicating focused and impactful collaborations.

The network also showcases a global presence, with contributors like the École Polytechnique Fédérale de Lausanne, Nagoya Institute of Technology, and the University of Oklahoma reflecting the international interest in secure healthcare systems. Institutions such as the Roski Eye Institute and the Brunel Institute for Bioengineering further highlight the interdisciplinary nature of the research, bridging fields like electronics, biomedical engineering, and computer science. Overall, the collaboration trends suggest that a few leading institutions dominate the network, while mid-tier and emerging players support its expansion. Strengthening connections among these institutions could enhance the global research ecosystem, driving innovation in secure healthcare systems using smart implants and biosensors.

6.3 Most Cited Journals in the Field

Figure 10. Bradford's law

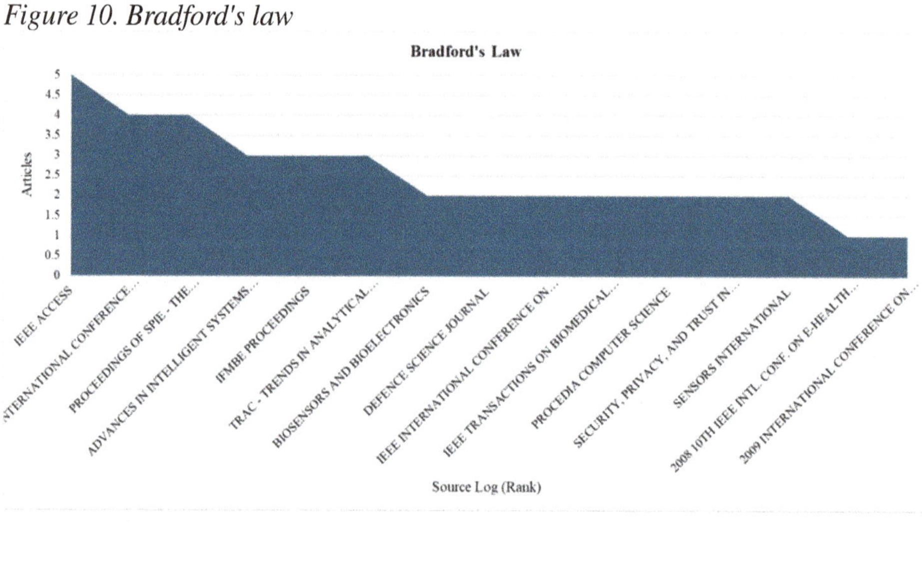

Bradford's Law as shown in Figure 10 provides a framework for analyzing the distribution of articles across journals and identifying core sources in a specific research domain. For the domain of Smart Implants and Biosensors in Secure Healthcare Systems, the dataset of Table 5 reveals that journals can be ranked based on their productivity, measured by the number of articles contributed. The rank, frequency, and cumulative frequency highlight the concentration of articles within a few core sources and their gradual decline across the remaining sources.

Table 5. Most cited journals

Source	Rank	Freq	cumFreq
IEEE ACCESS	1	5	5
ACM INTERNATIONAL CONFERENCE PROCEEDING SERIES	2	4	9
PROCEEDINGS OF SPIE - THE INTERNATIONAL SOCIETY FOROPTICAL ENGINEERING	3	4	13
ADVANCES IN INTELLIGENT SYSTEMS AND COMPUTING	4	3	16
IFMBE PROCEEDINGS	5	3	19
TRAC - TRENDS IN ANALYTICAL CHEMISTRY	6	3	22
BIOSENSORS AND BIOELECTRONICS	7	2	24
DEFENCE SCIENCE JOURNAL	8	2	26
IEEE INTERNATIONAL CONFERENCE ON COMMUNICATIONS	9	2	28
IEEE TRANSACTIONS ON BIOMEDICAL ENGINEERING	10	2	30
PROCEDIA COMPUTER SCIENCE	11	2	32
SECURITY, PRIVACY, AND TRUST IN WBANS AND E-HEALTHCARE	12	2	34
SENSORS INTERNATIONAL	13	2	36
2008 10TH IEEE INTL. CONF. ON E-HEALTH NETWORKING, APPLICATIONS AND SERVICE, HEALTHCOM 2008	14	1	37
2009 INTERNATIONAL CONFERENCE ON ULTRA MODERN TELECOMMUNICATIONS AND WORKSHOPS	15	1	38
2010 IEEE INTERNATIONAL CONFERENCE ON BIOINFORMATICS AND BIOMEDICINE WORKSHOPS, BIBMW 2010	16	1	39

The highest-ranking journal as shown in Table 4, IEEE Access (Rank 1), contributes the most articles (5), establishing it as the most influential source in this field. As the rank increases, the frequency of articles decreases, with journals ranked 2nd (ACM International Conference Proceeding Series) and 3rd (Proceedings of SPIE) each contributing four articles. These high-ranking journals collectively account for a significant share of the cumulative frequency, reaching 13 articles by Rank 3. This marks the end of the core zone, where most research contributions are concentrated.

The middle zone begins with sources ranked 4th to 6th, including Advances in Intelligent Systems and Computing, IFMBE Proceedings, and TRAC - Trends in Analytical Chemistry, each contributing three articles. The cumulative frequency reaches 22 articles by the 6th rank. This zone reflects moderately influential journals that still hold importance but contribute fewer articles than the core sources.

The peripheral zone includes journals with lower ranks (Rank 7 and beyond), such as Biosensors and Bioelectronics and Defence Science Journal, which contribute only two articles each. By Rank 15, journals like the 2008 10th IEEE International

Conference on E-Health Networking contribute just one article, bringing the cumulative frequency to 39. This diminishing trend in article contributions across ranks is also evident in the visual representation, which shows a steep initial drop in the number of articles followed by a slower decline as the rank increases.

The analysis of rank, frequency, and cumulative frequency aligns with Bradford's Law, emphasizing the importance of core journals while acknowledging the contributions of middle and peripheral sources. For researchers, focusing on high-frequency journals like IEEE Access and ACM International Conference Proceeding Series ensures access to the majority of relevant studies. At the same time, consulting peripheral journals may provide valuable niche insights to support innovative or interdisciplinary research.

6.4 Leading Institutions and Their Contributions

The analysis of over 10 leading institutions as shown in Figure 11 reveals a worldwide push in Smart Implants and Biosensors for Secure Healthcare Systems. The most productive university is the Massachusetts Institute of Technology (MIT), USA with 17 articles published proving the institute's contribution to the development of interdisciplinary science. Qilu University of Technology, China has published 11 articles and indicates China's increasing interest in healthcare technology.

Nine papers were contributed by Central South University, China, Purdue University, USA, and the University of Michigan-Shanghai Jiao Tong University Joint Institute, a U.S.-China venture highlighting the role of international collaboration Five papers were contributed by École Polytechnique Fédérale de Lausanne, Switzerland, and Lawrence Livermore National Laboratory, USA, which underlines the research output of technologically developed countries.

The Brazilian Universidade Federal do Rio de Janeiro published 7 articles and emphasizes Latin American experience, while BAM Federal Institute for Materials Research and Testing from Germany and Hubei University from China published 6 articles each and focus on regional experience. Such intercontinental integration demonstrates the advancements in the field of protected healthcare, with the United States, China, and other globe partners.

Figure 11. Leading institutions

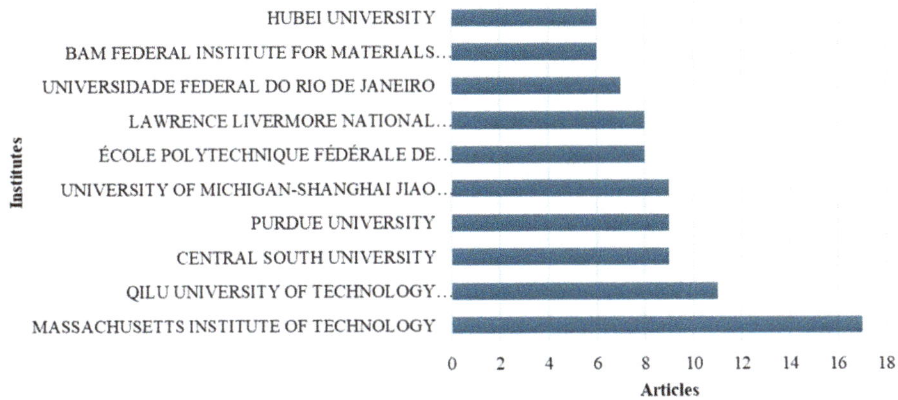

7. THEMATIC AREAS AND RESEARCH METHODOLOGIES (RQ3)

7.1 Identification of Major Research Themes

Thematic analysis shown in Figure 12 identifies and categorizes significant research themes within the domain of smart implants and biosensors for secure healthcare systems. The thematic clusters are organized based on their relevance (centrality) and development (density) into four quadrants: Motor Themes, Niche Themes, Emerging or Declining Themes, and Basic Themes. These clusters provide insights into the current focus and future potential of various topics in this field.

Motor themes, located in the upper-right quadrant, are both highly relevant and well-developed. Authentication emerges as a leading research focus, with high centrality (6.282) and density (672.696). This theme revolves around ensuring secure access to healthcare systems and devices, emphasizing the importance of user identity verification and data protection. Similarly, Monitoring (centrality: 4.394, density: 629.924) is critical for real-time health tracking using smart implants and biosensors, enabling better patient management. Bio-Sensors (centrality: 6.070, density: 605.058) represent another key focus, advancing sensor technologies for precise data collection and enhanced healthcare outcomes.

Figure 12. Thematic map

Niche themes, located in the upper-left quadrant, are well-developed but less central to the field. Simulation (centrality: 0, density: 505) explores virtual modeling and testing of healthcare systems, providing a foundation for understanding the functionality of smart implants. Biometrics (centrality: 0, density: 604.167) investigates the use of unique biological identifiers, such as fingerprints and retinal scans, to enhance system security. Similarly, Bio Sensors (centrality: 0, density: 540.625) emphasize specialized mechanisms for accurate data acquisition. These themes indicate deep research interest but have yet to achieve broader integration into mainstream healthcare systems.

Themes in the lower-left quadrant are either emerging areas of interest or declining topics. Human Body Communication (HBC) (centrality: 0, density: 450) explores the development of communication protocols for body-based networks, which could potentially improve data exchange between implants. Education (centrality: 0.917, density: 482.143) reflects the integration of biosensor technology into academic and training programs, supporting knowledge dissemination and innovation in healthcare technologies.

Basic themes, found in the lower-right quadrant, are highly relevant but underdeveloped. Security (centrality: 15.655, density: 443.730) is a core research focus, addressing the critical need for safeguarding sensitive health data from cyber threats. Internet of Things (IoT) (centrality: 14.945, density: 487.800) emphasizes the connectivity of smart devices in healthcare systems, paving the way for interoperability and seamless data integration. Blockchain (centrality: 1.920, density: 415.278) introduces decentralized security protocols, offering promising solutions to enhance data privacy and integrity.

7.2 Thematic shift over time

The evolution of research themes in smart implants and biosensors is illustrated in Figure 13, depicting a thematic progression from 1973 to 2025. This shift demonstrates how research priorities have transitioned from foundational concepts to more sophisticated and integrated technologies.

- **Early Research Themes (1973–2014)**

During the initial phase, research was predominantly centered on biosensors, healthcare, and implantable devices. These themes laid the groundwork for utilizing biosensors to monitor metabolites and other biological parameters, emphasizing their application in healthcare systems. The focus on human aspects also highlighted the foundational need to understand the interaction between technology and biology.

- **The expansion of applications is focused on the period between 2015 and 2019.**

From 2015-2019, the research area expanded and involve more area like network security, diagnosis and mobile devices. During this period, biosensors were incorporated into diagnosis and mobile health care systems. Surgically inserted smart devices appeared as the next subgroup of implants because of the developments in the implantable technologies. Electrocardiography integration pointed at biosensors usage in cardiovascular health assessment.

- **New Technologies and Directions (2020–2022)**

Analyzing the trend of the published research themes, it is possible to state that from 2020 to 2022 they were published more frequently focused on the specificity and the utilization of technologies. Thus, bio sensors and diagnosis were further explored, as well as other areas such as energy efficiency underlining the importance of long lasting and sustainable solutions in healthcare. Cryptography and healthcare were the two sub topics that depicted the increasing concern to protect the identity of the patients and the integrity of the systems in health care facilities while implants (surgical) was among the major themes, which underscore the progress made in the implant technology.

- **Present Trends and Future Developments (2023-2025)**

In the most recent period, research has move to higher and integrated systems. Human remains a major theme, which is a testimony to the fact that people have never lost their focus on the importance of human-oriented design and usability. New trends defined as Internet of Things and blockchain suggest the use of decentralized and integrated approaches in data protection and performance. There has also emerged e-learning which proposes incorporating biosensor technologies into educational systems for training of healthcare workers. Network security and bio sensors remains important, which reiterates the need to protect the privacy of health information and develop biosensors more.

Figure 13. The evolution of research themes

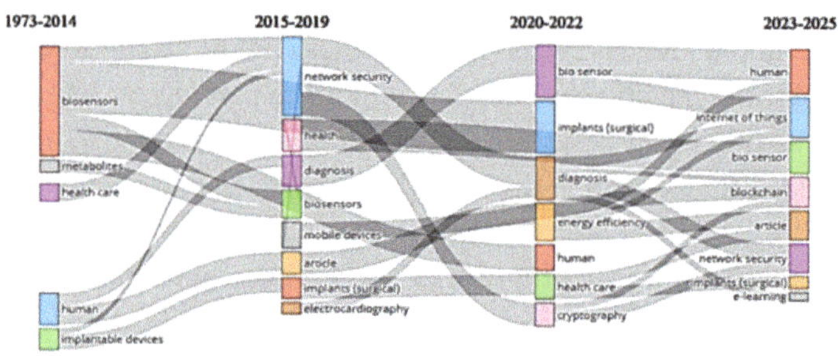

7.3 Keyword Analysis

The keyword analysis in Table 6 reflects the evolving focus of research in Smart Implants and Biosensors for Secure Healthcare Systems from 2014 to 2024. The trends show significant growth in keywords such as “Implants (Surgical)” and “Biosensors”, which are consistently dominant. The usage of “Implants (Surgical)” increased from 8 in 2014 to 41 in 2024, while “Biosensors” saw steady growth from 20 to 39 during the same period, highlighting a sustained interest in biomedical technology.

Keywords related to “Human” and “Humans” also saw a marked increase, with “Human” rising from 12 in 2014 to 37 in 2024 and “Humans” from 7 to 21. This suggests a growing emphasis on human-centric healthcare solutions. Similarly, “Implantable Devices” increased from 9 in 2014 to 30 in 2024, reflecting advancements in implant technologies.

The significance of cybersecurity in healthcare is evident from the rise of "Network Security" and "Computer Security", with the former increasing from 2 in 2014 to 27 in 2024 and the latter from 8 to 22. This growth underscores the importance of protecting sensitive medical data and devices.

"Internet of Things (IoT)" shows a rapid rise, particularly after 2018, growing from 0 in 2014 to 25 in 2024. This highlights the increasing integration of IoT in healthcare. Keywords like "Health Care" and "Article" also show steady increases, indicating consistent documentation and interest in broader healthcare topics.

Table 6. Keyword analysis

Year	2014	2015	2016	2017	2018	2019	2020	2021	2022	2023	2024
Implants (Surgical)	8	11	18	20	23	27	30	34	36	40	41
Biosensors	20	20	24	26	28	29	31	32	34	37	39
Human	12	12	13	13	14	16	21	24	26	30	37
Implantable Devices	9	11	15	17	19	20	22	24	26	28	30
Network Security	2	5	11	13	13	16	19	20	20	23	27
Health Care	8	8	11	11	12	15	17	18	22	24	26
Internet Of Things	0	0	2	2	6	10	11	12	16	24	25
Article	9	9	11	11	12	14	17	19	20	21	24
Computer Security	8	8	9	9	11	11	11	16	17	22	22
Humans	7	7	8	8	9	10	12	14	15	18	21

Figure 14. Keyword co-occurrence network

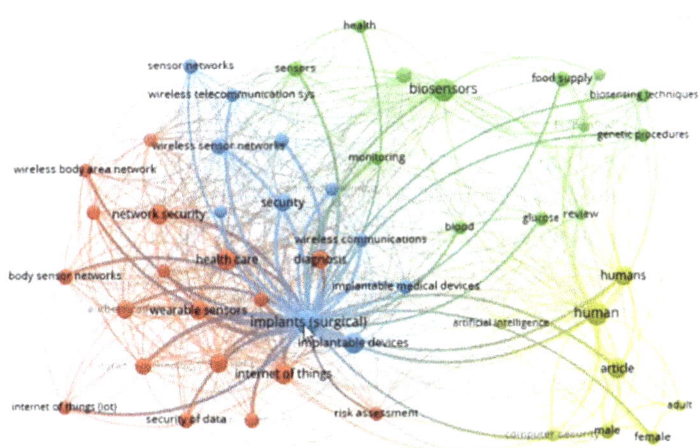

The keyword co-occurrence network depicted in Figure 14 showcases the interconnected nature of research themes in the field of smart implants and biosensors. Each node in the network represents a keyword, with the size of the nodes indicating the frequency of use, while the edges connecting them signify their co-occurrence in research publications. The clustering of keywords reveals distinct thematic areas within the field.

- **Biosensors and Health Applications**

The green cluster is dominated by keywords such as biosensors, health, and monitoring, highlighting the pivotal role of biosensors in healthcare applications. This cluster also connects to specialized themes like glucose review and food supply, reflecting the growing use of biosensors in managing chronic conditions like diabetes and ensuring food safety. The dense network within this cluster underscores the widespread integration of biosensors into diagnostic and monitoring systems, emphasizing their importance in advancing healthcare practices.

- **Security and Network Integration**

The red cluster centers around security-related terms such as network security, internet of things (IoT), and wearable sensors, pointing to the critical role of secure communication in healthcare systems. Keywords like wireless body area networks and security of data highlight concerns related to safeguarding sensitive medical information in interconnected systems. This cluster reveals the increasing reliance on IoT and wearable technologies in healthcare while addressing the challenges of ensuring robust cybersecurity.

- **Human-Centric and Ethical Considerations**

The yellow cluster focuses on human-centered themes, with terms like human, humans, and article signifying research that prioritizes patient needs, ethical concerns, and the social implications of technological advancements. The presence of artificial intelligence and genetic procedures within this cluster suggests an ongoing effort to integrate cutting-edge technologies while addressing their impact on individuals and society. This cluster underscores the importance of aligning technological innovations with ethical and human-centered principles.

- **Bridging Themes and Interdisciplinary Connections**

Keywords such as implants (surgical), health care, and diagnosis occupy central positions in the network, acting as bridges between the clusters. Their strong interconnections with both security-focused and health-oriented themes indicate the interdisciplinary nature of research in this field. These bridging keywords reflect the integration of diverse technologies and methodologies to address complex healthcare challenges.

8. RESEARCH GAPS AND FUTURE DIRECTIONS(RQ4)

The domain of smart implants and biosensors in secure healthcare systems has made significant progress, yet several gaps remain that must be addressed to unlock its full potential. Below are key gaps and suggestions for future research (Patil & Patil, 2023).

8.1 Key Gaps in Current Research

- **Cybersecurity Integration**: Limited integration of advanced cybersecurity measures with biosensor networks and implantable devices (Patil et al., 2023). IoT-based systems and wireless communication are widely used, but their vulnerability to cyber threats is underexplored.
- **Interoperability**: Absence of standardized protocols for system interoperability, leading to fragmented implementations.
- **Ethical and Privacy Concerns:** Insufficient research on handling sensitive patient data, particularly in cross-border healthcare systems (Santosh & Gaur, 2021; Li et al., 2022).
- **Scalability and Cost-Effectiveness**: Current studies often focus on high-end solutions, ignoring the feasibility of deployment in resource-constrained environments.
- **Emerging Technology Integration**: Lack of practical frameworks for the integration of AI and blockchain in smart implants and biosensors.

8.2 Under-researched Areas

- **Energy Efficiency**: Real-time energy efficiency remains underexplored. Optimizing battery life and reducing energy consumption are crucial for the sustainability of implantable devices.
- **Biocompatibility**: Limited research on the biocompatibility and long-term performance of materials used in smart implants, particularly for diverse patient demographics (Singh et al., 2024).

- **Personalized Healthcare Solutions**: A general lack of focus on personalized solutions based on biosensor data analytics, with most research being application-centric (Hulsen, 2021).

8.3 Suggestions for Future Research Directions

- **Cybersecurity Frameworks**: Develop robust frameworks integrating encryption, authentication, and anomaly detection for implantable devices and biosensors. Explore the integration of blockchain and AI to enhance security and real-time data analysis (Gupta et al., 2024).
- **Scalability and Cost-Effective Solutions**: Design scalable, cost-effective solutions suitable for low-resource settings. Consider self-powered biosensors using energy-harvesting technologies like piezoelectric materials or biofuel cells.
- **Material Development**: Advance biocompatible materials with enhanced durability and reduced rejection rates (Naresh & Lee, 2021; Abdulbari & Basheer, 2017).
- **Interdisciplinary Approaches**: Encourage collaboration across fields such as materials science, data analytics, and clinical sciences for holistic solutions.
- **Ethical Considerations**: Focus on transparent data governance models and patient-centric designs to ensure equitable and secure healthcare systems (Misera et al., 2024; Bhatia et al., 2024)

By addressing these gaps and pursuing the outlined directions, the field of smart implants and biosensors can achieve significant breakthroughs in secure healthcare systems.

9. CONCLUSION

This bibliometric analysis provides an in-depth exploration of the research landscape surrounding smart implants, biosensors, and security in healthcare systems from 2005 to 2024. The study reveals significant growth in the volume of research, highlighting influential authors, journals, and institutions, while also identifying key thematic areas and research methodologies. Despite promising advancements in the field, critical gaps remain, particularly in the integration of robust security measures and the need for long-term clinical studies. The analysis also points to the limited collaboration among authors, indicating a potential for greater interdisciplinary engagement in addressing these challenges. Future research should focus on bridging these gaps by fostering collaboration across disciplines and ensuring the development of secure, scalable, and cost-effective solutions. Emphasis on security protocols, personalized healthcare solutions, and biocompatible materials will be pivotal in advancing the practical application of smart implants and biosensors. By addressing these gaps, researchers can unlock the full potential of these transformative technologies, ultimately improving patient outcomes and ensuring the safety and efficacy of smart healthcare systems.

REFERENCES

Abdulbari, H. A., & Basheer, E. A. (2017). Electrochemical biosensors: Electrode development, materials, design, and fabrication. *ChemBioEng Reviews*, *4*(2), 92–105. DOI: 10.1002/cben.201600009

Abounassar, E. M., El-Kafrawy, P., & Abd El-Latif, A. A. (2022). Security and interoperability issues with Internet of Things (IoT) in healthcare industry: A survey. In *Security and Privacy Preserving for IoT and 5G Networks* (pp. 159–189). Techniques, Challenges, and New Directions. DOI: 10.1007/978-3-030-85428-7_7

Abyzova, E., Dogadina, E., Rodriguez, R. D., Petrov, I., Kolesnikova, Y., Zhou, M., Liu, C., & Sheremet, E. (2023). Beyond tissue replacement: The emerging role of smart implants in healthcare. *Materials Today. Bio*, *24*, 100784. DOI: 10.1016/j.mtbio.2023.100784 PMID: 37731959

Alam, F., Ahmed, M. A., Jalal, A., Siddiquee, I., Adury, R. Z., Hossain, G. M. M., & Pala, N. (2024). Recent progress and challenges of implantable biodegradable biosensors. *Micromachines*, *15*(4), 475. DOI: 10.3390/mi15040475 PMID: 38675286

Andreu-Perez, J., Leff, D. R., Ip, H. M., & Yang, G. Z. (2015). From wearable sensors to smart implants—Toward pervasive and personalized healthcare. *IEEE Transactions on Biomedical Engineering*, *62*(12), 2750–2762. DOI: 10.1109/TBME.2015.2422751 PMID: 25879838

Batista, E., Moncusi, M. A., López-Aguilar, P., Martínez-Ballesté, A., & Solanas, A. (2021). Sensors for context-aware smart healthcare: A security perspective. *Sensors (Basel)*, *21*(20), 6886. DOI: 10.3390/s21206886 PMID: 34696099

Bhatia, D., Paul, S., Acharjee, T., & Ramachairy, S. S. (2024). Biosensors and their widespread impact on human health. *Sensors International*, *5*, 100257. DOI: 10.1016/j.sintl.2023.100257

Bussooa, A., Neale, S., & Mercer, J. R. (2018). Future of smart cardiovascular implants. *Sensors (Basel)*, *18*(7), 2008. DOI: 10.3390/s18072008 PMID: 29932154

Cho, M., Han, J. K., Suh, J. N., Kim, J. J., Ryu, J. R., Min, I. S., Sang, M., Lim, S., Kim, T. S., Kim, K. B., Kang, K., Hwang, K., Kim, K., Hong, E.-B., Nam, M.-H., Kim, J., Song, Y. M., Lee, G. J., Cho, I.-J., & Yu, K. J. (2024). Fully bioresorbable hybrid opto-electronic neural implant system for simultaneous electrophysiological recording and optogenetic stimulation. *Nature Communications*, *15*(1), 45803. DOI: 10.1038/s41467-024-45803-0 PMID: 38448437

Choi, Y. S., Jeong, H., Yin, R. T., Avila, R., Pfenniger, A., Yoo, J., Lee, J. Y., Tzavelis, A., Lee, Y. J., Chen, S. W., Knight, H. S., Kim, S., Ahn, H.-Y., Wickerson, G., Vázquez-Guardado, A., Higbee-Dempsey, E., Russo, B. A., Napolitano, M. A., Holleran, T. J., & Rogers, J. A. (2022). A transient, closed-loop network of wireless, body-integrated devices for autonomous electrotherapy. *Science, 376*(6592), 1006–1012. DOI: 10.1126/science.abm1703 PMID: 35617386

Deore, S., Bachche, R., Bichave, A., & Patil, R. (2021). EHR-Sec: A blockchain-based security system for electronic health. In *International Conference on Soft Computing and Signal Processing* (pp. 295–303). Springer Nature Singapore.

Deore, S., Bachche, R., Bichave, A., & Patil, R. (2022). Review on applications of blockchain for electronic health records systems. In *Second International Conference on Image Processing and Capsule Networks: ICIPCN 2021* (pp. 609–616). Springer International Publishing. DOI: 10.1007/978-3-030-84760-9_52

Duttagupta, S., Marin, E., Singelée, D., & Preneel, B. (2023). HAT: Secure and practical key establishment for implantable medical devices. In *Proceedings of the ACM Conference on Data and Application Security and Privacy*. DOI: 10.1145/3577923.3583646

Ghafar-Zadeh, E. (2015). Wireless integrated biosensors for point-of-care diagnostic applications. *Sensors (Basel), 15*(2), 3236–3261. DOI: 10.3390/s150203236 PMID: 25648709

Gil, B., Ip, H., Kassanos, P., Lo, B., Yang, G. Z., & Anastasova, S. (2022). Smart implanted access port catheter for therapy intervention with pH and lactate biosensors. *Materials Today. Bio, 15*, 100298. DOI: 10.1016/j.mtbio.2022.100298 PMID: 35634169

Guo, Y., Hao, Z., Zhao, S., Gong, J., & Yang, F. (2020). Artificial intelligence in health care: Bibliometric analysis. *Journal of Medical Internet Research, 22*(7), e18228. DOI: 10.2196/18228 PMID: 32723713

Gupta, K., Mishra, V., & Makkar, A. (2024). A global cybersecurity standardization framework for healthcare informatics. *IEEE Journal of Biomedical and Health Informatics*, 1–8. DOI: 10.1109/JBHI.2024.3467179 PMID: 39321006

Gupta, N., Renugopalakrishnan, V., Liepmann, D., Paulmurugan, R., & Malhotra, B. D. (2019). Cell-based biosensors: Recent trends, challenges and future perspectives. *Biosensors & Bioelectronics, 141*, 111435. Advance online publication. DOI: 10.1016/j.bios.2019.111435 PMID: 31238280

Heo, Y. J., & Takeuchi, S. (2013). Towards smart tattoos: Implantable biosensors for continuous glucose monitoring. *Advanced Healthcare Materials*, *2*(1), 43–56. DOI: 10.1002/adhm.201200167 PMID: 23184404

Hu, C., Wang, L., Liu, S., Sheng, X., & Yin, L. (2024). Recent development of implantable chemical sensors utilizing flexible and biodegradable materials for biomedical applications. *ACS Nano*, *18*(5), 3969–3995. DOI: 10.1021/acsnano.3c11832 PMID: 38271679

Hulsen, T. (2021). Challenges and solutions for big data in personalized healthcare. In *Big Data in Psychiatry & Neurology* (pp. 69–94). Academic Press. DOI: 10.1016/B978-0-12-822884-5.00016-7

Jalali, M. S., Razak, S., Gordon, W., Perakslis, E., & Madnick, S. (2019). Health care and cybersecurity: Bibliometric analysis of the literature. *Journal of Medical Internet Research*, *21*(2), e12644. DOI: 10.2196/12644 PMID: 30767908

Kasoju, N., Remya, N. S., Sasi, R., Sujesh, S., Soman, B., Kesavadas, C., Muraleedharan, C. V., Varma, P. H., & Behari, S. (2023). Digital health: Trends, opportunities and challenges in medical devices, pharma and bio-technology. *CSI Transactions on ICT*, *11*(1), 11–30. DOI: 10.1007/s40012-023-00380-3

Kaveti, R., Lee, J., Youn, J., Jang, T. M., Han, W. B., Yang, S., Shin, J. W., Ko, G., Kim, D. J., & Han, S. (2023). Soft, long-lived, bioresorbable electronic surgical mesh with wireless pressure monitor and on-demand drug delivery. *Advanced Materials*, •••, 36. PMID: 37770105

Kwarteng, E., & Cebe, M. (2022). *A survey on security issues in modern implantable devices: Solutions and future issues*. arXiv. https://arxiv.org/

Ledet, E. H., Liddle, B., Kradinova, K., & Harper, S. (2018). Smart implants in orthopedic surgery, improving patient outcomes: A review. *Innovation and Entrepreneurship in Health*, *5*, 41–51. DOI: 10.2147/IEH.S133518 PMID: 30246037

Li, F., Ruijs, N., & Lu, Y. (2022). Ethics & AI: A systematic review on ethical concerns and related strategies for designing with AI in healthcare. *AI*, *4*(1), 28–53. DOI: 10.3390/ai4010003

Li, J., Liu, J., Wu, Z., Shang, X., Li, Y., Huo, W., & Huang, X. (2023). Fully printed and self-compensated bioresorbable electrochemical devices based on galvanic coupling for continuous glucose monitoring. *Science Advances*, *9*(22), eadi3839. DOI: 10.1126/sciadv.adi3839 PMID: 37467335

Li, S., Lu, D., Li, S., Liu, J., Xu, Y., Yan, Y., Rodriguez, J., Bai, H., Avila, R., Kang, S., Ni, X., Luan, H., Guo, H., Bai, W., Wu, C., Zhou, X., Hu, Z., Pet, M. A., Hammill, C. W., & Rogers, J. A. (2024). Bioresorbable, wireless, passive sensors for continuous pH measurements and early detection of gastric leakage. *Science Advances*, *10*(16), 10. DOI: 10.1126/sciadv.adj0268 PMID: 38640247

Liu, J., Liu, N., Xu, Y., Wu, M., Zhang, H., Wang, Y., Yan, Y., Hill, A., Song, R., Xu, Z., Park, M., Wu, Y., Ciatti, J. L., Gu, J., Luan, H., Zhang, Y., Yang, T., Ahn, H.-Y., Li, S., & Rogers, J. A. (2024). Bioresorbable shape-adaptive structures for ultrasonic monitoring of deep-tissue homeostasis. *Science*, *383*(6687), 1096–1103. DOI: 10.1126/science.adk9880 PMID: 38452063

Misera, J., Melchert, J., & Bork-Hüffer, T. (2024). Biosensing and biosensors—Terminologies, technologies, theories and ethics. *Geography Compass*, *18*(11), e70007. DOI: 10.1111/gec3.70007 PMID: 39600691

Mohammadi, A. T., Mohammad Taheri, S. A., Karamouz, M., & Sarhaddi, R. (2024). *Rising innovations: Revolutionary medical and dental breakthroughs revolutionizing the healthcare field*. Nobel Sciences.

Naresh, V., & Lee, N. (2021). A review on biosensors and recent development of nanostructured materials-enabled biosensors. *Sensors (Basel)*, *21*(4), 1109. DOI: 10.3390/s21041109 PMID: 33562639

Nidhya, R., Kumar, M., Maheswar, R., & Pavithra, D. (2022). Security and privacy issues in smart healthcare system using Internet of Things. In *IoT-Enabled Smart Healthcare Systems* (pp. 63–85). Services and Applications. DOI: 10.1002/9781119816829.ch4

Patil, R. Y. (2024). A secure privacy preserving and access control scheme for medical Internet of Things (MIoT) using attribute-based signcryption. *International Journal of Information Technology : an Official Journal of Bharati Vidyapeeth's Institute of Computer Applications and Management*, *16*(1), 181–191. DOI: 10.1007/s41870-023-01569-0

Patil, R. Y., Karati, A., Patil, Y., & Bannore, A. (2023). Reliable data sharing in medical cyber physical system using fog computing. In *Intelligent Edge Computing for Cyber Physical Applications* (pp. 67–83). Academic Press. DOI: 10.1016/B978-0-323-99412-5.00007-1

Patil, R. Y., Karati, A., & Patil, Y. H. (2024). A signcryption with identity-based authentication for secure EHR sharing in IoMT utilizing ECC. *International Journal of Information Technology : an Official Journal of Bharati Vidyapeeth's Institute of Computer Applications and Management, 16*(8), 5133–5148. DOI: 10.1007/s41870-024-01963-2

Patil, R. Y., & Patil, Y. H. (2023). A proxy signcryption scheme for secure sharing of industrial IoT data in fog environment. *International Journal on Computer Science and Engineering, 26*(2), 118–128.

Patil, R. Y., Patil, Y. H., & Bannore, A. (2024). Cybersecurity in medical Internet of Things: Issues, challenges, and future directions. In *Security Implementation in Internet of Medical Things* (pp. 87–110). CRC Press.

Raymond, D. A., Apetorgbor, M., Kumar, P., & Goureshettiwar, P. (2024). IoT-enabled smart implants: Revolutionizing post-surgical care. In *2024 2nd International Conference on Sustainable Computing and Smart Systems (ICSCSS)* (pp. 430–435). IEEE.

Rooney, L., Rimpiläinen, S., Morrison, C., & Nielsen, S. L. (2019). *Review of emerging trends in digital health and care: A report by the Digital Health and Care Institute.* Digital Health and Care Institute.

Santosh, K. C., & Gaur, L. (2021). Privacy, security, and ethical issues. In Santosh, K. C., & Gaur, L. (Eds.), *Artificial Intelligence and Machine Learning in Public Healthcare: Opportunities and Societal Impact* (pp. 65–74). Academic Press. DOI: 10.1007/978-981-16-6768-8_8

. Sarwal, R., Prasad, U., Gopal, K. M., Kalal, S., Kaur, D., Kumar, A., Regy, P. V., & Sharma, J. (2021). *Investment opportunities in India's healthcare sector.*

Sharma, A., Badea, M., Tiwari, S., & Marty, J. L. (2021). Wearable biosensors: An alternative and practical approach in healthcare and disease monitoring. *Molecules (Basel, Switzerland), 26*(3), 748. DOI: 10.3390/molecules26030748 PMID: 33535493

Sheng, H., Jiang, L., Wang, Q., Zhang, Z., Lv, Y., Ma, H., Bi, H., Yuan, J., Shao, M., Li, F., Li, W., Xie, E., Liu, Y., Xie, Z., Wang, J., Yu, C., & Lan, W. (2023). A soft implantable energy supply system that integrates wireless charging and biodegradable Zn-ion hybrid supercapacitors. *Science Advances, 9*(46), 9. DOI: 10.1126/sciadv.adh8083 PMID: 37967195

. Singh, A. B., Khandelwal, C., & Dangayach, G. S. (2024). Advancements in healthcare materials: Unraveling the impact of processing techniques on biocompatibility and performance. *Polymer-Plastics Technology and Materials*, 1–37.

Zhang, Y., Rytkin, E., Zeng, L., Kim, J. U., Tang, L., Zhang, H., Mikhailov, A., Zhao, K., Wang, Y., Ding, L., Lu, X., Lantsova, A., Aprea, E., Jiang, G., Li, S., Seo, S. G., Wang, T., Wang, J., Liu, J., & Rogers, J. A. (2025). Millimetre-scale bioresorbable optoelectronic systems for electrotherapy. *Nature*, *640*(8057), 77–86. DOI: 10.1038/s41586-025-08726-4 PMID: 40175757

Chapter 6
Bioorganometallic Polymers as Next-Generation Biodegradable and Bioactive Biomaterials

Shamim Ahmad Khan
http://orcid.org/0009-0007-5921-2955
Shibli National College, Azamgarh, India

ABSTRACT

Bioorganometallic polymers represent a new generation of novel biomaterials which combine metal centers and organic polymer backbones to realize novel physicochemical and biological performance. Such materials have become the subject of interest as next generation biodegradable and bioactive materials in biomedical applications by combining polymer versatility with the catalytic, electronic and bioactive properties of organometallic compounds. Addition of biologically relevant metals like iron, zinc, titanium and ruthenium improves antimicrobial, biocompatibility and controlled degradation. Tunable nature enables them to be controlled carefully with respect to mechanical strength, degradation rates and biological interactions. They can therefore be broadly utilized in drug delivery, tissue engineering scaffolds, biosensors, antimicrobial coating and cancer treatment. The developments in synthesis and bioorganometallic chemistry are still allowing designing of multifunctional biomaterials with better performance and safety.

DOI: 10.4018/979-8-2600-0060-1.ch006

1. INTRODUCTION

Biomaterials have been central to recent advances in biomedical engineering because they can interface with biological systems for therapeutic or diagnostic applications. These materials are widely used in medical devices, tissue engineering scaffolds, drug delivery systems, implants and regenerative medicine (Menagadevi et al., 2026). Over the past several decades, materials science, polymer chemistry and biotechnology have advanced significantly, enabling the development of novel biomaterials that mimic natural biological conditions and serve as cellular scaffolds. The primary objective of biomaterials research is to develop biocompatible materials that are safe and integrate with biological tissues without adverse immune reactions (Datta et al., 2020). As healthcare technologies evolve rapidly, advanced biomaterials that are more functional, biodegradable and bioactive have become increasingly in demand.

Bioactive and biodegradable biomaterials are of great importance in regenerative medicine and tissue engineering. Biodegradable items can be broken down in the body into non-toxic products, so they do not require surgical removal once their therapeutic purpose is served (Kalpana Manivannan et al., 2024). The property is particularly useful in implants like temporary implants, drug delivery vehicles and tissue scaffolds. Bioactive biomaterials, in turn, are also intended to interact directly with biological tissues and induce specific cellular reactions, including cell adhesion, cell growth, differentiation and tissue regeneration (Table 1). These substances can activate biological responses and promote healing mechanisms, making them useful in contemporary therapeutic practices (Sindhi et al., 2025).

Table 1. Conventional vs bioorganometallic polymers

Property	Conventional Polymers	Bioorganometallic Polymers
Bioactivity	Low	High
Antimicrobial activity	Limited	Excellent
Stimuli responsiveness	Limited	High
Catalytic properties	Absent	Present
Cost	Low	High
Toxicity risk	Low	Moderate

Although there have been advances in polymer-based biomaterials, there are a few limitations of conventional polymeric materials when employed in biomedical devices. Several conventional synthetic polymers are known to be limited in bioactivity and do not have the capacity to communicate with biological systems actively

(Pires et al., 2023). Besides, other polymers exhibit poor mechanical performance, low degradation or insufficient stability in physiological conditions. In some other instances, degradation products of polymers can lead to inflammation or toxicity, which can undermine the efficacy and safety of biomedical devices (Abdulsalam et al., 2025). Moreover, traditional polymers are usually not antimicrobial, which exposes medical implants to infection and biofilms (Negut et al., 2022). Such shortcomings have encouraged scientists to consider new classes of hybrid materials that are more functional. Bioorganometallic polymers have become one of the most promising types of multifunctional biomaterials in recent years; they combine the benefits of organic polymers with the specific function of metal-containing compounds (Carraher, 1981; Constantino et al., 2023). Bioorganometallic polymers are composite materials where metal atoms or metal complexes are embedded in polymer chains or networks. The presence of centers on metals adds further functionalities, including catalytic activities, antimicrobial properties, redox-responsiveness and electronic conductivity. These properties allow one to create high-performing materials that can conduct various biological and therapeutic tasks at the same time (Emmanuel et al., 2026).

The addition of metals like iron, zinc, copper, titanium and silver into the polymer framework enables researchers to create materials with adjustable physicochemical characteristics as well as enhanced biological functionality. Indicatively, metal ions can amplify the antimicrobial action, stimulate tissue regeneration and control drug release (Tousian et al., 2024). Moreover, these hybrid materials can be used in many biomedical applications owing to the polymer framework, which offers structural flexibility, biocompatibility and processability. Because of this, bioorganometallic polymers are receiving more and more interest in drug delivery, tissue engineering, biosensing and regenerative medicine (Hudi ă & Gălă eanu, 2023). Figure 1 represents the structure and concept of bioorganometallic polymers.

Figure 1. General structure and concept of bioorganometallic polymers integrating metal centers with organic polymer frameworks.

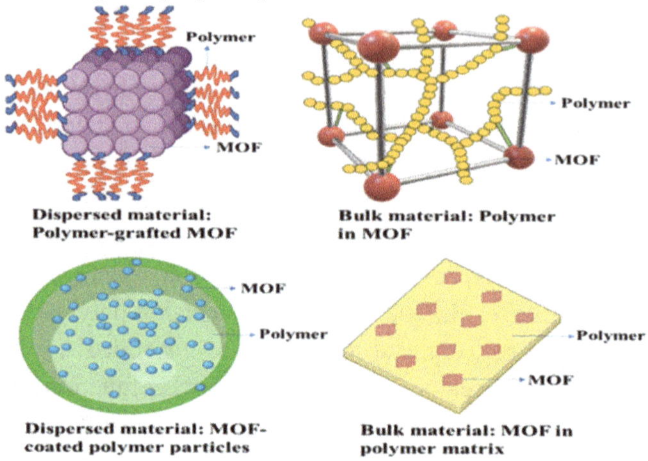

Reproduced with permission from Rajeev and Panicker (2025)

Despite significant advances in biodegradable polymers such as PLA, PGA, PCL, chitosan and alginate, many conventional systems still suffer from drawbacks, including limited bioactivity, inadequate antimicrobial performance, low responsiveness to biological stimuli and limited multifunctionality. These limitations have spurred interest in bioorganometallic polymers, which integrate metal-mediated functionality with polymeric versatility. However, a comprehensive understanding of their synthesis strategies, biodegradation mechanisms, bioactivity and biomedical applications remains incomplete (Abdulsalam et al., 2025). Therefore, this chapter critically reviews recent developments in bioorganometallic polymers and highlights their potential as next-generation biodegradable and bioactive biomaterials. Despite these advances, many existing biomaterials primarily provide structural support with limited biological functionality. This limitation has created a need for multifunctional materials that combine controlled biodegradation, bioactivity and therapeutic performance, making bioorganometallic polymers promising candidates for biomedical applications.

2. FUNDAMENTALS OF BIOORGANOMETALLIC POLYMERS

Bioorganometallic polymers are a type of hybrid material that combines the structural flexibility of organic polymers with the functional behavior of metal-containing compounds. The materials have attracted a lot of interest in materials science and biomedical engineering as they possess physicochemical and biological characteristics that are not common among traditional polymers (Rauf et al., 2023). The use of metal centers in polymer frameworks allows scientists to create multi-functional materials that are more catalytically active, antimicrobial, mechanically strong and biologically compatible (Zheng et al., 2025).

2.1 Definition and Classification

Organometallic polymers are polymeric substances that include metal atoms directly bonded to carbon atoms in the polymer backbone or side-chains. Metal-carbon bonds found in these polymers confer unusual chemical reactivity and enable a wide range of functionalities to be incorporated into the structure of the polymers (Qu et al., 2022). Organometallic polymers are interesting catalysts, electronics and advanced materials applications have been extensively studied owing to their distinctive electronic and chemical properties. A specialized category of organometallic polymers that are used in biological and biomedical applications is bioorganometallic polymers. These materials use biocompatible metals in polymer systems that are capable of interacting with biological environments in a safe manner. Iron, zinc, copper, titanium and silver are popular due to their ability to offer antimicrobial activity, catalytic functionality in biological systems and enhanced cell and biomolecule interactions (Tilden et al., 2024).

Bioorganometallic polymers can be categorized into three broad types depending on how the metal components will be organized within the polymer matrix, as represented in Table 2.

- Main-chain organometallic polymers are polymers that have metal atoms as part of the chain structure. Metal atoms are switched with organic groups in these systems to create repeating metal-ligand structures or metal-carbon structures. Since the metal centers are embedded in the backbone, these polymers tend to exhibit special electronic, catalytic, as well as structural characteristics (Cué-Sampedro & Sánchez-Fernández, 2024).
- Side-chain organic metallometallic polymers are composed of organic polymer backbones, to which metal complexes are linked as pendant groups on the side-chains. In this structure, the backbone would offer mechanical stability and the metal-bearing side groups would offer functional properties including catalytic activity, antimicrobial effects and redox behavior. Polymer properties can also be modified more easily through this structure (Qu et al., 2022).
- Organometallic polymer networks of cross-linked type are achieved through the use of metal ions as cross-linking agents between several polymer chains. The systems produce three-dimensional networks with enhanced mechanical

strength and structural stability. Metal centers in the network can also add new capabilities, including being responsive to environmental stimuli and other biological performance capabilities (Kim et al., 2021).

Table 2. Classification of bioorganometallic polymers with examples and applications

Type of Polymer	Structural Characteristics	Example Metal Components	Typical Applications	Ref.
Main-chain organometallic polymers	Metal atoms integrated into polymer backbone	Fe, Ti, Ru	Catalysis, electronic biomaterials	(Nobahar et al., 2026)
Side-chain organometallic polymers	Metal complexes attached to polymer side-chains	Cu, Zn, Ag	Drug delivery, antimicrobial materials	(Mohamed et al., 2025)
Cross-linked organometallic networks	Metal ions act as cross-linkers in polymer network	Ca, Zn, Ti	Tissue engineering scaffolds, biomaterials	(Nobahar et al., 2026)

The choice of structure for a bioorganometallic polymer depends on its intended biomedical application. Main-chain polymers typically offer superior catalytic and electronic properties, whereas side-chain polymers provide greater flexibility for functionalization. Cross-linked networks are particularly suitable for tissue engineering applications due to their excellent mechanical stability and structural integrity.

2.2 Chemical Structure and Properties

Bioorganometallic polymers mainly possess their characteristics due to the presence of metal-carbon bonds and metal center-polymer matrix interactions. Metal-carbon bonds are both chemically stable and reactive and have an impact on the electronic structure and catalytic activity of the polymer. Functional groups of the polymer can also be reacted with metal ions by coordination bonding, by electrostatic interactions or by covalent bonds. The interactions influence the structural organization of the polymer, its mechanical strength and its degradation (Nobahar et al., 2026; Sun et al., 2024).

The introduction of metal centers is likely to increase mechanical characteristics like tensile strength and stability since the metal coordination can strengthen the polymer structure. Certain bioorganometallic polymers are also electrically conducting and redox-active and therefore find application in biosensors and biomedical electronic devices. The other essential characteristic is catalytic activity. Metal ions are catalysts for a wide range of chemical or biological reactions. Bioorganometallic polymers can therefore be involved in biochemical reactions or help in the controlled release of drugs (Emmanuel et al., 2026). Their catalytic characteristics render them useful in designing responsive or smart biomaterials.

2.3 Biological Interactions

The interface between bioorganometallic polymers and biological systems is essential in the case of biomedical applications. The materials should be able to react with proteins, enzymes and cells without being toxic or triggering adverse immune reactions. Biomolecules can be able to interact with metal centers in the polymer either via coordination or electrostatic interactions. These interactions can affect the activity of enzymes, control biochemical pathways or improve the responses of cells. In other instances, polymers created by incorporating metal may also resemble natural biological systems and engage in catalytic reactions that also occur in living organisms (Iacob, 2025).

Bioorganometallic polymers are capable of affecting cellular activities, including cell adhesion, cell proliferation and cell differentiation. There are certain metal ions that are important in the biological system and among them, zinc and copper may promote tissue growth and regeneration when they are present in biomaterials. Nevertheless, it is necessary to preserve biocompatibility. The choice of metal ions and polymer components has to be as non-toxic as possible and degrade safely in the body. The biocompatible metals and biodegradable polymer building blocks are usually designed in such a way that the materials decompose into non-toxic products once their task is complete (Liang et al., 2017). Beyond cell adhesion and proliferation, the biological performance of bioorganometallic polymers is influenced by protein adsorption, immune responses and the controlled release of metal ions. Interactions between metal centers and biological molecules can modulate cellular signaling pathways involved in tissue repair and regeneration. However, excessive release of metal ions can induce oxidative stress or inflammatory responses, highlighting the importance of maintaining a balance between bioactivity and biocompatibility.

3. SYNTHESIS AND FABRICATION STRATEGIES

Bioorganometallic polymers are developed based on novel synthesis and fabrication methods that allow fine control of the composition, structure, as well as functionality. Because these materials consist of metal and polymer frameworks, their synthesis needs to retain their stability, biocompatibility and functional performance. Contemporary polymer chemistry provides various viable methods, such as coordination polymerization, controlled radical polymerization, sol-gel processing, self-assembly and more innovative methods, including the 3D printing of materials and nanotechnology-based methods (Figure 2) (Mankaev & Karlov, 2023).

Figure 2. Synthetic pathways used for preparing bioorganometallic polymers: (a) Schematic of ATRP and Surface-Initiated ATRP (SI-ATRP). (b) Schematic representation of the sol-gel processing. (c) Advanced fabrication methods.

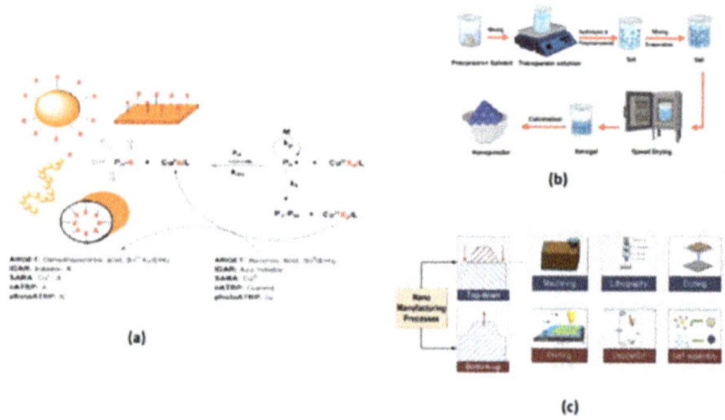

(a) Reproduced with permission from Hui et al. (2014); (b) Reproduced with permission from El-Khawaga et al. (2023); (c) Reproduced with permission under Creative Commons CC BY license from Z. Fan et al. (2022)

3.1 Coordination Polymerization

This involves the joining of the various separated components into a single structure. Coordination polymerization is a common method for the production of bioorganometallic polymers. In this process, the metal ions or complexes are used to react with organic ligands in order to form coordinate bonds to create polymeric structures that have metal centers embedded in them. These metal atoms serve as points of connection between organic molecules into long structures. These polymers are held together by metal-ligand interactions, which determine their structure and stability (Abtahi et al., 2025). The use of transition metals, including iron, copper, zinc, titanium and ruthenium, is because of their stable bond formation with functional group-containing ligands, including carboxylates, amines, phosphines and imidazoles. This allows the creation of well-organized polymer networks with adjustable properties. The major benefit of the coordination polymerization is the versatility of the structure. Researchers can form linear, branched and three-dimensional structures by the choice of appropriate metal ions and ligands. The properties that can be optimized using this control are mechanical strength, porosity, catalytic activity and biodegradability (Yoshinari & Konno, 2023). Also, a large number of coordination polymers are stimuli-responsive and therefore can respond to changes in the environment, including pH, temperature and ionic strength. These features are most useful for biomedical uses such as controlled drug delivery and responsive biomaterials.

3.2 Controlled Radical Polymerization

Radical polymerization methods have been subjected to controlled polymerization and this has revolutionized the production of functional polymers due to the ability to determine the particular molecular weight, architecture and composition with accuracy. These techniques are particularly effective in the incorporation of metal-bearing functional groups into polymer chains and maintaining homogeneity (Grishin, 2023). The main method in this category is Atom Transfer Radical Polymerization (ATRP). It uses transition metal catalysts, usually copper complexes, to control the growth of the polymer by mechanisms of reversible deactivation and activation. This is a regulated process that generates polymers with a high level of narrow molecular weight distributions and predefined structures. ATRP is a widely used method for producing side-chain organometallic polymers in which metal complexes are introduced either during or after polymerization (Dadashi-Silab & Matyjaszewski, 2020).

Another important technique is Reversible Addition-Fragmentation Chain Transfer (RAFT) polymerization. It has chain transfer agents in it to regulate the growth of the polymers, which enables the exact control of polymer size and structure. RAFT can be used with a variety of monomers and reaction environments, allowing it to be used in biomedical applications. It also allows introducing functional groups that can subsequently liaise with metal ions to create customized bioorganometallic polymer systems. These methods assist the architecture of complex forms of polymers like block, graft and star-shaped polymers. These structures are especially handy in drug delivery systems and nanostructured biomaterials whereby controlled molecular organization is critical (Martínez et al., 2022).

3.3 Sol–Gel and Self-Assembly Techniques

Another technique that is essential in the production of hybrid bioorganometallic materials is the sol-gel process. In this method, metal alkoxides or organometallic precursors are reacted by hydrolysis and condensation reactions to create a three-dimensional inorganic network. The sol-gel technique enables the incorporation of metal ions onto polymer matrices to give materials of special structural and functional properties. Sol–gel processing has benefits that include low processing temperatures, high purity and the creation of porous structures (Barrino, 2024). These characteristics render it suitable for drug delivery, biosensing and tissue engineering. The methods of self-assembly also have a great effect in assembling nanostructured bioorganometallic polymers. Self-assembly refers to the process where molecules spontaneously form ordered structures upon non-covalent interactions, which can be either hydrogen bonding, electrostatic forces or coordination of metal. The interactions between metal and ligand trigger the development of nanoscale structures in these systems, including micelles, vesicles and polymeric nanoparticles (J. Zhang et al., 2024).

The small size and high surface area of these nanostructures are important benefits for biomedical applications because they increase the interactions of the nanostructures with biological systems. They have specific applications in targeted drug delivery, imaging and biosensing. The sol-gel process combined with self-assembly allows the production of hybrid materials that are stronger in mechanical properties, catalytically active and have regulated biodegradation (Abdullah et al., 2025).

3.4 Advanced Fabrication Methods

The new technologies have brought about new fabrication methods for bioorganometallic polymers. Out of them, 3D printing and bioprinting have become potent techniques for creating personalized biomedical assemblies. 3D printing can also be used to create polymeric materials in precise three-dimensional shapes in a layer-by-layer process. Researchers can create printable bioinks that include metal-containing polymers to create customized products that have mechanical and biological properties. These scaffolds also have the ability to imitate natural tissue structures and facilitate cell growth, which is useful in regenerative medicine (Kumar et al., 2021). Bioprinting is an expansion of this idea, in which living cells are incorporated into biomaterials to form tissue-like structures. Bioorganometallic polymers are used as supporting matrices, which offer structural stability and functional performance (Saini et al., 2021).

The fabrication methods that are based on nanotechnology also increase the scope of design. Nanoparticle synthesis, nanofiber fabrication and electrospinning are techniques that allow the creation of materials with regulated nanoscale structures. The nanostructured polymers can be used to increase the capacity of drug loading, improve cell-material interactions and provide targeted therapeutic effects (Anusiya & Jaiganesh, 2022). Synthesis and fabrication strategies play a crucial role in determining the structural organization, metal distribution, degradation behavior and biological performance of bioorganometallic polymers. Therefore, understanding these fabrication methods is essential for enhancing the biodegradability and bioactivity of these materials for biomedical applications.

4. BIODEGRADABILITY AND BIOACTIVITY

Advanced biomaterials to be used in biomedical applications demand materials that are functional and compatible with biological systems. Biodegradability and bioactivity are some of the most significant features of next-generation biomaterials (Oleksy et al., 2023). These properties allow the materials to respond to the tissues effectively and safely degrade once their intended purpose is accomplished. The polymers of bioorganometallic type are of particular interest since, due to the possibility of adjusting their structure, it is possible to control their rates of degradation and biological reactions. Multifunctionality of these materials may be attained by incorporating metal centers into the biodegradable polymer frameworks, such as therapeutic, catalytic and antimicrobial properties (Emmanuel et al., 2026; Kayed & Almeataq, 2023).

4.1 Mechanisms of Biodegradation

Biodegradation is the process of decomposing materials into small units through biological or chemical reactions. Biodegradable polymers are more desirable in biomedical applications since they help minimize the use of surgical procedures for removal and reduce accumulation in the body with time. The bioorganometallic polymers' degradation behaviour is determined by their composition, structure and environment (Kurowiak et al., 2023). Hydrolytic degradation is one such common mechanism, in which water molecules break chemical bonds in the polymer. Polymers with hydrolysable bonds like esters, amides, carbonates and others degrade slowly in the physiological environment, as represented in Figure 3. Hydrolysis can also influence metal–ligand bonds in bioorganometallic systems and metal ions are released in a controlled manner, which is quite beneficial in drug delivery, where gradual degradation of material can be used to control therapeutic release (H. Fan et al., 2025).

Another significant pathway is enzymatic degradation, which is catalyzed by enzymes like proteases, esterases and lipases. These enzymes are selective in the breakdown of polymer chains in mild physiological conditions. In bioorganometallic polymers, the organic backbone, as well as some metal-coordinated structures, can be targeted, which enables degradation to be done in a more controlled and specific manner (Rosato et al., 2022). The degradation also depends on environmental factors. pH, oxidative stress, temperature and variations in biological fluids have the ability to increase or slow down the degradation processes. As an example, inflamed tissues are likely to be acidic, where hydrolysis will increase, whereas redox-active metal centers may be influenced by oxidative environments (Mittal et al., 2014; Tumilaar et al., 2024). The knowledge of such influences assists in the design of materials whose degradation profiles are predictable.

4.2 Bioactive Properties

Along with biodegradability, bioorganometallic polymers are also highly bioactive and can therefore interact with the biological system and have the ability to mediate cellular activities. One of them is antimicrobial activity. Silver, copper and zinc metal ions may prevent microorganisms by breaking cell membranes, forming reactive oxygen species or disrupting metabolism. These metals are also effective in protecting against infections and hence are used in medical equipment, implants and dressings when incorporated into the polymers. (Smola-Dmochowska et al., 2023). These polymers are also catalytic and redox active. The possibility of involving metal centers in biochemical reactions allows metal centers to be used in drug activation or controlled therapeutic release that is controlled. Materials that are redox responsive can be used to react to physiological conditions and enable targeted and stimuli-responsive treatment approaches (Kothawade & Shende, 2024). Moreover, the bioorganometallic polymers facilitate the interactions and regeneration of tissues. Key metal ions such as calcium, zinc and magnesium are involved in cellular metabolism and cellular signaling. They have the ability to increase cell proliferation, differentiation and tissue repair when incorporated in biomaterials. An example would be the ability of zinc-based materials to promote bone regeneration and copper-containing systems to promote angiogenesis. Applications of these properties are largely useful in regenerative medicine as well as tissue engineering (Su et al., 2019).

4.3 Biocompatibility and Safety

In spite of all these benefits, the safety of bioorganometallic polymers is a major concern for clinical application. Biocompatibility is the property of a material that allows that material to work without any toxicity, inflammation and immune response. Metal and polymer materials used are crucial to safety consequences. Huge effect of cytotoxicity is of major concern because excess release of metal ions can damage cells and tissues (Bernard et al., 2018). Thus, it is necessary to manage the level of concentration and discharge of the metal ions. Iron, zinc, calcium and titanium are biocompatible metals that are sometimes used since they are less toxic and are also naturally found in some parts of biological systems. In addition to cytotoxicity, metal hypersensitivity and allergic immune reactions can occur in susceptible individuals. Certain metal-containing biomaterials may cause localized inflammation following implantation. Therefore, the careful selection of metal components and thorough assessment of biocompatibility are essential for the safe clinical use of bioorganometallic polymers. The strategies of controlled release also increase safety. In properly designed systems, the release of metal ions is gradual with the degradation of the polymer and the system remains biologically active without being accumulated in a harmful manner. This has found special use in antimicrobial therapy and tissue regeneration where a constant flow of bioactive ions promotes healing (Amirtharaj Mosas et al., 2022). Altogether, bioorganometallic polymers are good alternatives for high-tech biomedical use due to their biodegradability, bioactivity and regulated safety. These polymers have a great future in medical technology due to the optimization of degradation behavior, biological interactions and material composition (Karimi et al., 2025).

Figure 3. Biodegradation mechanisms of polymers.

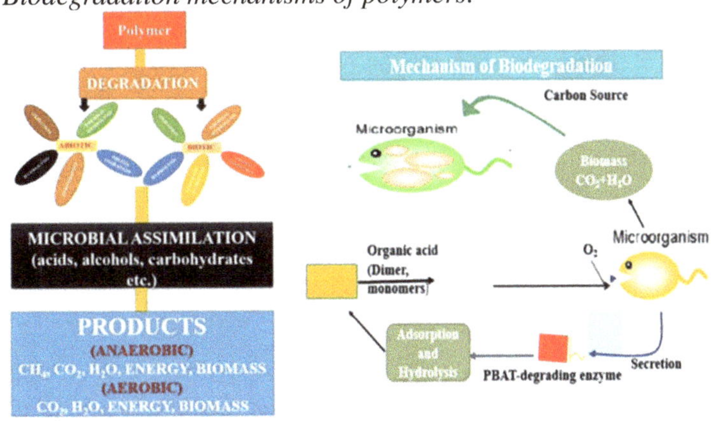

Reproduced with permission from Kanwal et al. (2022)

5. BIOMEDICAL APPLICATIONS

The bioorganometallic polymers have attracted significant interest over the past few years because of their multifunctional nature and capability to combine both the merits of organic polymers and metal-based components. They are a very versatile compound in biomedical use due to their distinctive physicochemical characteristics, such as catalytic activity, antimicrobial effects, controlled degradability and stimulus responsiveness, as represented in Figure 4 (R. A. Omer et al., 2025). Such hybrid materials can be designed to respond to biological systems in a controlled and useful way and therefore have applications in drug delivery, tissue engineering, antimicrobial therapies and biosensing technologies. The idea of the addition of biologically active metal ions to polymer matrices gives the chance to design materials which would not only have structural functions but also be active in the therapeutic and diagnostic process (Srivastava & Azad, 2023).

Figure 4. Biomedical applications of bioorganometallic polymers.

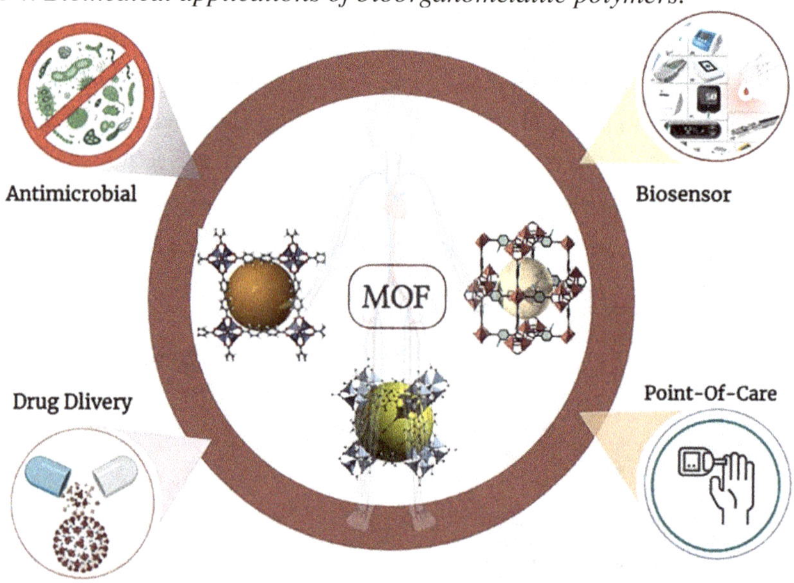

Reproduced with permission from Rezaee et al. (2022)

5.1 Drug Delivery Systems

Advanced drug delivery systems are one of the most promising biomedical uses of bioorganometallic polymers. The classical approaches to drug delivery are usually limited by a lack of good drug stability, quick degradation, non-specificity and unwanted side effects. Another solution to these problems is offered through bioorganometallic polymers, as they allow targeted and controlled drug delivery (Ezike et al., 2023). Some of the major characteristics of these materials are that they enable the control of drug release mechanisms. There is a possibility of the polymer matrix containing therapeutic molecules that can slowly be released as the polymer degrades or in response to environmental stimuli. The metal centers in the structure of the polymer can also be used to control the release of drugs using catalytic or redox-responsive release methods (Herdiana et al., 2022). As an example, metal-ligand interactions can entrap drug molecules in the polymer framework and liberate them within a particular physiological stimulus such as a pH alteration, temperature and redox potential. This is a type of controlled release method, whereby the therapeutic agents are administered at the required rate and place, which enhances the effectiveness of the treatment and minimizes side effects (Benny et al., 2024).

The other notable benefit of bioorganometallic polymers is that they can be used in targeted delivery of drugs with metal-responsive polymers. Some metal ions have the ability to react to certain biological signals or environmental conditions in diseased tissues. For example, the chemical environment of tumor tissues can be highly distinct, including acidic pH and high oxidative stress. It is possible to use these conditions to target drugs to the target site by using metal-responsive polymer systems (Tabasi et al., 2021). Also, magnetic metal nanoparticles incorporated in polymer matrices can be directed by using external magnetic fields, which means that they will enable the specific targeting of drugs in particular tissues or organs. These techniques find application especially in cancer treatment and precision medicine, where personalized treatment is likely to have a great impact on patient outcomes (Shi et al., 2025). Ferrocene- and iron-based polymer systems have shown particular promise for redox-responsive and targeted drug delivery applications.

5.2 Tissue Engineering and Regenerative Medicine

The other significant use of bioorganometallic polymers includes tissue engineering and regenerative medicine that involves the use of biomaterials to repair, replace or regenerate damaged tissues (Socci et al., 2023). Tissue engineering generally makes use of scaffolds that imitate the natural extracellular matrix and offer structural support to the growth and development of the tissues as well as cells (Mangani et al., 2025). The bioorganometallic polymers may be constructed in order to form bone and soft tissue scaffold structures. Tissue regeneration processes can be improved by incorporating metal ions that are biologically active, like calcium, magnesium, zinc and copper, in polymer scaffolds (Farjaminejad et al., 2024). An example of this is the use of calcium-containing polymer scaffolds that can promote mineralization and osteogenesis of bone and copper-containing materials that can promote angiogenesis or the creation of new blood vessels, which are required to heal tissues (Khalkhali et al., 2025).

Moreover, the biodegradable property of these materials enables the scaffold to degrade over time as new tissue grows and no surgical intervention is necessary to remove the scaffold (Mishra & Singh, 2025). The most recent development in 3D printing and nanotechnology has strengthened the promise of bioorganometallic polymers in tissue engineering. These technologies enable the production of very fine tissue scaffold structures that are very close to the ones found in nature. Consequently, bioorganometallic polymer scaffolds are gaining popularity in the research and clinical practice of regenerative medicine (Laird et al., 2021).

5.3 Antimicrobial Materials

Medical implant and wound infections are still a significant healthcare issue. A good alternative to this is bioorganometallic polymers that have a metal-based antimicrobial effect. Specific metal ions such as silver, copper and zinc are highly recognized in terms of their antimicrobial action. These metal ions, when added to polymer matrices, can offer continuous protection against bacterial and fungal infections (Jiang et al., 2024). The antimicrobial effect of these materials is usually achieved through interruption of the microbial cell membrane, the formation of reactive oxygen species and disruption of the normal metabolic processes of the microorganisms. The formation of biofilms is also discouraged by metal ions, a leading cause of persistent infections of medical devices (Sharma et al., 2023).

The bioorganometallic polymers are highly applicable in the field of wound healing and implant coating due to these properties. Antimicrobial polymers can be used in wound dressings to avoid infection and assist in the healing of wounds and tissue regeneration (Alven et al., 2022). Similarly, bioorganometallic polymers can be used to coat implantable medical devices like orthopedic implants, catheters and prosthetics to minimize the chances of post-operative infection. The antimicrobial metal ions are designed to be released gradually to provide definite protection and prevent extreme toxicity to the nearby tissues (Negut et al., 2022). The antimicrobial activity of these materials has been reported against common pathogens such as *Staphylococcus aureus* and *Escherichia coli.*

5.4 Biosensors and Diagnostic Systems

There is also extensive research on the use of bioorganometallic polymers in the design of advanced biosensors and diagnostic systems. The special electronic and catalytic characteristics of metal centers render bioorganometallic polymers the best materials in sensor design (P. K. Omer et al., 2025). Polymer sensors based on metals are capable of detecting a wide variety of biomolecules, such as glucose, proteins, nucleic acids and enzymes. As an example, polymers containing metallic elements can be used to facilitate electron transfer reactions to increase the sensitivity and accuracy of electrochemical biosensors (Abhishek et al., 2023).

Biomolecule and pathogen detection is also another application of the technology. Given that bioorganometallic polymers can be functionalized with particular recognition molecules like antibodies or enzymes, they may be used to selectively bind particular target molecules. The interaction of the target molecule with the sensor surface can produce some type of detectable signal to the metal centers in the polymer, which may include electrical, optical and catalytic reactions (Lahiri & Basu, 2024). Consequently, bioorganometallic polymer-based biosensors are currently being fabricated to be used in early disease detection, real-time health-related measurements and point-of-care testing. Altogether, metal functionality combined with polymeric materials has provided new possibilities in designing multifunctional biomedical systems (Table 3). The role of bioorganometallic polymers in the future of biomedical innovation is constantly increasing in terms of drug delivery technologies, regenerative medicine, antimicrobial therapies and diagnostic tools (Casini & Pöthig, 2024).

Table 3. Biomedical applications of bioorganometallic polymers with metal components and functions

Biomedical Application	Metal Component	Function of Metal	Example Use	Ref.
Drug delivery systems	Fe, Au	Stimulus-responsive release	Targeted cancer therapy	(Jia et al., 2021)
Tissue engineering scaffolds	Ca, Zn, Mg	Bone regeneration and cell growth	Bone tissue scaffolds	(Grewal et al., 2026)
Antimicrobial materials	Ag, Cu	Antibacterial and antifungal activity	Wound dressings and implant coatings	(Ul Haq & Krukiewicz, 2023)
Biosensors and diagnostics	Pt, Au, Cu	Catalytic and electronic properties	Detection of biomolecules	(Lino et al., 2022)

6. CHALLENGES AND LIMITATIONS

- The problem of metal toxicity, structural stability, cost of production and regulatory approval is also a significant issue that has to be considered closely when developing the material design. The issue of metal toxicity is one of the major difficulties of bioorganometallic polymers. Even though some metal ions like iron, zinc, magnesium and calcium are naturally found in biological systems and are deemed biocompatible, some of them may have toxic properties when they are in high concentrations (Casini & Pöthig, 2024). Overrelease of metal ions in the polymer matrix may cause cytotoxicity, oxidative stress and cellular damage. In this case, the use of such metals as copper, silver and cadmium can disrupt cell metabolism and enzymatic activity if their content surpasses the limits of safe physiological concentrations. Thus, close regulation of the rates of metal incorporation and release should be taken to guarantee that bioorganometallic polymers will not cause any harm to biomedical applications (Egorova & Ananikov, 2017). Furthermore, variations in individual patient responses may influence the safety profile of metal-containing biomaterials, emphasizing the need for personalized biocompatibility assessment.

- The second limitation is also associated with the stability of polymer-metal complexes. The strength and stability of metal-ligand interactions in the polymer structure are critical to the performance and functionality of bioorganometallic polymers. Such interactions can be destabilized in the biological environment by changes in pH, temperature, enzyme activity and competing ions. This can cause early degradation, uncontrolled release of metal ions or it can also cause the loss of material functionality. Therefore, scientists have to develop polymer systems that exhibit unperturbed coordination bonds and optimized architectures to ensure their structural integrity even in physiological conditions (Ronga et al., 2023).
- The large-scale bioorganometallic polymer manufacturing is also associated with cost and scalability issues. The manufacture of these hybrid materials can require complicated chemical reactions, special catalysts and pure metal sources; thus, these materials may cost more to produce. Also, there are more sophisticated methods of fabrication like controlled radical polymerization and the synthesis of nanostructured materials, which involve more complex equipment and technical skills. These can inhibit the commercial scale of bioorganometallic polymers, especially for mass medical and industrial use (Reynes et al., 2024).
- Finally, regulatory and safety issues are also extremely important in the development of biomedical materials. Bioorganometallic polymers have to pass a series of tests to ensure that they are safe, effective and stable before being applied in clinical practice. The regulatory agencies need a lot of data about the biocompatibility, toxicity, products and environmental impact of degradation. The metal components can also complicate the approval process because of the possible anxiety about tissue accumulation of metal and long-term health impacts could be raised (Bernard et al., 2018; Tian et al., 2022). Consequently, both preclinical and clinical research is required to determine the safety and functionality of such materials. In addition, large-scale manufacturing, batch-to-batch consistency and cost-effective production remain important challenges for the clinical translation of bioorganometallic polymers.

7. FUTURE PERSPECTIVES AND EMERGING TECHNOLOGIES

- The sphere of bioorganometallic polymers is developing quite fast and new technologies should help to improve and increase the design, functionality and biomedical uses of those polymers considerably. The new directions in computational technology, nanotechnology and sustainable materials science are providing new opportunities in creating a new generation of biomaterials that are safer and more efficient (R. A. Omer et al., 2025).

- Among the most promising trends is the use of artificial intelligence in the design of biomaterials. The large datasets associated with polymer chemistry, metal coordination and biological interactions can be analyzed with AI and machine learning algorithms in order to predict the behavior and properties of new materials. The methods can be used to develop bioorganometallic polymers with ideal properties like regulated degradation rates, greater biocompatibility and increased therapeutic efficacy (Otyepka et al., 2025).
- The other trend is the incorporation of bioorganometallic polymers with nanotechnology and smart biomaterials. Nanostructured biomaterials have excellent surface area, mechanical characteristics and biological interactions. Using metal-containing polymers in nanoscale assemblies like nanoparticles, nanofibers and nanogels, it is possible to obtain multifunctional assemblies that can be used in targeted drug delivery, responsive therapeutic systems and advanced biosensing platforms (Eltaib, 2025; G. Zhang et al., 2020). Particularly, smart biomaterials, which react to environmental factors such as pH, temperature and redox, are promising in precision medicine and targeted therapeutic delivery.
- Biomaterial development is also being changed by the use of digital modeling and predictive materials science. The use of computational modeling tools enables scientists to model the interactions of molecules between polymers, metallic ions and biological molecules. Such simulations assist in foreseeing mechanical characteristics, degradation patterns and biological reactions so that the optimization of material organizations becomes more effective (Xue et al., 2021). Predictive modeling is also effective in developing personalized biomedical materials in response to a certain medical condition or patient need.

Figure 5. Future directions in bioorganometallic biomaterials.

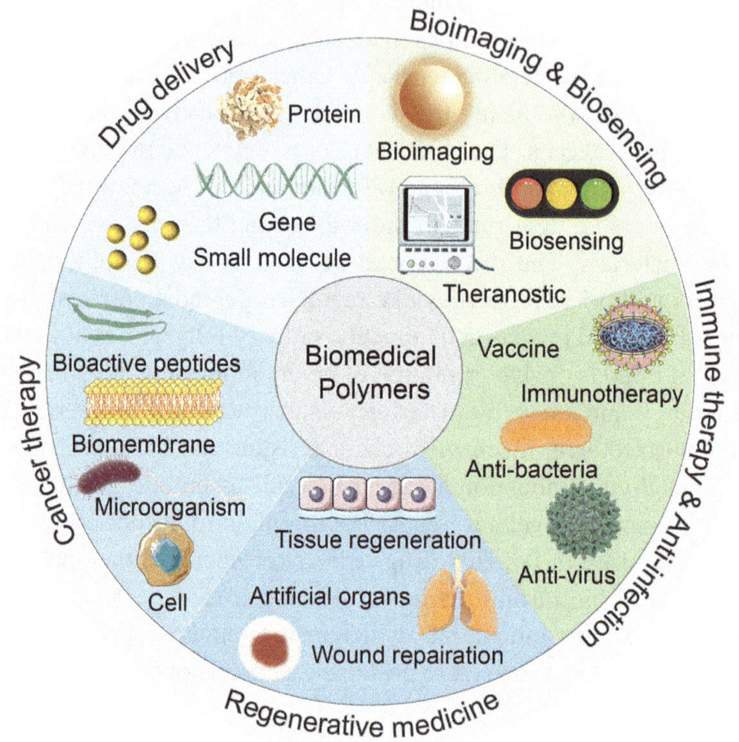

Reproduced with permission from Chen et al. (2022)

- Finally, there is an increasing trend towards sustainable development of biomaterials. To minimize the environmental impact of biomaterial synthesis, researchers are looking into the possibilities of eco-friendly synthesis, renewable sources of polymers and biodegradable metal complexes (Vinchurkar et al., 2025). Sustainable strategies support green chemistry and responsible biomaterial development (Vinchurkar et al., 2025). Future advances in personalized biomaterials and AI-assisted material design are expected to further accelerate the clinical translation of bioorganometallic polymers.

8. CONCLUSION

Bioorganometallic polymers are an emerging class of advanced biomaterials that integrate the structural versatility of organic polymers with the unique functional properties of metal centers. This hybrid combination offers tunable mechanical characteristics, controlled biodegradability, catalytic activity, antimicrobial performance and stimuli-responsive behavior, making these materials highly attractive for biomedical applications. Their potential has been demonstrated in drug delivery systems, tissue engineering scaffolds, antimicrobial coatings and biosensing platforms, where the controlled release of bioactive metal ions can enhance therapeutic efficacy and tissue regeneration. Incorporating biologically essential metals such as calcium, magnesium, zinc and iron further enhances bioactivity. This enables the creation of multifunctional biomaterials capable of fulfilling both structural and therapeutic roles. Recent advancements in nanotechnology, smart biomaterials, computational modeling and additive manufacturing have accelerated the design of next-generation bioorganometallic polymer systems offering superior performance and precision. Despite these promising developments, challenges related to metal toxicity, long-term biocompatibility, regulatory approval and large-scale production remain significant hurdles to clinical use. Overcoming these limitations will require interdisciplinary research involving chemists, materials scientists, biomedical engineers and clinicians. Continued innovation and rigorous evaluation are expected to facilitate the safe and effective integration of bioorganometallic polymers into biomedical and regenerative medicine applications in the future.

Declaration

The authors used AI-assisted tools solely for language editing, grammar correction and improving the readability of the manuscript. All scientific content, interpretations and conclusions were developed, reviewed and approved by the authors, who take full responsibility for the work.

REFERENCES

Abdullah, M., Obayedullah, M., Shariful Islam Shuvo, M., Abul Khair, M., Hossain, D., & Nahidul Islam, M. (2025). A review on multifunctional applications of nanoparticles: Analyzing their multi-physical properties. *Results in Surfaces and Interfaces*, *21*, 100635. DOI: 10.1016/j.rsurfi.2025.100635

Abdulsalam, L., Abubakar, S., Permatasari, I., Lawal, A. A., Uddin, S., Ullah, S., & Ahmad, I. (2025). Advanced Biocompatible and Biodegradable Polymers: A Review of Functionalization, Smart Systems, and Sustainable Applications. *Polymers*, *17*(21), 2901. DOI: 10.3390/polym17212901 PMID: 41228661

Abhishek, N., Verma, A., Singh, A., Vandana, , & Kumar, T. (2023). Metal-conducting polymer hybrid composites: A promising platform for electrochemical sensing. *Inorganic Chemistry Communications*, *157*, 111334. DOI: 10.1016/j.inoche.2023.111334

Abtahi, S., Hendeniya, N., Mahmud, S. T., Mogbojuri, G., Iheme, C. L., & Chang, B. (2025). Metal-Coordinated Polymer–Inorganic Hybrids: Synthesis, Properties, and Application. *Polymers*, *17*(2), 136. DOI: 10.3390/polym17020136 PMID: 39861209

Alven, S., Peter, S., Mbese, Z., & Aderibigbe, B. A. (2022). Polymer-Based Wound Dressing Materials Loaded with Bioactive Agents: Potential Materials for the Treatment of Diabetic Wounds. *Polymers*, *14*(4), 724. DOI: 10.3390/polym14040724 PMID: 35215637

Amirtharaj Mosas, K. K., Chandrasekar, A. R., Dasan, A., Pakseresht, A., & Galusek, D. (2022). Recent Advancements in Materials and Coatings for Biomedical Implants. *Gels (Basel, Switzerland)*, *8*(5), 323. DOI: 10.3390/gels8050323 PMID: 35621621

Anusiya, G., & Jaiganesh, R. (2022). A review on fabrication methods of nanofibers and a special focus on application of cellulose nanofibers. *Carbohydrate Polymer Technologies and Applications*, *4*, 100262. DOI: 10.1016/j.carpta.2022.100262

Barrino, F. (2024). Hybrid Organic–Inorganic Materials Prepared by Sol–Gel and Sol–Gel-Coating Method for Biomedical Use: Study and Synthetic Review of Synthesis and Properties. *Coatings*, *14*(4), 425. DOI: 10.3390/coatings14040425

Benny, A., Kalathiparambil Rajendra Pai, S. D., Pinheiro, D., & Chundattu, S. J. (2024). Metal organic frameworks in biomedicine: Innovations in drug delivery. *Results in Chemistry*, *7*, 101414. DOI: 10.1016/j.rechem.2024.101414

Bernard, M., Jubeli, E., Pungente, M. D., & Yagoubi, N. (2018). Biocompatibility of polymer-based biomaterials and medical devices – regulations, in vitro screening and risk-management. *Biomaterials Science*, *6*(8), 2025–2053. DOI: 10.1039/C8BM00518D PMID: 29968869

Carraher, C. E.Jr. (1981). Organometallic polymers. *Journal of Chemical Education*, *58*(11), 921. DOI: 10.1021/ed058p921

Casini, A., & Pöthig, A. (2024). Metals in Cancer Research: Beyond Platinum Metallodrugs. *ACS Central Science*, *10*(2), 242–250. DOI: 10.1021/acscentsci.3c01340 PMID: 38435529

Chen, W.-H., Chen, Q.-W., Chen, Q., Cui, C., Duan, S., Kang, Y., Liu, Y., Liu, Y., Muhammad, W., Shao, S., Tang, C., Wang, J., Wang, L., Xiong, M.-H., Yin, L., Zhang, K., Zhang, Z., Zhen, X., Feng, J., & Chen, X. (2022). Biomedical polymers: Synthesis, properties, and applications. *Science China. Chemistry*, *65*(6), 1010–1075. DOI: 10.1007/s11426-022-1243-5 PMID: 35505924

Constantino, V. R. L., Figueiredo, M. P., Magri, V. R., Eulálio, D., Cunha, V. R. R., Alcântara, A. C. S., & Perotti, G. F. (2023). Biomaterials Based on Organic Polymers and Layered Double Hydroxides Nanocomposites: Drug Delivery and Tissue Engineering. *Pharmaceutics*, *15*(2), 413. DOI: 10.3390/pharmaceutics15020413 PMID: 36839735

Cué-Sampedro, R., & Sánchez-Fernández, J. A. (2024). Supramolecular systems and their connection with metal–organic structures. *Frontiers in Chemistry*, *12*, 1468916. Advance online publication. DOI: 10.3389/fchem.2024.1468916 PMID: 39564433

Dadashi-Silab, S., & Matyjaszewski, K. (2020). Iron Catalysts in Atom Transfer Radical Polymerization. *Molecules (Basel, Switzerland)*, *25*(7), 1648. DOI: 10.3390/molecules25071648 PMID: 32260141

Egorova, K. S., & Ananikov, V. P. (2017). Toxicity of Metal Compounds: Knowledge and Myths. *Organometallics*, *36*(21), 4071–4090. DOI: 10.1021/acs.organomet.7b00605

El-Khawaga, A. M., Zidan, A., & El-Mageed, A. I. A. A. (2023). Preparation methods of different nanomaterials for various potential applications: A review. *Journal of Molecular Structure*, *1281*, 135148. DOI: 10.1016/j.molstruc.2023.135148

Eltaib, L. (2025). Polymeric Nanoparticles in Targeted Drug Delivery: Unveiling the Impact of Polymer Characterization and Fabrication. *Polymers*, *17*(7), 833. DOI: 10.3390/polym17070833 PMID: 40219222

Emmanuel, E. O., Idumah, C. I., Ekweonu, C. D., Madu, I., & Ikebudu, C. J. (2026). Optimization and functionalization of hybridized ferrocene polymeric nanoarchitectures and multifunctional applications. *Hybrid Advances, 12*, 100618. DOI: 10.1016/j.hybadv.2026.100618

Ezike, T. C., Okpala, U. S., Onoja, U. L., Nwike, C. P., Ezeako, E. C., Okpara, O. J., Okoroafor, C. C., Eze, S. C., Kalu, O. L., Odoh, E. C., Nwadike, U. G., Ogbodo, J. O., Umeh, B. U., Ossai, E. C., & Nwanguma, B. C. (2023). Advances in drug delivery systems, challenges and future directions. *Heliyon, 9*(6), e17488. Advance online publication. DOI: 10.1016/j.heliyon.2023.e17488 PMID: 37416680

Fan, H., Wang, H., Tian, Z., Shi, J., He, W., Qi, D., Guo, B., & Xu, J. (2025). Study on Hydrolytic Degradation of Polyester and Polyamide in Basic Solutions at High Temperatures. *Polymers, 17*(23), 3090. DOI: 10.3390/polym17233090 PMID: 41374779

Fan, Z., Hu, X., & Gao, R. X. (2022). Indirect Measurement Methods for Quality and Process Control in Nanomanufacturing. *Nanomanufacturing and Metrology, 5*(3), 209–229. DOI: 10.1007/s41871-022-00148-4

Farjaminejad, S., Farjaminejad, R., Hasani, M., Garcia-Godoy, F., Abdouss, M., Marya, A., Harsoputranto, A., & Jamilian, A. (2024). Advances and Challenges in Polymer-Based Scaffolds for Bone Tissue Engineering: A Path Towards Personalized Regenerative Medicine. *Polymers, 16*(23), 3303. DOI: 10.3390/polym16233303 PMID: 39684048

Grewal, N. S., Verma, M., Kumar, S., Khan, A. R., Kumar, K., & Sharma, N. (2026). Advancements in Mg-based scaffolds for bone tissue engineering: From design strategies to clinical perspectives. *Journal of Magnesium and Alloys, 102010*, 102010. Advance online publication. DOI: 10.1016/j.jma.2026.102010

Grishin, D. F. (2023). Methods of controlled radical polymerization in the synthesis of functional polymers and macromolecular structures. *Russian Chemical Bulletin, 72*(6), 1285–1298. DOI: 10.1007/s11172-023-3904-0

Herdiana, Y., Wathoni, N., Shamsuddin, S., & Muchtaridi, M. (2022). Drug release study of the chitosan-based nanoparticles. *Heliyon, 8*(1), e08674. DOI: 10.1016/j.heliyon.2021.e08674 PMID: 35028457

Hudi ă, A., & Gălă eanu, B. (2023). Polymer Materials for Drug Delivery and Tissue Engineering. *Polymers, 15*(14), 3103. DOI: 10.3390/polym15143103 PMID: 37514492

Hui, C. M., Pietrasik, J., Schmitt, M., Mahoney, C., Choi, J., Bockstaller, M. R., & Matyjaszewski, K. (2014). Surface-Initiated Polymerization as an Enabling Tool for Multifunctional (Nano-)Engineered Hybrid Materials. *Chemistry of Materials*, *26*(1), 745–762. DOI: 10.1021/cm4023634

Iacob, A.-T. (2025). Biomedical Application of Polymeric Materials. *Polymers*, *17*(17), 2401. DOI: 10.3390/polym17172401 PMID: 40942319

Jia, R., Teng, L., Gao, L., Su, T., Fu, L., Qiu, Z., & Bi, Y. (2021). <p>Advances in Multiple Stimuli-Responsive Drug-Delivery Systems for Cancer Therapy</p>. *International Journal of Nanomedicine*, *16*, 1525–1551. DOI: 10.2147/IJN.S293427 PMID: 33658782

Jiang, H., Li, L., Li, Z., & Chu, X. (2024). Metal-based nanoparticles in antibacterial application in biomedical field: Current development and potential mechanisms. *Biomedical Microdevices*, *26*(1), 12. DOI: 10.1007/s10544-023-00686-8 PMID: 38261085

Kalpana Manivannan, R., Sharma, N., Kumar, V., Jayaraj, I., Vimal, S., & Umesh, M. (2024). A comprehensive review on natural macromolecular biopolymers for biomedical applications: Recent advancements, current challenges, and future outlooks. *Carbohydrate Polymer Technologies and Applications*, *8*, 100536. DOI: 10.1016/j.carpta.2024.100536

Kanwal, A., Zhang, M., Sharaf, F., & Li, C. (2022). Polymer pollution and its solutions with special emphasis on Poly (butylene adipate terephthalate (PBAT)). *Polymer Bulletin*, *79*(11), 9303–9330. DOI: 10.1007/s00289-021-04065-2

Karimi, M., Asefnejad, A., Noshadi, B., Iranmanesh, F., & Nasri, P. (2025). An Overview of Biodegradable Polymers: Revolutionizing Biomedical Applications by Increasing Stability and Biocompatibility for Biomaterial Applications from Lumen-Apposing Metal Stents to Dental Implants. *Nanochemistry Research*, *10*(2). Advance online publication. DOI: 10.22036/NCR.2025.02.05

Kayed, S. F., & Almeataq, M. S. (2023). Photocatalytic Activity and Thermal Stability of Hybrid Metal–Polymer-Coordinated Complexes Derived from Gallic Acid and Ethylenediamine. *Langmuir*, *39*(30), 10445–10452. DOI: 10.1021/acs.langmuir.3c00869 PMID: 37458686

Khalkhali, P., Omidi, M., Masson-Meyers, D. S., Akbari, B., Dehghan, M. M., Aminianfar, H., Farzad-Mohajeri, S., Mansouri, V., Nikpasand, A., & Tayebi, L. (2025). Promoting Angiogenesis/Osteogenesis by a New Copper/Magnesium Hydroxide Hybrid Nanoparticle: In Vitro and In Vivo Investigation. *Journal of Biomedical Materials Research. Part A*, *113*(1), e37855. DOI: 10.1002/jbm.a.37855 PMID: 39815692

Kim, E. E., Kononevich, Yu. N., Anisimov, A. A., Buzin, M. I., Vasil'ev, V. G., Korlyukov, A. A., Ionov, D. S., Khanin, D. A., Shtykova, E. V., Volkov, V. V., & Muzafarov, A. M. (2021). Cross-linked polymer networks based on polysiloxane and nickel β-diketonate precursors. *Reactive & Functional Polymers, 164*, 104896. DOI: 10.1016/j.reactfunctpolym.2021.104896

Kothawade, S., & Shende, P. (2024). Coordination bonded stimuli-responsive drug delivery system of chemical actives with metal in pharmaceutical applications. *Coordination Chemistry Reviews, 510*, 215851. DOI: 10.1016/j.ccr.2024.215851

Kumar, P., Rajak, D. K., Abubakar, M., Ali, S. G. M., & Hussain, M. (2021). 3D Printing Technology for Biomedical Practice: A Review. *Journal of Materials Engineering and Performance, 30*(7), 5342–5355. DOI: 10.1007/s11665-021-05792-3

Kurowiak, J., Klekiel, T., & Będziński, R. (2023). Biodegradable Polymers in Biomedical Applications: A Review—Developments, Perspectives and Future Challenges. *International Journal of Molecular Sciences, 24*(23), 16952. DOI: 10.3390/ijms242316952 PMID: 38069272

Lahiri, H., & Basu, K. (2024). Supramolecular Sensing Platforms: Techniques for In Vitro Biosensing. *ChemEngineering, 8*(4), 66. DOI: 10.3390/chemengineering8040066

Laird, N. Z., Acri, T. M., Chakka, J. L., Quarterman, J. C., Malkawi, W. I., Elangovan, S., & Salem, A. K. (2021). Applications of nanotechnology in 3D printed tissue engineering scaffolds. *European Journal of Pharmaceutics and Biopharmaceutics, 161*, 15–28. DOI: 10.1016/j.ejpb.2021.01.018 PMID: 33549706

Liang, Y., Li, L., Scott, R. A., & Kiick, K. L. (2017). 50th Anniversary Perspective: Polymeric Biomaterials: Diverse Functions Enabled by Advances in Macromolecular Chemistry. *Macromolecules, 50*(2), 483–502. DOI: 10.1021/acs.macromol.6b02389 PMID: 29151616

Lino, C., Barrias, S., Chaves, R., Adega, F., Martins-Lopes, P., & Fernandes, J. R. (2022). Biosensors as diagnostic tools in clinical applications. *Biochimica et Biophysica Acta (BBA) -. Revisiones en Cáncer, 1877*(3), 188726. DOI: 10.1016/j.bbcan.2022.188726 PMID: 35367530

Mangani, S., Vetoulas, M., Mineschou, K., Spanopoulos, K., Vivanco, M., Piperigkou, Z., & Karamanos, N. K. (2025). Design and Applications of Extracellular Matrix Scaffolds in Tissue Engineering and Regeneration. *Cells, 14*(14), 1076. DOI: 10.3390/cells14141076 PMID: 40710329

Mankaev, B. N., & Karlov, S. S. (2023). Metal Complexes in the Synthesis of Biodegradable Polymers: Achievements and Prospects. *Materials (Basel)*, *16*(20), 6682. DOI: 10.3390/ma16206682 PMID: 37895663

Martínez, I. V., Ek, J. I., Ahn, E. C., & Sustaita, A. O. (2022). Molecularly imprinted polymers via reversible addition–fragmentation chain-transfer synthesis in sensing and environmental applications. *RSC Advances*, *12*(15), 9186–9201. DOI: 10.1039/D2RA00232A PMID: 35424874

Mishra, K. J., & Singh, D. N. K. (2025). Biodegradable Polymers-Based Scaffolds for Wound Healing and Tissue Engineering. [IJST]. *International Journal of Science and Technology*, *2*(3), 135–145. DOI: 10.70558/IJST.2025.v2.i3.241096

Mittal, M., Siddiqui, M. R., Tran, K., Reddy, S. P., & Malik, A. B. (2014). Reactive Oxygen Species in Inflammation and Tissue Injury. *Antioxidants & Redox Signaling*, *20*(7), 1126–1167. DOI: 10.1089/ars.2012.5149 PMID: 23991888

Mohamed, H. I., ElKhawaga, H. A., Abd El-Lateef, H. M., Ibrahim, S. M., Reda, L. M., & Metwally, A. M. (2025). Development of novel coordination polymeric metal complexes with potent antimicrobial and antitumor activity. *European Polymer Journal*, *232*, 113944. DOI: 10.1016/j.eurpolymj.2025.113944

Negut, I., Bita, B., & Groza, A. (2022). Polymeric Coatings and Antimicrobial Peptides as Efficient Systems for Treating Implantable Medical Devices Associated-Infections. *Polymers*, *14*(8), 1611. DOI: 10.3390/polym14081611 PMID: 35458361

Nobahar, A., Passos, H., Schaeffer, N., Sosa, F. H. B., & Coutinho, J. A. P. (2026). Engineered Biopolymer Architectures for Selective Metal Ions Extraction. *Advanced Sustainable Systems*, *10*(4), e70457. DOI: 10.1002/adsu.70457

Oleksy, M., Dynarowicz, K., & Aebisher, D. (2023). Advances in Biodegradable Polymers and Biomaterials for Medical Applications—A Review. *Molecules (Basel, Switzerland)*, *28*(17), 6213. DOI: 10.3390/molecules28176213 PMID: 37687042

Omer, P. K., Abdulkareem, E. I., Omer, R. A., & Rashid, R. F. (2025). A review of organometallic compounds as versatile sensors in environmental, medical, and industrial applications. *Reviews in Inorganic Chemistry*, *45*(2), 397–409. DOI: 10.1515/revic-2024-0055

Omer, R. A., Sdiq, A., Qader, A. F., Salih, M., Abdulkareem, E., Ismail, H., Rashid, R., & Raheja, U. (2025). Organometallic Nanoparticles: Catalysts for Sustainable Chemistry and Advanced Nanotechnology. *Chemical Review and Letters*, *8*(3), 612–627. DOI: 10.22034/crl.2025.504643.1536

Otyepka, M., Pykal, M., & Otyepka, M. (2025). Advancing materials discovery through artificial intelligence. *Applied Materials Today*, *47*, 102981. DOI: 10.1016/j.apmt.2025.102981

Pires, P. C., Mascarenhas-Melo, F., Pedrosa, K., Lopes, D., Lopes, J., Macário-Soares, A., Peixoto, D., Giram, P. S., Veiga, F., & Paiva-Santos, A. C. (2023). Polymer-based biomaterials for pharmaceutical and biomedical applications: A focus on topical drug administration. *European Polymer Journal*, *187*, 111868. DOI: 10.1016/j.eurpolymj.2023.111868

Qu, R., Suo, H., Gu, Y., Weng, Y., & Qin, Y. (2022). Sidechain Metallopolymers with Precisely Controlled Structures: Synthesis and Application in Catalysis. *Polymers*, *14*(6), 1128. DOI: 10.3390/polym14061128 PMID: 35335458

Rajeev, S., & Panicker, U. G. (2025). Polymer-Metal Organic Frameworks: Recent Advances in Synthesis Strategies and Applications. *Journal of Inorganic and Organometallic Polymers and Materials*, *35*(7), 5157–5188. DOI: 10.1007/s10904-025-03616-8

Rauf, U., Shabir, G., Bukhari, S., Albericio, F., & Saeed, A. (2023). Contemporary Developments in Ferrocene Chemistry: Physical, Chemical, Biological and Industrial Aspects. *Molecules (Basel, Switzerland)*, *28*(15), 5765. DOI: 10.3390/molecules28155765 PMID: 37570735

Reynes, J. F., Leon, F., & García, F. (2024). Mechanochemistry for Organic and Inorganic Synthesis. *ACS Organic & Inorganic Au*, *4*(5), 432–470. DOI: 10.1021/acsorginorgau.4c00001 PMID: 39371328

Rezaee, T., Fazel-Zarandi, R., Karimi, A., & Ensafi, A. A. (2022). Metal-organic frameworks for pharmaceutical and biomedical applications. *Journal of Pharmaceutical and Biomedical Analysis*, *221*, 115026. DOI: 10.1016/j.jpba.2022.115026 PMID: 36113325

Ronga, L., Varcamonti, M., & Tesauro, D. (2023). Structure–Activity Relationships in NHC–Silver Complexes as Antimicrobial Agents. *Molecules (Basel, Switzerland)*, *28*(11), 4435. DOI: 10.3390/molecules28114435 PMID: 37298911

Rosato, A., Romano, A., Totaro, G., Celli, A., Fava, F., Zanaroli, G., & Sisti, L. (2022). Enzymatic Degradation of the Most Common Aliphatic Bio-Polyesters and Evaluation of the Mechanisms Involved: An Extended Study. *Polymers*, *14*(9), 1850. DOI: 10.3390/polym14091850 PMID: 35567020

Saini, G., Segaran, N., Mayer, J. L., Saini, A., Albadawi, H., & Oklu, R. (2021). Applications of 3D Bioprinting in Tissue Engineering and Regenerative Medicine. *Journal of Clinical Medicine, 10*(21), 4966. DOI: 10.3390/jcm10214966 PMID: 34768485

Sharma, S., Mohler, J., Mahajan, S. D., Schwartz, S. A., Bruggemann, L., & Aalinkeel, R. (2023). Microbial Biofilm: A Review on Formation, Infection, Antibiotic Resistance, Control Measures, and Innovative Treatment. *Microorganisms, 11*(6), 1614. DOI: 10.3390/microorganisms11061614 PMID: 37375116

Shi, K., Nokhodchi, A., & Ghafourian, T. (2025). Magnetic microscale polymeric nanocomposites in drug delivery: Advances and challenges. *Drug Discovery Today, 30*(1), 104276. DOI: 10.1016/j.drudis.2024.104276 PMID: 39736462

Sindhi, K., Pingili, R. B., Beldar, V., Bhattacharya, S., Rahaman, J., & Mukherjee, D. (2025). The role of biomaterials-based scaffolds in advancing skin tissue construct. *Journal of Tissue Viability, 34*(2), 100858. DOI: 10.1016/j.jtv.2025.100858 PMID: 39827732

Smola-Dmochowska, A., Lewicka, K., Macyk, A., Rychter, P., Pamuła, E., & Dobrzyński, P. (2023). Biodegradable Polymers and Polymer Composites with Antibacterial Properties. *International Journal of Molecular Sciences, 24*(8), 7473. DOI: 10.3390/ijms24087473 PMID: 37108637

Socci, M. C., Rodríguez, G., Oliva, E., Fushimi, S., Takabatake, K., Nagatsuka, H., Felice, C. J., & Rodríguez, A. P. (2023). Polymeric Materials, Advances and Applications in Tissue Engineering: A Review. *Bioengineering (Basel, Switzerland), 10*(2), 218. DOI: 10.3390/bioengineering10020218 PMID: 36829712

Srivastava, A., & Azad, U. P. (2023). Nanobioengineered surface comprising carbon based materials for advanced biosensing and biomedical application. *International Journal of Biological Macromolecules, 253*, 126802. DOI: 10.1016/j.ijbiomac.2023.126802 PMID: 37690641

Su, Y., Cockerill, I., Wang, Y., Qin, Y.-X., Chang, L., Zheng, Y., & Zhu, D. (2019). Zinc-Based Biomaterials for Regeneration and Therapy. *Trends in Biotechnology, 37*(4), 428–441. DOI: 10.1016/j.tibtech.2018.10.009 PMID: 30470548

Sun, Y., Ma, J., Ahmad, F., Xiao, Y., Guan, J., Shu, T., & Zhang, X. (2024). Bimetallic Coordination Polymers: Synthesis and Applications in Biosensing and Biomedicine. *Biosensors (Basel), 14*(3), 117. DOI: 10.3390/bios14030117 PMID: 38534224

Tabasi, H., Babaei, M., Abnous, K., Taghdisi, S. M., Saljooghi, A. S., Ramezani, M., & Alibolandi, M. (2021). Metal–polymer-coordinated complexes as potential nanovehicles for drug delivery. *Journal of Nanostructure in Chemistry, 11*(4 (December 2021)), 501–526. Advance online publication. DOI: 10.1007/s40097-021-00432-7

Tian, J., Song, X., Wang, Y., Cheng, M., Lu, S., Xu, W., Gao, G., Sun, L., Tang, Z., Wang, M., & Zhang, X. (2022). Regulatory perspectives of combination products. *Bioactive Materials, 10*, 492–503. DOI: 10.1016/j.bioactmat.2021.09.002 PMID: 34901562

Tilden, J. A. R., Doud, E. A., Montgomery, H. R., Maynard, H. D., & Spokoyny, A. M. (2024). Organometallic Chemistry Tools for Building Biologically Relevant Nanoscale Systems. *Journal of the American Chemical Society, 146*(44), 29989–30003. DOI: 10.1021/jacs.4c07110 PMID: 39468851

Tousian, B., Khosravi, A. R., Ghasemi, M. H., & Kadkhodaie, M. (2024). Biomimetic functionalized metal organic frameworks as multifunctional agents: Paving the way for cancer vaccine advances. *Materials Today. Bio, 27*, 101134. DOI: 10.1016/j.mtbio.2024.101134 PMID: 39027676

Tumilaar, S. G., Hardianto, A., Dohi, H., & Kurnia, D. (2024). A Comprehensive Review of Free Radicals, Oxidative Stress, and Antioxidants: Overview, Clinical Applications, Global Perspectives, Future Directions, and Mechanisms of Antioxidant Activity of Flavonoid Compounds. *Journal of Chemistry, 2024*(1), 5594386. DOI: 10.1155/2024/5594386

Ul Haq, I., & Krukiewicz, K. (2023). Antimicrobial approaches for medical implants coating to prevent implants associated infections: Insights to develop durable antimicrobial implants. *Applied Surface Science Advances, 18*, 100532. DOI: 10.1016/j.apsadv.2023.100532

Vinchurkar, K., Bukke, S. P. N., Jain, P., Bhadoria, J., Likhariya, M., Mane, S., Suryawanshi, M., Veerabhadrappa, K. V., Eftekhari, Z., & Onohuean, H. (2025). Advances in sustainable biomaterials: Characterizations, and applications in medicine. *Discover Polymers, 2*(1), 2. DOI: 10.1007/s44347-025-00014-8

Xue, K., Wang, F., Suwardi, A., Han, M.-Y., Teo, P., Wang, P., Wang, S., Ye, E., Li, Z., & Loh, X. J. (2021). Biomaterials by design: Harnessing data for future development. *Materials Today. Bio, 12*, 100165. DOI: 10.1016/j.mtbio.2021.100165 PMID: 34877520

Yoshinari, N., & Konno, T. (2023). Multitopic metal–organic carboxylates available as supramolecular building units. *Coordination Chemistry Reviews, 474*, 214850. DOI: 10.1016/j.ccr.2022.214850

Zhang, G., Zhang, L., Rao, H., Wang, Y., Li, Q., Qi, W., Yang, X., Su, R., & He, Z. (2020). Role of molecular chirality and solvents in directing the self-assembly of peptide into an ultra-pH-sensitive hydrogel. *Journal of Colloid and Interface Science*, *577*, 388–396. DOI: 10.1016/j.jcis.2020.05.087 PMID: 32497920

Zhang, J., Zhou, L., Jiang, X., Hu, W., & Yang, R. (2024). Self-Assembly of Metabolite Nanostructures toward Functional Biomaterials. *ACS Materials Letters*, *6*(2), 674–696. DOI: 10.1021/acsmaterialslett.3c01337

Zheng, Z., Cui, J., Wu, S., Cao, Z., & Cao, P. (2025). Engineering Metal-Organic Frameworks for Enhanced Antimicrobial Efficacy: Synthesis Methodologies, Mechanistic Perspectives, and Versatile Applications. *Journal of Functional Biomaterials*, *16*(9), 353. DOI: 10.3390/jfb16090353 PMID: 41003425

Chapter 7
Advanced Biomaterials for Tissue Repair and Regeneration:
A Circular Economy Approach Utilizing Tilapia Skin

Soumin Ghosh
JIS University, India

Soumyabrata Roy
JIS University, India

Dibyarupa Pal
http://orcid.org/0000-0002-8564-1485
JIS University, India

ABSTRACT

The surging demand for sustainable biocompatible materials from renewable resources has drawn attention towards Tilapia skin waste for tissue repair and generation. Tilapia has abundant source of type I collagen, structurally similar to mammalian collagen with outstanding biocompatibility, low immunogenicity, and no cultural or health restrictions. In order to treat burns wounds, tilapia skin maintains its bioactive qualities for cell adhesion, proliferation, and controlled inflammatory responses. Modern fabrication techniques employ hydrogels, electrospun membranes, and porous scaffolds for skin regeneration. Co-functional bio active molecules improves mechanical stability, moisture retention and improves antimicrobial properties. With

DOI: 10.4018/979-8-2600-0060-1.ch007

the advent of clinical translation, these affordable biomaterials made from tilapia exhibit huge waste utilization potential to be incorporated in circular economy approach for healthcare and environmental purposes.

1. INTRODUCTION

1.1 Brief contextualization of Tissue Engineering imperatives

Tissue regeneration is the process by which damaged or missing tissues are repaired, replaced or restored to their normal biological function. An alternative process to conventional reconstructive techniques, tissue engineering provides substitutes for tissues or organs, acting as a tissue resource in cases of scarcity, such as in elderly patients and burn victims. Due to the dearth of donor tissues and the inability of certain tissues to regenerate like cardiac tissue and components of the nervous system, tissue engineering techniques have emerged as a long term solution to these challenges. However limited success has been reported for treating complex organs, but successful clinical results has been reported for some target organs like skin, cartilage, urinary bladder and trachea. Advancement in tissue engineering fields have satisfactorily fulfilled the increasing demands for intricate organ and tissue models in the biomedical sector. Furthermore, innovative methodologies and tissue sources continually emerge to enhance existing tissue engineering initiatives by integrating pathway engineering for cell differentiation regulation, nanoscale bioactive agent patterning, and non-invasive imaging techniques (Zorlutuna et al., 2013; Mandrycky et al., 2017; Meretsky et al., 2024).

Tissue engineering is fundamentally organized into two main strategies: top-down and bottom-up techniques. The top-down approach utilizes additive manufacturing techniques to construct three-dimensional scaffolds by sequentially adding material, resulting in porous structures with customized architecture to facilitate tissue development. In this approach, viable cells are then introduced or 'seeded' on or within the scaffold. A primary advantage of top-down approaches is their adaptability, enabling the utilization of diverse processing materials and the creation of scaffolds with tailored mechanical properties appropriate for the desired application. This approach is constrained by several limitations, including insufficient vascularization within the construct, challenges in attaining a uniform distribution of various cell types, and the resulting inability to achieve tissue-specific cell densities, all of which impede the functional complexity of the engineered tissue. Conversely, bottom-up methodologies entail the concurrent assembly of scaffolding materials, cells, and frequently bioactive substances into modular building units of diverse forms and dimensions. Methods include hydrogel encapsulation, self-assembled cell aggregation, cell sheet engineering, and three-dimensional bioprinting, facilitate the creation of structures with improved functional complexity. Scaffolds possessing an optimal morphology and substantial mechanical, physical, and biological properties are essential. The 3D printing of antibacterial scaffolds possessing enough mechanical strength and superior biocompatibility is a promising approach to address microbial infection challenges. However, Bottom-up techniques have gained significance due to their enhanced ability for precise spatial control over cell arrangement, enabling the final architecture to accurately replicate the structural and organizational complexity of natural tissues. These sort of advanced tissue engineering builds engineered scaffolds according to the patient's specific requirement of tissue shape and structures (Whitely et al., 2018; Sultan and Mathew, 2018; Rivera-Briso and Serrano-Aroca, 2018; Liu et al., 2022; Dubey et al., 2023).

Effective tissue engineering is essential for designing a bioengineered environment that encourages cellular activity and tissue repair (Figure 1). This can be accomplished by using specialized scaffolds that facilitate cell growth, specialization, and the controlled delivery of growth-promoting signals. These scaffolds replicate the body's natural support structure, providing a framework for new tissue formation. Growth factors are crucial in this system, as they trigger cellular communication, stimulate blood vessel growth (angiogenesis), and support tissue reorganization. The development of new blood vessels is especially important, as it delivers oxygen and nutrients to implanted cells, ensuring their survival and integration with surrounding tissues. By strategically combining scaffolds, living cells, and bioactive compounds, tissue engineering offers a promising solution for repairing or replacing damaged organs and restoring normal function. This multidisciplinary approach paves the way for breakthroughs in regenerative therapies and in the treatment of severe tissue damage.

Tissue regeneration relies on two key components: cellular building blocks and the extracellular matrix (ECM) (Vilaça-Faria, et al., 2024).

1.2 Cellular Building Blocks

The cellular building blocks denote the population of viable cells that facilitate the creation, maintenance, and repair of tissue structures. This encompasses tissue-specific parenchymal cells that execute particular functions like cardiomyocytes in the heart or hepatocytes in the liver, alongside multipotent stem cells and progenitor cells that facilitate regeneration by developing into specialized cell types.

1) Fibroblasts: responsible for generating extracellular matrix and the primary source of collagen

2) Mesenchyme cells: responsible for generation of various lineages of cells in tissues and controls inflammation.
3) Keratinocytes: Mediates epithelialization of skin.

The efficacy of tissue repair in regenerative therapies depends on the sourcing, viability, and integration of biological components. Regardless of whether sourced from the patient's own body (autologous), a donor (allogeneic), or differentiated from induced pluripotent stem cells (iPSCs), these cellular components must achieve effective cell-to-cell communication, endure the inflammatory milieu of an injury site, and conform to the structural guidance of the extracellular matrix to restore functional tissue. The cellular building blocks constitute further different cell types:

Stem Cells: Undifferentiated cells (e.g., embryonic, mesenchymal, or induced pluripotent stem cells) that can develop into specialized cell types.

Progenitor Cells: Partially differentiated cells that contribute to tissue repair (e.g., satellite cells in muscles).

Differentiated Cells: Mature cells that maintain and restore tissue function (e.g., fibroblasts in skin repair).

1.3 Extracellular Matrix (ECM): Nature's Blueprint

The extracellular matrix (ECM) is a complex three-dimensional structure composed of structural proteins, including fibrillar collagen, glycoproteins and elastin, with polysaccharides such as hyaluronan and proteoglycans. The ECM serves as a dynamic scaffolding facilitating critical cellular processes, including attachment, migration, and intracellular signalling, in addition to providing mechanical support. According to O'Brien (2011), a scaffold is a type of biomaterial that can facilitate cell attachment, proliferation, and differentiation, which in turn can result in the development of new tissue. Scaffolds are designed to mimic the shape and physiology of cells in living organisms. They have specific purposes and possess desirable qualities such as a high surface-to-volume ratio, biodegradability, mechanical strength, flexibility, and the ability to promote cell adhesion. It also functions as a reservoir for growth factors such as vascular endothelial growth factor (VEGF) and transforming growth factor-beta (TGF-β) which are released in response to cellular signals to modulate proliferation, differentiation, and tissue remodelling. The intricate interplay between cells and their ECM is crucial for successful tissue regeneration, as the mutual communication between these components dictates the fate of both indigenous and transplanted cells. Consequently, recent advancements in regenerative medicine have focused on developing biomimetic materials and decellularized scaffolds that replicate the biochemical and biophysical properties of the natural extracellular matrix. These engineered platforms aim to harness the instructional capabilities of the extracellular matrix, so enhancing the efficacy of regenerative therapies and improving outcomes in complex clinical scenarios such as substantial tissue loss or organ failure (Kaur, and Reinhardt, 2015; Karamanos, et al., 2021; Dzobo et al., 2023; Liu et al., 2023).

The extracellular matrix (ECM) affects cell mitogenesis and chemotaxis, guides cell differentiation, and stimulates beneficial host tissue remodelling responses. The three-dimensional ultrastructure, surface topology, and compositional characteristics of the ECM likely influence these intricate effects. These intrinsic characteristics work together to provide a bioactive environment that promotes cellular infiltration and actively directs the phenotypic behaviour of recruited cells towards regenerative outcomes rather than fibrotic scar formation. Research indicates that leftover cellular material within biological scaffolds reduces or completely negates the advantageous benefits of tissue remodelling linked to these materials when implanted in vivo. The existence of residual DNA, cellular membranes, and intracellular antigens can elicit detrimental inflammatory reactions, compromising the host's capacity to initiate an effective regeneration response. Thus, tissue processing methods especially decellularization techniques are vital factors in achieving clinical success (Kim et al., 2017; Yao et al., 2019; Brouki et al., 2025).

1.4 Decellularized Extracellular Matrix (dECM) Scaffolds: The Rationale

A three-dimensional porous scaffold can be built using biodegradable materials, which can be derived from either natural sources or manufactured polymers. For example decellularized tilapia skin was used to create a low cost and immune compatible granular hydrogel using microfluidics (Hodde, 2002; Im et al., 2024). Scaffolds made of decellularized extracellular matrix (dECM) produced from real tissues have gained popularity due to their special biomimetic characteristics. These dECM scaffolds can facilitate the generation of new tissues for restoration and augment the body's intrinsic healing capacity extending beyond conventional tissue transfers and transplantation techniques. The effectiveness of ECM-derived scaffolds is thus significantly reliant on the meticulous removal of cellular components while maintaining the structural integrity, biochemical composition, and three-dimensional architecture of the natural ECM. Attaining this equilibrium continues to be a primary issue in the advancement of biologic scaffold materials for regenerative medicine applications for diverse organs, including skin, nerve, bone, heart, liver, lung, and kidney (Crapo et al., 2011; Kabirian et al., 2020; Gadre et al., 2024).

1.5 Collagen

Collagen is the most abundant protein of the mammals (25% of the total protein) with a characteristic triple-helical structure, predominant in connective tissues, bones, skin and cartilage. The protein has exceptional tensile strength, consisting of three polypeptide chains and cell binding features, widely employed as a biomaterial in creating scaffolds that imitate natural tissue (Rappu et al., 2019; Cao et al., 2024). Recombinant human collagen represents a promising, safer alternative to animal-derived collagen for tissue engineering. While challenges in scalable, cost-effective production remain, ongoing advances in genetic engineering, biomaterial design, and 3D bioprinting position RHC as a transformative biomaterial for regenerative medicine. The four predominant varieties are: type I (present in skin, tendons, bone, and cornea), type II (found in cartilage), type III (located in blood vessels and skin), and type IV (associated with basement membranes). This diversity facilitates improved biomimetics through the selection of collagen types specific to the target tissue. Type I bovine collagen has been utilized to create porous scaffolds that enhance angiogenesis, nutrient transport, and cell migration—essential for tissue survival. These scaffolds have effectively replicated human endometrium when populated with stromal or organoid cells, exhibiting responsiveness to hormonal stimulation. In corneal engineering, rat tail collagen I scaffolds provide robust mechanical strength, excellent transparency, elevated water content, and gradual degradation (extendable through UV cross-linking), facilitating the proliferation of epithelial cells and keratocytes for corneal tissue substitutes with prospective transplantation applications.In vitro studies have recombinant produced different types of collagens from isolated genes, including types I, II, III, and V collagen. Compared to natural collagen, recombinantly produced collagen achieves proper post-translational modifications, including hydroxylation and glycosylation. Additionally, collagen is biocompatible, exhibit low antigenicity and high biodegradability, mostly derived from animal tendon tissues like pigs, bovines and rat (Avila Rodríguez et al., 2018). But apprehensions about transmission of zoonotic diseases and religious constraints have opined for alternative and safer collagen sources, like from fish, sponges and jellyfish. Huge prospect lies in aquaculture industry where processing of fish generates heads, fins, viscera, scales, bones and skin waste. Though most Fish collagen has diminished proline and hydroxyproline content required for mechanical strength and conformation stability; exception lies in tilapia, a warm water fish with comparable amino acids to that of mammalian sources (Jafari et al., 2020).

2. TILAPIA SKIN AS AN OPPORTUNITY

2.1 Structural biochemical composition and advantages of Tilapia Skin

Among the various marine species, Nile tilapia (*Oreochromis niloticus*), a tropical freshwater fish, has gained attention recently. The tilapia fish collagen scaffold has many benefits like it replicates human collagen and helps in natural healing by providing a stable, biocompatible environment for new tissue formation (Table 1). The fish has high denaturation temperature of its type I collagen (35 °C), the highest among all reported fish collagen (Sugiura et al., 2009). A high denaturation temperature is desirable as the collagen would not denature easily during processing. Tilapia collagen was first identified as a potential biomaterial by Sugiura et al (2009), who reported that tilapia collagen sponges exhibited similar biosorption rates and inflammatory responses to porcine collagen sponges. Ever since, tilapia collagen has been proven to support corneal regeneration (Yuan et al., 2014), oral-maxillofacial bone regeneration (Tang and Saito, 2015) and skin regeneration (Zhou et al., 2016). Currently, tilapia is farmed in more than 100 countries due to its fast breeding, high protein content, adaptability in a wide range of culture systems, and high resistance to disease (Fitzsimmons, 2011). The wide geographical distribution and constant availability of tilapia allow the large scale production of tilapia-derived biomaterials economically. It is one of the most farmed fish globally making it an ideal resource for developing tilapia collagen biomedical products due to its abundant resource and low cost (Lau et al., 2019; Huang et al., 2024; Zimba et al., 2024).

Table 1. Highlights the Tilapia skin: Composition, properties and biomimetic advantages

Features of Tilapia skin	Characteristics	Remarks on Biomimetic advantage	References
Collagen content and denaturation temperature	Type I collagen (abundant); high denaturation temperature (35°C)	Low immunogenicity, less immune rejection. Similar skin absorption rate as porcine skin.	Sugiura et al., 2009
Amino acid profile	Comparable to mammalian collagen; high proline and hydroxyproline	Absence of zoonotic disease risk and dissemination	Jafari et al., 2020; Avila Rodríguez et al., 2018
Bioactive componenets	Omega-3 PUFAs (EPA, DHA); glycosaminoglycans	Religious and cultural acceptability	Tozetto et al., 2025
Structural organization	Dermal matrix with aligned collagen fibers; high porosity	Highly structurally similar to mammalian skin, quick healing properties	Lau et al., 2019; Huang et al., 2024

3. CIRCULAR ECONOMY FRAMEWORK: WASTE TO RESOURCE

3.1 Aquaculture waste problem

Tilapia can grow in a variety of salinities, from freshwater to seawater, can withstand acidic and alkaline conditions, as well as low oxygen (2 mg/l) and high ammonia (50 mg/l). The global aquaculture business generates millions of tonnes of fish processing waste annually, with tilapia skin accounting for 30-45% of the fish's weight as discarded byproduct like skin, scales, and bone (Moraiatypical s et al., 2020; Mundra et al., 2023; Zimba et al., 2024; Biyada & Urbonavičius, 2025). Traditional disposal methods used are landfilling and incineration which have environmental and economic consequences. Circular principle advocates transforming such wastes into valuable resources reducing ecological impact, nutrient recovery and regenerating new revenue streams. The EU's Circular Economy Action Plan (European Commission, 2020b) promotes this transition from linear to circular model, aiming to increase renewable energy, waste resource re-use, and valorisation of by-products (Table 2).

3.2 Waste valorization: The circular economy approach

Figure 1. Schematic diagram showing the process of Tilapia skin collagen extraction to application

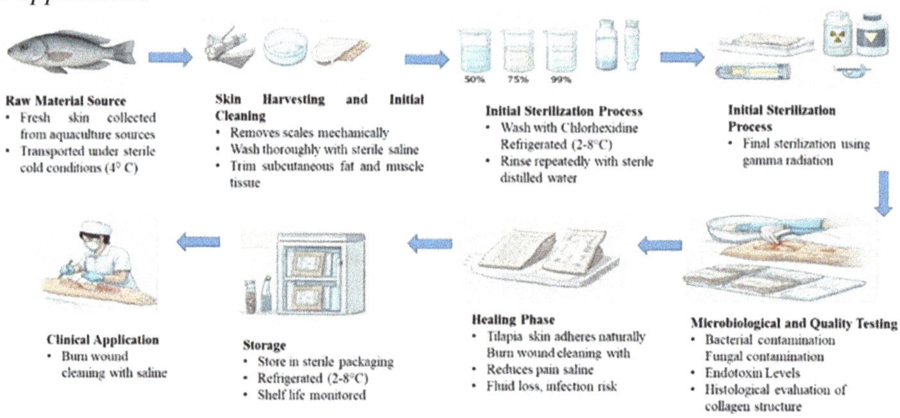

Table 2. The circular economy benefits of Tilapia skin valorization

Different aspects	Impact on society and Environment	References
Tilapia waste	Diversion of dumping into landfills and waterways	Zimba et al., 2024
Reduction in carbon foorprint	Lower compared to livestock farming	Biyada and Urbonavičius, 2025
Processing cost	Less with minimal processing ways	Lau et al., 2019
Revenue generation and rural development	Value addition in pharmaceutical, biomedical and cosmetics sector; new job openings	Munshi et al., 2025
Waste turned to product	Integrated closed loop system; waste turned into medical asset	Shang et al., 2025

4. EXTRACTION AND PROCESSING TECHNOLOGIES

4.1 Collagen extraction from Tilapia skin

Tilapia biomaterial is modified into tilapia collagen membrane by creating electrospun fibre. Electrospinning is recognized as a method for producing nanofibrous three-dimensional (3D) scaffolds that replicate the extracellular matrix (ECM) milieu for tissue engineering (TE). In contrast to conventional densely-packed two-dimensional (2D) nanofibrous membranes, three-dimensional (3D) electrospun nanofiber scaffolds offer enhanced spatial control, providing adequate porosity and a 3D environment that closely resembles in vivo conditions, with similar optimized properties like injectability, compressibility, and bioactivity. During scaffold manufacturing, tilapia fish collagen is initially removed and prepared for electrospinning by formulating a 50% wt. solution of the collagen in 0.1 M acetic acid. A 20% polyvinyl alcohol (PVA) solution is independently produced in deionized water at 80°C and agitated until uniform. Upon preparation of both solutions, they are combined in a 50:50 (v/v) ratio at ambient temperature, agitated to achieve uniformity, and thereafter stored at 4°C for 24 hours to facilitate the release of any entrapped air bubbles. The resultant collagen/PVA solutions are subsequently employed for electrospinning to produce nanofiber scaffolds. These scaffolds possess significant biomedical uses, especially in tissue engineering, where they function as wound dressings, templates for skin regeneration, or matrices for tendon repair, utilizing the biocompatibility and structural advantages of collagen generated from tilapia.

Burn wound healing therapy has evolved using Nile tilapia fish skin as less dressing changes are required compared to conventional method. A study indicated that the decellularized tilapia dermal matrix enhanced cellular differentiation and mineralization of stem cells. Such Fish-derived matrix are abundant in omega-3 polyunsaturated fatty acids and has demonstrated promising outcomes in the treatment of chronic diabetic foot ulcers (DFUs), necrotic and angiodermatitis wounds. The omega-3 fatty acids in the decellularized tilapia accelerated the healing process by enhancing cytokine production of the inflammatory phase. Acellular fish skin matrix has also shown to markedly reduce both the wound depth in patients with complex non-healing ulcers. Moreover, the fish skin showed great adherence to the wound bed and this decreased the amount of analgesics used throughout the entire treatment. In another clinical and pre-clinical experiments indicate that acellular Tilapia skin reduces the healing time of superficial burns to 10–11 days, compared to 2 weeks with conventional treatment approaches and Deep burns often heal in roughly 21 days, compared to almost 3 weeks. Treating with tilapia fish skin was more sustainable and likewise reduced the budgetary impact of the therapy when compared to more conventional methods (Tozetto et al., 2025; Munshi et al., 2025).

4.2 Decellularization of Tilapia skin

The protocol of dECM initiates with the extraction and purification from washed tilapia scales, to eliminate surface contaminants. Non-collagenous proteins were subsequently removed by NaCl solution and demineralization by using hydrochloric acid solution to eliminate calcium and phosphorus minerals. The extracted collagen was subjected to dialysis for purification and when the yield was sufficient, freeze-dried, for biomedical use (Bi et al., 2019). Tilapia skin collagen sponge was developed into electrospun nanofibers for rapid wound dressing. In another investigation, tannic acid was incorporated into electrospun Tilapia fibre for more stability and application in food packaging. Furthermore, the 3D morphology governs cellular interactions and facilitates proliferation, migration, and differentiation of cells for matrix remodelling (Zhou et al., 2016; Vilches et al., 2025). Decellularized Tilapia skin collagen is also cross-linked in presence of glutaraldehyde to develop innovative hybrid membranes. Tang and Saito, 2015 reported use of such hybrid biomaterials to induce the formation of odontoblast-like cells and osteoblasts from human mesenchyme stem cells. Significant improvement in the odontoblast cell attachment was observed compared to control and reports also highlight that tilapia collagen can modulate immune cell activity, induce osteogenic differentiation and promote wound healing.

Decellularization is the procedure of eliminating cellular constituents from tissues or organs to produce biological scaffolds comprised of extracellular matrix (ECM). The principal objective is to maintain the ECM's intricate composition, three-dimensional ultrastructure, and bioactivity while removing biological components that may elicit a detrimental immune response. Decellularization typically involves a combination of chemical, biologic, and physical agents. The choice depends on tissue density and the desired properties of the final scaffold.

The application of Nile Tilapia (*Oreochromis niloticus*) skin in the treatment of burns and wounds has garnered interest owing to its structural, histological, and mechanical resemblance to human skin, in addition to its accessibility, affordability, safety, and significant biocompatibility, notably its type I collagen concentration. Sterilization is a crucial step in preparing tilapia skin for therapeutic application, as contamination by aquatic bacteria such *Pseudomonas aeruginosa, Aeromonas hydrophila,* and *Candida parapsilosis* presents infection hazards. Traditional sterilization techniques employed for mammalian skin grafts such as ethylene oxide gas, irradiation, and chemical agents exhibit constraints. Ethylene oxide generates hazardous residues, whereas irradiation may affect collagen integrity by inducing cross-linking or peptide bond cleavage. Common disinfectants such as chlorhexidine and povidone-iodine demonstrate efficacy against a wide range of bacteria, although they may influence tissue characteristics. Silver nanoparticles have being investigated as an antibacterial option. An optimal sterilization technique must significantly diminish microbial load while preserving the structural integrity, biochemical composition, and biocompatibility of the skin graft. This study seeks to assess the effects of three distinct sterilization methods differing in duration and technique on microbial count, collagen characteristics, and histological structure, due to the limited research on sterilization protocols specifically for fish skin. Moreover Tilapia skin can reinforce various other biopolymers to form biocomposites for applications, in the area of bone and skin regeneration. Table 3 demonstrates the various decellularization techniques applied to get a low immunogenic trigger during grafting process (Mohammed et al., 2025; Martinez et al., 2025).

Table 3. Depicting decellularization process using different agents

Agents	Examples	Mechanism	Effects on ECM
Chemical Agents			
Acids and Bases	Peracetic acid, Sodium hydroxide	Hydrolytic degradation of biomolecules; cell lysis.	damaged collagen, growth factors eradicated, and reduce mechanical strength
Hypo/Hypertonic Solutions	Saline solutions	Osmotic shock causes cell lysis	Effectively lyses cells but does not fully remove cellular residues; minimal ECM disruption.

continued on following page

Table 3. Continued

Agents	Examples	Mechanism	Effects on ECM
Detergents	Ionic (SDS, Deoxycholate), Non-ionic (Triton X-100, Zwitterionic (CHAPS)	Solubilize cell membranes; disrupt DNA-protein interactions	SDS: Most effective for cell ultrastructure disruption and growth factors removal. Triton X-100: Better for thinner tissues; less disruptive but may remove GAGs.
Solvents	Alcohols, Acetone	Dehydration; lipid removal	Effective for dense tissues but can crosslink and precipitate proteins like collagen
Biologic Agents			
Enzymes	Trypsin, Nucleases (DNase), Dispase	Cleave specific peptide bonds or nucleic acids.	Prolonged exposure to Trypsin can disrupt collagen and elastin; nucleases can remove DNA
Chelating Agents	EDTA, EGTA	Disrupt cell adhesion to ECM by binding metal ions.	Typically used with enzymes; ineffective alone
Physical Agents			
Freeze-Thaw	-	cells lysed by Intracellular ice crystals	Minimal loss of ECM protein with ultrastructure disruption
Pressure	Hydrostatic pressure, Perfusion	bursts cells; forces agents through tissue	Can disrupt ECM if pressure is excessive
Electroporation	Non-thermal irreversible electroporation	Electrical pulses destabilize cell membranes	Preserves ECM morphology but limited by tissue size and immune-mediated clearance

4.3 Biomaterial Formats for Fabrication

The Tilapia skin can be transformed into several biomaterial forms that have potential uses in tissue engineering and regenerative medicine. Composite materials, xenografts, hydrogels (including granular hydrogels), electrospun nanofibers, acellular dermal matrices, collagen sponges, and collagen-gelatin blends are among these formats. There are distinct benefits to using each format, which are based on structural features, bioactivity, and therapeutic relevance.

4.3.1 Hyaluronic acid hydrogels

HA is naturally found in connective tissues, a component of the extracellular matrix playing a critical role in wound healing, joint lubrication, and tissue hydration. Considered as the simplest glycosaminoglycan consisting of Glucuronic acid and N-acetylglucosamine, HA is extracted from different sources like vitreous humor, umbilical cord, synovial fluid and rooster combs. HA regulates the activity of immune cells, helps to resolve inflammation and prevents the development of chronic inflammation that would hinder the process of recovery. It also promotes the migration and proliferation of important cells like fibroblasts and keratinocytes, essential for wound closure and tissue regeneration. It also promotes angiogenesis, the process of new blood vessel development, which is critical for supplying oxygen and nutrients to the healing tissue. Furthermore, HA promotes the development of the ECM, which helps to shrink the wound and prevent tissue scarring. Being hygroscopic in nature, the capacity to hold onto moisture makes it an ideal healing medium and reduces the likelihood of scab formation (Ding et al., 2022; Chang et al., 2024). HA has antimicrobial properties that help prevent infection, a critical factor in wound healing. By enhancing cell functions, maintaining hydration, and promoting tissue remodelling, hyaluronic acid accelerates healing in both acute and chronic wounds. HA is also recognized as a potential biomaterial for the development of wound dressing. There are various reports that emphasized the wound healing efficacy of HA in stimulating epithelial and mesenchymal cell differentiation and migration, therefore improving collagen deposition and angiogenesis.

A hyaluronic acid (HA) hydrogel scaffold is a 3D, porous structure made from hyaluronic acid, a naturally occurring glycosaminoglycan designed to mimic the properties of natural tissue for biomedical applications such as tissue engineering, wound healing, and drug delivery. This scaffold combines the hydrophilic, biocompatible, and biodegradable properties of HA with the structural benefits of a hydrogel, creating an ideal environment for cell growth, regeneration, and controlled release of bioactive molecules. To enhance the mechanical properties, stability, and controlled degradation of the HA hydrogel scaffold, crosslinking agents are often used. Common crosslinkers include 1,4-butanediol diglycidyl ether (BDDE), glutaraldehyde, or genipin (Reddy et al., 2015; Li et al., 2020). Crosslinking forms a network of bonds between the HA molecules, resulting in a more robust and durable structure, while still maintaining its water-retention and porous characteristics. The hyaluronic acid solution is mixed with the crosslinking agent and then subjected to conditions that facilitate gelation. These conditions include high or low temperature changes or UV irradiation, depending on the nature of the crosslinking method. The HA solution then forms a gel-like structure, which can be shaped into scaffolds using Molds or 3D printing techniques. To create a highly porous structure, the HA hydrogel can be freeze-dried. This process removes water from the gel and leaves behind a scaffold with interconnected pores, which are essential for cell migration, nutrient exchange, and tissue growth. The final HA hydrogel scaffold is sterilized through methods such as gamma irradiation or ethanol sterilization to ensure it is safe for medical and clinical applications (Chen et al., 2023).

4.3.2 Alginate derived hydrogels and blends

Alginate is a natural, anionic and hydrophilic polymer commercially produced by various marine algae of Phaeophyceae family and some bacteria (*Pseudomonas or Azotobacter*). Chemically consist of thousand blocks of β-d-mannuronic acid and α-l-guluronic acid monomers connected via 1→4 linkage. It is a versatile polymer, highly biocompatible, easy to handle, nontoxic, economic affordability and with gelation properties. Different forms of Alginate-based biomaterials are developed like hydrogels, microcapsules, foams and sponges, using different fabrication techniques. Nonetheless, alginate hydrogels has great importance for biomedical applications in drug delivery system and as the supporting scaffold matrix for tissue repair and regeneration. It is used in the form of injectable hydrogels, microcapsules, fibre, and in making anisotropic 3D scaffolds. Alginate is found in different quality from low, medium, or high viscosity and thus tendency to forms a stable hydrogel in the presence of polyvalent cations (Mg^{2+}, Ca^{2+}), also varies. As a natural polysaccharide, alginate is hygroscopic shows a pH-dependent anionic character and interacts with cationic polyelectrolytes and proteoglycans, making it a convenient system for the loading of cells. Their ability to form a supportive 3D structure, retain moisture, and degrade naturally makes them an excellent choice for promoting tissue regeneration and controlled release of therapeutic agents. Like osteochondral tissue repair, bone and, cartilage regeneration, reconstruction of the corneal endothelium has been achieved by fabricating alginate in different forms like hydroxyapatite alginate-based scaffolds, alginate-gelatin-calcium composite hydrogels, methacrylamide gelatin/alginate hydrogels, alginate/hydroxypropyl chitosan hydrogels. It also serves as an exceptional component for creating bioinks, either alone or in synergism, used in 3D printed scaffolds (Sarker et al., 2017; Lee et al., 2012; Farshidfar et al., 2023; Hu et al., 2024).

Alginate is a commonly used polymer of natural origin for the design and fabrication of drug delivery systems (DDS) due to its well-known advantageous properties. The presence of active functional (carboxylic and hydroxyl) groups allows a wide range of possibilities for biochemical, chemical, or enzymatic modification necessary for the design of various functional and effective DDSs. The design of a DDS containing alginate as a single/combined polymer requires the control of important properties such as porosity, mechanical strength, swelling behaviour, degradation, and cell and/or bioactive agent loading and release properties according to the desired drug release profile (sustained/controlled), drug administration route (oral, transdermal, parenteral, pulmonary, localized/targeted) as well as to the challenges that must be overcome by bioactive molecules (Suhail, et al., 2021; Kavand et al., 2024).

4.3.3 Silk Fibroin scaffold as electrospun membranes

Silk fibroin (SF), a naturally occurring biopolymer generated from *Bombyx mori* silkworm, has been used as a suture in humans for a long time. The easily accessible and exceptional properties have led to its exploitation in biomedical research domains. These properties include biocompatibility, gradual degradation, sterilizability, and great mechanical properties (Sun et al., 2021). SF has the ability to be transformed into tissue-engineered scaffolds, such as films, hydrogels, sponges, and fibers, with remarkably reproducible forms (Rockwood et al., 2011). According to Yang et al. (2014), Liu et al. (2019), and Ullah and Chen (2020), silk proteins can be mixed with other materials to improve or attain certain qualities, such as biomedical traits like cell adhesion and biocompatibility. Scaffolds made of SF mimic the natural tissue's extracellular matrix and provide a physical framework for cells and vasculature to interact with, while also supporting, the development of new tissues. Cells and vasculature travel to scaffolds after implantation and develop into their openings. Granulation tissue creation involves the ongoing degradation of the scaffold and its eventual replacement by fully formed new tissue. A silk fibroin scaffold is a three-dimensional framework obtained from the *Bombyx mori* silkworm. Extensively utilized in tissue engineering, it integrates superior mechanical strength with biocompatibility and biodegradability to mimics the cellular extracellular matrix (ECM), thus facilitating cell proliferation and tissue regeneration.

4.3.4 Chitosan conjugated dressings

SF-based dressings are also conjugated with chitosan, a biocompatible and biodegradable polysaccharide, presenting antimicrobial and haemostatic activity. More importantly, chitosan also accelerates wound healing. Guang et al., 2015 examined a chitosan/SF (CS/SF) composite scaffold in skin wound healing in Sprague–Dawley rats. In vitro examination revealed antimicrobial properties of this dressing after 24 h of a culture with the inhibition against *E. coli*, *S. aureus*, *Pseudomonas aeruginosa*, and *Monilia albicans*. Scaffolds made of SF alone did not cause inhibition of microbial growth. In vivo study of a CS/SF composite scaffold on Sprague-Dawley rats showed promotion of skin wound healing without signs of infection and completely healed skin injuries after 21 days. Silver nanoparticles (AgNPs) can be an influential component of fibroin dressings due to their broad antimicrobial properties.

5. BIOMEDICAL APPLICATIONS OF TILAPIA SKIN

As an eco-friendly and cost-effective substitute for conventional scaffolds obtained from mammals, tilapia skin acellular dermal matrix (TADM) has demonstrated exceptional efficacy in the management of acute skin wounds. Since tilapia skin is often thrown away as aquaculture waste, these results are in line with the concepts of the circular economy, which aim to turn an environmental burden into a medicinal asset by reusing and recycling materials. A hybrid scaffold might be created by combining electrospun nanofibers with collagen from tilapia skin. This would allow the scaffold to take advantage of both the mechanical strength and adjustable breakdown of SF and the inherent bioactivity and cell signalling capabilities of tilapia collagen (Ibrahim et al., 2020).

5.1 Burn Treatment and Wound Healing

Biological dressings made of tilapia skin have shown promising results in treating burns and chronic wounds. Research shows that compared to conventional therapies, using tilapia skin speeds healing, lessens discomfort, and decreases infection rates. This cost-efficient strategy has proved beneficial in both human and veterinary medicine, thanks to the skin's high collagen concentration and structural resemblance to human skin. Collagen scaffolds made of freeze-dried tilapia skin were just as effective as commercially available collagen dressings made of pig or cow skin in healing excisional wounds in mice.

5.2 Bone regeneration

Earlier studies have shown that collagen composites made from tilapia skin, especially when mixed with hydroxyapatite, are very biocompatible, degrade at suitable rates, and have beneficial osteogenic characteristics for use in bone tissue engineering. Oral and maxillofacial bone abnormalities can be effectively treated using these composites (Shang et al., 2025).

5.3 Corneal Repair

Tilapia skin has demonstrated promising outcomes in the restoration of eyesight in dogs suffering from corneal ulcers and serious eye injuries. This highlights the adaptable feature of Tilapia skin in ophthalmic applications (Melo et al., 2024).

5.4 Periodontal and dental applications

Periodontal bone defects were repaired by hydrogel and scaffolding approaches where Tilapia skin was applied as occlusive dressing in the palate area. Dentin-pulp complexes are soft tissue which were repaired by injectable formulations in Tilapia platforms (Sridhar et al., 2024).

5.5 Clinical Case studies with advanced tissue engineering

The literary evidence supports the use of Tilapia skin xenografts in wound care with efficacy and encouraging signs of advanced clinical trials. However, this evidence is limited by small sample sizes, inconsistent methodologies, and a lack of long-term follow-up data. Systematic reviews and meta-analyses are the main sources of this evidence which are cited below as successful cases of clinical trials:

Dong et al., 2020 developed conformable hydrogels made of hyaluronic acid (HA) and adipose-derived stem cells to treat burn injuries. The storage modulus increased considerably within three minutes to signal the gelation point, indicating that higher polymer concentrations accelerated the hydrogels' rapid gelation at ambient temperature. *In vivo* studies on a mouse model of severe second-degree burns revealed that the hydrogels significantly accelerated wound closure and reduced scar formation.

Ying et al., 2019 developed in situ collagen-hyaluronic acid hydrogels utilizing human microvascular endothelial cells (HMEC) and fibroblasts (COS-7). The swelling capacity of the hydrogels increased progressively over 72 hours, attaining 45%, as indicated by the swelling index. This swelling efficiency facilitates cell outgrowth by exchanging nutrients and signalling molecules between cells cultured in the HA hydrogel and those at the wound site. Following a three-hour incubation with the hydrogels at 37 °C, approximately 47% of *S. aureus* and 55% of *E. coli* were eradicated, illustrating the antibacterial efficacy of the hydrogels. COS-7 and HMEC encapsulated within these hydrogels exhibited substantial cellular proliferation. Thus wounds treated with hydrogels had a superior healing process compared to commercial medication.

Wang et al., 2019 formulated in situ hyaluronic acid (HA) hydrogels that encapsulate plasmid DNA encoding vascular endothelial growth factor (VEGF) for the treatment of burn wounds. *In vivo* tests utilizing Sprague Dawley mice indicated that the hydrogels expedited healing in splinted burn sites by suppressing inflammation, enhancing microvascular growth, and exhibiting biocompatibility.

Makvandi et al., 2019 developed hydrogels composed of hyaluronic acid (HA) infused with silver nanoparticles (AgNPs). Rheological assessment verified superior mechanical capabilities and a gelation temperature close to physiological levels, facilitating straightforward local application. Cytotoxicity studies demonstrated that the hydrogels exhibited significant biocompatibility with L929 cells. Antibacterial assays demonstrated substantial growth inhibition of B. subtilis and E. coli. In vitro wound healing models indicated that AgNP-loaded hydrogels expedited wound closure and repair relative to the control.

Da Silva et al., 2017 demonstrated a novel strategy for diabetic wound healing by leveraging integration of human adipose stem cells within hyaluronic acid to orchestrate neo- vascularization, neo-innervation and inflammation. In contrast to conventional bulk hydrogels, the hybrid approach enhanced porosity, improved oxygen permeability and fluid absorption, all critical factors in exudate-prone diabetic wounds. The approach also emphasized on controlled inflammatory response and transition to successful nerve regeneration. The study demonstrates that while stem cell secretome modulation can enhance epidermal differentiation and macrophage polarization, but excessive tilting towards anti-inflammatory stage can hamper nerve generation. Thus crucial balance in wound healing is necessary with timely inflammation to recruit and guide the repair cells. The stem cell-hydrogel constructs improvises on stabilizing cell retention, biocompatibility, and structural support and maintain neurotropic factor gradients for successful wound healing.

Karimi et al., 2022 investigated the in vivo regenerative effects of a daphne mucronate (DM)-loaded alginate scaffold (1.25 g alginate in 100 mL distilled water) for wound healing purposes. They found that alginate-based scaffolds loaded with 1.25 g DM/100 g alginate achieved controlled release of daphne mucronate, as a rich source of herbal antioxidants, and accelerated the tissue regeneration process. The daphne mucronate-loaded alginate based scaffolds prompted fast expression and high levels of EGF, bFGF and MMP13 genes essential for the wound regeneration process and substituted pathological tissues with normal, physiological forms.

Monavari et al., 2023 designed an alginate dialdehyde-gelatin hydrogel reinforced by lysozyme and silica-calcium nanoparticles and loaded by cerium to obtain hydrogel scaffolds for bone tissue engineering. Scaffolds containing 5% alginate were used. Obtained in vitro results showed that the hydrogel scaffolds improved proliferation, adhesion and differentiation of pre-osteoblast cells due to the excellent biocompatibility of alginate dialdehyde-gelatin hydrogel, the bioactivity of the cerium and silica-calcium nanoparticles, and the stabilizing effect of lysozyme on the scaffold structure. Additionally, the scaffolds provided a release of lysozyme throughout the 21-day immersion time and decreased the proliferation of MG63 osteosarcoma cells due to the effect of lysozyme. All obtained results supported the convenience of the scaffolds for bone tissue engineering able to induce bone regeneration, address infection and inhibit tumor formation.

Zhang et al., 2019 conducted a study comparing two distinct electrospun silk fibroin membranes against a control group treated with a commercial wound dressing (3MTM TegadermTM). A poly-dopamine modified electrospun SF membrane (PESF) was created, while the second SF dressing remained unmodified (ESF). Full-thickness skin wound models were established on Sprague-Dawley rats. The authors demonstrated that all treatments significantly improved wound healing. They disclosed that the poly-dopamine-modified SF dressing exhibited a considerably improved performance ($p < 0.001$) compared to the other dressings throughout the healing phase. Moreover, wounds managed with both types of SF membranes exhibited reduced inflammatory infiltration on the third postoperative day compared to the control group. Full re-epithelialization was noted in all groups on the fourteenth day. The highest efficacy in skin regeneration was noted in the poly-dopamine-modified silk fibroin dressing group. This study demonstrated that SF membrane.

Hodgkinson et al., 2014 showed that SF scaffolds with fibre diameters in the range of 250–300 nm significantly promoted the proliferation of primary human dermal fibroblasts (PHDFs) and upregulated expression of collagen I and III compared to scaffolds with larger fibres. This nanoscale architecture appears to better mimic the native extracellular matrix, thereby enhancing fibroblast activity, a critical component of granulation tissue formation. Moreover, *ex vivo* studies using human full-thickness skin models revealed that SF scaffolds improved re-epithelialization by accelerating keratinocyte migration, a finding that aligns with the enhanced re-epithelialization observed in the Wj-MSC-SF-treated groups in Millán-Rivero's in vivo model.

Guo et al., 2021 investigated two types of silk fibroin (SF) scaffolds, namely random silk fibroin scaffolds (RSSs) and aligned silk fibroin scaffolds (ASSs), and their effects on vascularization and wound healing in female Sprague-Dawley rats. In vivo findings shown that both dressing types facilitate epithelialization in the full-thickness wound model. Throughout the investigation, wound healing was continuously more rapid in the ASS group compared to the RSS group, with a significant difference observed on day 14 ($p < 0.01$). Both types of dressings markedly expedited wound healing in comparison to the control (untreated) group. Moreover, both materials facilitated neovascularization in the wounds, which is crucial for efficient wound healing.

A noteworthy study was performed by Millán-Rivero et al., 2019 with umbilical cord Wharton's jelly mesenchyme stem cells (Wj-MSCs-SF) seeded on electrospun silk fibroin scaffolds to enhance cutaneous wound healing in a murine excisional wound splinting model. In this study, control wounds were covered with the commercial dressing Linitul® and with the SF scaffolds. Histological examination showed that wound healing was comparable in the Linitul® and SF scaffold groups. The wounds treated with the Wj-MSCs-SF caused a delay in the wound healing process, but inflammatory infiltration with polymorphonuclear neutrophils (PMN) and macrophages was significantly reduced compared to the SF scaffold group. Infiltration with T CD3+ cells was also decreased in the Wj-MSCs-SF group compared to the SF scaffold group. Furthermore, groups treated with SF scaffolds alone and with the addition of Wj-MSCs in the first two weeks post-injury had significantly ($p < 0.001$) increased vascular surface compared to the Linitul® group, which is crucial in effective skin regeneration.

Patil et al., 2015 investigated nano silver-loaded SF (NSF) wound healing activity in male Wistar rats. The authors compared NSF gel with optimized SF gel, silver nanoparticulate gel, Carbopol gel, a positive control (Soframycin gel), and a negative control (untreated wounds). After 15 days of the experiment, the wounds treated with NSF gel had the highest healing rate compared to the other groups and were significant ($p < 0.05$) compared to the Soframycin group.

Chouhan et al., 2018 examined an SF hydrogel and its influence on skin regeneration in third-degree burn wounds in Wistar albino rats. The SF hydrogel significantly accelerated wound healing compared to untreated wounds. Histological examination showed early development of granulation tissue and rapid reepithelialisation in treated wounds. By day 14 post-treatment, the wound area in the SF hydrogel-treated wounds was 40%, while the area of untreated wounds was 80%, which represented significantly ($p \leq 0.01$) accelerated wound healing in the treated wounds comp red to the control wounds. Furthermore, SF-treated wounds had higher vessel density compared to untreated wounds. In addition, treatment with the SF dressing revealed increased expression of CD163, a marker of the M2 type of macrophage. It means that the SF-treated wounds were characterized by the infiltration of M2 macrophages, which have a beneficial effect on wound healing by enhancing skin remodelling and regeneration.

Clinical trials are subjected to very less number of individuals and are subjected to constraints.

6. POTENTIAL CHALLENGES

Vascularization: Adequate vascularization of the engineered tissue is essential to create a sustainable functioning tissue. But promoting effective vascularization is often a daunting task that leads to considerable failures. Combination of various growth factors has been used to stimulate angiogenesis, the process of formation of new blood vessels. Although vascular endothelial growth factor (VEGF) is often applied to stimulate angiogenesis in artificially regenerating tissues with promising outcomes, various tissue specific challenges reappear and call for additional insights.

Immune rejection: Immune system of the host is a challenging barrier to the prospect of tissue engineering. Self-non-self-discrimination power of the innate and adaptive immunity is particularly challenging for xenogeneic biomaterials (from animals). The adverse effect is circumvented by selecting biomaterials with mild immunogenicity, use of immunosuppressive drugs and modification of the materials to render them less immunogenic. Standardization: Strict quality control is required because the skin composition of tilapia changes with age, food, and farming circumstances (Ghosh et al., 2023).

Restoring full function: Restoration of full function though considered impossible, achieving a sufficient degree is challenging as seen in skin grafts where hair follicles and sweat glands are difficult to regenerate.

Cost and mechanical properties: Escalating cost and disagreeable mechanical properties limits the selection of many ubiquitous biomaterials.

Clinical validation: Medical devices made from tilapia skin are still in their early stages and need more clinical validation, in contrast to silk-based items which have long histories of regulatory approval. Even though tilapia collagen are biodegradable, it is important to ensure complete resorption without adverse effects when managing their end-of-life.

7. FUTURE PROSPECTS AND REGULATORY STATUS

The capacity to transform aquaculture by-products into valuable therapeutic agents in line with the circular economy principles gives tilapia skin, a lot of hope for the future. Advanced cross-linking techniques that improve stability, biocompatibility, and cell adhesion are being developed to circumvent current mechanical strength limitations of Tilapia. In parallel the advent to make new composite materials like Tilapia-collagen hydroxyapatite demonstrates excellent osteogenic and biocompatible properties. Pre-clinical successes has further progressed but challenges lies in standardization of extraction and processing protocols, conducting larger-scale clinical trials to prove their safety and efficacy over the long periods, and scaling up production from waste streams.

Regulatory status of Tilapia skin as a biomaterial is at promising advanced level, but is still region-specific where Brazil is at the forefront. Phase II and Phase III randomized controlled trials in Brazil with 115 patients showed substantial benefits compared to standard silver sulfadiazine treatment, putting the product in a prime position to become the first nationally studied animal skin registered for clinical use by the National Health Surveillance Agency (ANVISA).

8. CONCLUSION

The use of tilapia skin as a cutting-edge biomaterial for repairing and regenerating tissues offers strategic solutions to convert fish waste into valorised product which fits well into circular economy. This method not only helps reduce pollution but also offers a biocompatible, inexpensive, and structurally appropriate scaffold for skin regeneration and wound healing by making use of a common aquaculture by-product. Tilapia skin's has collagen-rich matrix, antibacterial characteristics, and low immunogenicity which draws negligible ethical concerns compared to mammalian biomaterials. Thus medical innovation can drive tilapia waste into skin-based technology, which promotes clinical efficacy and environmental stewardship in a closed-loop system.

REFERENCES

Avila Rodríguez, M. I., Rodríguez Barroso, L. G., & Sánchez, M. L. (2018). Collagen: A review on its sources and potential cosmetic applications. *Journal of Cosmetic Dermatology*, *17*(1), 20–26. DOI: 10.1111/jocd.12450 PMID: 29144022

Bi, C., Li, X., Xin, Q., Han, W., Shi, C., Guo, R., & Zhong, J. (2019). Effect of extraction methods on the preparation of electrospun/electrosprayed microstructures of tilapia skin collagen. *Journal of Bioscience and Bioengineering*, *128*(2), 234–240. DOI: 10.1016/j.jbiosc.2019.02.004 PMID: 30833145

Brouki Milan, P., Masoumi, F., Biazar, E., Zare Jalise, S., & Mehrabi, A. (2025). Exploiting the potential of decellularized extracellular matrix (ECM) in tissue engineering: A review study. *Macromolecular Bioscience*, *25*(1), 2400322. DOI: 10.1002/mabi.202400322 PMID: 39412772

Cao, L., Zhang, Z., Yuan, D., Yu, M., & Min, J. (2024). Tissue engineering applications of recombinant human collagen: A review of recent progress. *Frontiers in Bioengineering and Biotechnology*, *12*, 1358246. DOI: 10.3389/fbioe.2024.1358246 PMID: 38419725

Chang, W., Chen, L., & Chen, K. (2024). The bioengineering application of hyaluronic acid in tissue regeneration and repair. *International Journal of Biological Macromolecules*, *270*, 132454. DOI: 10.1016/j.ijbiomac.2024.132454 PMID: 38763255

Chen, H., Xue, H., Zeng, H., Dai, M., Tang, C., & Liu, L. (2023). 3D printed scaffolds based on hyaluronic acid bioinks for tissue engineering: A review. *Biomaterials Research*, *27*(1), 137. DOI: 10.1186/s40824-023-00460-0 PMID: 38142273

Chouhan, D., Lohe, T. U., Samudrala, P. K., & Mandal, B. B. (2018). In situ forming injectable silk fibroin hydrogel promotes skin regeneration in full thickness burn wounds. *Advanced Healthcare Materials*, *7*(24), 1801092. DOI: 10.1002/adhm.201801092 PMID: 30379407

Commission of the European Communities. (2009). Communication from the Commission to the European Parliament, the Council, the European Economic and Social Committee and the Committee of the Regions: A Strategy for ICT R&D and Innovation in Europe: Raising the Game (Vol. 116). Office for Official Publications of the European Communities. A new Circular Economy Action Plan. For a cleaner and more competitive Europe (2020), pp. 1-19.

Crapo, P. M., Gilbert, T. W., & Badylak, S. F. (2011). An overview of tissue and whole organ decellularization processes. *Biomaterials*, *32*(12), 3233–3243. DOI: 10.1016/j.biomaterials.2011.01.057 PMID: 21296410 da Silva, L. P., Santos, T. C., Rodrigues, D. B., Pirraco, R. P., Cerqueira, M. T., Reis, R. L., & Marques, A. P. (2017). Stem cell-containing hyaluronic acid-based spongy hydrogels for integrated diabetic wound healing. Journal of Investigative Dermatology, 137(7), 1541-1551. DO DOI: 10.1016/j.jid.2017.02.976

de Souza Melo, M., da Silva, S. M., Neto, A. E. V., Vieira, N. C. G., de Moraes, M. E. A., Júnior, E. M. L., & de Moraes Filho, M. O. (2024). Advances in veterinary ophthalmology: Acellular protein matrix from tilapia skin (Oreochromis Niloticus) in the recovery of corneal ulcers in dogs and cats. *LUMEN ET VIRTUS*, *15*(39), 2031–2045. DOI: 10.56238/levv15n39-035

Ding, Y. W., Wang, Z. Y., Ren, Z. W., Zhang, X. W., & Wei, D. X. (2022). Advances in modified hyaluronic acid-based hydrogels for skin wound healing. *Biomaterials Science*, *10*(13), 3393–3409. DOI: 10.1039/D2BM00397J PMID: 35575243

Dong, Y., Cui, M., Qu, J., Wang, X., Kwon, S. H., Barrera, J., & Gurtner, G. C. (2020). Conformable hyaluronic acid hydrogel delivers adipose-derived stem cells and promotes regeneration of burn injury. *Acta Biomaterialia*, *108*, 56–66. DOI: 10.1016/j.actbio.2020.03.040 PMID: 32251786

Dubey, A., Vahabi, H., & Kumaravel, V. (2023). Antimicrobial and biodegradable 3D printed scaffolds for orthopedic infections. *ACS Biomaterials Science & Engineering*, *9*(7), 4020–4044. DOI: 10.1021/acsbiomaterials.3c00115 PMID: 37339247

Dzobo, K., & Dandara, C. (2023). The extracellular matrix: Its composition, function, remodeling, and role in tumorigenesis. *Biomimetics*, *8*(2), 146. DOI: 10.3390/biomimetics8020146 PMID: 37092398

Farshidfar, N., Iravani, S., & Varma, R. S. (2023). Alginate-based biomaterials in tissue engineering and regenerative medicine. *Marine Drugs*, *21*(3), 189. DOI: 10.3390/md21030189 PMID: 36976238

Fitzsimmons, K. E. (2011). An assessment of the luminescence sensitivity of Australian quartz with respect to sediment history. *Geochronometria*, *38*(3), 199–208. DOI: 10.2478/s13386-011-0030-9

Gadre, M., Kasturi, M., Agarwal, P., & Vasanthan, K. S. (2024). Decellularization and their significance for tissue regeneration in the era of 3D bioprinting. *ACS Omega*, *9*(7), 7375–7392. DOI: 10.1021/acsomega.3c08930 PMID: 38405516

Ghosh, B., Sánchez-Velazco, D. F., Prem, P., Ali, A. S. M., Dominique, O., Shah, S. A., & Obeysekera, D. (2023). Use of Nile tilapia fish skin in treatment for burn victims. *International Journal of Surgery. Global Health*, *6*(5), e0257. DOI: 10.1097/GH9.0000000000000257

Guang, S., An, Y., Ke, F., Zhao, D., Shen, Y., & Xu, H. (2015). Chitosan/silk fibroin composite scaffolds for wound dressing. *Journal of Applied Polymer Science*, *132*(35), app.42503. Advance online publication. DOI: 10.1002/app.42503

Guo, P., Du, P., Zhao, P., Chen, X., Liu, C., Du, Y., & Lv, G. (2021). Regulating the mechanics of silk fibroin scaffolds promotes wound vascularization. *Biochemical and Biophysical Research Communications*, *574*, 78–84. DOI: 10.1016/j.bbrc.2021.08.026 PMID: 34438350

Hodde, J. (2002). Naturally Occurring Scaffolds for Soft Tissue Repair and Regeneration. *Tissue Engineering*, *8*(2), 295–308. DOI: 10.1089/107632702753725058 PMID: 12031118

Hodgkinson, T., Yuan, X. F., & Bayat, A. (2014). Electrospun silk fibroin fiber diameter influences in vitro dermal fibroblast behavior and promotes healing of ex vivo wound models. *Journal of Tissue Engineering*, *5*, 2041731414551661. DOI: 10.1177/2041731414551661 PMID: 25383171

Hu, T., Fang, J., Shen, Y., Li, M., Wang, B., Xu, Z., & Hu, W. (2024). Advances of naturally derived biomedical polymers in tissue engineering. *Frontiers in Chemistry*, *12*, 1469183. DOI: 10.3389/fchem.2024.1469183 PMID: 39635576

Huang, J. Y., Wong, T. Y., Tu, T. Y., Tang, M. J., Lin, H. H., & Hsueh, Y. Y. (2024). Assessment of tilapia skin collagen for biomedical research applications in comparison with mammalian collagen. *Molecules (Basel, Switzerland)*, *29*(2), 402. DOI: 10.3390/molecules29020402 PMID: 38257315

Ibrahim, A., Hassan, D., Kelany, N., Kotb, S., & Soliman, M. (2020). Validation of three different sterilization methods of Tilapia skin dressing: Impact on microbiological enumeration and collagen content. *Frontiers in Veterinary Science*, *7*, 597751. DOI: 10.3389/fvets.2020.597751 PMID: 33426019

Im, P., Shin, H., & Kim, J. (2024). Tilapia-derived granular hydrogel as a 3D scaffold promoting rapid wound healing. Biomacromolecules, 25(2), 1153-1161. doi/.DOI: 10.1021/acs.biomac.3c01137

Jafari, H., Lista, A., Siekapen, M. M., Ghaffari-Bohlouli, P., Nie, L., Alimoradi, H., & Shavandi, A. (2020). Fish collagen: Extraction, characterization, and applications for biomaterials engineering. *Polymers*, *12*(10), 2230. DOI: 10.3390/polym12102230 PMID: 32998331

Kabirian, F., & Mozafari, M. (2020). Decellularized ECM-derived bioinks: Prospects for the future. *Methods (San Diego, Calif.)*, *171*, 108–118. DOI: 10.1016/j.ymeth.2019.04.019 PMID: 31051254

Karamanos, N. K., Theocharis, A. D., Piperigkou, Z., Manou, D., Passi, A., Skandalis, S. S., & Onisto, M. (2021). A guide to the composition and functions of the extracellular matrix. *The FEBS Journal, 288*(24), 6850–6912. DOI: 10.1111/febs.15776 PMID: 33605520

Karimi, S. A., Abdolmaleki, A., Sadeghi, A., Naseri, S., Ayni, M., & Gholami, M. R. (2022). Drug delivery system through alginate dermal scaffold loaded with hydroalcoholic extract of Daphne Mucronata improves dermal excisional wound healing: An experimental research. *Regenerative Engineering and Translational Medicine*, *8*(4), 523–534. DOI: 10.1007/s40883-022-00252-3

Kaur, J., & Reinhardt, D. P. (2015). Extracellular matrix (ECM) molecules. In Stem cell biology and tissue engineering in dental sciences (pp. 25-45). Academic Press

Kavand, A., Noverraz, F., & Gerber-Lemaire, S. (2024). Recent advances in alginate-based hydrogels for cell transplantation applications. *Pharmaceutics*, *16*(4), 469. DOI: 10.3390/pharmaceutics16040469 PMID: 38675129

Khunmanee, S., Jeong, Y., & Park, H. (2017). Crosslinking method of hyaluronic-based hydrogel for biomedical applications. *Journal of Tissue Engineering*, *8*, 2041731417726464. DOI: 10.1177/2041731417726464 PMID: 28912946

Kim, B. S., Kim, H., Gao, G., Jang, J., & Cho, D. W. (2017). Decellularized extracellular matrix: A step towards the next generation source for bioink manufacturing. *Biofabrication*, *9*(3), 034104. DOI: 10.1088/1758-5090/aa7e98 PMID: 28691696

Lau, C. S., Hassanbhai, A., Wen, F., Wang, D., Chanchareonsook, N., Goh, B. T., & Teoh, S. H. (2019). Evaluation of decellularized tilapia skin as a tissue engineering scaffold. *Journal of Tissue Engineering and Regenerative Medicine*, *13*(10), 1779–1791. DOI: 10.1002/term.2928 PMID: 31278852

Lee, K. Y., & Mooney, D. J. (2012). Alginate: Properties and biomedical applications. *Progress in Polymer Science*, *37*(1), 106–126. DOI: 10.1016/j.progpolymsci.2011.06.003 PMID: 22125349

Li, S., Pei, M., Wan, T., Yang, H., Gu, S., Tao, Y., & Xiao, P. (2020). Self-healing hyaluronic acid hydrogels based on dynamic Schiff base linkages as biomaterials. *Carbohydrate Polymers*, *250*, 116922. DOI: 10.1016/j.carbpol.2020.116922 PMID: 33049836

Li, Y., Wang, H., Li, J., Zhang, N., Xu, B., Li, Y., & Ge, B. (2024). Self-assembly and cross-linking preparation of tilapia-skin-derived collagen/alginate hydrogels for efficient wound repairing. *Polymer Engineering and Science*, *64*(5), 2146–2156. DOI: 10.1002/pen.26682

Liu, N., Zhang, X., Guo, Q., Wu, T., & Wang, Y. (2022). 3D bioprinted scaffolds for tissue repair and regeneration. *Frontiers in Materials, 9*, 925321. DOI: 10.3389/fmats.2022.925321

Liu, S., Yu, J. M., Gan, Y. C., Qiu, X. Z., Gao, Z. C., Wang, H., Chen, S.-X., Xiong, Y., Liu, G.-H., Lin, S.-E., McCarthy, A., John, J. V., Wei, D.-X., & Hou, H. H. (2023). Biomimetic natural biomaterials for tissue engineering and regenerative medicine: New biosynthesis methods, recent advances, and emerging applications. *Military Medical Research, 10*(1), 16. DOI: 10.1186/s40779-023-00448-w PMID: 36978167

Makvandi, P., Ali, G. W., Della Sala, F., Abdel-Fattah, W. I., & Borzacchiello, A. (2019). Biosynthesis and characterization of antibacterial thermosensitive hydrogels based on corn silk extract, hyaluronic acid and nanosilver for potential wound healing. *Carbohydrate Polymers, 223*, 115023. DOI: 10.1016/j.carbpol.2019.115023 PMID: 31427021

Mandrycky, C., Phong, K., Zheng, Y. (2017) Tissue engineering toward organ-specific regeneration and disease modeling. MRS Commun. Sep;7(3):332-347. .DOI: 10.1557/mrc.2017.58

Martinez, A. T. T., Zuluaga-Vélez, A., Sepúlveda-Arias, J. C., Santa, J. F., & Buitrago-Sierra, R. (2025). Development of biocompatible scaffolds of collagen fibers from Tilapia scales (Oreochromis niloticus) modified with PVA. *Fibers and Polymers, 26*(3), 995–1009. DOI: 10.1007/s12221-024-00803-1

Meretsky, C. R., Polychronis, A., Liovas, D., & Schiuma, A. T. (2024). Advances in tissue engineering and its future in regenerative medicine compared to traditional reconstructive techniques: A comparative analysis. *Cureus, 16*(9). Advance online publication. DOI: 10.7759/cureus.68872 PMID: 39376883

Millán-Rivero, J. E., Martínez, C. M., Romecín, P. A., Aznar-Cervantes, S. D., Carpes-Ruiz, M., Cenis, J. L., & García-Bernal, D. (2019). Silk fibroin scaffolds seeded with Wharton's jelly mesenchymal stem cells enhance re-epithelialization and reduce formation of scar tissue after cutaneous wound healing. *Stem Cell Research & Therapy, 10*(1), 126. DOI: 10.1186/s13287-019-1229-6 PMID: 31029166

Mohammed, M. R. (2025). Recent Advances in Fish Scale-Based Polymer Biocomposite Scaffolds for Tissue Engineering. *Biomedical Materials & Devices*. Advance online publication. DOI: 10.1007/s44174-025-00621-w

Monavari, M., Medhekar, R., Nawaz, Q., Monavari, M., Fuentes-Chandía, M., Homaeigohar, S., & Boccaccini, A. R. (2022). A 3D printed bone tissue engineering scaffold composed of alginate dialdehyde-gelatine reinforced by lysozyme loaded cerium doped mesoporous silica-calcia nanoparticles. *Macromolecular Bioscience, 22*(9), 2200113. DOI: 10.1002/mabi.202200113 PMID: 35795888

Morais, C. A. R. S., Santana, T. P., Santos, C. A., Passetti, R. A. C., Melo, J. F. B., de Macedo, F. D. A. F., & Del Vesco, A. P. (2020). Effect of slaughter weight on the quality of Nile tilapia fillets. *Aquaculture (Amsterdam, Netherlands), 520*, 734941. DOI: 10.1016/j.aquaculture.2020.734941

Mundra, S., Shetty, P. J., & Shetty, N. J. (2024). Tilapia fish waste: An asset for tissue engineering–A review. *Journal of Indian Society of Periodontology, 27*(6), 568–577. DOI: 10.4103/jisp.jisp_395_22 PMID: 38434505

Munshi, M. H., Saha, R., Kamruzzaman, M., Akter, M., Al Rakib Shikder, A., & Repon, M. R. (2025). Recent Advancements in Biomaterial Applications: A Review. *International Journal of Environmental Research, 19*(6), 244. DOI: 10.1007/s41742-025-00908-2

O'Brien, F. J. (2011). Biomaterials & scaffolds for tissue engineering. *Materials Today, 14*(3), 88–95. DOI: 10.1016/S1369-7021(11)70058-X

Patil, S., George, T., & Mahadik, K. (2015). Green synthesized nanosilver loaded silk fibroin gel for enhanced wound healing. *Journal of Drug Delivery Science and Technology, 30*, 30–36. DOI: 10.1016/j.jddst.2015.09.001

Rappu, P. M., Salo, A., Myllyharju, J., & Heino, J. (2019). Role of prolyl hydroxylation in the molecular interactions of collagens. *Essays in Biochemistry, 63*(3), 325–335. DOI: 10.1042/EBC20180053 PMID: 31350381

Reddy, N., Reddy, R., & Jiang, Q. (2015). Crosslinking biopolymers for biomedical applications. *Trends in Biotechnology, 33*(6), 362–369. DOI: 10.1016/j.tibtech.2015.03.008 PMID: 25887334

Rivera-Briso, A. L., & Serrano-Aroca, Á. (2018). Poly(3-Hydroxybutyrate-Co-3-Hydroxyvalerate): Enhancement Strategies for Advanced Applications. *Polymers, 10*(7), 732. DOI: 10.3390/polym10070732 PMID: 30960657

Rockwood, D. N., Preda, R. C., Yücel, T., Wang, X., Lovett, M. L., & Kaplan, D. L. (2011). Materials fabrication from Bombyx mori silk fibroin. *Nature Protocols, 6*(10), 1612–1631. DOI: 10.1038/nprot.2011.379 PMID: 21959241

Ruiz-Alonso S., Lafuente-Merchan M., Ciriza J., et al. Tendon tissue engineering: cells, growth factors, scaffolds and production techniques, J. Contr. Release 333 (2021) 448–486, .DOI: 10.1016/j.jconrel.2021.03.040

Sarker, B., Zehnder, T., Rath, S. N., Horch, R. E., Kneser, U., Detsch, R., & Boccaccini, A. R. (2017). Oxidized alginate-gelatin hydrogel: A favorable matrix for growth and osteogenic differentiation of adipose-derived stem cells in 3D. *ACS Biomaterials Science & Engineering*, *3*(8), 1730–1737. DOI: 10.1021/acsbiomaterials.7b00188 PMID: 33429654

Shang, Y., Li, Y., Peng, C., & Ma, S. (2025, November 28). Tilapia skin collagen composite: A promising candidate in the modulation of bone regeneration. *Dental Materials Journal*, *44*(6), 705–715. DOI: 10.4012/dmj.2025-042 PMID: 41207685

Sridhar, L. P., Jagannathan, R., Veeraraghavan, V. P., El-Sherbiny, M., Asfari, J., Balaji, T. M., & Alluhaibi, S. T. (2024). Nile Tilapia Skin-Derived Materials for Dental Application. *Dental Hypotheses*, *15*(2), 33–34. DOI: 10.4103/denthyp.denthyp_13_24

Sugiura, N., Awaji, T., Masuda, S., Toyoda, T., Igarashi, H., Ishikawa, Y., Ishii, M., & Kimoto, M. (2009). Potential for decadal predictability in the North Pacific region. *Geophysical Research Letters*, *36*(20), 2009GL039787. DOI: 10.1029/2009GL039787

Suhail, M., Khan, A., Rosenholm, J. M., Minhas, M. U., & Wu, P.-C. (2021). Fabrication and characterization of diclofenac sodium loaded hydrogels of sodium alginate as sustained release carrier. *Gels (Basel, Switzerland)*, *7*(1), 10. DOI: 10.3390/gels7010010 PMID: 33514036

Sultan, S., & Mathew, A. P. (2018). 3D printed scaffolds with gradient porosity based on a cellulose nanocrystal hydrogel. Nanoscale, 10(9), 4421-4431. DOIhttps:// DOI: 10.1039/C7NR08966J

Sun, W., Gregory, D. A., Tomeh, M. A., & Zhao, X. (2021). Silk fibroin as a functional biomaterial for tissue engineering. *International Journal of Molecular Sciences*, *22*(3), 1499. DOI: 10.3390/ijms22031499 PMID: 33540895

Tang, J., & Saito, T. (2015). Effect of type I collagen derived from tilapia scale on odontoblast-like cells. *Tissue Engineering and Regenerative Medicine*, *12*(4), 231–238. DOI: 10.1007/s13770-014-0114-8

Tozetto, R., De Macedo, J. B., Da Silva, T. L. M., Ventura, A. C. T., Beltrame, F. L., & Ferrari, P. C. (2025). Evaluation of Nile Tilapia (Oreochromis niloticus) Skin Peptides for Wound Healing: A Systematic and Meta-Analysis Review. *ACS Biomaterials Science & Engineering*, *12*(1), 53–70. DOI: 10.1021/acsbiomaterials.5c01219 PMID: 41329667

Vilaça-Faria, H., Noro, J., Reis, R. L., & Pirraco, R. P. (2024). Extracellular matrix-derived materials for tissue engineering and regenerative medicine: A journey from isolation to characterization and application. *Bioactive Materials*, *34*, 494–519. DOI: 10.1016/j.bioactmat.2024.01.004 PMID: 38298755

Vilches, J. L., da Silva Martin, C., Sanches, A. O., & Malmonge, J. A. (2025). Tannic acid crosslinking in Tilapia gelatin microfibers produced by Solution Blow Spinning. *Journal of Food Engineering*, *397*, 112574. DOI: 10.1016/j.jfoodeng.2025.112574

Wang, P., Huang, S., Hu, Z., Yang, W., Lan, Y., Zhu, J., Hancharou, A., Guo, R., & Tang, B. (2019). In situ formed anti-inflammatory hydrogel loading plasmid DNA encoding VEGF for burn wound healing. *Acta Biomaterialia*, *100*, 191–201. DOI: 10.1016/j.actbio.2019.10.004 PMID: 31586729

Whitely, M., Cereceres, S., Dhavalikar, P., Salhadar, K., Wilems, T., Smith, B., Mikos, A., & Cosgriff-Hernandez, E. (2018). Improved in situ seeding of 3D printed scaffolds using cell-releasing hydrogels. *Biomaterials*, *185*, 194–204. DOI: 10.1016/j.biomaterials.2018.09.027 PMID: 30245387

Yao, Q., Zheng, Y. W., Lan, Q. H., Kou, L., Xu, H. L., & Zhao, Y. Z. (2019). Recent development and biomedical applications of decellularized extracellular matrix biomaterials. *Materials Science and Engineering C*, *104*, 109942. DOI: 10.1016/j.msec.2019.109942 PMID: 31499951

Ying, H., Zhou, J., Wang, M., Su, D., Ma, Q., Lv, G., & Chen, J. (2019). In situ formed collagen-hyaluronic acid hydrogel as biomimetic dressing for promoting spontaneous wound healing. *Materials Science and Engineering C*, *101*, 487–498. DOI: 10.1016/j.msec.2019.03.093 PMID: 31029343

Zhang, Y., Lu, L., Chen, Y., Wang, J., Chen, Y., Mao, C., & Yang, M. (2019). Polydopamine modification of silk fibroin membranes significantly promotes their wound healing effect. *Biomaterials Science*, *7*(12), 5232–5237. DOI: 10.1039/C9BM00974D PMID: 31596279

Zhou, T., Wang, N., Xue, Y., Ding, T., Liu, X., Mo, X., & Sun, J. (2016). Electrospun tilapia collagen nanofibers accelerating wound healing via inducing keratinocytes proliferation and differentiation. *Colloids and Surfaces. B, Biointerfaces*, *143*, 415–422. DOI: 10.1016/j.colsurfb.2016.03.052 PMID: 27037778

Zimba, B. L., Rwiza, M. J., & Sauli, E. (2024). Utilizing tilapia fish skin biomaterial for burn wound dressing: A systematic review. *Scientific African*, *24*, e02245. DOI: 10.1016/j.sciaf.2024.e02245

Zorlutuna, P., Vrana, N. E., & Khademhosseini, A. (2013). The expanding world of tissue engineering: The building blocks and new applications of tissue engineered constructs. *IEEE Reviews in Biomedical Engineering*, *6*, 47–62. DOI: 10.1109/RBME.2012.2233468 PMID: 23268388

Chapter 8
Sustainable Transience in Personalized Implants With Regenerative Bioelectronics and Green Materials

L. B. Muralidhar
http://orcid.org/0000-0003-3453-613X
Jain (Deemed to be) University, India

K. P. Sheeba
http://orcid.org/0009-0003-4707-5392
Jain (Deemed to be) University, India

ABSTRACT

This chapter advances sustainable transience as a unifying paradigm for next-generation personalised implants that heal, monitor, and disappear without ecological residue. Moving beyond durability-centred biomaterials, it integrates regenerative medicine, biodegradable bioelectronics, and circular green materials science to reconceptualise implants as temporally intelligent socio-technical systems. Drawing on global health, sustainability, and innovation data, it identifies conceptual paradoxes linking longevity, scalability, digital governance, and environmental accountability. The proposed Sustainable Transient Implant (STI) Framework aligns regeneration–degradation kinetics, life-cycle assessment, AI-enabled monitoring, and equity-oriented design. By bridging theory, practice, and regulation, the chapter reframes implantable medicine as clinically effective, digitally integrated, and ecologically responsible.

DOI: 10.4018/979-8-2600-0060-1.ch008

INTRODUCTION: DESIGNING FOR DISAPPEARANCE IN THE AGE OF REGENERATIVE TECHNOLOGIES

Traditionally, implantable medicine has prioritised longevity, but many clinical applications require only temporary solutions. This is particularly true of bone repair, nerve regrowth, and post-surgical stabilisation, where a device is intended to assist, guide, and/or monitor tissue healing and is then left in place after healing is complete. The bioresorbable scaffolds, transient electronics, and biodegradable metals, on the other hand, reimagine disappearance as a design goal for clinical success over the longer term, without causing the undesirable appearance of failure, spoiling the materials, or requiring revision procedures (Fiocchi et al., 2020).

What is important, then, is not degradation per se, but whether degradation exceeds repair, mechanical demand, and patient metabolic rate. In tissue engineering, premature resorption may compromise structural support. At the same time, excessive persistence may inhibit remodelling and promote a fibrotic response in the surrounding tissue (encapsulation), thereby reducing biological integration (Giannokopoulos et al., 2020). In the case of transient bioelectronics, these concepts can be extended to allow for short-term sensing, stimulation, and monitoring, but not for extraction, and for short-term signal stability, data security, and post-device accountability, complementary new requirements are introduced (Hong et al., 2020). This is a good example of how conductive bioresorbable nerve scaffolds can merge regenerative guidance and electroactive function. They, however, represent a greater design limitation, as no material can optimise bioactivity, conductivity, strength, and predictable resorption simultaneously (Li et al., 2023).

Therefore, for the transient implant design, one cannot only focus on optimising a single property, but must optimise its design by multiple criteria over time. For biodegradable magnesium implants, for example, a challenge is controlling corrosion via coatings or alloying without sacrificing bioactivity and mechanical performance (Kim et al., 2019). Shifting that will achieve clinical credibility also demonstrates regulatory progress toward approval of bioresorbable metallic trauma devices, but clinical examples remain limited due to differences in degradation across anatomic sites, biochemical environments, loading conditions, and individual metabolic profiles (European Medicines Agency, 2018).

Personalisation is thus a key element of 'safe disappearance '. Digital modelling and digital twins can assist in this regard by simulating dissolution, tissue integration, mechanical transfer, and sensor performance over time as the design space evolves, rather than treating these as fixed properties of the material. This digital layer also raises ethical questions regarding consent, ownership, transparency, cybersecurity, and the absence of data after implant degradation (Stahl et al., 2018). Sustainability should also be considered a fact-based, not a presumed, requirement, not based on biodegradability. While a device that dissolves in vivo may not be biodegradable, it may have significant environmental burdens associated with its extraction, manufacturing, in vivo sterilisation, packaging, transport, procurement and clinical monitoring, or conversely, avoid the burdens of in vivo revision surgery (Eckelman & Sherman, 2020; Kuper & Blomsma, 2024).

The Sustainable Transient Implant model is thus more of an organising model, as it combines four interdependent axes: functional-regenerative alignment, lifecycle accountability, digital governance, and equity-oriented implementation. Functional-regenerative alignment is a question of whether there is a time window for healing (temporary mechanical or electrical support). Lifecycle accountability challenges the question of comparative environmental evidence to support disappearance. How the data from sensors, AI-aided design, and digital twins are clinically valid, secure and ethically controlled are questions raised by digital governance. Equity-oriented implementation focuses on questions of the production, monitoring, and accessibility of transient implants across a variety of health care systems. These axes allow the three sustainability, degradability, and digital monitoring concepts to be treated as a single translational design problem rather than as three key concepts.

The chapter translates from theory to conceptual tensions, the empirical cases, methods, regulation, ethics and implementation. It assesses transient implants by considering which functions are transient, what time frame of degradation is clinically acceptable, what information is available once the implant is gone, which phase of implant life is most burdensome, and whether the implant can be afforded and monitored in the intended setting. Programmed biodegradation intelligence, in this regard, is the application of materials engineering, sensor feedback and computational prediction of implant degradation and dysfunction timing and safe degradation. The beauty of transient implants is that they are not only invisible but can also incorporate the clinical requirements for therapeutic duration, biological repair, and environmentally responsible, accountable digital control.

THEORETICAL FOUNDATIONS: BRIDGING BIOMEDICAL ENGINEERING, SYSTEMS ECOLOGY, AND DIGITAL HEALTH

The theoretical aspect of the chapter is not to examine regenerative medicine, systems ecology, and digital health as distinct entities, but to demonstrate how these three areas must be integrated in designing implants that are transient, degradable, clinically informative, and environmentally responsible. The theory of classical implants is guided by inertness, stability, and long-term mechanical reliability, which have arisen from reconstructive models that rely on fixation and replacement. This model cannot be used when treatment success involves regeneration rather than continuous substitution (Fiocchi et al., 2020).

The benefit of the implant in regenerative applications is as a temporary support. It should aid in tissue repair, but not until the tissue can sustain itself. This perspective on the materiality of implants is one of time, relationships and processes. Understands that the body is a dynamic composition of biological, technological, and ecological relations, not a closed system (Braidotti, 2013). To this end, regenerative medicine and other cutting-edge technologies also contribute. Healing cannot be broken down into cells, scaffolds and signalling processes, as they all have to work together, with mechanical load, biochemical gradients, scaffold architecture and degradation state all playing a role in healing. Mechanotransduction plays a key role, as cells respond to stiffness, strain, degradation by-products, and porosity, which affect their differentiation and matrix formation (Giannokopoulos et al., 2020).

Thus, the degradation rate of the scaffolds must be comparable to the healing rate. Gradually degrading the material can delay remodelling processes, and degrading too rapidly can cause the loss of structural and mechanical cues before tissue organisation is successful. Conditional assistance – the implant should act in a guiding manner, but not replace an endogenous regeneration – is the most defendable principle. The most defensible principle is conditional assistance, which refers to the implant acting in a guiding manner rather than replacing endogenous regeneration (Fiocchi et al., 2020). The principle is the same for transient bioelectronics, which are sensors and stimulators that operate for a short period. These create a monitoring or electrical function within a specified time window after the function, and the function is reduced thereafter. (For example, in the area of peripheral nerve repair, they are conductive bioresorbable scaffolds (CBSs) (Hong et al., 2020; Li et al., 2023).

STI thus considers implants to be temporary cyber-physical-ecological systems rather than materials. There is another constraint: Sustainability. That is, the product needs to be biodegradable, but not just that; it also needs to benefit the environment. The life-cycle evidence for environmental claims in the healthcare sector is significant, as it involves many resources and emissions across the extraction, production, sterilisation, transport, clinical use, and avoided revision stages (Eckelman & Sherman, 2020; Kuper & Blomsma, 2024). This is supported by digital twins and AI-driven materials discovery, which model degradation and regenerative responses, device performance, and environmental effects at both the biological and operational scales.

However, the literature is only of questionable merit in favour of STI. It indicates that controlled degradation, transient sensing, and lifetime assessment are possible, but does not prove that they can be optimised simultaneously in a single device without compromise (Kim et al., 2019; Stahl et al., 2018). Mechanical timing, degradation safety, data governance and lifecycle burden are all factors that make a difference in a transient becoming sustainable. This results in the paradox of the chapter: The implant has to be maintained for long enough to heal the wound and be gone again soon enough to avoid chronic burden and burden of data; it needs to generate evidence without causing an informational risk and, finally, it needs to prove sustainability only when evidence exists on the wider system.

CONCEPTUAL GAPS AND PARADOXES

The paradoxes of Sustainable Transient Implants (STI) stem from each one of the four STI axes: functional-regenerative alignment creates the paradoxes of persistence and disappearance, lifecycle accountability creates the paradoxes of biodegradability and sustainability, digital governance creates the paradoxes of disappearing hardware and persistent data, and equity-oriented implementation creates the paradoxes of personalised optimisation and scalable access. These tensions demonstrate that one cannot define and explain STI with the traditional design of implants.

The first contradiction concerns performance over time. A transient implant will need to function for a sufficiently long period to establish tissues, monitor recovery, or stimulate tissues. However, it must also resorb before it becomes a chronic biological or ecological burden. Therefore, the success of a clinical application will not be reliant on degradation, but on degradation within a therapeutic window that matches tissue repair. Mechanical support may be lost before regeneration can be established, and remodelling may be delayed, resulting in foreign-body reactions and inflammation. Data from magnesium orthopaedic implants and bioresorbable vascular scaffolds demonstrate that the same classes of materials can have different clinical performance depending on the anatomical location, loading and healing profiles of the patient; therefore, degradation kinetics should be considered a clinical-performance variable and not a material property (Fiocchi et al., 2020; Kim et al., 2019).

In the case of transient electronics, conductors, encapsulations, and power systems need to maintain the fidelity of the electronic signals throughout a limited therapeutic window and be designed to dissolve safely (Hong et al., 2020). However, one remaining limitation is the lack of modelling of the degradation chemistry, electronic decay, and tissue remodelling as a single interconnected system. The second paradox concerns the dichotomy between personalisation and scalability. While patient-specific anatomy, metabolism, immune response, biomechanical loading, and healing rate can offer a means of achieving therapeutic precision, they also make the manufacturing, regulatory validation, and cost control of these products more complex, and the distribution of these devices more difficult to achieve equitably.

Although materials discovery and computational design may be faster with AI, the industrial dilemma of reliably manufacturing customised transient devices at scale remains to be solved. The third paradox concerns environmental credibility. While it is possible to avoid removal surgery, this does not necessarily mean these are sustainable, as it does not account for the impacts of synthesis, sterilisation, packaging, transport, procurement and supply-chain (Eckelman & Sherman, 2020). However, the ecological benefits of renewable origin and degradability are not always positive, as clinical benefits may be outweighed by the intensity of production and the consequences of disposal. (Kuper & Blomsma, 2024).

The fourth paradox concerns persistence on the computer. The concept of digital twins and sensor-enabled implants can help with adaptive therapy and predictive modelling. The information they generate can still be stored in records, algorithms, and cloud systems even when the device is gone (Li et al., 2023). The levels of consent, transparency, cybersecurity, and ownership and stewardship frameworks are still less developed than technical capabilities (European Medicines Agency, 2018; Stahl et al., 2018). In addition to these tensions, it is an equity issue because the advanced transient implants are still stuck in high-capacity health research facilities in the West and East Asian area and may be unaffordable to other health systems with fewer resources, who are left sidelined from what is being touted as 'sustainable'. In fact, STI needs an integrated theory that understands and addresses transient implants as data-intensive, environmentally impactful, and socio-technical systems that are not monodisciplinary but rather a mix of socio-technical and biomedical.

EMPIRICAL LANDSCAPE: GLOBAL TRENDS AND COMPARATIVE PERSPECTIVES

In the empirical world of the sustainable transient implant, it is an impermanence that is more important than the novelty of the technology—whether a device disappears or not is not the issue; the issue is whether it disappears when it should and not when it should not, whether there is enough evidence to support it, whether it can be defended environmentally, and whether it can be institutionally controlled. Examples like RemeOs™, coated magnesium implants, BVS, and conductive nerve scaffolds demonstrate a lack of uniformity: clinical need and engineering innovation are outpacing evidence of regulatory, ecological and implementation (Kim et al., 2019; Fiocchi et al., 2020; Li et al., 2023; Kuper & Blomsma, 2024).

As health expenditure rises, population ages and the number of degenerative, traumatic and reconstructive needs grows, implants that hold tissue in place, promote healing and minimize revision surgeries are critical to supporting the demand for healthcare, but they also add material and carbon responsibilities for healthcare (Eckelman & Sherman, 2020). Despite the benefits of biodegradable implants, they have yet to be proven sustainable because life-cycle assessment (LCA) is often not integrated into the design and validation process, procurement, or the comparison of products in use (Kuper & Blomsma, 2024).

There is a clear pattern of asymmetry of evidence. Routine adoption does not occur until there is regulatory approval for the device class, the redesign of the scaffold demonstrates the feasibility of resorption, and monitoring value outweighs validation, cybersecurity and cost burdens in a transient electronic system (Fiocchi et al., 2020; Hong et al., 2020; Stahl et al., 2018). Pathways also differ across regions. While the European policy discourse seems more comfortable linking the concepts of biodegradability to those of safety, environmental accountability and material governance, the evidence at programme level is less complete; the United States is more attuned to evidence of safety, performance and comparative substantiation, which favour selective entry, but may limit multifunctional or personalised designs (European Medicines Agency, 2018).

RemeOs™ is a significant translational marker of transition. This approval is the first for biodegradable metals expected to meet current safety and efficacy criteria for selected indications. At the same time, it represents a brief period of fixation with a bioresorbable magnesium alloy, followed by controlled corrosion after the bone has healed (Kim et al., 2019). Its more general teaching, however, depends on indication-specific fixation targets, strategies for corrosion control, imaging follow-up, and clinician guidance, as well as evidence correlating degradation kinetics with functional outcomes and disappearance.

All of these implementations (coated magnesium implants, Bioresorbable Scaffold, conductive nerve platforms) have a defined therapeutic window, a period during which mechanical and/or electrical function can decline safely, and a post-implant protocol in which degradation products, residual risk, and patient monitoring are assessed. All of these implementations have a defined therapeutic window, a period in which mechanical and/or electrical function can decline safely, and a post-implant protocol in which the degradation products, residual risk and patient monitoring are monitored (Kim et al., 2019; Li et al., 2023; Hong et al., 2020). This is complemented by an Asian perspective – particularly from Japan – which offers a clinically mature but provisional evidence base: in terms of vascular outcomes and orthopaedic outcomes, it is clear that the success of degradable implants depends on the type of lesion, anatomical location, loading circumstances and the long-term behaviour of the material.

Although the coatings have multilayer structures, they may help prevent corrosion and maintain bioactivity; however, it is not clear whether they are necessary for safety and reliability, as biodegradability does not imply safety or reliability (Kim et al., 2019). The transient neural interfaces and therapeutic monitoring systems have greater scientific potential due to their combination of short-term sensing and therapeutic support with a lack of explantation, but at the same time, the predictability, reproducibility, cost and data governance of these systems are a cause for concern (Hong et al., 2020; Li et al., 2023). Failure of the polymer is also a valuable lesson: if the polymer resorption rate exceeds local tolerance, the by-products of this process may be acidic and irritate the tissue (Fiocchi et al., 2020). Implantable devices, then, can be sustainable only if they disappear in synergy with the mechanical, biological, economic and ecological considerations.

OPERATIONALISING THE SUSTAINABLE TRANSIENT IMPLANT (STI) FRAMEWORK

This section implements the STI framework as a series of steps in the design, validation, regulatory, and procurement processes, rather than as an introductory concept. It aims to fill the gaps between regenerative medicine, biodegradable electronics, digital health and sustainable materials research. Regenerative engineering defines how biomaterials, cells, and signalling environments facilitate tissue regeneration but seldom accounts for environmental externalities or the governance challenges posed by digitally mediated implants. Lifecycle and circular-economy approaches, on the other hand, reinforce ecological accountability in healthcare procurement but are incomplete unless they account for the mechanobiological and cyber-physical conditions that influence clinical success (Kuper & Blomsma, 2024; Eckelman & Sherman, 2020).

STI is an alignment test that is not descriptive. Therefore, it serves as a cross-domain test of alignment. It questions the coherence of regenerative performance, programmed degradation, digital accountability, lifetime sustainability, and the feasibility of implementation within a single therapeutic pathway (Fiocchi et al., 2020; Hong et al., 2020; Stahl et al., 2018). The framework is based on regenerative engineering, systems theory, socio-technical systems theory, and lifecycle sustainability theory to view implant success as a temporally coordinated socio-technical process. Personalisation is also included, with the proviso that no personalisation will be allowed that could compromise validity, reproducibility, traceability, or affordability. So it is STI, first tests therapeutic timing, then degradation burden, then digital governance, and then finally the possibility of scaling up the implementation.

Figure 1. STI decision sequence for transient implant implementation

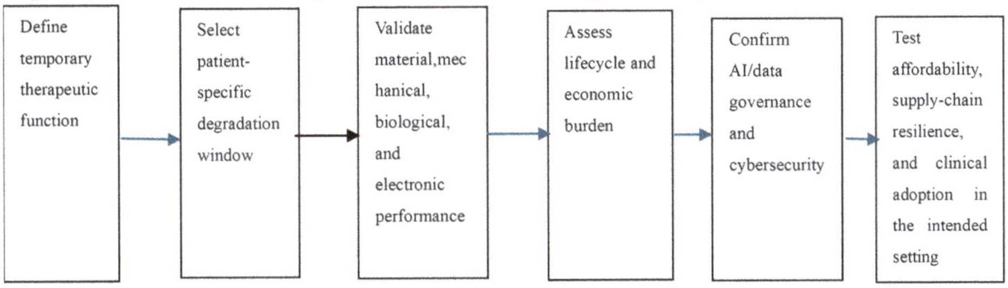

The first axis of STI is working and related to the "coupling of regeneration and degradation. In traditional implant design, material degradation is considered a desirable behaviour, provided mechanical support does not fail prematurely. STI places greater weight on the design of temporal alignment. The structural support needs to be removed at a rate that coincides with the tissue-specific healing kinetics; that is, once the endogenous tissue can take over load-bearing and biological functions, the structural support should be reduced (Fiocchi et al., 2020). This is a very important requirement, as the trajectory varies across sites depending on age, inflammation, and biomechanical conditions. The same principle applies to devices that are bio-electronic in nature, but of short duration. The activity of sensors, electrodes, and wireless interfaces should not be limited to attachment to degradable scaffolds. Instead, it should be synchronised with the biological events in which the scaffold is involved, such as nerve conduction return or the closure of the window of opportunity for post-surgical complications (Hong et al., 2020).

In this context, degradation is not an endpoint of a degradation therapy, but rather an active design variable in therapy. The field is still hindered by the lack of reliable predictive models linking tissue remodelling, material dissolution, and electronic failure, so temporal co-ordination remains more wishful thinking than reality (Li et al., 2023). The second STI axis relates to the ecological part and takes biodegradability into account in the lifecycle sustainability. Many studies in the literature suggest that in vivo resorption is beneficial for the environment, but this alone is not sufficient. While temporary implants can provide a cost of up to one million dollars for 30 years of use, they can still involve carbon-intensive production, dangerous precursor chemistry and waste in the supply chain, allowing for the potential for savings in avoiding removal surgery (Eckelman & Sherman, 2020).

Lifecycle assessment should thus be introduced into the design as a brief that uses evidence, specifies a functional unit, comparator, and system boundary, and provides justification for a higher manufacturing burden (Eckelman & Sherman, 2020; Kuper & Blomsma, 2024). This translates to a criterion for procurement and regulation. It also differentiates between renewable, biobased, and biodegradable materials because while ecological origin is not enough to ensure clinical suitability, it is not sufficient to ensure that the material will degrade in a benign manner (Fiocchi et al., 2020).

Table 1. Implementation constraints for sustainable transient implants

Domain	Key Implementation Question	Main Construct	Required Evidence
Personalisation	Can degradation be matched to patient healing?	Variability in anatomy, load, metabolism, and inflammation	Imaging, risk stratification, validated design envelope
AI infrastructure	Can model output be trusted during care?	Dataset bias, cybersecurity, model drift, interoperability	Validation dataset, audit trail, change-control plan
Economics	Can the device scale clinically?	Upfront cost, reimbursement, training, workflow disruption	Total-cost comparison, avoided revision/explantation evidence
Geopolitics	Can it be accessed outside high-resource systems?	Supply-chain dependence, regulatory capacity, and local monitoring limits	Local serviceability, procurement feasibility, post-market surveillance
Sustainability	Is disappearance environmentally meaningful?	Manufacturing burden, sterilisation, packaging, digital energy use	Lifecycle assessment against a durable-device comparator

The third one of STI's axes is digital and connects smart monitoring with the question of data ethics. When transient implants are used to provide temporary intelligence to assist in healing surveillance, degradation forecasting, and prediction of remaining therapeutic life (Hong et al., 2020), they add value. The implant is not a 'dissolving object' then. Its material behaviour, sensing function, and degradation pathway should be monitored or modelled together. The therapeutic function should not be degraded during the recovery period and should be degraded only when continued support is no longer clinically required. STI translates this into the concept of Programmed Biodegradation Intelligence, which merges the capabilities of artificial intelligence and prediction, along with feedback from sensors and material-response modelling, to inform when and how to cease an implant's activity and safely break it down. This is a new form of regenerative engineering that uses chemically programmed and cyber-physically interpreted cases of degradation.

However, if it is to be adopted routinely, then digital infrastructure needs to be as robust as that for medical devices. Robustness, Reproducibility, Explainability and Clinical Relevance are important criteria for the validation of predictive models for degradation forecasting, Adaptive monitoring and decision support. Data integrity is also a critical design factor because during critical recovery moments, sensor degradation, loss of communication, device failure, software updates, and interoperability issues can jeopardise data reliability. In such scenarios involving the interaction of transient implants with hospital systems, wireless networks, cloud systems, and digital-twin environments, cybersecurity is also a key concern. To safeguard clinical trust, secure authentication, encryption, audit trails, and surveillance should be established to prevent data manipulation by unauthorised parties (Stahl et al., 2018).

To successfully deploy an AI tool in practice, validated degradation datasets, secure sensor data pipelines, EHR compatibility, responsibility for model updates, and software change-control procedures for future updates must be defined. Digital sustainability needs to be evaluated as well, as storage, transmission, cloud computing, retraining, and redundant monitoring carry environmental and financial costs that must be clinically justified(U.S. Food and Drug Administration, 2025; Stahl et al., 2018).

The fourth "STI axis" is equity, reflecting the fact that, despite advances in innovation, implants are still generally found in high-income clinical settings. Access is considered part of sustainability at STI. Devices which can be simplified, adapted for low-resource environments or manufactured using controlled additive manufacturing processes are thus at the core rather than the periphery. They rely on the performance of very uneven surgical, sterilisation, procurement, follow-up and digital infrastructures to be of value on a global scale. While additive manufacturing can help reduce waste and enable localised production, it requires feedstock validation, quality control, and skilled personnel. STI thus combines regenerative engineering, sustainability science, digital governance, economic feasibility, and equity within a unified framework for implementing a lifecycle approach.

METHODOLOGICAL INNOVATIONS

A progressive methodological transition towards sustainable transient implants entails considering disappearance as a coupled phenomenon across the chemical, mechanical, biological, electronic, and environmental fields. The conventional methods used to evaluate implants are not well adapted to this goal, since they were designed for implants with long-term durability, with key considerations including static biocompatibility, structural stability, and long-term mechanical integrity. Rather, transient systems are useful for their transitory nature, focusing on the evolution of interactions among degradation chemistry, tissue regeneration, load transfer, and local biochemical response—and, if applicable, the fidelity of the signal, its decay, and the reliability of its sensing—over time (Hong et al., 2020).

Degradation has to be measured as a multidimensional performance criterion, not only in terms of mass loss, corrosion rate and conductivity loss. Clinically relevant assessment should be correlated with changes in strength, pH, inflammation, tissue remodelling, electrical function, degradation products, recovery measures, and environmental measures (Kim et al., 2019). This will produce a degradation-performance map of therapeutic windows identified for physiological loading, pH changes, immune response, and mechanical loading changes. The same applies to the fragmentation of conductors, encapsulation failure, and degradation or loss of the substrates' sensing and stimulation abilities (Zhang et al., 2025; Roshi & Wang, 2025).

The challenge is epistemic and technical because there is still a separation of cultures of validation, timeframes of investigation, and levels of evidence between biomaterials, electronics, clinical medicine and research on sustainability. Digital-twin methods could provide some relief to this division, but if they are not implemented carefully. For a clinically useful twin for transient implants, patient anatomical data, material properties, degradation rates and longitudinal physiological data would assume a pivotal role in estimating the remaining functional life in real conditions, particularly in the presence of age-related, anatomical-site-related, inflammatory-state-related, and metabolic-profile-related changes in material behaviour.

However, these models are still limited by the limited availability of longitudinal degradation datasets, by the lack of interoperability between biological and engineering data streams, and by the lack of clear rules for successful disappearance. They involve the use of patient-specific data, including healing, inflammatory, pH, loading, corrosion or polymer dissolution data, signal decay and outcomes, as well as interpretable predictions, which explain uncertainty, failure thresholds and when support can safely be lowered (Hong et al., 2020).

Achieving interoperability, cyber security and regulatory acceptance are also key. A digital twin needs to incorporate material data, imaging, sensor outputs, clinical records, and post-market surveillance, and maintain privacy, data integrity, and auditability. During prediction, sensor degradation, loss of communication, and partial dissolution of the device can occur, all of which can introduce missing or biased data when the most valuable prediction is needed. Thus, in addition to physical safety, computational models for predicting degradation and healing need to be robust, explainable, and clinically valid and are a crucial component of the regulatory evaluation of implants. Therefore, in addition to physical safety, computational models for predicting degradation and healing must also be robust, explainable and clinically valid, and are a critical part of the regulatory evaluation of an implant (Stahl et al., 2018; U.S. Food and Drug Administration, 2025).

Another parallel priority is to develop more robust MIIVICs. The differences between in-body and accelerated bench test behaviour are often attributed to the latter failing to mimic most of the fluid dynamics, immune system responses, enzymatic processes and non-uniform loading that occur in vivo, as seen in scaffolds, magnesium implants and transient bioelectronics, respectively (Kim et al., 2019; Li et al.,

2023). Few examples of weak links in the laboratory exist, such as the early redesigns of BRSs that relied on prematurely resorbed scaffolds as indicators of control, when in reality the tissue-level outcomes of success were fibrosis, inflammation, remodelling, and nerve conduction (Fiocchi et al., 2020).

Artificial Intelligence can facilitate materials discovery and extend structure-property exploration. It should, however, primarily treat sustainability indicators as a set of coequal constraints alongside other design properties, such as bioactivity, resorption profile, toxicity, manufacturability, and electronic performance. However, this potential is constrained by the availability of ecological indicators in materials libraries, the limited scope of evidence on degradation, and the often biased nature of that evidence. Therefore, socio-technical LCA will need to be coupled with design rather than applied as an afterthought. Environmental advantage cannot be inferred from in vivo resorption alone.

The functional unit should be defined within a scope of LCA, which involves establishing a scope and selecting a comparator, as well as setting boundaries for extraction, precursor synthesis, manufacturing, coating/integration of electronics, sterilisation, packaging, transport, implantation, avoided removal, degradation by-products and accessory end-of-life (Eckelman & Sherman, 2020; Kuper & Blomsma, 2024). This method does not assume biodegradability as a proxy for sustainability, as would be done in a traditional design (Eckelman & Sherman, 2020; Kuper & Blomsma, 2024). A validation pathway should therefore be a mix of the three types of testing, namely, a practical validation pathway that should include bench testing, simulated-use testing, clinical follow-up, and lifecycle modelling. All of these layers, put together, would minimise ambiguous sustainability claims and help inform better regulatory, hospital, and manufacturer decisions (Kim et al., 2019; Hong et al., 2020; U.S. Food and Drug Administration, 2023).

GOVERNANCE, REGULATION, AND INSTITUTIONAL ECOSYSTEMS

In their role, regulation and governance remain key challenges to the success of sustainable transient implants, as these technologies challenge the assumptions on which the oversight of implantable medical devices has been built. Existing systems were created largely for the durable artefacts for which material identities are stable, therapeutic uses are known, and the risks of use after the product's production and marketing are well known. Bioresorbable and bioelectronic implants are other implants that test each condition. They have a material identity that evolves throughout their lifespan; their clinical value may depend on the synergy among material, electronic, and software functions; their risks extend beyond degradation and are present throughout the degradation pathway up to the implant phase. The regulatory challenge is thus not just about categorising a new material as an implant, but also about defining an implant that, over time, ceases to exist, regulating an implant after it has physically changed, and assessing the therapeutic/ecological life cycle of an implant (European Medicines Agency, 2018).

Existing assessment frameworks in the USA and EU serve as a starting point for assessment, with a focus on safety, effectiveness, performance testing and validation, risk management, conformity testing and post-market surveillance. Approval of Bioretec's RemeOs™ trauma screw demonstrates that a bioresorbable metallic implant can meet today's regulatory requirements. It should instead be interpreted as evidence of the feasibility, and not proof that such a more complex system already exists that can control the other types of devices, such as those that combine biomaterials, electronics, software, artificial intelligence and patient-specific design (European Medicines Agency, 2018).

For the transient implants, the degradation dynamics, digital function and life-cycle performance need to be considered, and regulatory models need to be developed that incorporate all of these factors. This need becomes evident in Europe, where the risk does not stop at the implant but extends as the device is gradually dissolved over time, along with its degradation products and the resorption rate of each patient. (European Medicines Agency, 2018). The most easily adaptable biodegradable devices for either jurisdiction are those similar to a traditional implant. More challenges arise when scaffolds, sensors, stimulators, algorithmic monitoring, and temporary data infrastructures form a unified therapeutic system. Classification is then substantive, not merely procedural, as it sets evidentiary thresholds, tests, approval prices, reimbursement, and surveillance obligations. Failure to classify matches technological reality may delay useful innovations, encourage "business as usual" design, and leave important risks insufficiently addressed (Hong et al., 2020; Eureka PatSnap, 2025).

This is because there is a need to go beyond the static biocompatibility and mechanical reliability of the standardisation. It should be used to assess the temporal variation in mass loss, corrosion and/or resorption, mechanical integrity, tissue response, electrical activity, and the safety of the by-product (Kim et al., 2019). The implant corrosion rate does not just determine the clinical value of a magnesium fixation implant, nor does the persistence of the signal determine the value of a bioresorbable electrode; with both types of implants, the value is determined by both time-dependent support and safe degradation and functional withdrawal (Kim et al., 2019).

The limited capacity of governance, laboratory infrastructure, technical expertise and post-market surveillance is further exacerbated in the Global South. This can result in overlooking local surgical conditions, monitoring capacity, waste systems, climate, procurement systems, and digital connectivity. Equity thus exists in addition to being distributive. Other vulnerabilities include concentrated supply chains, specialised alloys, rare coatings, medical polymers, semiconductor components, sterilisation systems, and proprietary software. A responsible STI pathway should be a fusion of the international community's evidence standards and adaptability in procurement, serviceability, training, and surveillance in the local context. However, access gaps can only be addressed at the regional level through validated materials, quality assurance, and reliable post-market monitoring (OECD, 2024; European Medicines Agency, 2018).

There are then three pathways of governance: incremental adaptation, hybrid regulatory innovation and 'transformative' circular regulation. There are, however, pragmatic but limited incremental adaptations; hybrid adaptations, better suited to multifunctional systems but requiring institutional capacity; and circular adaptations, which are most successful because they incorporate lifecycle sustainability, supply chain transparency, and ecological accountability into the regulatory process. However, it is politically challenging (Stahl et al., 2018; Kuper & Blomsma, 2024; Eckelman & Sherman, 2020). The main idea is that no architecture can govern sustainable transient implants. Their disappearance should be considered the purpose, burden of proof and regulatory requirement of the device.

ETHICAL, LEGAL, AND CULTURAL DIMENSIONS

Ethical, legal, and cultural issues play a key role in sustainable implants; they significantly impact the legitimacy, adoption, and access to implants. With the emergence of biodegradable implants, the traditional principles of safety, efficacy, and informed consent no longer suffice when considering the ethics of these new devices and their data-generating and computational functions. A broader governance structure must be put in place, as bodily intervention now intersects with information persistence, proprietary control, cultural interpretation, and environmental equity issues. An odd thing about the material implant is that some of the technology may disappear. However, the data and the legal and supply-chain infrastructure remain after the treatment is over. Hence, the need for an exact description of the end of an implant for the sustainable transience. Data sovereignty is particularly challenging with this issue. In the context of adaptive and personalised care, finite therapeutic windows for sensing, wireless communication, and predictive analytics of electrophysiological signals, healing trajectories, and device performance can be achieved through transient implants (Hong et al., 2020).

In bio-medical contexts, however, the issues of ownership, access, reuse, control and secondary exploitation of biological information remain unanswered, particularly in the context of neuro-ICTs (Stahl et al., 2018). The impermanence of the material may give the false impression of its informational impermanence. Cloud platforms and analytics systems, repositories, and commercial data pipelines can be left behind as the implant dissociates. Data sovereignty should therefore be understood as ongoing control over information generated by temporal implants, such as restrictions on data use and control over aggregated information from other datasets used to train AI systems. This concern is also a factor in informed consent. Patients deserve information not only on the risks of implantation and how long the implant will last, but also on how and when it will stop working, how long the information will remain on the implant, and how and when the algorithm will be involved in monitoring and/or intervening (Eureka PatSnap, 2025).

Consent should thus be process-based, not a single consent, as a single consent cannot account for functions that occur in sequence and the long-lasting consequences of data. The legal aspect concerns striking a balance between protecting intellectual property and maintaining an open infrastructure that promotes transparency, interoperability, and accessibility. While IP can be used to recoup translation expenses, it can also act as a barrier to diffusion. If there is excessive enclosure, it increases prices and restricts scientific review. Open-source discovery infrastructures provide, for the time being, an alternative by sharing databases, collaborating pre-competitively, and offering distributed modelling. There is another level of interpretation. It is often assumed that disappearance is a good thing in biomedical discourse.

However, what is considered good and bad varies across cultures, depending on their conceptions of respect for the body, contamination, trust, and technological validity. Braidotti's posthuman perspective can help conceptualise embodiment as relational rather than biologically closed. (Braidotti, 2013). Last, but not least, environmental justice should be part of the ethical assessment, as the less waste there is at the end of its use, the more waste there is at the beginning of its extraction and manufacture (Eckelman & Sherman, 2020; Kuper & Blomsma, 2024).

TOWARD A TYPOLOGY OF SUSTAINABLE TRANSIENCE

A typology of sustainable transience is required, as traditional classifications of biomaterials are typically based on composition, degradation time, or the indication for which they are used, which do not address how a temporary implant is both biologically and environmentally responsible, and digitally responsible. These older taxonomies work on a material basis, considering transience as a characteristic of material objects. A better way to think of it, however, is as a socio-technical design approach in which tissue repairs, material vanishes, environmental burden, and data governance are configured. Four different types of transience can be identified:

functional transience, eco-optimised transience, digitally augmented transience and regenerative-circular convergence. The most established functional transience is that of the functional transience. It excludes implants intended to be temporary, as they would be resorbed after sufficient recovery, thereby preventing the need for secondary removals in certain indications and reducing the overall burden of foreign bodies (Fiocchi et al., 2020).

Examples of such devices include bioresorbable scaffolds and orthopaedic fixation devices, which require a concerted action among resorption kinetics, tissue regeneration, and load transfer to be successful (Kim et al., 2019). It is significant in history because of the legitimisation of disappearance clinically. The drawback is that its environmental impact is seldom part of the design, and its digital capacity is treated as an afterthought. Eco-optimised transience takes this a step further, using not only the safe in vivo resorption but also feedstocks, manufacturing processes, degradation products, and life-cycle burdens as a basis for judging implants. This shift is significant, as healthcare's climate impact is substantial, and life-cycle assessment is increasingly relevant to device assessment and procurement (Eckelman & Sherman, 2020; Kuper & Blomsma, 2024). It is important because it counters the notion that biodegradability equals sustainability and because it demands proof of environmental claims rather than their symbolism of vanishing (Roshi & Wang, 2025).

However, this classification can become technocratic if it fails to consider affordability, extraction politics, and digital infrastructure. The next generation of degradable implants is digitally augmented transience, which includes the ability to sense, stimulate, monitor or predict. In the model, the material and cyber-physical transience exist. The clinically relevant data are collected by bioresorbable conductors and substrates, as well as electrodes, and are then washed out after treatment (Hong et al., 2020; Zhang et al., 2025). This transition is exemplified by post-surgical monitoring and peripheral nerve regeneration, particularly in the context of integrated structural and electronic functions on a single temporary platform, as in conductive BPSs (Li et al., 2023).

This sophistication, however, introduces new tensions, as data infrastructures are likely to persist even as hardware is replaced, leaving software responsibilities unsettled and creating cybersecurity challenges and governance liabilities (Stahl et al., 2018; Eureka PatSnap, 2025). The regenerative-circular convergence category is the most integrated. It entails planning for environmental protection, safe degradation, limited digital monitoring, and tissue regeneration simultaneously during implant design. This approach includes the synchronisation of degradation and healing, the selection of materials based on end-of-life considerations, and the selective use of digital systems to monitor, predict, and adaptively shut down (Kuper & Blomsma, 2024).

The ability to discover materials with AI tools and to provide open computational infrastructure to support it may facilitate this model approach, which can leverage multi-objective optimisation for resorption, bioactivity, signal stability, manufacturability, and environmental indicators. It is an integrating concept with very difficult practical application because it requires the optimisation of several criteria across the biological, digital, ecological, and regulatory systems. The typology should not be interpreted as a ladder of complexity and sustainability, but rather as a way to convey the different levels of complexity across the various site types. The typology is therefore not a ladder of complexity and sustainability but rather a means of communicating the complexity of the different site types.

Although predictable resorption and low cost are key, functional transience may be a superior choice compared to perceived transience. Eco-optimised transience is not adequate when clinical performance is subpar. Regenerative-circular convergence is not the final goal but rather an optimal state, which may or may not be achieved (Hong et al., 2020). This typology also makes sense for the research agenda. The field does not necessarily strive to include more advanced devices, but rather to determine how much transience is appropriate for a specific clinical context, evidence base, infrastructure, and environmental goal.

Future studies should therefore go beyond the materials optimisation approach and develop a more holistic programme that establishes a "holistic" interconnection between ontology, method and context. Theoretically, it is important to try to transcend the paradigms of biomaterials based on inertness, durability, and passive compatibility. Temporary support, controlled degradation, and patient-specific fit are already valued in regenerative medicine and urgently need stronger connections to environmental and digital governance (Fiocchi et al., 2020). The post-humanist ontologies of regeneration help in this regard, as they consider implants to be temporary actors in tissue repair, material metamorphosis, digital management, and ecological consequences, rather than external bodies inserted into enclosed bodies (Braidotti, 2013).

However, it is not engineering evidence that can be replaced by ontological refinement. It needs to be related to mechanobiological information on how changes in interactions among scaffold properties (stiffness, porosity, biochemical gradients, degradation by-products, tissue response) influence scaffold performance (Giannokopoulos et al., 2020). Another theoretical priority is the integration of systems immunology/material science. Inflammation, metabolism, and repair pathways are patient-specific and thus affect degradation outcomes, as well as biocompatibility and foreign-body reaction.

The method would enhance predictions for each body, every stage of disease, and every location, and place greater pressure on the scalability of design and manufacturing. This would improve predictions for each body, across disease states and at the anatomical level, and increase the pressure to design and manufacture at scale. The third priority concerns the medical device sector of the circular bioeconomy. However, there are only limited rules governing circularity for resorbable implants, as there is no post-use recovery. Transience, measured by life cycle accounting, is minimized but not at any cost; it is a desired outcome, not a foregone one, as concerns are given to safer feedstocks, benign degradation products and reduced revision burdens, and not attempting to assume the properties of biobased or biodegradable materials (Kuper & Blomsma, 2024; Eckelman & Sherman, 2020).

Three directions are crucial regarding the method. While traceability has been widely recognised as a potential solution, there is a need for systematic research on how it can be applied as a governance measure for opaque supply chains of biodegradable metals, coatings, polymers and embedded electronics, and what its current potential is (Medical supply chain integrated with blockchain and IoT, 2023). Real-time degradation analytics should also play an important role, as temporary sensors and conductive scaffolds can monitor in real time the regeneration process in the limited therapeutic window; corrosion and signal loss, tissue integration and residual function can be modelled in real time, preferably using digital-twin architectures, as suggested by Hong et al. (2020), Li et al., 2023, and Zhang et al. (2025). Data rights, cybersecurity, consent and algorithmic transparency must be part of this advance, as if material disappears, so does persistent information (Stahl et al., 2018; Eureka PatSnap, 2025).

Comparative regulatory ethnographies are also required to account for the different diffusion patterns of similar technologies in Europe and the United States, as well as in resource-constrained countries like Japan and countries that have different institutional configurations, such as safety, sustainability, value, affordability and infrastructural integration (European Medicines Agency, 2018). Last but not least, the future research agenda should also include aspects of an ageing society, manufacturing, and low-resource healthcare, all of which are climate-dependent. In the case of limited infrastructure, these pressures call for simplified, standardised, cost-disciplined implants that minimise the burden of revision, help decarbonise processes, and support reductions in consumption and pollution. It can then be said that the starring role, the determining one, is that of establishing an immunologically informed science of disappearance, a digitally accountable science of disappearance, a materially traceable science of disappearance, and a socially distributive science of disappearance.

IMPLICATIONS

The implications for sustainable temporary implants are far-reaching: what an implant is, what it looks like, and what institutional structures are needed to regulate it are all being redefined. From a theoretical point of view, several publications have recently emerged, revealing that biomaterials are no longer considered inert therapeutic devices whose therapeutic activity can be summarised as biocompatibility and mechanical resistance. Instead, they can more credibly be understood as socio-ecological systems with a temporal scope where the mechanisms of function depend on the interplay between degradation dynamics and tissue repair, digital monitoring, supply chain origin and its downstream impacts on the ecosystem (Roshi & Wang, 2025).

This change is significant, as it introduces a processual approach to biomaterials theory, in contrast to the static approach used in classical implant design, and shifts attention to therapeutic value. The concept of the implant is that it is not inserted into the body but enters a dynamic body of change and interaction: biological, informational, and ecological. This is an important point to consider, as a posthumanist account, such as Braidotti's (2013), offers a conceptual language for thinking about the relationally constituted body, device, and environment in ways that depart from their ontological separation.

However, the significance of this possibility of the abstract goes beyond the abstract itself. The beauty of this is that it offers insight into the rationale for assessing transient implants, which is not as easily explained by local clinical outcomes. They are also based on the distribution of both risk and material burden, as well as data ownership, across much wider socio-technical systems. It is this convergence of the best theoretical promise that the field holds for achieving analytical convergence in regenerative medicine, sustainability science and digital health.

They include models of tissue guidance and temporally calibrated healing, regenerative medicine, the division of life cycles, among others, as well as an ecological accounting approach, sustainability science, predictive modelling, sensing, and adaptive intervention (Giannokopoulos et al., 2020; Eckelman & Sherman, 2020). However, some paradoxes remain unclarified by this synthesis. A device can be clinically effective but unaffordable. It can be technologically sophisticated but poorly validated, and it can be bioactive but inaccessible to people outside high-resource health care systems (Stahl et al., 2018; Kuper & Blomsma, 2024). However, in theory, it is not a more united, tension-free model; rather, it is a more realistic model that can analyse trade-offs that were not apparent in previous implementation paradigms.

This will have significant real-world impact as it needs to be integrated into the implant design process, with LCA, AI-driven modelling, and clinical validation. Integrating life-cycle assessment and modelling, with the help of AI, into the implant design process, rather than using them as an a posteriori evaluative tool, is a major need. However, current evidence suggests that biodegradability is not enough to consider sustainability as a criterion since other factors such as extraction, synthesis, sterilisation, logistics, and complexity of the device are also significant factors in the impact of the product on the environment (Eckelman & Sherman, 2020). This is particularly emphasised by Kuper and Blomsma (2024), who reveal that procurement decisions increasingly demand a cradle-to-grave assessment rather than a functional assessment alone.

For implant manufacturers, it will be essential to overcome the challenges of mechanical strength, bioactivity, and degradation control, and to embrace the concepts of energy intensity, feedstock quality, packaging burden, and avoiding design revisions. In cases where biological response, corrosion behaviour and electronic endurance must all be harmonised, AI-driven materials discovery can facilitate this change by enabling faster, more accurate screening of material combinations to predict performance in the target scenario.

The feasibility and cost-effectiveness of sustainable transient implants have to be assessed based on the cost of care, not solely on the unit price. In addition to potentially higher initial costs, these devices might be more expensive due to the use of specialised materials and coatings, additive manufacturing, imaging-based planning, software validation, cyber security maintenance, and post-market monitoring. The economic value may still exist if they reduce secondary removal surgery, shorten the follow-up period, prevent revision surgery, and reduce long-term inflammation or waste management burden. For this reason, clinical uptake will depend on whether there are systems in place that reward avoiding a procedure, as well as on life-cycle benefits and better purchase value than other, cheaper, permanent implants. In addition, manufacturing repeatability, sterilisability, supply and accessibility, clinician training, and integration into the existing surgical workflow also play a part in scalability. In low and middle-income settings, affordability should be considered a design constraint, and simpler models with simpler sensing capabilities may be more sustainable than more technologically advanced implants, which may not be affordable or able to be scaled up for procurement or maintenance, and monitored at scale (Kuper & Blomsma, 2024)

However, the possibilities of AI are limited. This can lead to blind spots in the design being reproduced due to data quality issues, model bias, low interoperability, and a closed black box of optimisation criteria. Likewise, a digital twin framework has the potential to improve predictive control of remaining functionality and degradation behaviour, but it is limited without reliable data capture and clinically interpretable outputs. In interdisciplinary training, many pressures make it essential. Competence will be required in the future practitioners in specific areas such as bioengineering, environmental science, data modelling, and ethics. Sustainable transient implants are not in any of these areas but rather at their intersection. Otherwise, the design team might create an appealing device for the environment but not for clinical and/or data-governance needs, or an effective device for clinical needs but not for the environment or data-governance needs.

These changes, theoretical and practical, lead to the policy implications. For sustainable transient implants to move beyond a niche idea and be taken seriously, procurement and regulation should provide clear incentives for implant design decisions that reduce ecological impact without compromising therapeutic value. Hospitals and public health systems have a significant impact on the formation of large-scale demand structures, a fact that is particularly important for green procurement. For manufacturers, the incentives to incorporate sustainability into design will be greater if purchase criteria include sustainability proof and resistance to degradation, and if sources of origin can be traced. The need to internalise sustainability in design will be stronger for manufacturers if purchase criteria include sustainability proof, resistance to degradation, and traceable sources of origin (Kuper & Blomsma, 2024; Eckelman & Sherman, 2020).

Meanwhile, more international standardisation of biodegradable electronics is needed. The current regulatory landscape remains uneven, with different weights assigned to evidentiary requirements, sustainability claims, and regulatory pathways across Europe, the United States, and Asia (European Medicines Agency, 2018). This fragmentation can slow diffusion and make compliance costly, especially when a medical implant is multifunctional, such as a scaffold, sensor, or resorbable electronic device. Greater coherence in standards would enhance comparability, safety assessment, and scalability at the international level, especially for multifunctional implants, including scaffolds, sensors, and resorbable electronics.

However, harmonisation should not be taken to the extent of flattening of regulations. It should be flexible from a contextual perspective across communication infrastructure levels: from affordability to basic infrastructure to technical sophistication. Thus, the deeper meaning of the success of the sustainable implant innovation will be rooted in the common notion. What matters with an implant is not that it lasts forever, but that it lasts for the right period of time, in the right manner and in the right moral context.

The implications of this return to the central claim of the chapter, namely, for the quality of their temporary function, their safety regarding disappearance, the accountability of their data practices, and the evidence supporting their environmental claims, should be noted.

CONCLUSION: TOWARD REGENERATIVE AND ECOLOGICALLY ACCOUNTABLE IMPLANT FUTURES

The literature review in this chapter demonstrates a paradigm shift between the two concepts of sustainable transience in implant medicine. In this new paradigm, implants are not thought of as inert and permanent implants. They should ultimately be viewed as adaptive systems that can control scaffold mechanics and degradation rates and dictate bioelectronic sensing and therapeutic decisions within clinically relevant parameters (Li et al., 2023; Zhang et al., 2025). The reframing, both theoretically, as bringing the three fields in the same field of analysis of regenerative medicine, sustainability science and digital health into the same field of analysis, is important, because it uncovers tensions which have not been sufficiently addressed.

The final result then circles back to the first STI paradigm mentioned. Only those implants that provide functional support timed to healing are successful, and those that do must be environmentally responsible; digital systems must be trustworthy even after the loss of the material; and they must not impede access to high-resource environments during execution. The framework is introduced at the beginning of the chapter. It continues throughout, providing STI with a consistent basis for future research, evaluation, and governance of biological performance, programmed degradation, environmental accountability, digital intelligence, and healthcare accessibility.

Material for optimal bioactivity may not be optimal for controlling resorption, whilst materials that disappear may leave unsustainable data footprints, and technologies touted as sustainable may leave an environmental frame (Eckelman & Sherman, 2020; Stahl et al., 2018). These contradictions are no stumbling block to the paradigm. Instead of innovation operating in silos, a system-level convergence of regenerative medicine, green materials science, and digital bioelectronics is required to enable future developments. Experience from other jurisdictions around the world, such as Europe, the USA and Asia, also indicates that there is a need for evidence to be harmonised, sustainability measures to be taken into account, and usability of clinical practice to be achieved while taking into consideration the non-superseding context of differences that exist between the jurisdictions in the field of costs, infrastructure and access (European Medicines Agency, 2018).

Perhaps the most powerful vision for the future is not just for implants to be biodegradable, but for each individual to have their own system that heals, monitors, and then disappears with minimal ecological footprint, revision rate, and informational density. In that respect, it is not the length of the life of the next generation of implants that will be judged, but rather their flawless finish.

REFERENCES

Braidotti, R. (2013). *The posthuman*. Polity Press.

Eckelman, M. J., & Sherman, J. D. (2020). *Health care's climate footprint: The health sector's contribution to global emissions*. Health Care Without Harm.

Eureka PatSnap. (2025, August 27). *Regulatory and ethical considerations for transient and biodegradable electronics*. https://eureka.patsnap.com/report-regulatory-and-ethical-considerations-for-transient-and-biodegradable-electronics

European Medicines Agency. (2018, November 27). *Medical devices*. European Medicines Agency. https://www.ema.europa.eu/en/human-regulatory-overview/medical-devices

Fiocchi, S., Prandi, C., & Section, A. (2020). Bioresorbable scaffolds: Current technology and future perspectives. *Frontiers in Cardiovascular Medicine*, *7*, 589571. DOI: 10.3389/fcvm.2020.589571 PMID: 33330651

Giannokopoulos, S., Pistono, C., & Willermet, C. (2020). Design, materials, and mechanobiology of biodegradable scaffolds for bone tissue engineering. *Current Opinion in Biomedical Engineering*, *3*(2), 91–104. DOI: 10.1016/j.cobme.2020.03.002

Hong, S., Seung, S., & Kim, D. (2020). Bioresorbable materials on the rise: From electronic components and physical sensors to in vivo monitoring systems. *Advanced Materials*, *32*(3), 1904567. DOI: 10.1002/adma.201904567

Kim, H., Park, J., & Lee, S. (2019). Biodegradable and bioactive orthopaedic magnesium implants with multilayered coatings. *ACS Applied Bio Materials*, *2*(7), 2940–2949. DOI: 10.1021/acsabm.9b00313

Kuper, M., & Blomsma, F. (2024). *Life-cycle assessment study for hospital medical-device purchasers*. European Circular Economy Stakeholder Platform. https://circulareconomy.europa.eu/platform/en/toolkits-guidelines/life-cycle-assessment-study-hospital-medical-device-purchasers

Li, X., Chen, L., & Zhang, Q. (2023). A bioresorbable and conductive scaffold integrating silicon membranes for peripheral nerve regeneration. *Advanced Healthcare Materials*, *12*(32), e2301859. DOI: 10.1002/adhm.202301859 PMID: 37750601

Stahl, B. C., Rainey, S., Harris, E., & Fothergill, B. T. (2018). The role of ethics in data governance of large neuro-ICT projects. *Journal of the American Medical Informatics Association : JAMIA*, *25*(8), 1099–1107. DOI: 10.1093/jamia/ocy040 PMID: 29767726

U.S. Food and Drug Administration. (2023). *Evidentiary expectations for 510(k) implant devices* (Draft guidance). U.S. Food and Drug Administration. https://www.fda.gov/media/171835/download

Xia, B., Liu, Y., Xing, Y., Shi, Z., & Pan, X. (2025). Biodegradable medical implants: Reshaping future medical practice. *Advancement of Science*, *12*(35), e2508014. DOI: 10.1002/advs.202508014 PMID: 40782070

Chapter 9
Transient Biomaterials and Implants:
Challenges in Forensic Identification and Evidence Interpretation

Baljeet Yadav
http://orcid.org/0000-0003-0624-6784
KR Mangalam University, India

Anurag Gupta
http://orcid.org/0009-0009-9931-6868
Banaras Hindu University, Varanasi, India

Sameer Saharan
http://orcid.org/0000-0002-7487-6297
Mody University of Science and Technology, India

Kumud Kant Awasthi
http://orcid.org/0000-0001-8701-7574
KR Mangalam University, India

Mahipal Singh Sankhla
http://orcid.org/0000-0002-7361-3786
KR Mangalam University, India

Garima Awasthi
http://orcid.org/0000-0002-0038-1240
KR Mangalam University, India

ABSTRACT

Transient biomaterials and implants represent a specialized frontier in regenerative medicine, designed to provide temporary structural support or physiological monitoring before degrading within the body. While these materials offer significant clinical benefits and enhance patient privacy by leaving no permanent traces, they present profound challenges for forensic identification and evidence interpretation. These temporal dynamics often result in the rapid deterioration of identification markers, such as manufacturing codes or bioactive coatings, potentially leaving the timeframe between implantation and death indeterminate. This chapter examines

DOI: 10.4018/979-8-2600-0060-1.ch009

the conceptual foundations and various types of transient biomaterials, including biocompatible devices and degradable scaffolds. It further evaluates the limitations of current forensic sampling protocols and analytical techniques, such as infrared spectroscopy, which may fail to distinguish individual traces from surrounding biological residues.

1. INTRODUCTION

Derived from joint considerations among myriad scientific domains, transient biomaterials, analogous to exoskeletons or shells, transitorily bolster living entities. Transient biomaterials, analogous to exoskeletons or shells, transitorily bolster living entities (La Mattina et al., 2020). These seem to posit significant challenges for forensic identification and subsequent evidence framing (Figure 1). Their initial forensic relevance and potential evidentiary weight shifts considerably until the very instances when testimony is sought. Even when being capable of establishing an evident link between a lived experience and the occurrence of a pertinent event, the appropriateness of advancing individual inferences concerning subsequent conduct is exceedingly debatable. The fragility of the counterevidence and the prerequisite for tracers enabling human access to a shared framework posits difficulties in establishing sufficient certainty to contextualise the already tenuous individual reasoning chain. Seizing specific attention from law enforcement during the forensic triage phase, these elements consequently represent an intriguing candidate for joint consideration among forensic disciplines, even though decisive literature analogue remains absent (Riehakainen et al., 2021).

Figure 1. Challenges in forensic identification and evidence interpretation of transient biomaterials and implants

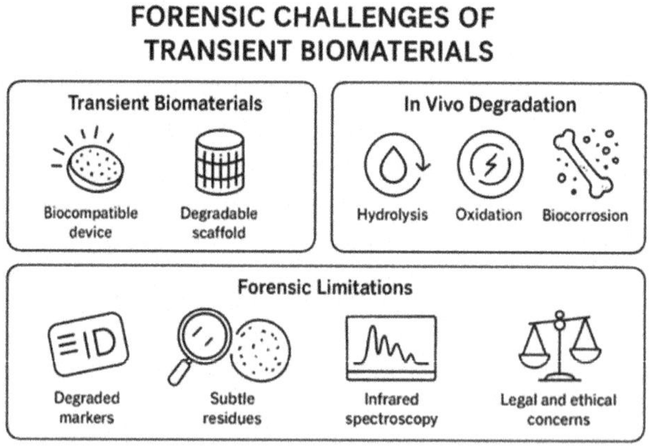

2. CONCEPTUAL FOUNDATIONS OF TRANSIENT BIOMATERIALS

Transience is a versatile and complex concept that applies across diverse fields of inquiry. For example, in art, it carries an aesthetic component; in philosophy, temporal transience denotes the paradox of existence; in life, it embodies the journey of birth, growth, decay, and death; and in personal experience, it conveys the notion of possessions in constant flux. Transient implants are defined by the conditions of their existence after implantation, such as a physiological environment conducive to their processing and/or a service period limited by programmed absorbance and degradation that puts them beyond usefulness within the body (Li et al., 2017). These materials of implantable devices consequently differ in kind and function from the more common implants that remain within the body permanently after the initial implantation process.

Whereas classic implants are hidden and non-detectable, transient implants, once the implanted device is removed or dissolved, can often become detectable, allowing the artisanal link to be retained yet verifiable, similar to the association observable from no dissolving or permanent materials (Cha et al., 2019). Consequently, the permanence versus transience property continues to retain relevance in later stages of the forensic analysis. Evidence of temporary implants such as orthopaedic and contraceptive devices, and bioactive surface treatments and markers, persists on the surface of extracted devices but can nevertheless become concealed through post-extraction alterations such as cleaning, corrosion, contamination, or environmental deposition (Bukhamseen et al., 2022).

3. FORENSIC IDENTIFICATION: PRINCIPLES AND CONSTRAINTS

Identification of human bodies is an important step in forensic practice (A. H. Bukhamseen et al., 2022). A number of modern surgical procedures involve the use of implantable devices, and orthopedic surgical implantable devices are particularly useful for the identification of human corpses. The identification information provided by orthopedic implants is essential, especially when other conventional methods such as DNA analysis, fingerprints, and dental charting are not applicable or in challenging circumstances. The evidentiary value of transient implants depends not only on material persistence but also on environmental exposure, biological interactions, and postmortem degradation dynamics

Identification primarily depends on the nature and extent of the manufacturing data, which depends on two different considerations: the information inscribed on the devices and the methods used for fabrication, that is, the design and manufacture. Marking of orthopedic implants does not only depend on the national impact of the forensic science, but also the requirements of the mandatory use of the unique code on orthopedic implants by the medical regulatory authority (Blessing & Lin, 2018).

Unlike conventional permanent implants, transient biomaterials present unique forensic complications because their structural integrity, chemical composition, and detectability progressively diminish over time. This temporal instability challenges the reproducibility of forensic examinations and complicates medico-legal interpretation during delayed postmortem investigations.

4. TYPES OF TRANSIENT BIOMATERIALS AND IMPLANTS

Temporary biomaterials can take different forms and offer varied functionality (Prakasam et al., 2017). While some restore tissue after injury, others provide mechanical strength and optimal architecture for regeneration. Transient implants include those designed for limited functionality, degrading after fulfilling their intended role. Implant-associated materials, such as coatings, fixation agents, and drug-releasing matrices, also allow transit. Non-implantable biomaterials apply transiently before tissue integration in forensics and beyond, as temporary biomaterials are designed to provide structural support and degrade after tissue repair and regeneration is achieved (Bonferoni et al., 2021).

Temporary biocompatible devices recover structural integrity and prevent further damage following tissue loss (Amini et al., 2011). These include meshes for hernia repair, mechanical supports for large bone defects, and splints for orthopaedic fractures. Total tissue recover requires gradually reduced implant functionality through bioreclamation or bioreplacement. Transitory tissue-restoration devices must degrade before the final recovery stage, ensuring patient safety. Temporary supportive devices and coatings that facilitate functional transit constitute auxiliary transitional biomaterials.

Altering surface properties to reduce adhesion between product and tissue assists device removal, speeds up recovery, and lowers the risk of post-removal adverse events (Figure 2). A variety of techniques, such as physical surface modifications, enable surface engineering to establish desirable properties, improving implant management. Time-release local-anesthetic matrices and other controlled drug-delivery biomaterials function as auxiliary transitional resources, as demonstrated by absorbable anesthetic matrices designed to provide sustained nerve blockade through gradual drug release (Wang et al., 2009).

Figure 2. Classification of transient biomaterials and implants based on functionality, degradability, and application context

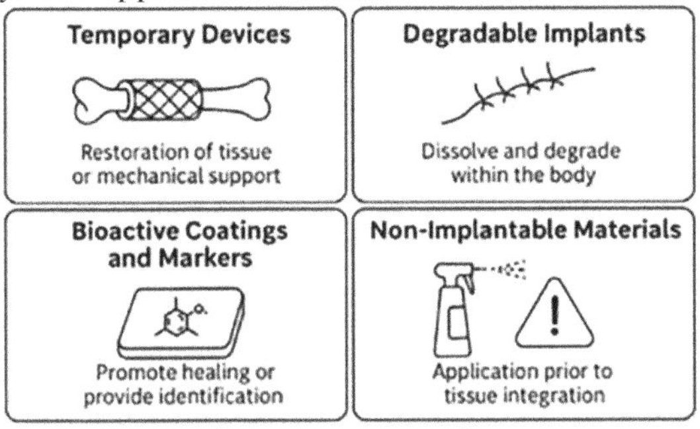

4.1. Temporary Biocompatible Devices

Temporary biocompatible devices, widely used in conventional medicine, also have a role in biometric applications like photo identification and biomedical tracking, as biodegradable and implantable electronic systems are designed for short-term monitoring of physiological signals and degrade naturally after use, eliminating the need for removal (Hosseini et al., 2021). Such materials seem to fade quickly as biometric devices become digital and more innovative. Nevertheless, some regions are still adopting temporary biocompatible prosthetics. Temporary prosthetics are often detachable and do not require any surgical or installation process. Temporary prosthetics of body parts like fingerprints, toes, and lips also modify bespoke biometric models or templates. Temporary biocompatible devices are either thermoset, thermoplastic, or viscoelastic polymeric materials used in biomedical applications and widely available in the market (Satchanska et al., 2024).

4.2. Degradable and Dissolving Implants

Dissolvable implants such as stitches, staples, and surface coatings are routinely employed in medicine. Dissolvable implants such as stitches are routinely employed in medicine (Pillai & Sharma, 2010). Recent advances in synthetic biomaterials allow complete elution of bioactive agents or drug-polymer constructs through dissolution.

Tuning the material composition enables control of release profiles between seconds and years. Fully biodegradable implants composed of poly-methylmethacrylate, poly-dl-lactide, copolymers, and polycaprolactone exhibit the additional capability of complete degradation and are reportedly used for maxillofacial, dental, orthopedic, and biodegradable electronic devices, among other applications. Selective laser-melted biodegradable scaffolds composed of poly(d,l-lactide) have shown potential for maxillofacial reconstruction (Smeets et al., 2017).

For soluble materials, utilization within a therapeutic window are decisive criteria to consider the devices as forensically significant. Implants would be considered irrelevant when release or degradation products are completely exhausted. Implant traceability through manufacturing codes, polymer signatures, or embedded radiopaque markers may assist forensic experts in linking recovered materials to medical records and patient histories.

From a forensic perspective, degradable implants may provide temporary identification markers during early postmortem intervals; however, prolonged environmental exposure may result in partial or complete loss of traceable characteristics

4.3. Bioactive Coatings and Markers

Implantable devices that are bioactive and transient have a significant role in implant marking and biocompatibility. A marker should be biocompatible, enabled on existing implants, non-harmful in pre- or post-degradation, and easily implementable into devices already approved for medical use. Several materials including platinum, tantalum, niobium, titanium, and ceramics, allow marking of implants. Titanium is the most accepted material. Methods include laser engraving, etching, anodisation, micro welding, bonding electronic microchips, and coating polymer markers. Commercial systems offering titanium-based marker systems prolong identification time. Tissue contacts markers need to withstand design, grinding, polishing, and sterilization processes (Colvenkar & Sv, 2022). After implantation, transient markers are exposed to human cell and tissue contact, which leads to degradation and loss of identification capability. For example, Titanium/bioactive glass coatings promote osseointegration of implants and allow bioactive temporary markers on degradable implants (Drnovsek et al., 2012).

Biomaterials applied onto the surface of titanium can become an identification marker. Partial-device coating does not hinder implant performance. Coatings need to be biocompatible and enable either temporary or permanent identification of the implant. Customisation of their design enables the coating systems to transduce specific identification markers and allow temporary identification during the degradation stage.

Bioactive coatings used in orthopedic and dental implants may retain microscopic or spectroscopic signatures even after partial degradation. Advanced analytical techniques such as SEM-EDS, FTIR spectroscopy, Raman spectroscopy, and micro-CT imaging can therefore support material characterization in compromised forensic samples. For enhanced transient marking of implants at a given active time window, titanium and biodegradable polycaprolactone have been investigated as identification systems (Smith et al., 2010).

4.4. Non-Implantable Transient Biomaterials in Forensics

The existing body of forensic literature on transient biomaterials addresses implants exclusively; yet, other significantly bioactive materials offer promise for forensic identification without the need for implantation. This group encompasses primarily non-implantable inorganic and organic biocompatible materials with notable forensic potential. They are principally classified as transient biomaterials because they undergo substantive changes over time and beyond any prospect of detection. Mechanisms for post-extraction transformation involve micro-scale particle detachment and active compound release from the matrix (Stewart et al., 2018). Forensic processes for conventional bulk materials and polymers should be amendable to the surface-dominant, particulate nature of most transient pertinent materials expanded matrices of greater surface area encompass enriched quantities of active functions ready for release and thus enhance exposure to the surrounding medium and bioactive particulate detachment.

Transitory materials exhibiting rapid changes compared to conventional implants tend to be less durable, since biodegradable biomaterials are engineered to degrade progressively in biological environments, reducing their long-term persistence (Prakasam et al., 2017). Transient bioactive coatings adhered superficially to implants often diminish to irretrievable levels prior to the implant's removal. The residue of other non-deposited materials yet remains viable for financial tracing or other identification (Bukhamseen et al., 2022).

5. POST-EXTRACTION AND DEGRADATION DYNAMICS

Although transient biomaterials and implants exhibit considerable variability in their degradation pathways and mechanisms, discernible patterns may still facilitate the determination of the extraction timeframe and other elements for the estimating time since discharge (TSD) (Das et al., 2012). Several relevant concepts arise from the study of post-extraction degradation dynamics for the sake of legal and forensic evaluation. After extraction from a living entity, transient substances invariably undergo post-mortem alterations and environmental exposure, regardless of their specific in vivo degradation features. The early adoption of a relevant strategy based on degradation systems allows the identification of time-frames for the recoverability and permanence of traces to support forensic analyses.

During in vivo utilization, temporary devices or biomedical structures degrade through pathways such as disintegration, solubilization, hydrolysis, oxidation, thermal processes, macromolecular chain scission, and biocorrosion. Biomaterials degrade through hydrolytic, thermal, oxidative, and biodegradation processes (Woodard & Grunlan, 2018).

The degradation behavior of transient biomaterials may assist forensic investigators in estimating implantation duration, post-surgical intervals, or approximate timelines of medical intervention. Chemical transformations associated with these modes can generate minute quantities of low-molecular-weight by-products, epithelial residues, or soluble substances, which may become incorporated into the surrounding material and offer additional clues to the analysis. Following extraction from the host, these initial compounds occupy a significant role in the post-mortem assessment of discoverable transient residues.

5.1. In Vivo Degradation Pathways

After implantation, biocompatible materials can undergo both bulk and surface degradation through hydrolytic or oxidative pathways (Prabhath et al., 2021). Hydrolysable polymers such as poly(α-hydroxy esters) are the most widely used degradable implants. In vivo degradation generally occurs via hydrolytic scission of ester bonds, accompanied by diffusion of small molecules (water or saline) through the polymer into the final degradation products. These small molecules remain essentially harmless if the material is biocompatible, and project systems with slow biodegradability are designed so that the initial degradation stage does not reach these nondangerous products. Degradation in situ is frequently monitored in a bioreactor or in vivo, despite the difficulty of access (Wang et al., 2017).

5.2. Postmortem Transformations and Environmental Effects

The continued need for biomaterials and biocompatible implants is driven by increasing demand in regenerative medicine, tissue engineering, and implantable biosensors for in-situ monitoring of biological markers (Klosterhoff et al., 2017). Transient biomaterials, also referred to as temporary biomaterials or biodegradable biomaterials/biologics, can degrade sequentially or on-demand upon stimulation by environmental or biological factors. Transient implants can help overcome issues associated with the use of permanent implants and allow for reduced surgical operations. Forensics plays an inevitable role in the activation of safety programs, without which it would be difficult to track people who have abandoned or lost implanted biomaterials.

Forensic analysis of transient biomaterials can help convey insight into the nature of the event and time elapsed before the photograph was taken. Investigation and study of the temporal activity of the previously mentioned biomaterials would provide useful data to police forces and improve understanding of the post-mortem period.

Transient devices and implants are subject to degradation both in vivo and post-mortem, while the surrounding environment also influences degradation and transformation (Singh et al., 2020). The degradation and transformation processes can occur simultaneously or sequentially, and after an extraction the material colures and chemical groups further evolve and depend on the environmental conditions. Environmental exposure and contamination impact the physical aspect of the materials, nevertheless certain “fingerprints” can survive due to the locality of these marks. After extraction and exposure to the surrounding environment, the transient and bioactive materials can undergo degradation, transformation and contaminination which varies with time. The colour alteration and contaminant deposition can provide potential clues about the temporal window and reject the so-called “post-mortem” activity.Interpretation of degradation stages should be conducted cautiously because physiological variables, infection, pH fluctuations, and environmental exposure can significantly alter expected degradation rates

5.3. Time-frames of Detectability

The FBI conducted an extensive study of forensic specimen age and explained the challenges and limitations in predicting the time of deposit. Estimation of age in forensic anthropology remains challenging due to biological variability and methodological limitations (Ubelaker & Khosrowshahi, 2019). From their extensive testing of numerous tissue, fluid, and staining types, it was concluded that it is currently impossible to predict the viable age of a specimen beyond a preliminary time-frame of a few hours or days. Predicting the age of a specimen after days remained “marred by low-resolution estimates” and analysis across these additional time-points yielded an “alternative time-since-deposition calculation” the FBI study ranked “unreliable.”

Establishing detectability windows for transient implants is essential for forensic timeline reconstruction, particularly in cases involving unidentified remains, medical malpractice investigations, and disaster victim identification. In many studies focusing on blood, other body fluids, saliva, and seminal strains only resolution up to about a week was attainable.

Predicting the age of transient materials post-extraction has remained an open research area, particularly as many newly introduced transient implants allow qualitative and quantitative detection only for limited time frames, owing to their gradual degradation and disappearance from biological environments (Riehakainen et al., 2021). Standardized degradation databases and validated reference models are currently limited, highlighting the need for interdisciplinary forensic-biomedical collaboration.

6. EVIDENCE COLLECTION, ANALYSIS, AND INTERPRETATION

Transient materials after surgical procedures are subject to a variety of sampling and analytical approaches based on their anticipated properties and timelines. Collection of residues is normally preceded by a careful visual inspection of the area for macroscopic clues. Recognizing the presence of trace evidence is a critical first step in the forensic process (Cubbage et al., 2023). If visible, standard techniques for sampling the materials involved should be employed, ensuring the chance for subsequent comprehensive analysis and forensic reconstruction. The availability of residue depends on the material involved and the duration since incorporation or post-partum elimination. Evidence recovery protocols for transient biomaterials should prioritize contamination control, moisture stabilization, low-temperature storage, and rapid documentation to minimize post-recovery degradation

Lesser adherence and greater extractability support longer interval sampling for higher biomaterial dimensions. Collection of puncture marks typically requires over a thousand hours of elapsed time following in-vivo implantation. Degraded residues can still enable reconstruction analysis and estimation of elapsed time, as micro-CT three-dimensional reconstruction allows quantitative evaluation of implant degradation over time (Xu et al., 2018.

6.1. Sampling Protocols for Transient Materials

For transient materials, the selection of an appropriate sampling protocol and sampling site is a defining factor for trace recoverability based on biomaterial degradation kinetics, subsequent modifications of biomaterial traces after manufacture and/or installation (e.g. machining marks, scarring, contouring), and the establishment of an observational timeframe after sample deposition or manufacturing, installation, or cleaning.

Transient biomaterials degrade after completing their intended function, leading to shorter persistence and reduced trace survival compared to traditional implants (Prakasam et al., 2017). However, it is possible to discern the use of certain biocompatible materials even 6 months post deposition or after tampering attempts. Recognizing or detecting post-exploitation traces after a relatively short timeframe will direct attention to particular actions or locations, down-scaling the evidentiary potential (Figure 3).

Both water-soluble and biodegradable formulations, once colocalized, should lead to the examination of sampling sites corresponding to possible direct contact. Upon control or partial removal, leftover traces often persist in sufficient quantity and/or quality for identification or sourcing characterization. Traces sometimes survive longer in protected locations such as footwear, headwear, and toss surfaces, provided the latter exhibit on/off cycles requiring biomaterial re-application. Trace DNA persistence is strongly influenced by surface characteristics and environmental exposure, with some surfaces enabling longer retention of biological material (Arsenault et al., 2024). Biodegradable materials may continue degrading after collection, time-sensitive chain-of-custody documentation becomes particularly important in preserving evidentiary reliability.

6.2. Analytical Techniques and Their Limitations

Biodegradable and transient implant materials present unique challenges in identification and interpretation, making them an important subject of ongoing research (Prakasam et al., 2017). To date, evidence on transient biomaterials cannot contribute heavily in showing identification parameters from a post-mortem point of view. The field of forensic science is foundationally set to acquire, analyze, preserve, and present evidence in any judicial setting. Transient materials are gained at pre-mortem and post-mortem scenarios; marks from post-mortem point do not showcase validity of evidence in any judicial meeting. Dealing with identification from transient materials is severely limited to conventional means of collecting and exhibiting a material. To acquire information about permanent post-mortem damages allow what limitations can and cannot be presented. A multi-modal analytical workflow combining radiological imaging, microscopy, elemental profiling, and spectroscopic analysis improves reliability in the forensic characterization of transient implant residues. Current forensic protocols remain limited by insufficient reference databases, lack of standardized degradation benchmarks, and variability in implant composition across manufacturers.

Figure 3. Forensic workflow for transient biomaterial evidence interpretation

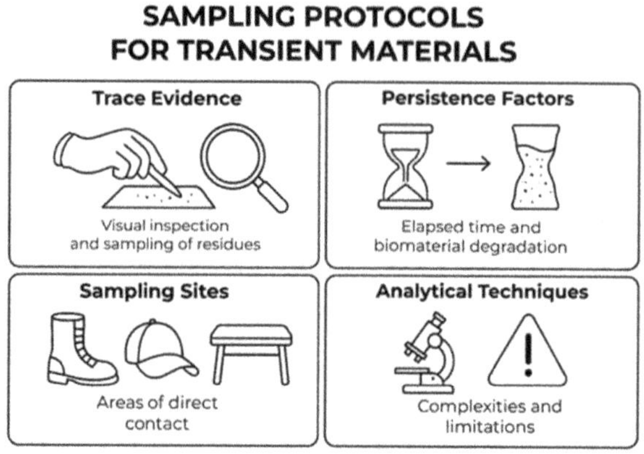

Figure 4. Schematic representation of sampling strategies, persistence factors, and analytical limitations associated with transient biomaterials in forensic investigations

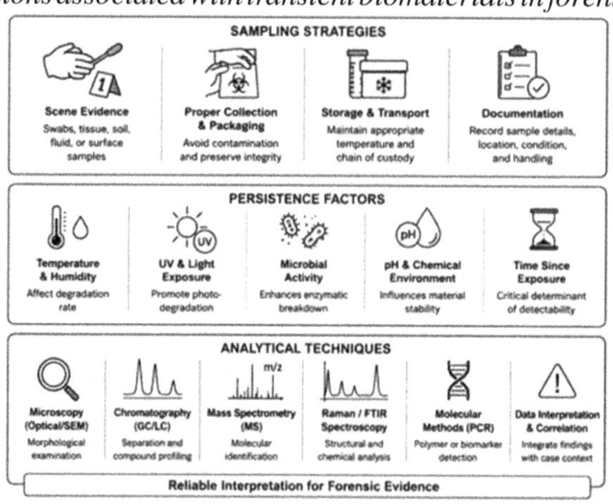

6.3. Documentation and Chain-of-Custody Considerations

The collection of evidence material requires proper documentation, which is essential to preserve the chain of custody from the first moment the investigator handles the material until its presentation in court (D'Anna et al., 2023). A chain of custody form should be filled out, containing the individual's signature, date, time of collection, a description of the collected material, and the initials of the person who collected the sample. This documentation complements the collection in a complementary manner. Furthermore, whenever a new person receives the sample, the chain of custody must be transferred and documented in accordance with the previously described steps.

The chain of custody must be prolonged for every reporting of aseptic sample extraction. It is important to now the property of the integrity of the sample and to preclude the intervention of/or changes induced by transferee, especially those who may deliver a contradictory report.

6.4. Interpretive Frameworks for Transient Evidence

The determination of degradation behavior and residual lifetime of transient biomaterials remains complex due to variability in material properties and environmental conditions (Li et al., 2017), posing potential challenges for forensic identification and evidence interpretation. The framework outlined here provides guidance on the time since extraction of biocompatible materials, often of interest when transient biomaterials are found in or on a corpse or in association with bodily fluids. Forensic scientists typically work backwards from the visible evidence, attempting to describe and interpret what may have been present initially. With transient biomaterials, however, identifying the starting point may be too speculative to provide a useful interpretation. Instead, it often suffices to establish a chronology of processing events and approximate time since the latest application when a material was still transferred. Forensic biologists are trained to analyse and interpret temporal transfer evidence and other transient materials, but different frameworks are available to assist scientific parameterization and judicial articulation. Forensic biologists analyze temporal transfer evidence and transient materials using structured frameworks (van Oorschot et al., 2021).

A specialised framework classifies biocompatible materials according to their medical, engineering or forensic relevance and provides detailed information on time since extraction, degradation routes, transfer processes, post mortem transformation, and relevant analytical techniques. Separation of medical, engineering and forensic types is also applied to classify deactivation and destruction techniques according to their forensic significance and the body fluid desired during the illicit transfer. A more general framework classifies evidence type according to physical phase homogeneous fluids, homogeneous solids, or heterogeneous mixtures without reference to function or application. Each scheme facilitates the choice of temperature sensors, sampling techniques, and models for transfer speed, and helps to articulate common features of active sampling theory, method, and subjective interpretation that arise in diverse forensic situations involving transient evidence.

Interpretation of transient biomaterial evidence should incorporate temporal degradation modeling, environmental context, medical history correlation, and analytical uncertainty assessment

7. LEGAL AND ETHICAL IMPLICATIONS

Since the introduction of computer numerically controlled (CNC) machines, the production of complex biocompatible implants that fit the human body has become easy. Computer-numerically controlled (CNC) machining has enabled the precise production of complex, patient-specific biocompatible implants (Honigmann et al., 2018). Together with these manufacturing advances, materials have been developed to produce transient implants that can dissolve or fully resorb to leave no traces in the body after a determined period. From these two conditions, some implantable or non-implantable transient biocompatible biomaterials are emerging in medicine and forensic science, allowing the implantation of transient materials into the biological body without leaving any trace which probably helps to protect the privacy of the patient. Courts may require demonstrable analytical reliability and validated methodology before accepting transient biomaterial evidence as scientifically admissible forensic evidence

7.1. Admissibility and Standard of Proof

The legal status of transient biomaterials and implants as forensic evidence is unsettled. Specimens that qualify as transient materials under the specification in Section 3.2 may or may not be admissible as evidence, depending on the surrounding circumstances. In general, to satisfy a court regarding the admissibility of transient evidence, the prosecution must prove three conditions. It must show that the evidence is relevant, legally obtained, and trustworthy.

Each condition is subject to a specific standard of proof. Relevance in forensic interpretation may be established by a preponderance of evidence, whereas higher evidentiary standards such as clear and convincing evidence are required for more definitive conclusions in legal and forensic decision-making (Testa & West, 2010). This requirement, while not a universal standard, is widespread in jurisdictions throughout the United States, including Nebraska. Transient biomaterials and implants fit the same evidentiary paradigm as other physical samples, tracing their evidentiary status across the full timeline from deposition to extraction.

7.2. Privacy, Consent, and Dual-Use Concerns

The use of transient biomaterials and transient implants raises issues related to privacy, consent, and dual-use applications analogous to concerns raised by other innovative technologies (Marinescu et al., 2025). Users of such materials, as well as individuals inadvertently exposed to them, may wish to maintain control over the information connected to their identity and may have legitimate objections to the collection and analysis of materials they did not select. Concerns over dual-use applications include the possibility that transient biomaterials or implants, while intended solely for therapeutic purposes, may nonetheless facilitate illicit activity.

Because transient biomaterials are designed for temporary use, adoption will afford a limited exposure window during which it may be possible to determine specific usage conditions. If those conditions are particularly sensitive for example, if particular installations derive from illicit activity the user may have a crucial interest in preventing subsequent identification. These materials therefore echo the privacy challenges already associated with other identifying forensic materials, such as DNA, ink, or natural materials. Like non-transient forensic markers, it may be possible to suppress disclosure of certain markers or materials without seriously impeding investigation.

The possibility that transient materials might be exploited to commit illicit activity raises legitimate concerns about dual-use applications. Such exploitation could occur by deliberately choosing, for instance, materials that specifically denote criminal activity. Certain transient procedures could also open up indirect possibilities for illicit activity, as when the very adoption of a therapy, either illicit in itself or easily mistaken for illicit activity, signals potential avenues for further inquiry. Therefore the increasing integration of smart biomaterials and traceable implant technologies raises important concerns regarding patient privacy, medical surveillance, and unauthorized forensic access to biomedical data.

7.3. Forensic Reporting and Expert Testimony

Expert witness reports compose a distinct but integral dimension of the forensic identification of transient implantable devices. The continued shifts in the fields of implantation, materials science, biomaterial interactions, and forensic particle analysis underscore the need for expert testimony to clarify and contextualize the findings of an examination of transient-material evidence. Such testimony seeks to fulfil the broader expectations of forensic reporting by establishing the identity of the evidence, the methods employed in the examination, the materials considered, and the limitations of the analysis performed; by indicating the relevance of the evidence to the case; and by addressing broader relevant questions. Since the transient biomaterial welding process can leave only a very subtle residue compared to many types of organic and inorganic microparticles found on commonly involved surfaces, the focus of expertise in such cases is frequently confined to bioresorbable implants, indicators of transient biomaterials in the absence of visible deposit characterisation, and the interpretation of photographic documentation of such deposits.

The identification and description of transient biomaterials and implants are subject to strict procedural requirements and success criteria, as medical findings such as implants can serve as reliable identifiers when appropriately documented and compared with antemortem records (Lemos et al., 2024). Re-extraction dynamics for a broad range of deposited transient materials, including organic and inorganic substances, as a function of still-life exposure duration at all common ambient temperatures in both indoor and outdoor settings have been characterised for uncoated and likewise untempered surfaces. Examples and models of the forensically relevant extraction or dissolution of transient biomaterials, on-time-to-visibility and early post-extraction formation of medium- to long-living sheds and deposits of still-identifiable colored biomaterial important particularly for screening purposes condensed reflection (including diffuse-specular and other optical effects), dormant expired substances, and other residual characteristics, have all been elucidated. Such an understanding allows the completion of a standardised, deterministic temporal framework for extraction, still-life watching, and other observables also documented for widely studied deposition and evidence-movement scenarios on glass, metals, and certain other non-bioresponsive surfaces. The continual emergence of novel non-resborbable yet transient implantable materials and their active incorporation into forensic-toolbox approaches underscore the need for additional expertise in these areas, alongside that emphasised above where all present-day formulations remain characterisable with readily obtained models and broader still-life iconography.

8. CASE STUDIES AND APPLIED METHODOLOGIES

Temporary or biodegradable materials have been increasingly considered in various fields of science, like agriculture, textile, coating, and biomaterials, attracted by the idea to materialize a sustainable globe (Dallaev et al., 2025). A temporary implant is typically defined as a material that should fulfill specific function(s) incomplete for a predefined time and is then absorbed or excreted. Temporary biomaterials can stimulate the native biological behaviours while stimulating the local tissue repair, and their rapid absorption rate can also increase the safety of load-bearing implants. In decomposed or skeletonized remains, partially preserved orthopedic or dental implants may provide critical identifiers through serial numbers, radiographic comparison, or manufacturer traceability records Forensic field has gained the benefit from these transient materials, as temporary installation non-consensually can occur at the surgical operation taking, removing traces afterward would cause a great challenge for forensic identification and determine the timeline of the operation. Several types of transient biomaterials or implant have been applied in forensic practice.

9. FUTURE DIRECTIONS AND RESEARCH GAPS

Bioabsorbable and bioactive materials designed to promote physiological processes at implantation sites have become a major focus of materials science research (Prakasam et al., 2017). Preliminary studies address transient biomaterials meant for elimination from human body. Despite their non-implantable nature, transient biomaterials preclude forensic identification. Transient biomaterials, precisely constrained in their intended location, seldom create removable markers. Future forensic research should focus on developing standardized degradation atlases, implant traceability systems, AI-assisted degradation prediction models, and validated protocols for transient biomaterial evidence interpretation.

"Closer collaboration among forensic scientists, biomaterial engineers, radiologists, and medico-legal investigators will be essential for establishing reproducible forensic standards

CONCLUSION

Notably, transient biomaterials and implants present complex challenges for forensic scientists. Their defining temporal dimension, in conjunction with diverse degradation patterns, severely constrains the availability of material suitable for forensic analysis. After implantation, identification markers rapidly deteriorate, and the ensuing inter-patient variability generates significant spatiotemporal uncertainty regarding the extant form and location of any potential extracts. Consequently, the time-frame between implantation and extraction (or death) may remain indeterminate. Subsequently, environmental exposure, coupled with the residual presence of co-extracts, further complicates the picture.

Once extracted, the absence of recognized standardized sampling schemata for non-classical biological materials calls for the formulation of specific protocols. Moreover, while conventional forensic methods (e.g., fluorescence, infrared spectroscopy) cover a broad range of materials, they often fail to distinguish individual traces. Where analysis succeeds, the predominance of trace-sized coupling agents typically returns inconclusive results, given that inclusion layers generally present no investigable identifiers. Alternative techniques targeting non-coupling-specific entities commonly prove destructive, eliminating any prospect for subsequent identification. The Brazilian scenario illustrates yet another limitation; stringent compulsory disclosure laws governing markers on transient implants inhibit meaningful traceability.

It may be concluded that the transient biomaterials represent a rapidly evolving challenge in forensic science because their evidentiary persistence is inherently time-dependent. Although these materials introduce complexities in recovery, preservation, and interpretation, they also offer emerging opportunities for implant-assisted identification and forensic timeline reconstruction. Advancements in analytical technologies, standardized forensic workflows, and interdisciplinary collaboration will be essential to ensure reliable medico-legal interpretation of transient biomaterial evidence.

REFERENCES

Amini, A. R., Wallace, J. S., & Nukavarapu, S. P. (2011). Short-term and long-term effects of orthopedic biodegradable implants. *Journal of Long-Term Effects of Medical Implants, 21*(2), 93–122. DOI: 10.1615/JLongTermEffMedImplants.v21.i2.10 PMID: 22043969

Arsenault, H., Kuffel, A., Daeid, N. N., & Gray, A. (2024). Trace DNA and its persistence on various surfaces: A long term study investigating the influence of surface type and environmental conditions - Part one, metals. *Forensic Science International. Genetics, 70,* 103011. DOI: 10.1016/j.fsigen.2024.103011 PMID: 38324952

Blessing, M. M., & Lin, P. T. (2018). Identification of Bodies by Unique Serial Numbers on Implanted Medical Devices. *Journal of Forensic Sciences, 63*(3), 740–744. DOI: 10.1111/1556-4029.13598 PMID: 28718881

Bonferoni, M. C., Caramella, C., Catenacci, L., Conti, B., Dorati, R., Ferrari, F., Genta, I., Modena, T., Perteghella, S., Rossi, S., Sandri, G., Sorrenti, M., Torre, M. L., & Tripodo, G. (2021). Biomaterials for Soft Tissue Repair and Regeneration: A Focus on Italian Research in the Field. *Pharmaceutics, 13*(9), 1341. DOI: 10.3390/pharmaceutics13091341 PMID: 34575417

Bukhamseen, A. H., Aldhameen, A. A., Alzayyat, N. T., Alqadeeb, M. M., Alotaibi, B. K., Aljohani, R. I., & Menezes, R. G. (2022). The use of orthopedic surgical devices for forensic identification: A systematic review. *Acta Biomedica, 93*(3), e2022082. DOI: 10.23750/abm.v93i3.12471 PMID: 35775772

Cha, G. D., Kang, D., Lee, J., & Kim, D. H. (2019). Bioresorbable Electronic Implants: History, Materials, Fabrication, Devices, and Clinical Applications. *Advanced Healthcare Materials, 8*(11), e1801660. DOI: 10.1002/adhm.201801660 PMID: 30957984

Colvenkar, S., & Sv, R. (2022). Denture Marking for Forensic Identification Using Laser-Marked Stainless Steel Quick Response (QR) Code. *Cureus, 14*(2), e22431. DOI: 10.7759/cureus.22431 PMID: 35371810

. Cubbage, H. R., Macey, C., & Scott, K. R. (2023). Macroscopic assessment of environmental trace evidence dynamics in forensic settings. *Science & justice: journal of the Forensic Science Society, 63*(3), 376–386. https://doi.org/DOI: 10.1016/j.scijus.2023.03.004

D'Anna, T., Puntarello, M., Cannella, G., Scalzo, G., Buscemi, R., Zerbo, S., & Argo, A. (2023). The Chain of Custody in the Era of Modern Forensics: From the Classic Procedures for Gathering Evidence to the New Challenges Related to Digital Data. *Healthcare (Basel)*, *11*(5), 634. DOI: 10.3390/healthcare11050634 PMID: 36900637

Dallaev, R., Papež, N., Allaham, M. M., & Holcman, V. (2025). Biodegradable Polymers: Properties, Applications, and Environmental Impact. *Polymers*, *17*(14), 1981. DOI: 10.3390/polym17141981 PMID: 40732859

Das, D., Zhang, Z., Winkler, T., Mour, M., Gunter, C., Morlock, M., Machens, H. G., & Schilling, A. F. (2012). Bioresorption and degradation of biomaterials. *Advances in Biochemical Engineering/Biotechnology*, *126*, 317–333. DOI: 10.1007/10_2011_119 PMID: 21975956

Honigmann, P., Sharma, N., Okolo, B., Popp, U., Msallem, B., & Thieringer, F. M. (2018). Patient-Specific Surgical Implants Made of 3D Printed PEEK: Material, Technology, and Scope of Surgical Application. *BioMed Research International*, *4520636*, 1–8. Advance online publication. DOI: 10.1155/2018/4520636 PMID: 29713642

Hosseini, E. S., Dervin, S., Ganguly, P., & Dahiya, R. (2021). Biodegradable Materials for Sustainable Health Monitoring Devices. *ACS Applied Bio Materials*, *4*(1), 163–194. DOI: 10.1021/acsabm.0c01139 PMID: 33842859

Klosterhoff, B. S., Tsang, M., She, D., Ong, K. G., Allen, M. G., Willett, N. J., & Guldberg, R. E. (2017). Implantable Sensors for Regenerative Medicine. *Journal of Biomechanical Engineering*, *139*(2), 0210091–02100911. DOI: 10.1115/1.4035436 PMID: 27987300

La Mattina, A. A., Mariani, S., & Barillaro, G. (2020). Bioresorbable Materials on the Rise: From Electronic Components and Physical Sensors to In Vivo Monitoring Systems. *Advanced Science (Weinheim, Baden-Wurttemberg, Germany)*, *7*(4), 1902872. DOI: 10.1002/advs.201902872 PMID: 32099766

Lemos, Y. V., Furtado, A. N., Lima, A. Z., Dionísio, A. S., Araújo, R. M., & Cunha, E. (2024). Human identification by medical findings in a forensic anthropology context. *Forensic Sciences Research*, *9*(3), owae041. Advance online publication. DOI: 10.1093/fsr/owae041 PMID: 39493280

Li, R., Wang, L., Kong, D., & Yin, L. (2017). Recent progress on biodegradable materials and transient electronics. *Bioactive Materials*, *3*(3), 322–333. DOI: 10.1016/j.bioactmat.2017.12.001 PMID: 29744469

Marinescu, M. R., Ionescu, O. N., Pachiu, C. I., Dinescu, M. A., Muller, R., & uchea, M. P. (2025). Next-Gen Healthcare Devices: Evolution of MEMS and BioMEMS in the Era of the Internet of Bodies for Personalized Medicine. *Micromachines*, *16*(10), 1182. DOI: 10.3390/mi16101182 PMID: 41156428

Pillai, C. K., & Sharma, C. P. (2010). Review paper: absorbable polymeric surgical sutures: chemistry, production, properties, biodegradability, and performance. *Journal of Biomaterials Applications*, *25*(4), 291–366. DOI: 10.1177/0885328210384890 PMID: 20971780

Prabhath, A., Vernekar, V. N., Vasu, V., Badon, M., Avochinou, J. E., Asandei, A. D., Kumbar, S. G., Weber, E., & Laurencin, C. T. (2021). Kinetic degradation and biocompatibility evaluation of polycaprolactone-based biologics delivery matrices for regenerative engineering of the rotator cuff. *Journal of Biomedical Materials Research. Part A*, *109*(11), 2137–2153. DOI: 10.1002/jbm.a.37200 PMID: 33974735

Prakasam, M., Locs, J., Salma-Ancane, K., Loca, D., Largeteau, A., & Berzina-Cimdina, L. (2017). Biodegradable Materials and Metallic Implants-A Review. *Journal of Functional Biomaterials*, *8*(4), 44. DOI: 10.3390/jfb8040044 PMID: 28954399

Riehakainen, L., Cavallini, C., Armanetti, P., Panetta, D., Caramella, D., & Menichetti, L. (2021). In Vivo Imaging of Biodegradable Implants and Related Tissue Biomarkers. *Polymers*, *13*(14), 2348. DOI: 10.3390/polym13142348 PMID: 34301105

Satchanska, G., Davidova, S., & Petrov, P. D. (2024). Natural and Synthetic Polymers for Biomedical and Environmental Applications. *Polymers*, *16*(8), 1159. DOI: 10.3390/polym16081159 PMID: 38675078

Singh, R., Bathaei, M. J., Istif, E., & Beker, L. (2020). A Review of Bioresorbable Implantable Medical Devices: Materials, Fabrication, and Implementation. *Advanced Healthcare Materials*, *9*(18), e2000790. DOI: 10.1002/adhm.202000790 PMID: 32790033

Smeets, R., Barbeck, M., Hanken, H., Fischer, H., Lindner, M., Heiland, M., Wöltje, M., Ghanaati, S., & Kolk, A. (2017). Selective laser-melted fully biodegradable scaffold composed of poly(d,l-lactide) and β-tricalcium phosphate with potential as a biodegradable implant for complex maxillofacial reconstruction: In vitro and in vivo results. *Journal of Biomedical Materials Research. Part B, Applied Biomaterials*, *105*(5), 1216–1231. DOI: 10.1002/jbm.b.33660 PMID: 27062073

Smith, C. M., Roy, T. D., Bhalkikar, A., Li, B., Hickman, J. J., & Church, K. H. (2010). Engineering a titanium and polycaprolactone construct for a biocompatible interface between the body and artificial limb. *Tissue Engineering. Part A*, *16*(2), 717–724. DOI: 10.1089/ten.tea.2009.0066 PMID: 19769529

Stewart, S. A., Domínguez-Robles, J., Donnelly, R. F., & Larrañeta, E. (2018). Implantable Polymeric Drug Delivery Devices: Classification, Manufacture, Materials, and Clinical Applications. *Polymers*, *10*(12), 1379. DOI: 10.3390/polym10121379 PMID: 30961303

. Testa, M., & West, S. G. (2010). Civil commitment in the United States. *Psychiatry (Edgmont (Pa.: Township))*, *7*(10), 30–40.

Ubelaker, D. H., & Khosrowshahi, H. (2019). Estimation of age in forensic anthropology: Historical perspective and recent methodological advances. *Forensic Sciences Research*, *4*(1), 1–9. DOI: 10.1080/20961790.2018.1549711 PMID: 30915413

van Oorschot, R. A. H., Meakin, G. E., Kokshoorn, B., Goray, M., & Szkuta, B. (2021). DNA Transfer in Forensic Science: Recent Progress towards Meeting Challenges. *Genes*, *12*(11), 1766. DOI: 10.3390/genes12111766 PMID: 34828372

Wang, C. F., Djalali, A. G., Gandhi, A., Knaack, D., De Girolami, U., Strichartz, G., & Gerner, P. (2009). An absorbable local anesthetic matrix provides several days of functional sciatic nerve blockade. *Anesthesia and Analgesia*, *108*(3), 1027–1033. DOI: 10.1213/ane.0b013e318193596a PMID: 19224820

Wang, J., Liu, L., Wu, Y., Maitz, M. F., Wang, Z., Koo, Y., Zhao, A., Sankar, J., Kong, D., Huang, N., & Yun, Y. (2017). Ex vivo blood vessel bioreactor for analysis of the biodegradation of magnesium stent models with and without vessel wall integration. *Acta Biomaterialia*, *50*, 546–555. DOI: 10.1016/j.actbio.2016.12.039 PMID: 28013101

Woodard, L. N., & Grunlan, M. A. (2018). Hydrolytic Degradation and Erosion of Polyester Biomaterials. *ACS Macro Letters*, *7*(8), 976–982. DOI: 10.1021/acsmacrolett.8b00424 PMID: 30705783

Xu, Y., Meng, H., Yin, H., Sun, Z., Peng, J., Xu, X., Guo, Q., Xu, W., Yu, X., Yuan, Z., Xiao, B., Wang, C., Wang, Y., Liu, S., Lu, S., Wang, Z., & Wang, A. (2018). Quantifying the degradation of degradable implants and bone formation in the femoral condyle using micro-CT 3D reconstruction. *Experimental and Therapeutic Medicine*, *15*(1), 93–102. DOI: 10.3892/etm.2017.5389 PMID: 29375677

Chapter 10
Scorpion Venom and Biodegradable Biomaterials:
Toward a New Biomedical Application

Adil El Housseini
http://orcid.org/0009-0007-9294-2685
Laboratory of Aquatic Systems, Marine, and Continental Environment, Faculty of Sciences, Ibn Zohr University, Agadir, Morocco

Khawla Rachmoune
http://orcid.org/0009-0001-6483-1060
Laboratory of Water Science, Microbial Biotechnology, and Sustainability of Natural Resources, Faculty of Science Semlalia, Cadi Ayyad University, Marrakech, Morocco

Sara El aimani
Faculty of Sciences, Mohammed V University, Morocco

Youssef El Joud
Laboratory of Biotechnology and Valorization of Natural Resources, Faculty of Sciences, Ibn Zohr University, Agadir, Morocco

Moulay Abdelmonaim El Hidan
Laboratory of Biotechnology and Valorization of Natural Resources, Ibn Zohr University, Agadir, Morocco.

El-Mustapha Laghzaoui
http://orcid.org/0000-0002-7776-5218
Higher Institute of Nursing Professions and Health Techniques of Marrakech, Morocco

Mehdi Ait Laaradia
http://orcid.org/0000-0003-2390-2659
Superior Institute of Skilled Professions and Techniques in Beni Mellal, Morocco

El Hassan El Mouden
Laboratory of Water Science, Microbial Biotechnology, and Sustainability of Natural Resources, Faculty of Science Semlalia, Cadi Ayyad University, Marrakech, Morocco

DOI: 10.4018/979-8-2600-0060-1.ch010

Mustapha Agnaou
Laboratory of Aquatic Systems, Marine, and Continental Environment, Faculty of Sciences, Ibn Zohr University, Agadir, Morocco

ABSTRACT

Scorpion venom is known for its richness in bioactive molecules, particularly peptides, which confer a wide range of remarkable biological activities. Notably, it exhibits antimicrobial, antitumor, and immunomodulatory effects. However, its clinical translation remains limited due to its poor stability, dose-dependent toxicity, and potential immunogenicity. In this context, biodegradable biomaterials have emerged as promising platforms to overcome these limitations. They enable precise, targeted, and sustained delivery of scorpion venom peptides. This chapter highlights an innovative approach based on the integration of scorpion venom into biodegradable biopolymer matrices for applications in modern medicine. By leveraging the unique properties of biodegradable biomaterials, this work outlines a roadmap for the development of next-generation clinical applications.

INTRODUCTION

Humans have long explored natural resources to meet their nutritional and health needs, leading to the discovery of a wide range of bioactive molecules with considerable therapeutic potential. Among these, animal venoms, particularly those from scorpions, are attracting growing interest due to their richness in peptides and proteins capable of modulating cellular processes with high specificity. The biomolecules exhibit diverse biological activities, including antimicrobial, anticancer, anti-inflammatory, and immunomodulatory effects, making them promising candidates for innovative biomedical applications (Tobassum et al., 2020; Mendes et al., 2023).

The venom of scorpions contains various biologically active components, such as ion channel-targeting toxins, enzymes, or low molecular weight peptides, which interact with membrane receptors and intracellular targets involved in pathological processes. This targeting specificity gives them potential for use in inhibiting tumor proliferation, disrupting microbial biofilms, and modulating immune responses (Mendes et al., 2023). However, in spite of these promising properties, the direct application of scorpion venom remains limited due to challenges such as dose-dependent toxicity, immunogenicity, and insufficient in vivo specificity, which constrain its safe clinical use (Huang et al., 2026).

To overcome these limitations, advanced delivery strategies are required to enhance the bioavailability, targeting, and controlled release of venom-derived compounds. In this context, biodegradable biomaterials have emerged as a promising platform. Derived from natural or synthetic polymers, these materials exhibit tunable physicochemical properties, excellent biocompatibility, and the ability to ensure sustained and localized delivery of therapeutic agents. Their application in regenerative medicine and tissue engineering has significantly improved the stability and efficacy of bioactive molecules while reducing systemic toxicity (Ayari-Riabi et al., 2016; Hoang et al., 2025).

The integration of scorpion venom-derived peptides into biodegradable polymeric matrices represents an innovative approach to harnessing their therapeutic potential while minimizing associated risks. Such systems can enhance peptide stability, preserve biological activity, and enable controlled release adapted to specific therapeutic needs (Sara et al., 2026). The strategy holds promise in various biomedical fields, notably targeted cancer therapy, tissue repair, anti-infective applications, and neurobiology. It should be noted that nanocarrier-based systems, such as PLGA nanoparticles and chitosan-based scaffolds, have shown promising results in improving drug delivery efficiency and therapeutic results (Majumder et al., 2019; Rachmoune et al., 2026).

Although progress has been made, several challenges remain, including limited clinical translation, potential toxicity at higher doses, and the need for standardized evaluation protocols. Few studies have systematically compared different biomaterial platforms for venom delivery, highlighting a critical gap in research aimed at optimizing these hybrid systems for clinical applications.

Therefore, this chapter aims to examine the biochemical composition of scorpion venom and its therapeutic potential, as well as to provide an overview of biodegradable biomaterials and their role in biomedical applications. Particular emphasis is placed on the integration of venom-derived biomolecules into biodegradable polymeric systems as a promising strategy for developing safe and effective therapeutic approaches.

I. BIOCHEMICAL COMPOSITION AND THERAPEUTIC POTENTIAL OF SCORPION VENOM

1. Characteristics and Composition of Venom

Scorpion venom can be defined as a heterogeneous substance composed of multiple components (Furtado et al., 2020). It can be collected using electrical or manual stimulation methods (Oukkache et al., 2013), or by maceration of the venom apparatus, the telson (Ozkan et al., 2006; de Roodt et al., 2019). Physically, it appears as a dark, viscous liquid with an acidic pH. Its chemical composition includes: mucus and mineral salts; macromolecules and enzymatic complexes (such as hyaluronidases and mucopolysaccharides); protease inhibitors; low molecular weight compounds such as histamine and serotonin; and peptide fractions (7–8 kDa) which, although they represent only 2–3% of the total venom mass, are the main agents responsible for its toxicity (Fuzita et al., 2015). Additionally, kinins and acetylcholinesterase have also been identified (Minutti-Zanella et al., 2021).

These protein-based components target ion channel receptors located on the membranes of excitable cells in the neuromuscular system (Quintero-Hernandez et al., 2013; Bhavya et al., 2016). Current research mainly focuses on dangerous species such as those belonging to the Buthidae family (Touloun, 2019). Toxins purified from these venoms are distinguished by their immunological and pharmacological specificities (Kalayci et al., 2016; Munawar et al., 2018; Saez & Herzig, 2019). Their action targets ion channels in nerve and muscle cells; to date, four categories of toxins have been identified based on their affinity for sodium, potassium, calcium, or chloride channels (Jiang et al., 2021; Díaz-García & Varela, 2021).

Mode of Action

The virulence of scorpion venom relies on several mechanisms of action, including a direct effect on cells, in addition to a strong inflammatory response accompanied by an increase in pro-inflammatory factors. Furthermore, scorpion venom disrupts the oxidant–antioxidant balance while generating large amounts of reactive oxygen species (ROS), leading to intense oxidative stress. Moreover, neurotoxins are capable of strongly stimulating the autonomic nervous system. This activation induces a massive release of neurotransmitters from the adrenal glands as well as the sympathetic and parasympathetic systems. This phenomenon is the primary driver of pathophysiological complications, potentially leading to multi-organ failure, particularly cardiorespiratory crises and severe neurological disorders (El Housseini et al., 2025).

2. Relevant Biological Activities

Scorpion venom exhibits a wide range of medically relevant properties, primarily due to the interaction of its peptides with cellular membranes, leading to the destruction of pathogenic agents.

a. Antimicrobial Activity

Antimicrobial peptides (AMPs) destroy microbial and cancer cells through membrane lysis, taking advantage of similarities in their membrane composition, unlike healthy eukaryotic cells. However, some studies suggest more complex mechanisms of action involving intracellular targets (Harrison et al., 2014). At sub-lethal doses, increased membrane permeability can paradoxically promote growth by facilitating nutrient uptake or by modulating cellular signaling. These toxins are classified based on their size (long-chain or short-chain) or their function (cytotoxic or neurotoxic) (Almaaytah & Albalas, 2014). Peptides lacking disulfide bridges represent a small fraction of the venom and exhibit diverse functions, although many remain poorly characterized. These low-molecular-weight molecules (13–50 amino acids) are often antimicrobial peptides or bradykinin-potentiating peptides (BPPs) (Almaaytah & Albalas, 2014). AMPs are typically cationic and amphipathic (2–5 kDa), acting against fungi and bacteria (Gram-positive and Gram-negative) by disrupting membrane integrity. While some peptides, such as hadrurin, display broad-spectrum activity, others, such as parabutoporin, are more selective (Arpornsuwan et al., 2014). The molecular diversity of scorpion venom, therefore, provides promising avenues for developing alternatives to conventional antibiotics in the context of rising antimicrobial resistance.

b. Antitumor Activity

Although venomous animals often evoke fear, their secretions serve as the basis for treatments for conditions such as hypertension and diabetes. In oncology, scorpion-derived peptides exert their effects through:

- Induction of apoptosis (via mitochondrial pathways and caspases)
- Arrest of the cell cycle
- Interaction with tumor ion channels
- Inhibition of metastasis and cell migration (Kwon et al., 2024)

c. Immunomodulatory Activity

Scorpion venom modulates the immune response by acting on immune cells (macrophages and lymphocytes) and cytokine secretion. This includes the upregulation or downregulation of factors such as TNF-α, IL-6, and IL-1β, thereby either enhancing antitumor activity or stabilizing inflammatory responses (El-Qassas et al., 2024).

d. Anti-inflammatory Activity

Certain molecules isolated from scorpion venom exhibit anti-inflammatory properties by:

- Inhibiting the NF-κB signaling pathway
- Reducing the production of pro-inflammatory mediators
- Limiting the recruitment of inflammatory cells (Arcos et al., 2025)

II. BIODEGRADABLE BIOMATERIALS: KEY CONCEPTS AND PROPERTIES

1. Overview of Biopolymers

Biopolymers constitute an important class of materials widely used in biomedicine due to their ability to interact in a controlled manner with biological systems. They can be defined as polymeric macromolecules of natural or synthetic origin, characterized by their biodegradability and biocompatibility. These two fundamental properties clearly distinguish them from conventional polymers and explain their increasing application in areas such as drug delivery, tissue regeneration, and the development of resorbable medical devices (Zarrintaj et al., 2023). These materials are capable of inducing diverse cellular responses, highlighting their significant biological relevance and their ability to modulate cellular behavior (Figure 1) (Zarrintaj et al., 2023).

Figure 1. Differential cellular responses induced by chitosan-based biopolymers (: enhancement; : reduction; : initiation)

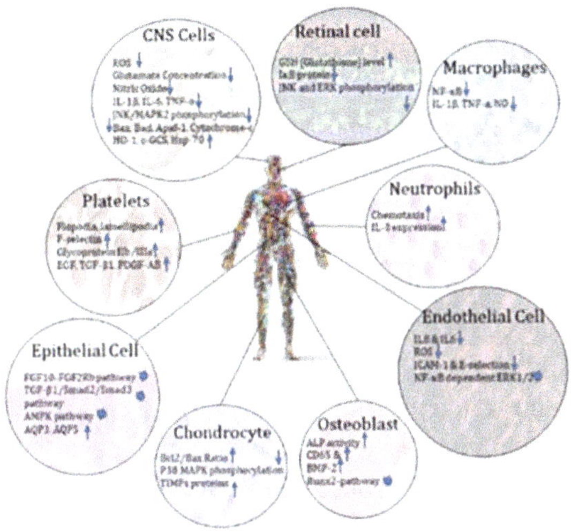

(Zarrintaj et al., 2023).

The use of biopolymers has historically evolved in three major phases. Originally, natural materials such as cellulose, collagen, and gelatin were used empirically for their mechanical and biological properties. This was followed by the development of biodegradable synthetic polymer materials such as polylactic acid (PLA), polyglycolic acid (PGA), and their copolymers (PLGA), which marked a breakthrough in the 1970s and 1980s, particularly due to their use in absorbable surgical sutures. Recently, advances in biotechnology and polymer chemistry have enabled the design of "smart" biomaterials that can respond to specific stimuli (pH, temperature, enzyme activity) (Sindhi et al., 2025).

Research has shown that the development of what are known as "smart" biomaterials depends on the possibility of precisely controlling the structural and functional properties of polymers, particularly through chemical modification or copolymerization, thereby enabling their behavior to be tailored in vivo (Narkar et al., 2022). On a structural level, biopolymers feature extremely diverse architectures, ranging from linear chain structures to branched or interconnected structures. Natural polymers, such as collagen, possess a complex hierarchical organization, ranging from the collagen triple helix to fibrillar structures, which lend them remarkable mechanical and biological properties. In turn, synthetic polymers offer greater consistency and precise control over parameters such as molar mass, molecular weight distribution, and crystallinity (Yin et al., 2022).

Research has shown that crystallinity has a significant impact on the mechanical strength and degradation rate of biopolymers: since crystalline regions are less accessible to water and enzymes, the degradation process is slowed down, whereas amorphous regions promote hydrolysis (Benito-Kaesbach et al., 2025). Biopolymers can be distinguished according to their origin into several broad categories. Certain ones are directly extracted from biomass, particularly from plant or animal sources, like alginate from brown algae, chitin from the exoskeletons of crustaceans, or cellulose of plant origin. Some are produced biotechnologically using microorganisms, such as polyhydroxyalkanoates (PHA), which are synthesized through bacterial fermentation. Lastly, a third category includes polymers obtained through chemical synthesis from bio-based monomers, such as polylactic acid (PLA), derived from the fermentation of sugars (Edo et al., 2025).

Studies have demonstrated that the origin of biopolymers plays a critical role not only in determining their physicochemical characteristics, such as crystallinity, molecular weight, and hydrophilicity, but also in influencing their biological behavior. In particular, natural biopolymers may contain residual impurities originating from extraction processes or their native biological environment, which can potentially trigger immune or inflammatory responses. Consequently, strict control of purification and production processes is essential to ensure the safety, reproducibility, and overall performance of biomaterials. The classification of natural-based materials, highlights the diversity of these biopolymers and their sources, providing a comprehensive framework for understanding their structural and functional variability (Figure 2) (Amponsah et al., 2025).

Figure 2. Classification of natural-based materials

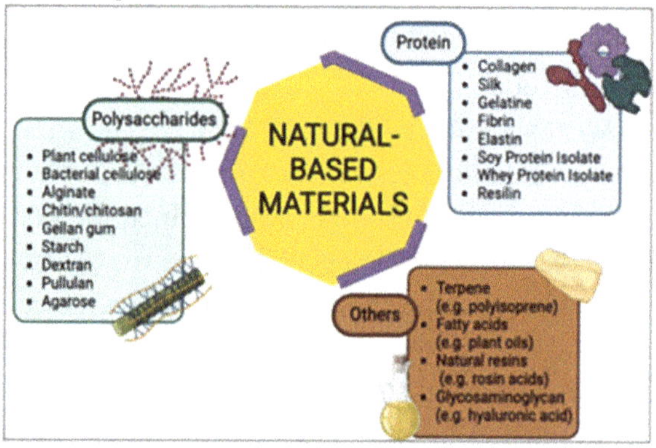

(Amponsah et al., 2025)

2. Properties of biopolymers

a) Biocompatibility

Biocompatibility extends beyond the absence of toxicity; it also encompasses a biomaterial's ability to promote specific cellular behaviors, such as cell adhesion, migration, and differentiation. Biocompatibility is highly dependent on the material's surface properties, including roughness, wettability, and the presence of functional groups (Jurak et al., 2021).

Studies have shown that the initial adsorption of plasma proteins at the surface of the biomaterial is a key event that determines the resulting cellular response. This coat of proteins influences cell adhesion via specific receptors such as integrins. In addition, natural biopolymers often contain intrinsic bioactive motifs (e.g., RGD sequences in collagen) which promote cell interaction. In stark contrast, most synthetic polymers require surface functionalization to enhance their bioactivity (Wronska et al., 2016).

b) Biodegradability controlled

Controlled biodegradability of biopolymers is a critical property in biomaterials engineering, making sure the material is completely absorbed over time in accordance with biological processes of tissue repair or regeneration. This phenomenon is driven by a complex interplay of physicochemical and biological mechanisms, involving both material-specific degradation processes and dynamic interactions with the physiological environment. Biopolymer degradation can be primarily classified into two broad categories: hydrolytic degradation and enzymatic degradation (Abdulsalam et al., 2025). Hydrolytic degradation, which is predominant in aliphatic polyesters such as polylactic acid (PLA), polyglycolic acid (PGA), and their copolymers (PLGA), involves the breaking of ester bonds under the action of water. This mechanism is heavily influenced by water diffusion within the polymer matrix, leading to the gradual hydrolysis of the macromolecular chains. This may occur homogeneously throughout the material's volume (bulk degradation) or at the surface (surface degradation), depending on the polymer's structure and permeability (Abdulsalam et al., 2025).

Similarly, enzymatic degradation, a characteristic of natural biopolymers such as polysaccharides (chitosan, alginate) and proteins (collagen, gelatin), involves the specific action of enzymes capable of cleaving the polymer's chemical bonds. This type of degradation is highly dependent on the biological microenvironment, particularly on enzyme concentration and the pathophysiological state of the tissue.

Unlike hydrolytic degradation, this mechanism is often more targeted and can be modulated by the presence of specific cells, such as macrophages or fibroblasts (Satchanska et al., 2024).

Biopolymer degradation rates are determined by a set of intrinsic physicochemical parameters. Molecular weight plays a central role, as low-molecular-weight polymers degrade more rapidly due to the presence of a greater number of chains ends susceptible to hydrolysis. Similarly, the hydrophilicity of the polymer directly influences water penetration into the matrix: hydrophilic polymers promote water absorption and thus accelerate hydrolysis reactions (Kamaly et al., 2016).

Cellularity is also a decisive factor. The crystalline regions, characterized by a dense and ordered molecular structure, are less accessible to water and enzymes, which slows their degradation. On the other hand, amorphous regions, which are more disordered, facilitate water diffusion and are therefore preferred for degradation. This structural heterogeneity can lead to non-uniform degradation of the material, affecting its mechanical properties over time (Bregado et al., 2019). Material properties, particularly size, shape, and surface area, also have a significant influence on degradation kinetics. The systems with a high surface area, such as nanoparticles or porous structures, undergo accelerated degradation due to increased exposure to water and enzymes. Conversely, massive or compact structures degrade more slowly, due to limited diffusion of degradation agents into the core of the material (Dilsad et al., 2024).

In addition to the inherent properties of the polymer, the physiological environment plays a crucial role in biodegradation. Factors such as pH, temperature, the presence of enzymes, and inflammatory conditions can significantly modulate degradation kinetics. In the case of an acidic environment, frequently observed in tumorous or inflammatory tissues, it can accelerate the hydrolysis of polyesters, while the infiltration of immune cells can increase local enzymatic activity (Abdulsalam et al., 2025). A major issue in the biodegradation of biopolymers is the accumulation of degradation products, particularly organic acids such as lactic acid and glycolic acid. This accumulation can lead to local acidification of the microenvironment, disrupting cellular homeostasis and potentially inducing an inflammatory response. This acidity can also further catalyze polymer degradation (autocatalytic effect), leading to an uncontrolled acceleration of the degradation process (De Souza and Gupta, 2024). At last, managing biodegradability is a major challenge in the design of biomaterials. Adjusting structural parameters (copolymerization, cross-linking, and the incorporation of additives) makes it possible to fine-tune the degradation kinetics to suit the intended therapeutic application. Too rapid degradation can compromise the material's mechanical properties, while too slow degradation can lead to undesirable accumulation and a prolonged inflammatory response (Mukherjee et al., 2023).

a) Non-toxicity

Biopolymer non-toxicity is a fundamental criterion for their use in biomedicine, encompassing not only the intrinsic safety of the material but also that of its degradation, reaction intermediates, and any impurities (El Housseini et al., 2026). The importance of this characteristic is heightened by the fact that biomaterials are often intended for long-term implantation or direct interaction with sensitive tissues (Dallaev et al., 2025).

Research has shown that the degradation of biodegradable polymers, particularly aliphatic polyesters such as PLA, PGA, and PLGA, leads to the formation of organic acids (lactic acid and glycolic acid), which can cause local acidification of the tissue environment. This lowering of pH can disrupt cellular homeostasis, affect enzymatic activity, and promote a local inflammatory response. In certain cases, excessive accumulation of these degradation products has been associated with indirect cytotoxicity, particularly through the activation of cellular stress pathways and the production of reactive oxygen species (ROS) (Lanao et al., 2013).

In addition, research has shown that certain biopolymers, particularly those derived from chemical synthesis processes, may contain residues of metal catalysts, organic solvents, or unreacted monomers. These contaminants can interact with cell membranes, altering their permeability and inducing cytotoxic or genotoxic effects. For this reason, traces of solvents such as dichloromethane or dimethylformamide, commonly used in the manufacture of nanoparticles, have been linked to adverse effects on cell viability (Dallaev et al., 2025).

Furthermore, several aspects of the size and shape of biomaterials, particularly in the case of nanoparticles, play a decisive role in their toxicity. Nanoparticles that are small in size (<100 nm) have an increased ability to penetrate cells and interact with intracellular organelles, particularly mitochondria and the nucleus, which can lead to disruption of cellular metabolism, DNA damage, and the induction of oxidative stress (Egbuna et al., 2021).

Assessment of non-toxicity employs a multidimensional approach that combines in vitro, ex vivo, and in vivo studies. In vitro tests include cytotoxicity assays (MTT, XTT, LDH) to evaluate cell viability, as well as genotoxicity tests (comet assay, micronucleus test) to detect DNA damage. These tests constitute an essential first step in the screening of biomaterials, although they do not always reflect the complexity of in vivo interactions (Lama et al., 2020).

Lastly, it is important to highlight that non-toxicity can be modulated through biomaterial engineering strategies. The surface modification (grafting of hydrophilic polymers, incorporation of anti-inflammatory agents, or antioxidants) can reduce the immune response and improve overall biocompatibility. Furthermore, the use of natural biopolymers or bio-based polymers can limit the risks associated with synthetic residues (Zhang et al., 2025).

b) Controlled release

Releasing bioactive molecules from biopolymers in a controlled way is a key focus in therapeutic systems engineering, enabling the optimization of pharmacological efficacy while minimizing side effects. This process is based on the complex interaction of several coupled physicochemical mechanisms, including the diffusion of the active molecule through the polymer matrix, the erosion or degradation of the polymer, and the swelling of the polymer network in the case of hydrophilic systems such as hydrogels. Studies have shown that these mechanisms can act simultaneously or sequentially, leading to nonlinear and often multiphase release profiles (Bajgiran et al., 2025).

In numerous polymeric systems, an initial rapid discharge, known as "burst release," is observed, corresponding to the diffusion of molecules that are weakly bound or adsorbed to the biomaterial surface. Generally, this phase is followed by a slower, prolonged release, driven by internal diffusion or the gradual degradation of the polymer. Studies have shown that this initial phase can be challenging in certain applications, particularly due to the risk of local drug overdose, but that it can also be leveraged to achieve a rapid therapeutic effect, followed by a prolonged maintenance of the active concentration (Gopinathan et al., 2014).

Diffusion is one of the primary mechanisms in systems where the polymer matrix remains comparatively stable and depends mainly on the pore size, the tortuosity of the polymer matrix, and the hydraulic size of the encapsulated molecule. It has been shown that increasing the porosity or pore size promotes the diffusion of active molecules, leading to a faster release. Conversely, a dense or highly cross-linked structure constrains the mobility of the molecules and slows their release (Hazra et al., 2021). Network density also plays a key role in controlling the release of the active ingredient, particularly in hydrogels. A high cross-linking density reduces the pore size of the polymer network, thereby limiting the diffusion of active molecules. Research has shown that modulating this density allows for precise adjustment of the delivery kinetics, ranging from rapid-release to sustained-release systems. The polymer's swelling in an aqueous environment increases the network's volume, facilitating molecular delivery and creating a dynamic release delivery profile (Ribeiro et al., 2025).

The interaction between the polymer and the active molecule is a key factor influencing drug release. These interactions can be electrostatic, hydrophobic, or mediated by hydrogen bonds. Research has shown that strong interactions between the polymer and the active molecule can significantly slow down drug release, requiring a preliminary dissociation process. By contrast, weak interactions promote rapid release. In the case of peptides, their three-dimensional structure, net charge, and hydrophilicity strongly influence their affinity for the polymer scaffold (Herdiana et al., 2022).

Additionally, conditions in the environment play a crucial role in the release kinetics. Studies have shown that changes in pH, temperature, or enzymatic composition can alter the polymer's structure and accelerate or slow down its degradation. In acidic environments, such as tumor or inflammatory tissues, the degradation of polyesters is generally accelerated, leading to a faster release of the encapsulated molecules. Conversely, the presence of specific enzymes can induce targeted degradation of natural polymers such as polysaccharides or proteins (Amoke et al., 2026).

In the particular case of peptides, controlled release presents additional challenges due to their physicochemical properties. Previous reviews have documented that the charge of peptides strongly influences their interaction with polymeric matrices, particularly in ionic systems such as chitosan- or alginate-based hydrogels (Liu et al., 2025). Additionally, their high hydrophilicity promotes rapid diffusion, which can complicate the control of their release. Furthermore, their tendency to undergo enzymatic degradation requires specific strategies to prolong their stability, such as encapsulation in nanoparticles or chemical modification (Baral et al., 2025).

Lastly, numerical modelling of drug release is an essential tool for understanding and predicting the performance of polymeric systems. Studies have shown that empirical and semi-empirical models, such as the Higuchi or Korsmeyer-Peppas models, can describe diffusion and erosion mechanisms and identify the dominant process. The development of these models is a key challenge in the field of drug delivery (Woodring et Ainslie, 2026).

III. USE OF VENOM PROTEIN PEPTIDES IN BIOMATERIALS

1. Encapsulation methods

Integrating venom peptides into biomaterials represents an innovative approach aimed at harnessing their exceptional biological properties (antimicrobial, anticancer, and neuroactive), while overcoming their inherent limitations, such as structural instability, susceptibility to enzymatic degradation, and low systemic bioavailability. Encapsulation thus emerges as an essential strategy for protecting these biomolecules and modulating their delivery in a controlled manner (Shankar et Unni, 2026).

Studies have shown that polymeric nanoparticles, particularly those based on PLGA, chitosan, or polycaprolactone, provide effective protection against proteolytic degradation by forming a physical barrier that limits protease access to the encapsulated peptides. Additionally, these nanoparticles facilitate cellular internalization via endocytosis mechanisms, enabling targeted intracellular delivery (Feczko, 2021). It has also been demonstrated that the size, surface charge, and functionalization of nanoparticles strongly influence their interaction with cell membranes and their biodistribution. Hydrogel injections represent another promising strategy, particularly well-suited for local administration. Their hydrated three-dimensional structure, which mimics the extracellular matrix, allows for the effective encapsulation of peptides while preserving their biological activity (Parveen et al., 2026). Studies have shown that these systems enable prolonged and localized release, thereby reducing systemic side effects. Furthermore, certain "smart" hydrogels are capable of responding to environmental stimuli (pH, temperature, enzymes), allowing for adaptive release of the peptides (Zhang et Wu, 2023).

Three-dimensional (3D) scaffolds, used in tissue engineering, offer a particularly promising platform for the integration of venom peptides. Studies have shown that these porous structures enable controlled spatiotemporal release of peptides, promoting cell regeneration and modulation of the tissue environment. Furthermore, peptides can be integrated directly into the scaffold matrix or immobilized on its surface, allowing for direct interaction with cells (Li et al., 2023).

However, despite these advances, several challenges remain, notably the preservation of the peptides' three-dimensional structure during encapsulation processes, as well as the precise control of their distribution within the biomaterial.

2. Loading techniques

The selection of the technique for loading venom peptides into biomaterials is a critical step that directly influences their stability, biological activity, and release profile. Every method has advantages and limitations that must be carefully considered based on the intended application. Physical adsorption relies on non-covalent interactions, such as electrostatic forces, hydrophobic interactions, and hydrogen bonds. The literature indicates that this method has the advantage of being simple, fast, and inexpensive, while preserving the native structure of the peptides (Ghodeif et al., 2025). However, the low binding energy results in rapid desorption, often leading to a significant initial release (burst release), thereby limiting its effectiveness for applications requiring sustained release.

Chemical binding, meanwhile, refers to the formation of covalent bonds between the peptide and the biomaterial. Studies have shown that this approach enables stable and long-lasting immobilization of peptides, reducing their diffusion and prolonging their activity. The strategy, however, has significant limitations, notably the risk of altering the peptide's three-dimensional conformation, which can lead to a partial or total loss of its biological activity. Consequently, the selection of conjugation sites and coupling agents is critical to preserving the functional integrity of the peptides (Varanko et al., 2020). Nanoparticle or microstructured encapsulation is a particularly effective approach for protecting peptides from adverse environmental conditions. Studies have shown that this technique helps maintain peptide stability, limit enzymatic degradation, and control their release over an extended period. However, it requires rigorous control of manufacturing parameters, such as temperature, pH, and the solvents used, to prevent peptide denaturation (Liu et al., 2025).

A major challenge in all these approaches lies in preserving the three-dimensional structure of the peptides, which is essential for their biological activity. Studies have shown that even minor changes in conformation can alter their interaction with biological targets, highlighting the importance of optimizing loading conditions (Wang et al., 2026).

3. Mechanism of release

Peptide release kinetics from biomaterials is a key factor in their therapeutic efficacy, as it determines the local concentration of the peptide and its duration of action. It results from the complex interaction between the properties of the biomaterial, those of the peptide, and the conditions of the biological microenvironment (Yu et al., 2026). Studies indicate that mathematical models, such as the Korsmeyer-Peppas model, are widely used to describe release mechanisms in polymeric systems. This model allows for the distinction between the relative contributions of diffusion and polymer erosion, depending on the diffusion exponent (n). However, in complex systems involving peptides, these models may have limitations, particularly due to specific interactions between the peptide and the polymeric matrix (Jahromi et al., 2020).

Biological microenvironments play a fundamental role in modulating peptide release. Studies have shown that acidic conditions, characteristic of many pathological environments (tumors, inflammation), can accelerate the degradation of polymer and promote a faster release of peptides (Liu et al., 2025). Similarly, the presence of proteolytic enzymes can not only degrade the biomaterial but also directly affect the peptides, reducing their activity. Inflammation can also be a key factor influencing release dynamics. Studies have shown that inflammatory conditions can alter vascular permeability, the composition of the extracellular matrix, and the activity of enzymes, which can significantly affect the diffusion and stability of peptides (Liu et al., 2025). Moreover, the production of free radicals at inflammatory sites can induce oxidative degradation of peptides.

Ultimately, the release kinetics must be tailored to the intended therapeutic application. In some cases, rapid release is desired to achieve an immediate effect, while in others, sustained release is necessary to maintain a stable therapeutic concentration. Studies have shown that optimizing this kinetics relies on a combination of strategies, including modification of the biomaterial, selection of the loading technique, and adaptation to pathophysiological conditions (Thang et al., 2023).

IV. EMERGING BIOMEDICAL APPLICATIONS

1. Ion channel targeting and delivery systems in cancer therapy

Scorpion venom, a complex cocktail of peptides and proteins perfected over millions of years of evolution (El Housseini et al., 2025), is currently being investigated for its remarkable anticancer properties (Rapôso, 2017). It represents a promising source of bioactive molecules for the development of targeted therapies. Scorpion venom peptides are distinguished by their high affinity for ion channels, whose overexpression is a common feature of numerous cancer cells (Ghodeif et al., 2025). These channels actively participate in the proliferation, migration, invasion, and metastatic potential of tumors (McFerrin & Sontheimer, 2006).

Among the most extensively investigated peptides, chlorotoxin (CTX derived from *Leiurus quinquestriatus*, was initially proposed as a chloride channel blocker. However, subsequent studies demonstrate that CTX's antitumor effects are primarily linked to its selective binding to MMP-2 in glioma cells, which suppresses tumor migration (Deshane et al., 2003; Soroccanu et al., 1998). This mechanistic understanding highlights significant limitations in the existing literature, as the theoretical assumptions surrounding ion channel targeting are often oversimplified. CTX has demonstrated superiority over traditional chemotherapeutic agents in enhancing tumor selectivity, positioning it as a standard reference for radiotherapy in malignant gliomas (MICAD Research Team, 2007). This has led to the development of TM-601, a synthetic analogue radiolabeled with iodine-131 (^{131}I-TM601), intended for imaging and radiotherapy in these tumors. Early-phase clinical trials (Phase I/II) have shown disease stabilization, suggesting potential clinical utility. However, the absence of Phase III trials indicates a significant translational gap. This stagnation may reflect challenges in the delivery process or limitations in the drugs themselves, such as therapeutic efficacy, regulatory hurdles, or manufacturing difficulties, issues that are often overlooked in preclinical studies.

In comparison to chloride channel-targeting peptides, potassium channel inhibitors such as charybdotoxin and margatoxin (MgTX) have demonstrated significant anticancer effects in vitro. Charybdotoxin blocks intermediate-conductance calcium-activated potassium channels (IKCa), thereby suppressing proliferation of prostate cancer cells (Parihar et al., 2003), while MgTX, a Kv1.3 inhibitor, induces apoptosis and inhibits migration in osteosarcoma cells (Lin et al., 2013). Notwithstanding these favorable outcomes, their translation is still slow compared to CTX. This discrepancy indicates that not every ion channel target is suitable for application to the clinic and could arise from differences in tissue distribution, toxicity profiles, or pharmacokinetics. Notably, the bulk of these investigations heavily utilize in *vitro* models and insufficiently validate these in clinically important in *vivo* systems, indicating a substantial gap in the literature.

Voltage-gated sodium channels (Nav), particularly the Nav1.5 subtype, are overexpressed in highly metastatic breast cancer cells and are associated with aggressive tumor behavior. Scorpion venom-derived peptides capable of specifically blocking this channel are the subject of active research. It synthesizes these advances and reports that Nav1.5 inhibition by certain venom peptides can lead to reduced cell proliferation and decreased expression of genes related to metastatic progression in breast cancer models (Ghodeif et al., 2025)

In a different context, TRP channels (Transient Receptor Potential), notably TRPC6 and TRPV2, are overexpressed in hepatocellular carcinoma (HCC) and contribute to tumor proliferation and migration. An innovative approach involves conjugating crude venom to nanoparticles to enhance its efficacy and specificity (Wang et al., 2024). Al-Asmari et al. (2017) demonstrated the effectiveness of a nanoliposomal delivery approach for scorpion venom against colorectal cancer cells, paving the way for similar applications in other tumor types. ghodeif and his team emphasize that nano-encapsulation or conjugation to nanoparticles improves peptide stability, bioavailability, and specificity while reducing toxicity, thus constituting a major strategy for exploiting the anticancer potential of venom peptides (Wang et al., 2024).

Furthermore, a novel peptide, androcin 18-1, has been identified from the venom transcriptome of the scorpion Androctonus bicolor. This peptide demonstrated antitumor activity against human U87 glioblastoma cells by inducing cell membrane disruption, accumulation of reactive oxygen species (ROS), and mitochondrial dysfunction, leading to apoptosis (Wang et al., 2024). Additionally, it suppresses cell migration by destabilizing the actin cytoskeleton and regulating MMPs/TIMPs expression (Wang et al., 2024).

Beyond their direct therapeutic potential, scorpion venom peptides also open innovative perspectives in diagnostic imaging. Recent work has shown that non-small cell lung cancers (NSCLC) overexpress several ion channels, including Nav1.7, ASIC channels, and nicotinic receptors α7 and α9. given the limitations of conventional imaging techniques, these authors propose using peptide toxins labeled via "click" chemistry as specific molecular imaging probes. Their preliminary results show that α-bungarotoxin (targeting α7/α9) and huwentoxin-IV (targeting Nav1.7) colocalize with the membrane of lung cancer cells, validating the concept of these peptides as biomarkers for diagnostic imaging (Baudat et al., 2023).

2. Wound healing and tissue engineering

Besides their antitumor effects mediated by selective modulation of ion channels, scorpion venom peptides possess pleiotropic biological properties that extend their therapeutic interest to other biomedical domains, particularly wound healing and tissue engineering. The modulation of wound healing is gaining increasing interest. Wound healing occurs in three overlapping phases: an inflammatory phase (hemostasis and immune cell recruitment), a proliferative phase (angiogenesis, fibroplasia, re-epithelialization), and a remodeling phase (collagen maturation) (Summer et al., 2026). The originality of scorpion venom peptides lies in their ability to act simultaneously or sequentially on these three phases, paving the way for the development of next-generation bioactive dressings (Figure 3B).

Figure 3. Therapeutic Integration of Scorpion Venom Peptides into Hydrogels for Cutaneous Tissue Repair

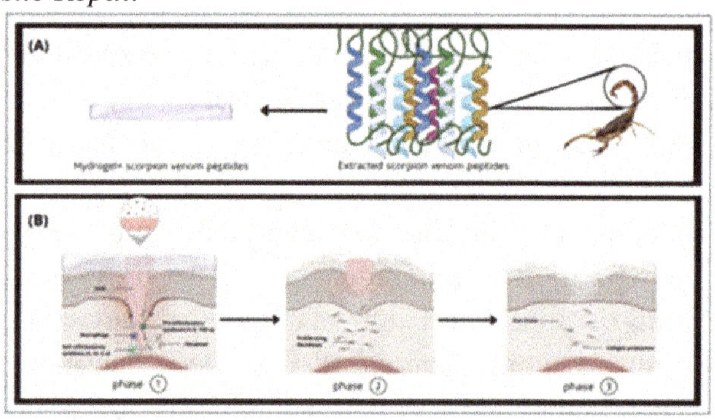

(A) *Biomaterial design schematic: bioactive peptides are extracted from scorpion venom and incorporated into a hydrogel matrix to produce a bioactive wound dressing intended for topical application.* **(B)** *Scorpion venom peptides intervene sequentially and complementarily during the cutaneous repair process. (Phase 1) Antimicrobial peptides ensure wound protection through bacterial destruction and infection prevention. (Phase 2) Anti-inflammatory peptides modulate the immune response by reducing the production of pro-inflammatory cytokines, notably TNF-α and IL-6, thereby normalizing the tissue microenvironment. (Phase 3) Pro-reparative peptides stimulate cell migration and proliferation, promote angiogenesis and collagen synthesis, leading to accelerated wound healing and the formation of higher-quality tissue.*

This capacity of scorpion venom peptides to act on the different phases of wound healing is not limited to cellular and tissue modulation; it is also accompanied by antimicrobial and immunomodulatory activity that reinforces their therapeutic potential. This antimicrobial activity helps overcome a major obstacle to wound healing bacterial contamination of wounds, which can lead to chronicity. Antimicrobial peptides (AMPs) from scorpion venom, initially synthesized by the animal to sterilize its own venom apparatus, are of major interest in this context (Shokhba et *al.*, 2026).

In this line of research, a recent study demonstrated the *in vivo* antibacterial efficacy of a novel scorpion venom peptide Smp24 conjugated with bimetallic nanomaterials. In a rat model of infected wound, the Smp24-nanomaterial conjugate significantly accelerated wound healing compared to control groups. Histological analysis revealed improved tissue reorganization, with reduced inflammatory infiltration and increased fibroblast and collagen density. This treatment also reduced bacterial load and modulated the immune response by decreasing pro-inflammatory cytokines (TNF-α, IL-6) and increasing anti-inflammatory cytokines (IL-4, IL-10) (Yang et *al.*, 2026). More recently, previous studies have identified the peptide derivative HP-H1K8 as a promising topical agent against methicillin-resistant *Staphylococcus aureus* (MRSA), opening the way for clinical applications in the treatment of infected wounds (Li et *al.*, 2020). Moreover, it was demonstrated that a scorpion-derived peptide and its analogues exhibited significant antibacterial activity *in vitro* and *in vivo* against various bacterial strains. Similarly, Yuan et al. (2022) developed a lysine-rich peptide, derived from a scorpion peptide, which showed potent antibacterial activity *in vitro* and *in vivo*, suggesting its potential for the treatment of skin infections (Daniele-Silva et *al.*, 2021). The cationic peptide Stigmurin, isolated from the venom of the Brazilian scorpion *Tityus stigmurus*, was structurally characterized by nuclear magnetic resonance (NMR). Its three-dimensional structure revealed essential features for its biological activity. *In vivo* tests demonstrated that Stigmurin possesses not only antibacterial activity but also significant wound healing activity, accelerating wound closure in animal models (Tan et *al.*, 2022).

In addition to their antimicrobial role, scorpion venom peptides are also involved in the regulation of the inflammatory response, a crucial step in wound healing. Although the inflammatory phase is necessary for wound debridement and immune cell recruitment, it can become deleterious if excessive or prolonged, particularly in chronic wounds. A major advance in understanding the mechanisms of action of scorpion peptides on wound healing (Krayem et *al.*, 2018). These researchers performed a proteomic analysis of exudates from chronic diabetic foot ulcers treated with a scorpion antimicrobial peptide (SAMP). The results revealed that SAMP treatment profoundly modulated the wound microenvironment by significantly decreasing the abundance of key pro-inflammatory cytokines (IL-1β, IL-6, MCP-1) and increasing the levels of proteins involved in tissue repair, notably fibronectin, which is essential for cell migration (Krayem et *al.*, 2018).

Furthermore, in addition to their antimicrobial and anti-inflammatory effects, certain scorpion venom peptides directly stimulate the anabolic processes of wound healing. They modulate angiogenesis by targeting integrins α5β1 and αvβ3, thereby promoting the formation of new blood vessels essential for supplying oxygen and nutrients to the repairing tissue (Krayem et *al.*, 2018). They also intervene in the cyclooxygenase-2 (COX-2) pathway, a key mechanism involved in tissue revascularization after injury (Correia et *al*., 2025). Notably, the anti-angiogenic effect of Hemilipin2 demonstrates that controlled modulation of angiogenesis, whether positive or negative, can be beneficial depending on the phase of wound healing, allowing fine regulation of the repair process (Jridi et *al*., 2017).

3. Scorpion venom biomedical devices

To date, no commercial bioactive dressing or implantable coating directly incorporates components derived from scorpion venom (Shokhba et *al*., 2026). Nevertheless, recent advances in biotechnology and biomaterials engineering have significantly strengthened interest in these biomolecules as next-generation therapeutic agents (Namiq et *al*., 2026). Current research is no longer limited to the extraction of crude venom but is oriented towards the identification, isolation, and valorization of specific bioactive peptides exhibiting promising pharmacological properties (Wan et *al*., 2017). Two major axes aim to transcend these properties to transform them into concrete and innovative therapeutic applications, particularly bioactive gels for cutaneous wound healing (Figure 3A) and the development of nanotechnologies for the encapsulation and targeted delivery of these peptides.

a. Bioactive gels for cutaneous wound healing

The application most directly related to the development of next-generation dressings concerns the formulation of bioactive gels. Researchers have developed gels based on Scorpion Venom Active Polypeptides (SVAP) specifically designed for the treatment of complex chronic wounds, such as diabetic or venous ulcers (Dai et *al*., 2008).

The efficacy of this approach has been demonstrated in preclinical studies where the venom gel significantly accelerated wound healing compared to a reference treatment, Epidermal growth Factor (EgF). The wound closure rate after 14 days reached 93.25% in the SVAP-treated group, compared to only 74.24% in the EgF group. The complete closure time was reduced to 21 days versus 35 days for the EgF-treated group. Histological analysis revealed that this acceleration was accompanied by a significant reduction in inflammatory cell infiltration, increased neovascularization, and improved collagen deposition, suggesting higher quality healing (Dai et *al*., 2008). Beyond its pro-healing effect, the gel possesses strong anti-inflammatory and antibacterial properties. Specific peptides, such as mucroporin (Conde et *al*., 2000) or scorpine (Tan et *al*., 2022), have been identified as being responsible for these antimicrobial effects, including against resistant bacterial strains.

Recent proteomic analysis of exudates from chronic wounds treated with a Scorpion Antimicrobial Peptide (SAMP) has elucidated the underlying molecular mechanisms. This study demonstrated that the treatment profoundly modulated the inflammatory microenvironment by significantly decreasing the abundance of key pro-inflammatory cytokines such as interleukin-1β (IL-1β), interleukin-6 (IL-6), and monocyte chemoattractant protein-1 (MCP-1). Simultaneously, SAMP increased the levels of proteins involved in tissue repair, notably fibronectin, which is essential for cell migration (Torres-Rêgo et *al.*, 2019). This dual action, both anti-inflammatory and pro-reparative, confirms the potential of scorpion venom peptides for the development of intelligent biomedical devices capable of acting directly on the wound microenvironment.

b. Nanotechnologies

To overcome the pharmacological obstacles inherent to the use of crude venom (notably its systemic toxicity and instability), research is turning to the use of nanotechnologies. The encapsulation of venom molecules into nanoparticles allows for the creation of safer and more effective coatings or delivery systems (Gláucia-Silva et *al.*, 2024; Huang et *al.*, 2026)

In this context, chitosan, a natural biodegradable polymer, emerges as a valuable ally. Researchers have developed cross-linked chitosan nanoparticles loaded with scorpion venom from the Brazilian species *Tityus stigmurus*. The results are remarkable, nano-encapsulation considerably improves the antimicrobial activity of the venom while reducing its toxicity. The minimum inhibitory concentration (MIC) against the bacterium *Staphylococcus aureus* was thus significantly reduced, and the nanoparticles demonstrated fungicidal activity against *Candida albicans* and *Candida tropicalis*, with satisfactory biocompatibility on fibroblasts (cell viability greater than 70%) (He et *al.*, 2025). Furthermore, a complementary approach consisted of encapsulating not the total venom, but a specific peptide, TistH, also derived from the venom of *Tityus stigmurus*. This chitosan nanoparticles loaded with the TistH peptide showed potentiated antifungal activity against *Candida* spp., as well as an ability to inhibit biofilm formation, a major factor in the chronicity of infected wounds (Gláucia-Silva et *al.*, 2024). This encapsulation strategy acts as a "shield" by protecting the active venom peptides from premature degradation and reducing their potential toxicity towards healthy cells. It thus paves the way for safer applications, whether for topical applications on the skin or in the longer term for systemic treatments (Wang et *al.*, 2012).

4. Neurobiology and personalized medicine

a. Ion channel-selective peptides

Extensive research has demonstrated that scorpion venom peptides exert diverse neurobiological effects primarily through specific interactions with voltage-gated ion channels (Lee et *al.*, 2000; Summer et *al.*, 2026). These peptides, evolved over millions of years, have developed remarkable selectivity for distinct ion channel subtypes, making them invaluable pharmacological tools and promising therapeutic leads (Muiruri & Zhong, 2023; Namiq et *al.*, 2026; Shokhba et *al.*, 2026). Scorpion venom neurotoxins can modulate neuronal excitability by targeting sodium (Nav), potassium (Kv), calcium (Cav), and chloride (ClC) channels, as well as transient receptor potential (TRP) channels, thereby influencing pain transmission, neurotransmitter release, and membrane potential regulation (Chen et *al.*, 2005; Qin et *al.*, 2025) (Table 1).

Table 1. Neurobiological effects and mechanisms of action of scorpion venom peptides on ion channels

Ion Channel Target	Toxin Type	Representative Peptides	Source (Species)	Mechanism of Action	Neurobiological Effect	References
Sodium Channels (Nav)	α-NaTx	AaH I, BmKI, Lqh2, Ts1, Ts2, Ts5, Ts6, Ts7	*Androctonus australis, Buthus martensii, Leiurus quinquestriatus, Tityus serrulatus*	inhibits channel inactivation, inducing prolonged channel opening	Neuronal hyperexcitability, intense pain	(Huang et al., 2026 ; Mouhat et al., 2008)
	β-NaTx	Cn2, Ts1, Ts2	*Centruroides noxius, Tityus serrulatus*	Hyperpolarizes the activation threshold, promoting facilitated channel opening	Spontaneous depolarization, increased neurotransmission	(Mendes et al., 2023 ; Mouhat et al., 2008)
	Bifunctional toxin	Ts1	*Tityus serrulatus*	Exerts dual modulatory effects on both activation and inactivation (α and β activities)	Mixed effect on activation and inactivation	(Mendes et al., 2023)
	Analgesic peptide	BmK AGAP	*Buthus martensii*	Suppresses MAPK signaling pathways, downregulates pro-inflammatory cytokine expression; antagonizes TRPV1 channels	Analgesic effect; potentiation of lidocaine effect	(Mouhat et al., 2008 ; Mouhat et al., 2004)
Potassium Channels (Kv)	Nav1.7-selective	(in development)	*Various Asian species*	Selectively blocks the voltage-gated sodium channel Nav1.7 implicated in pain transmission	Inhibition of pain transmission	(Xin et al., 2025)
	α-KTx	Charybdotoxin, Kaliotoxin, Margatoxin, BmKTX, Vm24	*Leiurus quinquestriatus, Buthus martensii, Vaejovis mexicanus, Androctonus mauretanicus, Centruroides margaritatus*	Occludes the channel pore via a functional dyad comprising a lysine and a hydrophobic/aromatic residue	Action potential prolongation, excitability modulation	(Huang et al., 2026 ; Mendes et al., 2023 ; Mouhat et al., 2008 ; Mouhat et al., 2004)
	γ-KTx	Ergtoxins	*Centruroides noxius*	Blocks ERG channels (Kv11.1) involved in cardiac repolarization	Prolongation of cardiac repolarization	(Mendes et al., 2023)
	κ-KTx	Hefutoxins	*Heterometrus fulvipes*	Blocks potassium channels independently of the CSαβ structural motif	Neuronal excitability modulation	
	BK Toxin	Martentoxin (MarTX)	*Buthus martensii*	Selectively blocks large-conductance calcium-activated potassium (BK) channels	Modulation of neurotransmitter release	(Dong et al., 2026)

continued on following page

Table 1. Continued

Ion Channel Target	Toxin Type	Representative Peptides	Source (Species)	Mechanism of Action	Neurobiological Effect	References
Calcium Channels (Cav)	RyR Agonists	Imperatoxin A (IpTxa)	*Pandinus imperator*	Activates ryanodine receptors, triggering intracellular calcium release	Intracellular calcium release, muscle contraction	(Huang et al., 2026 ; Mendes et al., 2023 ; Mouhat et al., 2008 ; Dong et al., 2026 ; Bourinet & Zamponi, 2017)
	RyR Stabilizers	Maurocalcine (MCa)	*Scorpio maurus palmatus*	Stabilizes the open conformation of ryanodine receptors	Calcium leak, contraction potentiation	
	T-type Inhibitors	Kurtoxin	*Parabuthus granulatus*	Blocks T-type calcium channels, specifically targeting Cav3.1 and Cav3.2 subtypes	Modulation of rhythmic neuronal excitability	
Chloride Channels (ClC)	ClC Blockers	Chlorotoxin (ClTx)	*Leiurus quinquestriatus*	Selectively inhibits chloride channels, disrupting ionic homeostasis	Inhibition of cell proliferation	
	Glioma-specific blockers	BmKCT	*Buthus martensii*	Targets chloride channels overexpressed in glioma cells	Selective inhibition of tumor cells	
TRP Channels	TRPV1 Modulators	WaTx	*Urodacus manicatus*	Prolongs TRPV1 channel opening, enhancing calcium influx	Hypersensitivity, neuronal depolarization	(Huang et al., 2026 ; Mendes et al., 2023 ; Mouhat et al., 2008)
	TRPV1 Activators	BmP01	*Mesobuthus martensii*	Activates TRPV1 channels under acidic conditions	Pain induction	
	TRPV1 Inhibitors	BmK AGAP	*Buthus martensii*	Antagonizes TRPV1 channels, blocking nociceptive signaling	Potentiation of analgesic effect	(Kampo et al., 2021)

Although scorpion venom peptides act on ion channels as their primary molecular targets, the specific mechanisms and physiological outcomes differ significantly depending on the channel subtype and the cellular context (Mouhat et *al.*, 2008; Summer et *al.*, 2026). In terms of their neurobiological effects, scorpion venom peptides primarily modulate neuronal activity by altering channel-gating properties, either delaying inactivation (α-toxins) or facilitating activation (β-toxins) of sodium channels (Huang et *al.*, 2026; Mendes et *al.*, 2023), or by occluding the pore of potassium channels through a functional dyad mechanism (Mouhat et *al.*, 2004; Mouhat et *al.*, 2008; Xin et *al.*, 2025). These interactions lead to profound changes in neuronal excitability, action potential propagation, and synaptic transmission (Dong et *al.*, 2026; He et *al.*, 2025). The therapeutic potential of these peptides extends beyond neurobiology. For example, α-toxins such as Ts1 and β-toxins including Cn2 have been extensively studied for their ability to modulate sodium channel activity, providing insights into channelopathies and pain mechanisms (Lee et *al.*, 2000; Kampo et *al.*, 2021). Potassium channel blockers like charybdotoxin and margatoxin selectively inhibit Kv1.3 channels, demonstrating immunosuppressive properties with potential applications in autoimmune diseases (Bourinet & Zamponi, 2017; Chow et *al.*, 2019; Tajti et *al.*, 2020) (Table 1). Chlorotoxin, derived from *Leiurus quinquestriatus*, specifically targets chloride channels overexpressed in glioma cells, thereby inhibiting tumor cell migration and invasion, highlighting the translational potential of these peptides in oncology (Tarcha et *al.*, 2017; Correia et *al.*, 2025).

Furthermore, emerging evidence reveals that certain scorpion venom peptides exhibit multimodal actions. BmK AGAP, isolated from *Buthus martensii*, not only blocks sodium channels to produce analgesic effects but also antagonizes TRPV1 channels, potentiating lidocaine-induced analgesia (Kampo et *al.*, 2021) (Table 1). Similarly, maurocalcine and imperatoxin A modulate ryanodine receptors, influencing intracellular calcium release and muscle contraction, with implications for muscle and cardiac disorders (Chen et *al.*, 2005; Mouhat et *al.*, 2004; Summer et *al.*, 2026).

b. Therapeutic applications of scorpion venom peptides in personalized medicine

Scorpion toxins offer promising perspectives for the development of targeted therapies in various pathological domains. In analgesia, the peptide BmK AGAP (*Buthus martensii*) inhibits TRPV1 channels and potentiates lidocaine's effect (Kampo et *al.*, 2021), enabling the development of selective analgesics without the side effects associated with opioids. This approach of ion channel modulation also finds application in Dravet syndrome, where the peptides Hj1a and Hj2a (*Hottentotta judaicus*) act as dual-action modulators on the mutated Nav1.1 channel (Chow et *al.*, 2019), providing a foundation for personalized epilepsy treatments.

In neurodegenerative diseases, the synthetic peptide SVHRSP (derived from Buthus martensii) exhibits neuroprotective effects in Parkinson's disease, improving motor dysfunction and attenuating neuronal damage by modulating the gut microbiota and inhibiting neuroinflammation (Chen et *al.*, 2025; Zhang et *al.*, 2023). Furthermore, the Kv1.3 channel, expressed on effector memory T lymphocytes, constitutes a key target in immunomodulation, as peptides BmKTX and Vm24 selectively block this channel with high affinity (Tajti et *al.*, 2020; Varga et *al.*, 2021; Veytia-Bucheli et *al.*, 2018). This strategy has been validated by dalazatide, which demonstrated a significant reduction in PASI score in 9 out of 10 psoriatic patients in a phase IB clinical trial (Tarcha et *al.*, 2017).

Finally, in precision oncology, the peptide BmKCT targets chloride channels overexpressed in glioma cells, inhibiting proliferation, invasion, and migration of tumor cells via down-regulation of MMP-2 expression (Wan et *al.*, 2017; Tan et *al.*, 2022; Ghodeif et *al.*, 2025).

VI. SCIENTIFIC AND TECHNOLOGICAL CHALLENGES

The considerable therapeutic potential of scorpion venom-derived peptides incorporated into biodegradable materials is limited by several scientific and technological challenges. One of the main issues is the stability of these peptides within polymer matrices. Due in part to their sensitivity to environmental factors such as pH, temperature, and enzymatic degradation, these biomolecules can undergo changes in their structure, leading to a loss of bioactivity (Sara et al., 2025). It is therefore essential to optimize encapsulation strategies and protective delivery systems to preserve their functionality.

An additional limitation is the potential immunogenicity of venom-derived peptides. As foreign molecules, they can trigger an immune response that reduces therapeutic efficacy and increases the risk of adverse effects, thereby limiting their long-term use. Similarly, the intrinsic toxicity of scorpion venom, which is dose-dependent, poses a major challenge. Some components may have beneficial effects at low concentrations but can become harmful at higher doses, thus requiring precise dose control and release kinetics (Vidya et al., 2021).

Interactions between peptides and biomaterial substrates must be carefully designed to ensure stability while enabling effective and targeted delivery. The variability in venom composition, which depends on the animal species, geographic origin, and extraction methods, further complicates standardization and reproducibility, two aspects that are essential for pharmaceutical development. The transition from preclinical research to clinical application has been limited. The majority of studies are limited to in vitro models or early-stage in vivo models, and clinical data are insufficient to validate safety and efficacy in humans. Additionally, regulatory constraints, production scalability, and profitability are major obstacles to the industrial and clinical implementation of these systems (Bates et al., 2024).

These challenges are best addressed through a multidisciplinary approach that integrates advances in molecular biology, materials science, pharmacology, and biomedical engineering. Future efforts should focus on improving peptide stability, reducing immunogenicity, optimizing delivery systems, and establishing standardized production and evaluation protocols to enable the successful translation of venom-derived biomaterial therapies into clinical practice.

CONCLUSION AND PERSPECTIVES

Due to its richness in bioactive peptides, scorpion venom may represent an exceptional natural source with remarkable therapeutic potential, including antimicrobial, anticancer, immunomodulatory, and anti-inflammatory activities, as well as other plausible biological effects. This makes it a promising avenue for the development of new therapeutic approaches in modern medicine.

However, despite these multiple activities, the direct use of venom in vital biological systems remains limited due to several constraints, ranging from venom stability to issues related to standardization and the control of its dose-dependent toxicity.

In this context, our chapter aims to fill this gap and address this challenge by proposing the integration of scorpion venom into biodegradable biopolymer matrices. This approach represents an innovative therapeutic strategy to optimize the efficacy of scorpion-derived peptides. Such integration could be exploited in various therapeutic fields, including anticancer therapy, wound healing, tissue engineering, and the development of anti-infective devices.

Subsequent research should aim to assess comparative *in vivo* studies of different matrix formulations, develop long-term safety and efficacy trials, test scalable GMP manufacturing processes, investigate synergistic venom-peptide combinations, and evolve towards other therapeutic applications, including autoimmune diseases and chronic pain. This collision between venom toxicology, biomaterials science, and nanotechnology is advancing this integration to the cutting edge of the next-generation precision medicine.

REFERENCES

Abdulsalam, L., Abubakar, S., Permatasari, I., Lawal, A. A., Uddin, S., Ullah, S., & Ahmad, I. (2025). Advanced biocompatible and biodegradable polymers: A review of functionalization, smart systems, and sustainable applications. *Polymers*, *17*(21), 2901. DOI: 10.3390/polym17212901 PMID: 41228661

Al-Asmari, A. K., Ullah, Z., Al Balowi, A., & Islam, M. (2017). In vitro determination of the efficacy of scorpion venoms as anti-cancer agents against colorectal cancer cells: A nano-liposomal delivery approach. *International Journal of Nanomedicine*, *12*, 559–574. DOI: 10.2147/IJN.S123514 PMID: 28144138

Almaaytah, A., & Albalas, Q. (2014). Scorpion venom peptides with no disulfide bridges: A review. *Peptides*, *51*, 35–45. DOI: 10.1016/j.peptides.2013.10.021 PMID: 24184590

Amoke, A., & Kalu, I. E. (2026). Advances in biodegradable polymeric materials: Focus on controlling factors and sustainable applications. *Next Research*, *7*, 101501. DOI: 10.1016/j.nexres.2026.101501

Arcos, S. S. S., Aguiar, M. R. D. C., Oliveira, J. D., Silva, M. R. D., Pimentel, I. D. O. C., Dos Anjos, N. G., Machado, G. H. R. S., Evangelista, K. B., Portaro, F. C. V., & Iwai, L. K. (2025). Scorpion venom as a source of cancer drugs: A comprehensive proteomic analysis and therapeutic potential. *International Journal of Molecular Sciences*, *26*(20), 9907. DOI: 10.3390/ijms26209907 PMID: 41155199

Arpornsuwan, T., Buasakul, B., Jaresitthikunchai, J., & Roytrakul, S. (2014). Potent and rapid antigonococcal activity of the venom peptide BmKn2 and its derivatives against different MALDI biotypes of multidrug-resistant *Neisseria gonorrhoeae*. *Peptides*, *53*, 315–320. DOI: 10.1016/j.peptides.2013.10.020 PMID: 24184420

Ayari-Riabi, S., Trimaille, T., Mabrouk, K., Bertin, D., Gigmes, D., Benlasfar, Z., & Elayeb, M. (2016). Polylactide conjugué au venin appliqué comme matériau biocompatible pour l'immunothérapie passive et active contre l'envenimation par scorpion. *Vaccine*, *34*(15), 1810–1815. DOI: 10.1016/j.vaccine.2016.02.030 PMID: 26902547

Bajgiran, N. K., Ahmadi, Y., Majidi, S., & Ahmadi, B. (2025). Smart Polymer-Based Membranes for Biomedical Applications: Mechanisms, Fabrication, and Advances in Controlled Drug Delivery and Tissue Engineering. *Materials Today. Communications*, •••, 114443.

Baral, K. C., & Choi, K. Y. (2025). Barriers and strategies for oral peptide and protein therapeutics delivery: Update on clinical advances. *Pharmaceutics*, *17*(4), 397. DOI: 10.3390/pharmaceutics17040397 PMID: 40284395

Bates, S. M., Evans, K. V., Delsing, L., Wong, R., Cornish, G., & Bahjat, M. (2024). Immune safety challenges facing the preclinical assessment and clinical progression of cell therapies. *Drug Discovery Today*, *29*(12), 104239. DOI: 10.1016/j.drudis.2024.104239 PMID: 39521331

Baudat, R., Benoit, E., Keck, M., Kessler, P., Truillet, C., & Servent, D. (2023). Les peptides de venins pour l'imagerie du cancer pulmonaire. *L'Actualité Chimique*, (490), 30–36.

Benito-Kaesbach, A., Beltrán-Sanahuja, A., Mathers, R. T., & Sanz-Lázaro, C. (2025). Understanding the degradation of bio-based polymers across contrasting marine environments using complementary analytical techniques. *Journal of Cleaner Production*, *524*, 146435. DOI: 10.1016/j.jclepro.2025.146435

Bhavya, J., Francois, N., More, N. S., & More, S. (2016). Scorpion toxin polypeptides as therapeutic agents: An overview. *Protein and Peptide Letters*, *23*(10), 848–859. DOI: 10.2174/0929866523666160630184635 PMID: 27397476

Bourinet, E., & Zamponi, G. W. (2017). Block of voltage-gated calcium channels by peptide toxins. *Neuropharmacology*, *127*, 109–117. DOI: 10.1016/j.neuropharm.2016.10.016 PMID: 27756538

Bregado, J. L., Secchi, A. R., Tavares, F. W., de Sousa Rodrigues, D., & Gambetta, R. (2019). Amorphous paracrystalline structures from native crystalline cellulose: A molecular dynamics protocol. *Fluid Phase Equilibria*, *491*, 56–76. DOI: 10.1016/j.fluid.2019.03.011

Chen, M., Zhang, Y., Hou, L., Zhao, Z., Tang, P., Sun, Q., & Wang, Q. (2025). SVHRSP protects against rotenone-induced neurodegeneration in mice by inhibiting TLR4/NF-κB-mediated neuroinflammation via gut microbiota. npj. *Parkinson's Disease*, *11*(1), 43. PMID: 40050294

Chen, T., Walker, B., Zhou, M., & Shaw, C. (2005). Molecular cloning of a novel putative potassium channel-blocking neurotoxin from the venom of the North African scorpion, Androctonus amoreuxi. *Peptides*, *26*(5), 731–736. DOI: 10.1016/j.peptides.2004.12.002 PMID: 15808902

Chow, C. Y., Chin, Y. K. Y., Walker, A. A., Guo, S., Blomster, L. V., Ward, M. J., Herzig, V., Rokyta, D. R., & King, G. F. (2019). Venom peptides with dual modulatory activity on the voltage-gated sodium channel NaV1.1 provide novel leads for development of antiepileptic drugs. *ACS Pharmacology & Translational Science, 3*(1), 119–134. DOI: 10.1021/acsptsci.9b00079 PMID: 32259093

Conde, R., Zamudio, F. Z., Rodríguez, M. H., & Possani, L. D. (2000). Scorpine, an anti-malaria and anti-bacterial agent purified from scorpion venom. *FEBS Letters, 471*(2-3), 165–168. DOI: 10.1016/S0014-5793(00)01384-3 PMID: 10767415

Correia, M. R., Han, S. W., Escalante, T., & Moreira, V. (2025). The role of the cyclooxygenase-2 pathway in tissue ischemia and revascularization following skeletal muscle injury induced by bothropic snake venom. *Microvascular Research, 157*, 104760. DOI: 10.1016/j.mvr.2024.104760 PMID: 39510245

da Silva, J. R., Castro-Amorim, J., Mukherjee, A. K., Ramos, M. J., & Fernandes, P. A. (2025). The application of snake venom in anticancer drug discovery: An overview of the latest developments. *Expert Opinion on Drug Discovery, 20*(3), 317–335. DOI: 10.1080/17460441.2025.2465364 PMID: 40012249

Dai, C., Ma, Y., Zhao, Z., Zhao, R., Wang, Q., Wu, Y., Cao, Z., & Li, W. (2008). Mucroporin, the first cationic host defense peptide from the venom of Lychas mucronatus. *Antimicrobial Agents and Chemotherapy, 52*(11), 3967–3972. DOI: 10.1128/AAC.00542-08 PMID: 18779362

Dallaev, R., Papež, N., Allaham, M. M., & Holcman, V. (2025). Biodegradable polymers: Properties, applications, and environmental impact. *Polymers, 17*(14), 1981. DOI: 10.3390/polym17141981 PMID: 40732859

Daniele-Silva, A., Rodrigues, S. D. C. S., Dos Santos, E. C. G., de Queiroz Neto, M. F., de Oliveira Rocha, H. A., da Silva-Júnior, A. A., & de Freitas Fernandes-Pedrosa, M. (2021). NMR three-dimensional structure of the cationic peptide Stigmurin from Tityus stigmurus scorpion venom: In vitro antioxidant and in vivo antibacterial and healing activity. *Peptides, 137*, 170478. DOI: 10.1016/j.peptides.2020.170478 PMID: 33359395

De Roodt, A. R., Lanari, L. C., Laskowicz, R. D., de Oliveira, V. C., Litwin, S., Calderon, L., & Lértora, E. (2019). Study on the obtaining of *Tityus trivittatus* venom in Argentina. *Toxicon, 159*, 5–13. DOI: 10.1016/j.toxicon.2018.12.004 PMID: 30611824

De Souza, F. M., & Gupta, R. K. (2024). Bacteria for bioplastics: Progress, applications, and challenges. *ACS Omega, 9*(8), 8666–8686. DOI: 10.1021/acsomega.3c07372 PMID: 38434856

Deshane, J., Garner, C. C., & Sontheimer, H. (2003). Chlorotoxin inhibits glioma cell invasion via matrix metalloproteinase-2. *The Journal of Biological Chemistry*, *278*(6), 4135–4144. DOI: 10.1074/jbc.M205662200 PMID: 12454020

Díaz-García, A., & Varela, D. (2021). Voltage-gated K^+/Na^+ channels and scorpion venom toxins in cancer. In *Ion channel signalling in cancer: From molecular mechanisms to therapeutics.*

Dilsad, A. M., Ahuja, A., Gupta, N., Bachala, S. K., & Rastogi, V. K. (2024). Analysing the impact of nanomaterials on the degradation behaviour of biopolymers: A comprehensive review. *European Polymer Journal*, *214*, 113189. DOI: 10.1016/j.eurpolymj.2024.113189

Dong, Y., Wang, J., Zhu, Y., Zhao, L., Zeng, Y., Tang, L., Xiao, Q., Cheng, J., Wang, C., & Tao, J. (2026). From venom peptides to neurotherapeutics: BmK defensins and short-chain peptides as modulators of ion channels. *Frontiers in Pharmacology*, *17*, 1754290. DOI: 10.3389/fphar.2026.1754290 PMID: 41693793

. Edo, G. I., Ndudi, W., Ali, A. M., Yousif, E., Jikah, A. N., Isoje, E. F., ... & Umar, H. (2025). Biopolymers: an inclusive review. *Hybrid advances, 9*, 100418.

Egbuna, C., Parmar, V. K., Jeevanandam, J., Ezzat, S. M., Patrick-Iwuanyanwu, K. C., Adetunji, C. O., Khan, J., Onyeike, E. N., Uche, C. Z., Akram, M., Ibrahim, M. S., El Mahdy, N. M., Awuchi, C. G., Saravanan, K., Tijjani, H., Odoh, U. E., Messaoudi, M., Ifemeje, J. C., Olisah, M. C., & Ibeabuchi, C. G. (2021). Toxicity of nanoparticles in biomedical application: Nanotoxicology. *Journal of Toxicology*, *2021*(1), 9954443. DOI: 10.1155/2021/9954443 PMID: 34422042

El Housseini, A., Aimrane, A., Youssef, E. L., Ait Laaradia, M., Laghzaoui, E. M., El Mouden, E. H., & Abou Oualid, J. (2025). Kinetics of brain oxidative stress in mice after envenomation by *Hottentotta gentili* (Pallary, 1924). *Toxicon*, •••, 108938. PMID: 41349899

El Housseini, A., Rachmoune, K., Ait Baba, A., Ait Hamdan, Y., Armadi, I., Harkat, C., . . . Agnaou, M. (2026). Secondary Metabolites: Therapeutic Tools for Tackling Neurodegenerative Diseases. In *Nanoparticles Agents and Secondary Metabolites in Neurodegenerative Diseases* (pp. 133-166). IGI Global Scientific Publishing.

El-Qassas, J., Abd El-Atti, M., & El-Badri, N. (2024). Harnessing the potency of scorpion venom-derived proteins: Applications in cancer therapy. *Bioresources and Bioprocessing*, *11*(1), 93. DOI: 10.1186/s40643-024-00805-0 PMID: 39361208

Feczko, T. (2021). Polymeric nanotherapeutics acting at special regions of body. *Journal of Drug Delivery Science and Technology*, *64*, 102597. DOI: 10.1016/j.jddst.2021.102597

Furtado, A. A., Silva, A. D., Silva-Júnior, A. A., & Fernandes-Pedrosa, M. F. (2020). Biology, venom composition, and scorpionism induced by Brazilian scorpion *Tityus stigmurus* (Thorell, 1876) (Scorpiones: Buthidae): A mini-review. *Toxicon*, *185*, 36–45. DOI: 10.1016/j.toxicon.2020.06.015 PMID: 32585220

Fuzita, F. J., Pinkse, M. W. H., Patané, J. S. L., Juliano, M. A., Verhaert, P. D. E. M., & Lopes, A. R. (2015). Biochemical, transcriptomic and proteomic analyses of digestion in the scorpion *Tityus serrulatus*: Insights into function and evolution of digestion in an ancient arthropod. *PLoS One*, *10*(1), 1–31. DOI: 10.1371/journal.pone.0123841 PMID: 25875018

Ghodeif, S. K., El-Fahla, N. A., Abdel-Rahman, M. A., & El-Shenawy, N. S. (2025). Arthropod venom peptides: Pioneering nanotechnology in cancer treatment and drug delivery. *Cancer Pathogenesis and Therapy*, *3*, E01–E17. PMID: 41541913

Ghodeif, S.K., El-Fahla, N.A., Abdel-Rahman, M.A., & El-Shenawy, N.S. (2026). rthropod venom peptides: Nanotechnology-enabled delivery and therapeutic potential in cancer treatment, Cancer Pathogenesis and Therapy, Volume 4, Issue 2, Pages 81-97, ISSN 2949-7132, DOI: 10.1016/j.cpt.2025.03.005

Gláucia-Silva, F., Torres, J. V. P., Torres-Rêgo, M., Daniele-Silva, A., Furtado, A. A., Ferreira, S. D. S., Chaves, G. M., Xavier-Júnior, F. H., Rocha Soares, K. S., Silva-Júnior, A. A., & Fernandes-Pedrosa, M. D. F. (2024). Tityus stigmurus-venom-loaded cross-linked chitosan nanoparticles improve antimicrobial activity. *International Journal of Molecular Sciences*, *25*(18), 9893. DOI: 10.3390/ijms25189893 PMID: 39337380

Gopinathan, N., Yang, B., Lowe, J. P., Edler, K. J., & Rigby, S. P. (2014). NMR cryoporometry characterisation studies of the relation between drug release profile and pore structural evolution of polymeric nanoparticles. *International Journal of Pharmaceutics*, *469*(1), 146–158. DOI: 10.1016/j.ijpharm.2014.04.018 PMID: 24726633

Harrison, P. L., Abdel-Rahman, M. A., Miller, K., & Strong, P. N. (2014). Antimicrobial peptides from scorpion venoms. *Toxicon*, *88*, 115–137. DOI: 10.1016/j.toxicon.2014.06.006 PMID: 24951876

Hazra, R. S., Dutta, D., Mamnoon, B., Nair, G., Knight, A., Mallik, S., Ganai, S., Reindl, K., Jiang, L., & Quadir, M. (2021). Polymeric composite matrix with high biobased content as pharmaceutically relevant molecular encapsulation and release platform. *ACS Applied Materials & Interfaces*, *13*(34), 40229–40248. DOI: 10.1021/acsami.1c03805 PMID: 34423963

He, D., Lei, Y., Qin, H., Cao, Z., & Kwok, H. F. (2025). Deciphering scorpion toxin-induced pain: Molecular mechanisms and ion channel dynamics. *International Journal of Biological Sciences*, *21*(7), 2921–2934. DOI: 10.7150/ijbs.109713 PMID: 40384871

Herdiana, Y., Wathoni, N., Shamsuddin, S., & Muchtaridi, M. (2022). Drug release study of the chitosan-based nanoparticles. *Heliyon*, *8*(1), e08674. DOI: 10.1016/j.heliyon.2021.e08674 PMID: 35028457

Hoang, C. N. M., Nguyen, S. H., & Tran, M. T. (2025). Nanoparticles in cancer therapy: Strategies to penetrate and modulate the tumor microenvironment–a review. *Smart Materials in Medicine*.

Huang, Y., Kamau, P. M., Wang, J., Gao, M., & Li, B. (2026). Scorpion Venom Neurotoxins: Molecular Diversity, Mechanisms, and Drug Scaffolds. *Toxins*, *18*(1), 25. DOI: 10.3390/toxins18010025 PMID: 41591171

Jahromi, L. P., Ghazali, M., Ashrafi, H., & Azadi, A. (2020). A comparison of models for the analysis of the kinetics of drug release from PLGA-based nanoparticles. *Heliyon*, *6*(2). PMID: 32140583

Jiang, D., Tonggu, L., Gamal El-Din, T. M., Banh, R., Pomès, R., Zheng, N., & Catterall, W. A. (2021). Structural basis for voltage-sensor trapping of the cardiac sodium channel by a deathstalker scorpion toxin. *Nature Communications*, *12*(1), 1–13. DOI: 10.1038/s41467-020-20078-3 PMID: 33397917

Jridi, I., Catacchio, I., Majdoub, H., Shahbazzadeh, D., El Ayeb, M., Frassanito, M. A., Solimando, A. G., Ribatti, D., Vacca, A., & Borchani, L. (2017). The small subunit of Hemilipin2, a new heterodimeric phospholipase A2 from Hemiscorpius lepturus scorpion venom, mediates the antiangiogenic effect of the whole protein. *Toxicon*, *126*, 38–46. DOI: 10.1016/j.toxicon.2016.12.001 PMID: 27940138

Jurak, M., Wiącek, A. E., Ładniak, A., Przykaza, K., & Szafran, K. (2021). What affects the biocompatibility of polymers? *Advances in Colloid and Interface Science*, *294*, 102451. DOI: 10.1016/j.cis.2021.102451 PMID: 34098385

Kalayci, S., Ustaoglu Iyigundogdu, Z., Yazici, M. A., Asutay, B., Demir, O., & Sahin, F. (2016). Evaluation of antimicrobial and antiviral activities of different venoms. *Infectious Disorders Drug Targets*, *16*(1), 44–53. DOI: 10.2174/1871526516666160108115307 PMID: 26743517

Kamaly, N., Yameen, B., Wu, J., & Farokhzad, O. C. (2016). Degradable controlled-release polymers and polymeric nanoparticles: Mechanisms of controlling drug release. *Chemical Reviews*, *116*(4), 2602–2663. DOI: 10.1021/acs.chemrev.5b00346 PMID: 26854975

Kampo, S., Cui, Y., Yu, J., Anabah, T. W., Falagán, A. A., Bayor, M. T., & Wen, Q. P. (2021). Scorpion Venom peptide, AGAP inhibits TRPV1 and potentiates the analgesic effect of lidocaine. *Heliyon*, *7*(12), e08612. DOI: 10.1016/j.heliyon.2021.e08560 PMID: 35005265

Krayem, N., Abdelkefi-Koubaa, Z., Marrakchi, N., Gargouri, Y., & Luis, J. (2018). Native and recombinant phospholipases A2 of Scorpio maurus venom glands impair angiogenesis by targeting integrins α5β1 and αvβ3. *International Journal of Biological Macromolecules*, *116*, 305–315. DOI: 10.1016/j.ijbiomac.2018.04.141 PMID: 29715557

Kwon, N. Y., Sung, H. K., & Park, J. K. (2024). Systematic review of the antitumor activities and mechanisms of scorpion venom in breast cancer. *Journal of Clinical Medicine*, *14*(9), 3181. DOI: 10.3390/jcm14093181 PMID: 40364211

Lama, S., Merlin-Zhang, O., & Yang, C. (2020). In vitro and in vivo models for evaluating the oral toxicity of nanomedicines. *Nanomaterials (Basel, Switzerland)*, *10*(11), 2177. DOI: 10.3390/nano10112177 PMID: 33142878

Lanao, R. P. F., Jonker, A. M., Wolke, J. G., Jansen, J. A., Van Hest, J. C., & Leeuwenburgh, S. C. (2013). Physicochemical properties and applications of poly (lactic-co-glycolic acid) for use in bone regeneration. *Tissue Engineering. Part B, Reviews*, *19*(4), 380–390. DOI: 10.1089/ten.teb.2012.0443 PMID: 23350707

Lee, D., Gurevitz, M., & Adams, M. E. (2000). Modification of synaptic transmission and sodium channel inactivation by the insect-selective scorpion toxin LqhαIT. *Journal of Neurophysiology*, *83*(3), 1181–1187. DOI: 10.1152/jn.2000.83.3.1181 PMID: 10712448

Li, S., Yu, Q., Li, H., Chen, M., Jin, Y., & Liu, D. (2023). Self-assembled peptide hydrogels in regenerative medicine. *Gels (Basel, Switzerland)*, *9*(8), 653. DOI: 10.3390/gels9080653 PMID: 37623108

Li, Z., Yuan, Y., Li, S., Deng, B., & Wang, Y. (2020). Antibacterial activity of a scorpion-derived peptide and its derivatives in vitro and in vivo. *Toxicon*, *186*, 35–41. DOI: 10.1016/j.toxicon.2020.07.028 PMID: 32768440

Lin, J., Zheng, Y., Chen, K., Huang, Z., Wu, X., & Zhang, N. (2013). Inhibition of FOXM1 by thiostrepton sensitizes medulloblastoma to the effects of chemotherapy. *Oncology Reports*, *30*(4), 1739–1744. DOI: 10.3892/or.2013.2654 PMID: 23912794

Liu, M., Svirskis, D., Proft, T., Loh, J., Yin, N., Li, H., Li, D., Zhou, Y., Chen, S., Song, L., Chen, G., Lu, W.-Y., Zhang, Z., Zhou, Z., Li, L., Huang, Y., Bunt, C., Sun, G., Harris, P. W. R., & Wen, J. (2025). Progress in peptide and protein therapeutics: Challenges and strategies. *Acta Pharmaceutica Sinica. B*, *15*(12), 6342–6381. DOI: 10.1016/j.apsb.2025.10.026 PMID: 41477339

Majumder, J., Taratula, O., & Minko, T. (2019). Nanocarrier-based systems for targeted and site specific therapeutic delivery. *Advanced Drug Delivery Reviews*, *144*, 57–77. DOI: 10.1016/j.addr.2019.07.010 PMID: 31400350

McFerrin, M. B., & Sontheimer, H. (2006). A role for ion channels in glioma cell invasion. *Neuron Glia Biology*, *2*(1), 39–49. DOI: 10.1017/S1740925X06000044 PMID: 16520829

Mendes, L. C., Viana, G. M. M., Nencioni, A. L. A., Pimenta, D. C., & Beraldo-Neto, E. (2023). Scorpion peptides and ion channels: An insightful review of mechanisms and drug development. *Toxins*, *15*(4), 238. DOI: 10.3390/toxins15040238 PMID: 37104176

MICAD Research Team. (2007). [^{131}I]-Chlorotoxin. In Molecular Imaging and Contrast Agent Database (MICAD). National Center for Biotechnology Information (NCBI). https://www.ncbi.nlm.nih.gov/books/NBK23197/

Minutti-Zanella, C., Gil-Leyva, E. J., & Vergara, I. (2021). Immunomodulatory properties of molecules from animal venoms. *Toxicon*, *191*, 54–68. DOI: 10.1016/j.toxicon.2020.12.018 PMID: 33417946

Mouhat, S., Andreotti, N., Jouirou, B., & Sabatier, J. M. (2008). Animal toxins acting on voltage-gated potassium channels. *Current Pharmaceutical Design*, *14*(24), 2503–2518. DOI: 10.2174/138161208785777441 PMID: 18781998

Mouhat, S., Mosbah, A., Visan, V., Wulff, H., Delepierre, M., Darbon, H., Grissmer, S., De WAARD, M., & Sabatier, J. M. (2004). The functional dyad of scorpion toxin Pi1 is not itself a prerequisite for toxin binding to the voltage-gated Kv1.2 potassium channels. *The Biochemical Journal*, *377*(1), 25–36. DOI: 10.1042/bj20030115 PMID: 12962541

Muiruri, K. P., & Zhong, J. (2023). Bioactive peptides from scorpion venoms: Therapeutic scaffolds and pharmacological tools. *Chinese Journal of Natural Medicines, 21*(1), 19–35. DOI: 10.1016/S1875-5364(23)60382-6 PMID: 36641229

Mukherjee, C., Varghese, D., Krishna, J. S., Boominathan, T., Rakeshkumar, R., Dineshkumar, S., Brahmananda Rao, C. V. S., & Sivaramakrishna, A. (2023). Recent advances in biodegradable polymers–properties, applications and future prospects. *European Polymer Journal, 192*, 112068. DOI: 10.1016/j.eurpolymj.2023.112068

Munawar, A., Ali, S., Akrem, A., & Betzel, C. (2018). Snake venom peptides: Tools of biodiscovery. *Toxins, 10*(11), 474. DOI: 10.3390/toxins10110474 PMID: 30441876

Namiq, S., Bouimeja, B., Abdessamad, E., Khadra, A., & Touloun, O. (2026). Pharmacological and therapeutic applications of scorpion venom peptides: A Review. *Journal of Natural Products Discovery, 4*(3), 45–58.

Narkar, A. R., Tong, Z., Soman, P., & Henderson, J. H. (2022). Smart biomaterial platforms: Controlling and being controlled by cells. *Biomaterials, 283*, 121450. DOI: 10.1016/j.biomaterials.2022.121450 PMID: 35247636

Oukkache, N., Chgoury, F., Lalaoui, M., Cano, A. A., & Ghalim, N. (2013). Comparison between two methods of scorpion venom milking in Morocco. *The Journal of Venomous Animals and Toxins Including Tropical Diseases, 19*(1), 5. DOI: 10.1186/1678-9199-19-5 PMID: 23849043

Ozkan, O., Adiguzel, S., Yakistiran, S., & Filazi, A. (2006). Study of the relationship between *Androctonus crassicauda* venom toxicity and telson size, weight and storage conditions. *The Journal of Venomous Animals and Toxins Including Tropical Diseases, 12*(2), 297–309.

Parihar, A. S., Coghlan, M. J., Gopalakrishnan, M., & Shieh, C. C. (2003). Effects of intermediate-conductance Ca^{2+}-activated K^{+} channel modulators on human prostate cancer cell proliferation. *European Journal of Pharmacology, 471*(3), 157–164. DOI: 10.1016/S0014-2999(03)01825-9 PMID: 12826234

Parveen, S., Bozal, S. B., Pober, J. S., & Saltzman, W. M. (2026). *Nanoparticle-hydrogel composites for precision medicine.* Cell Biomaterials. DOI: 10.1016/j.celbio.2026.100372

Qin, H., Luo, S., Zuo, W., Cao, Z., Assaraf, Y. G., & Kwok, H. F. (2025). Targeted eradication of glioblastoma via venom decapeptide-conjugated dendrimers: Inducing nuclear translocation and autophagic cell death. *Journal of Controlled Release, 383*, 113780. DOI: 10.1016/j.jconrel.2025.113780 PMID: 40315960

Quintero-Hernández, V., Jiménez-Vargas, J. M., Gurrola, G. B., Valdivia, H. H., & Possani, L. D. (2013). Scorpion venom components that affect ion-channel function. *Toxicon, 76*, 328–342. DOI: 10.1016/j.toxicon.2013.07.012 PMID: 23891887

Rachmoune, K., Mabsor-Zgandaoui, S., Bousaid, M., Armadi, I., El Housseini, A., Ait Hamdan, Y., & Gharssalaoui, A. (2026). Chitosan-based biopolymers for radioprotection: Mechanisms and biomedical applications in radiotherapy—A review. *International Journal of Biological Macromolecules, 348*, 150798. DOI: 10.1016/j.ijbiomac.2026.150798 PMID: 41663018

Rapôso, C. (2017). Scorpion and spider venoms in cancer treatment: State of the art, challenges, and perspectives. *Journal of Clinical and Translational Research, 3*(2), 233. DOI: 10.18053/jctres.03.201702.002 PMID: 30873475

Rodríguez de la Vega, R. C., Schwartz, E. F., & Possani, L. D. (2010). Mining on scorpion venom biodiversity. *Toxicon, 56*(7), 1155–1161. DOI: 10.1016/j.toxicon.2009.11.010 PMID: 19931296

Saez, N. J., & Herzig, V. (2019). Versatile spider venom peptides and their medical and agricultural applications. *Toxicon, 158*, 109–126. DOI: 10.1016/j.toxicon.2018.11.298 PMID: 30543821

Satchanska, G., Davidova, S., & Petrov, P. D. (2024). Natural and synthetic polymers for biomedical and environmental applications. *Polymers, 16*(8), 1159. DOI: 10.3390/polym16081159 PMID: 38675078

Shankar, P. R., & Unni, A. (2026). Deadly cures: Unlocking anticancer potential of reptile, amphibian, and arthropod venoms through molecular innovation and nanotechnology. *Next Nanotechnology, 9*, 100378. DOI: 10.1016/j.nxnano.2026.100378

Shokhba, A. S. M., Omran, M. A. E. D. A., Abdel-Rahman, M. A., El-Sayed, G. S., & Abdel-Nabi, I. M. (2026). In vivo Antibacterial Efficacy of a Novel Scorpion Venom Peptide Smp24 Conjugated with Bimetallic Nanomaterials. *Egyptian Journal of Chemistry, 69*(6), 599–608.

Sindhi, K., Pingili, R. B., Beldar, V., Bhattacharya, S., Rahaman, J., & Mukherjee, D. (2025). The role of biomaterials-based scaffolds in advancing skin tissue construct. *Journal of Tissue Viability, 34*(2), 100858. DOI: 10.1016/j.jtv.2025.100858 PMID: 39827732

Summer, M., Bibi, R., & Ali, S. (2026). Pharmacological Potential of Scorpion Venom: From Molecular Targets, Peptide Chemistry to Clinical Translation. *Revista Brasileira de Farmacognosia, 36*(1), 35. DOI: 10.1007/s43450-026-00739-y

Tajti, G., Wai, D. C., Panyi, G., & Norton, R. S. (2020). The voltage-gated potassium channel KV1.3 as a therapeutic target for venom-derived peptides. *Biochemical Pharmacology, 181*, 114146. DOI: 10.1016/j.bcp.2020.114146 PMID: 32653588

Tan, Z., Yu, Z., Xu, X., Meng, L., Yu, M., Tao, R., Wu, Y., & Zhu, Z. (2022). Proteomic analysis of exudates from chronic ulcer of diabetic foot treated with scorpion antimicrobial peptide. *Mediators of Inflammation, 2022*(1), 5852786. DOI: 10.1155/2022/5852786 PMID: 36225537

Tarcha, E. J., Olsen, C. M., Probst, P., Peckham, D., Muñoz-Elías, E. J., Kruger, J. G., & Iadonato, S. P. (2017). Safety and pharmacodynamics of dalazatide, a Kv1.3 channel inhibitor, in the treatment of plaque psoriasis: A randomized phase 1b trial. *PLoS One, 12*(7), e0180762. DOI: 10.1371/journal.pone.0180762 PMID: 28723914

Thang, N. H., Chien, T. B., & Cuong, D. X. (2023). Polymer-based hydrogels applied in drug delivery: An overview. *Gels (Basel, Switzerland), 9*(7), 523. DOI: 10.3390/gels9070523 PMID: 37504402

Tobassum, S., Tahir, H. M., Arshad, M., Zahid, M. T., Ali, S., & Ahsan, M. M. (2020). Nature and applications of scorpion venom: An overview. *Toxin Reviews, 39*(3), 214–225. DOI: 10.1080/15569543.2018.1530681

Torres-Rêgo, M., Gláucia-Silva, F., Soares, K. S. R., De Souza, L. B. F. C., Damasceno, I. Z., dos Santos-Silva, E., & de Freitas Fernandes-Pedrosa, M. (2019). Biodegradable cross-linked chitosan nanoparticles improve anti-Candida and antibiofilm activity of TistH, a peptide identified in the venom gland of the Tityus stigmurus scorpion. *Materials Science and Engineering C, 103*, 109830. DOI: 10.1016/j.msec.2019.109830 PMID: 31349502

Touloun, O. (2019). Liste actualisée et commentée de la faune scorpionique du Maroc (Arachnida: Scorpiones). *Revista Iberica de Aracnologia, 34*, 126–132.

Varanko, A., Saha, S., & Chilkoti, A. (2020). Recent trends in protein and peptide-based biomaterials for advanced drug delivery. *Advanced Drug Delivery Reviews, 156*, 133–187. DOI: 10.1016/j.addr.2020.08.008 PMID: 32871201

Varga, Z., Tajti, G., & Panyi, G. (2021). The Kv1.3 K^+ channel in the immune system and its "precision pharmacology" using peptide toxins. *Biologia Futura, 72*(1), 75–83. DOI: 10.1007/s42977-021-00071-7 PMID: 34554500

Veytia-Bucheli, J. I., Jiménez-Vargas, J. M., Melchy-Pérez, E. I., Sandoval-Hernández, M. A., Possani, L. D., & Rosenstein, Y. (2018). Kv1.3 channel blockade with the Vm24 scorpion toxin attenuates the $CD4^{+}$ effector memory T cell response to TCR stimulation. *Cell Communication and Signaling : CCS*, *16*(1), 45. DOI: 10.1186/s12964-018-0257-7 PMID: 30107837

Vidya, V., Achar, R. R., Himathi, M. U., Akshita, N., Kameshwar, V. H., Byrappa, K., & Ramadas, D. (2021). Venom peptides–a comprehensive translational perspective in pain management. *Current Research in Toxicology*, *2*, 329–340. DOI: 10.1016/j.crtox.2021.09.001 PMID: 34604795

Wan, T., Li, L., Zhu, Z., Liu, S., Zhao, Y., & Yu, M. (2017). Scorpion venom active polypeptide may be a new external drug of diabetic ulcer. *Evidence-Based Complementary and Alternative Medicine : eCAM*, *2017*(1), 5161565. DOI: 10.1155/2017/5161565 PMID: 29234410

Wang, K., Nguyen, T., Gao, Y., Guo, R., Fan, C., Liao, H., Li, J., Chai, J., Xu, X., Gong, Y., & Chen, X. (2024). Androcin 18-1, a novel scorpion-venom peptide, shows a potent antitumor activity against human U87 cells via inducing mitochondrial dysfunction. *Insect Biochemistry and Molecular Biology*, *170*, 104137. DOI: 10.1016/j.ibmb.2024.104137 PMID: 38759703

Wang, X., Wang, K., Fang, Y., Zhang, Y., Yi, L., Li, X., Zhao, Q., Zhu, X., Cai, S., & Wan, L. (2026). Strategies for overcoming multiple barriers of oral administration of protein and peptide therapeutics. *Materials Today. Bio*, *37*, 102763. DOI: 10.1016/j.mtbio.2026.102763 PMID: 41585434

Wang, X., Zhang, L., & Chen, Y. (2012). Scorpion venom peptides: A promising source for drug development. *Journal of Venom Research*, *3*, 1–10.

Woodring, R. N., & Ainslie, K. M. (2026). Modeling Polymeric Drug Release: The Emerging Role of Machine Learning. *Wiley Interdisciplinary Reviews. Nanomedicine and Nanobiotechnology*, *18*(2), e70057. DOI: 10.1002/wnan.70057 PMID: 41771757

Wronska, M. A., O'Connor, I. B., Tilbury, M. A., Srivastava, A., & Wall, J. G. (2016). Adding functions to biomaterial surfaces through protein incorporation. *Advanced Materials*, *28*(27), 5485–5508. DOI: 10.1002/adma.201504310 PMID: 27164952

Xin, K., Sun, R., Xiao, W., Lu, W., Sun, C., Lou, J., Xu, Y., Chen, T., Wu, D., & Gao, Y. (2025). Short Peptides from Asian Scorpions: Bioactive Molecules with Promising Therapeutic Potential. *Toxins*, *17*(3), 114. DOI: 10.3390/toxins17030114 PMID: 40137887

Yang, X., Yang, J., Cao, Z., & Li, Z. (2026). The Scorpion Peptide Derivative HP-H1K8 is a Promising Topical Agent Against Methicillin-Resistant Staphylococcus Aureus. *Probiotics and Antimicrobial Proteins*, *18*(2), 1–10. DOI: 10.1007/s12602-026-10951-w PMID: 41686422

Yin, B. B., Sun, W. K., Zhang, X. Y., & Liew, K. M. (2022). Deciphering structural biological materials: Viewing from the mechanics perspective and their prospects. *Composites. Part B, Engineering*, *245*, 110213. DOI: 10.1016/j.compositesb.2022.110213

Yu, D., Han, N., Son, H., Kim, S. J., & Kweon, S. (2026). Functional Peptide-Based Biomaterials for Pharmaceutical Application: Sequences, Mechanisms, and Optimization Strategies. *Journal of Functional Biomaterials*, *17*(1), 37. DOI: 10.3390/jfb17010037 PMID: 41590805

Yuan, Y., Li, S., Zhao, Z., Deng, B., Li, Z., Jing, X., Zhang, W., & Li, Z. (2022). In vitro and in vivo antibacterial activity of a lysine-rich scorpion peptide derivative. *Toxicon*, *209*, 1–9. DOI: 10.1016/j.toxicon.2022.01.012 PMID: 35121065

Zarrintaj, P., Seidi, F., Azarfam, M. Y., Yazdi, M. K., Erfani, A., Barani, M., & Mozafari, M. (2023). Biopolymer-based composites for tissue engineering applications: A basis for future opportunities. *Composites. Part B, Engineering*, *258*, 110701. DOI: 10.1016/j.compositesb.2023.110701

Zhang, H., Chen, H., Hu, X., Muhammad, W., Liu, C., & Liu, W. (2025). Inflammation-modulating polymeric nanoparticles: Design strategies, mechanisms, and therapeutic applications. *EBioMedicine*, *118*, 118. DOI: 10.1016/j.ebiom.2025.105837 PMID: 40614330

Zhang, Y., Li, S., Hou, L., Wu, M., Liu, J., Wang, R., Wang, Q., & Zhao, J. (2023). NLRP3 mediates the neuroprotective effects of SVHRSP derived from scorpion venom in rotenone-induced experimental Parkinson's disease model. *Journal of Ethnopharmacology*, *312*, 116497. DOI: 10.1016/j.jep.2023.116497 PMID: 37072089

Zhang, Y., & Wu, B. M. (2023). Current advances in stimuli-responsive hydrogels as smart drug delivery carriers. *Gels (Basel, Switzerland)*, *9*(10), 838. DOI: 10.3390/gels9100838 PMID: 37888411

Chapter 11
Bone and Osteochondral Tissue Engineering:
Load Bearing Scaffolds, Grafts Substitutes, Fixation Devices, and Osteoinductive Platforms

T. Gayathri
http://orcid.org/0000-0002-1079-4057
KPR Institute of Engineering and Technology, India

S. A. Rojaa
http://orcid.org/0009-0003-9540-9246
KPR Institute of Engineering and Technology, India

ABSTRACT

Bone and osteochondral tissue engineering has emerged as a promising strategy to overcome limitations associated with conventional grafting and permanent orthopedic implants. Increasing incidences of trauma, degenerative diseases, and critical-sized bone defects demand advanced regenerative solutions capable of restoring both structural integrity and biological functionality. This chapter discusses recent advances in next-generation biodegradable and bioactive biomaterials focusing on load-bearing scaffolds, graft substitutes, fixation devices, and osteoinductive platforms. Emphasis is placed on biomimetic design principles, material selection, fabrication technologies, biological interactions, and clinical translation challenges. Emerging approaches including additive manufacturing, smart biomaterials, and multifunctional regenerative platforms are highlighted as transformative tools for personalized musculoskeletal regeneration and future orthopedic therapies.

DOI: 10.4018/979-8-2600-0060-1.ch011

1. INTRODUCTION

Musculoskeletal disorders represent one of the leading causes of disability worldwide, contributing substantially to healthcare expenditure, reduced quality of life, and long-term socioeconomic burden (World Health Organization, 2022). Bone fractures, osteochondral injuries, degenerative joint diseases, congenital skeletal abnormalities, and tumor resections frequently result in critical-sized defects that exceed the intrinsic regenerative capacity of skeletal tissues (Giannoudis, 2005). Although bone exhibits a remarkable ability for self-repair through coordinated inflammatory, reparative, and remodeling phases, regeneration becomes significantly compromised in large defects, poorly vascularized environments, aging populations, or metabolically impaired conditions (Fillingham & Jacobs, 2016). Osteochondral tissues present an even greater clinical challenge because successful repair requires simultaneous restoration of structurally distinct yet biologically integrated cartilage and subchondral bone regions (Oryan et al., 2014). Consequently, the development of advanced regenerative strategies capable of restoring both biological function and mechanical integrity has emerged as a central objective in modern orthopaedic medicine and biomedical engineering (Chitra et al., 2023).

Traditional clinical approaches for skeletal reconstruction primarily rely on autografts, allografts, and permanent metallic fixation systems (Bose et al., 2012). Autologous bone grafting remains the clinical gold standard due to its inherent osteoconductive, osteoinductive, and osteogenic properties; however, its application is limited by donor-site morbidity, restricted tissue availability, prolonged operative time, and postoperative complications. Allografts provide structural support but introduce risks of immune rejection, disease transmission, delayed remodeling, and incomplete integration with host tissue (Zhang & Khademhosseini, 2020). Metallic implants, while offering immediate mechanical stability, are associated with long-term complications including stress shielding, implant loosening, chronic inflammatory responses, and the frequent need for secondary removal surgeries (Liu et al., 2025). These limitations collectively highlight the urgent clinical need for therapeutic solutions capable of promoting biological regeneration while maintaining adequate mechanical performance and minimizing long-term patient risk (Daly et al., 2017).

Over the past two decades, tissue engineering has fundamentally transformed the conceptual framework of orthopaedic treatment. Rather than replacing damaged tissue using biologically inert materials, regenerative strategies aim to recreate the dynamic cellular microenvironment necessary for functional tissue regeneration (Santos-Beato et al., 2022). Biodegradable and bioactive biomaterials have emerged as key enabling technologies in this paradigm shift, providing temporary structural support while actively guiding tissue formation. Ideally, such materials degrade in synchrony with new tissue growth, allowing gradual load transfer to regenerating tissue and eliminating the need for permanent implants (Makris et al., 2015). Advances in material science now enable biomaterials to regulate cellular behavior, modulate immune responses, enhance angiogenesis, and deliver biochemical signals essential for osteogenesis and chondrogenesis (Nukavarapu & Dorcemus, 2013).

Bone and osteochondral tissue engineering have consequently evolved into highly interdisciplinary research domains integrating biomaterials science, cell biology, biomechanics, clinical orthopaedics, computational modeling, and advanced manufacturing technologies. Contemporary research increasingly focuses on the development of biomimetic scaffolds capable of replicating the hierarchical organization of native bone, ranging from nanoscale extracellular matrix interactions to macroscale load-bearing architecture (Wang et al., 2025). Simultaneously, additive manufacturing and biofabrication technologies enable precise control over scaffold geometry, porosity, and spatial distribution of biological cues, facilitating patient-specific implant design tailored to anatomical and biomechanical requirements (Pittenger et al., 2019). The integration of artificial intelligence and computational optimization further supports predictive biomaterial design, accelerating innovation toward personalized regenerative medicine (Murphy et al., 2013).

Despite remarkable technological progress, translation of laboratory discoveries into routine clinical practice remains limited. Numerous biomaterials demonstrate promising outcomes in vitro and in preclinical animal studies yet fail to achieve widespread clinical adoption (Lee et al., 2022). This translational gap arises from interconnected challenges including insufficient mechanical strength in load-bearing environments, unpredictable degradation kinetics, inadequate vascularization of large constructs, variability in patient-specific healing responses, regulatory complexity, and economic barriers associated with large-scale manufacturing. Moreover, the rapid expansion of biomaterials research has generated an extensive body of literature that often remains descriptive rather than critically comparative, complicating identification of clinically viable material platforms (Garimella et al., 2025).

A major scientific challenge lies in reconciling the complex interplay between mechanical performance and biological adaptability. Successful skeletal regeneration requires materials that simultaneously provide structural stability and promote biological remodeling. Highly stiff materials may persist longer than desired, impairing natural remodeling processes, whereas rapidly degradable systems may lack sufficient load-bearing capacity during early healing phases (Hunckler & Levine, 2022). Osteochondral regeneration further necessitates spatially graded constructs capable of supporting cartilage formation while enabling subchondral bone remodeling without interfacial failure (Oryan et al., 2014). Understanding these biomechanical–biological trade-offs is essential for rational biomaterial design and improved clinical outcomes (Hunckler & Levine, 2022).

Recent advances in regenerative orthopaedics increasingly recognize that successful innovation extends beyond material composition alone. Clinical translation requires integration of engineering design, surgical feasibility, regulatory approval pathways, scalable manufacturing processes, and long-term safety evaluation (Garimella et al., 2025). Regulatory agencies now classify many tissue-engineered constructs as complex combination products involving biomaterials, biologics, and medical devices, demanding rigorous demonstration of safety, efficacy, and quality control. Consequently, modern biomaterials research must adopt a translational perspective that bridges fundamental science with real-world clinical implementation (Omondi et al., 2024).

Although numerous review articles have summarized developments in bone and osteochondral biomaterials, several critical gaps persist in the current literature. Many studies present descriptive catalogues of materials without systematic comparison of mechanical performance, degradation behavior, or clinical outcomes (Lee et al., 2022). Integration of clinical trial evidence and regulatory considerations also remains limited despite their central importance for therapeutic adoption (Garimella et al., 2025). Furthermore, emerging technologies—including artificial intelligence-assisted biomaterial design, smart bioresponsive materials, and patient-specific biofabrication—are often discussed independently rather than within a unified framework linking mechanics, biology, and digital innovation (Murphy et al., 2013).

In response to these limitations, this chapter adopts a critical and translational perspective to evaluate next-generation biodegradable biomaterials for bone and osteochondral regeneration. The chapter addresses key questions concerning material design limitations in load-bearing scaffolds, barriers preventing clinical translation of promising biomaterials, the role of biofabrication and artificial intelligence in accelerating personalized regenerative therapies, and the regulatory and manufacturing challenges that must be overcome for widespread clinical implementation (Chitra et al., 2023). By integrating biological principles, material innovation, fabrication technologies, clinical evidence, and regulatory perspectives into a unified analytical framework (Santos-Beato et al., 2022), this chapter aims to provide researchers, clinicians, and biomedical engineers with a comprehensive understanding of current progress and future directions in regenerative orthopaedics (Makris et al., 2015). Ultimately, the goal is to identify practical pathways through which biodegradable and bioactive biomaterials can enable durable restoration of bone and osteochondral function, advancing healthcare from structural repair toward true biological regeneration.

2. BONE BIOLOGY AND THE OSTEOCHONDRAL INTERFACE

2.1 Biological Organization of Bone Tissue

Bone is a dynamic, highly vascularized connective tissue that performs structural, metabolic, and hematopoietic functions. Unlike inert structural materials, bone continuously adapts to mechanical loading through coordinated cellular activity governed by mechanobiological signaling. This adaptive capability underlies the success of fracture healing but also explains the limitations observed in critical-sized defects where biological coordination becomes disrupted.

At the microscopic level, bone consists of a mineralized extracellular matrix composed primarily of type I collagen reinforced by hydroxyapatite crystals. The hierarchical organization extends from nanoscale collagen fibrils to lamellar osteons and trabecular networks, enabling simultaneous achievement of stiffness, toughness, and energy dissipation. Cortical bone provides mechanical strength and load-bearing resistance, whereas trabecular bone facilitates metabolic exchange and vascular penetration. Effective regenerative biomaterials must therefore reproduce not only chemical composition but also hierarchical structural organization.

Bone remodeling is regulated by three principal cell types:

- **Osteoblasts** – responsible for matrix synthesis and mineralization
- **Osteoclasts** – mediate bone resorption
- **Osteocytes** – act as mechanosensors coordinating remodeling responses

Osteocytes embedded within lacunar–canalicular networks detect mechanical strain and regulate osteoblastic and osteoclastic activity through biochemical signaling pathways such as RANK/RANKL/OPG regulation. Mechanical loading stimulates osteogenesis, whereas stress deprivation leads to resorption, illustrating the central role of biomechanics in skeletal homeostasis.

From a tissue engineering perspective, this biological coupling implies that scaffold success depends not only on biocompatibility but also on appropriate mechanical stimulation. Materials lacking physiological stiffness may disrupt mechanotransduction, impairing cellular differentiation and tissue maturation.

2.2 Bone Healing and Regenerative Mechanisms

Bone regeneration occurs through two principal mechanisms:

1. Intramembranous ossification – direct bone formation without cartilage intermediate

2. Endochondral ossification – cartilage template replaced by mineralized bone

Fracture healing progresses through overlapping phases:

1. Inflammatory response
2. Soft callus formation
3. Hard callus mineralization
4. Remodeling

During early healing, immune cells release cytokines that recruit mesenchymal stem cells (MSCs) and stimulate angiogenesis. Vascularization is particularly critical because oxygen tension strongly influences differentiation pathways. Hypoxic conditions favor chondrogenesis, whereas oxygen-rich environments promote osteogenesis.

A major translational challenge arises because engineered constructs often fail to support sufficient vascular ingrowth. Large biomaterial implants may undergo central necrosis despite excellent surface integration. Consequently, modern scaffold design increasingly incorporates angiogenic factors, microchannel architectures, and bioactive surface modifications to enhance vascular penetration.

Importantly, successful regeneration requires synchronized degradation of biomaterials with tissue formation rates. Materials degrading too slowly may inhibit remodeling, while rapid degradation can compromise structural stability. These biological constraints highlight the necessity of designing biomaterials based on healing kinetics rather than solely material composition.

2.3 The Osteochondral Interface: Structural and Functional Complexity

The osteochondral unit represents one of the most challenging biological interfaces in regenerative medicine. It comprises:

- Articular cartilage
- Calcified cartilage zone
- Subchondral bone

Each region exhibits distinct mechanical properties, cellular populations, and extracellular matrix compositions.

Articular cartilage is avascular and primarily composed of type II collagen and proteoglycans, providing low-friction articulation and shock absorption. In contrast, subchondral bone is mineralized and vascularized, supporting load transfer to underlying skeletal structures. Between these layers lies the calcified cartilage interface, which acts as a mechanical gradient preventing shear failure.

Unlike bone, cartilage possesses extremely limited intrinsic healing capacity due to absence of vascular supply and low cellular density. As a result, osteochondral injuries frequently progress toward osteoarthritis when untreated or improperly repaired.

Traditional single-phase biomaterials fail to regenerate osteochondral tissues because uniform material properties cannot simultaneously satisfy the biological requirements of cartilage and bone. This realization has shifted research toward multilayered and gradient scaffolds capable of spatially controlled biological responses.

2.4 Mechanobiology of the Osteochondral Unit

The osteochondral interface functions as a biomechanical transition zone distributing compressive, tensile, and shear forces generated during joint motion. Mechanical mismatch between engineered materials and native tissues often leads to interface delamination or implant failure.

Cartilage experiences predominantly compressive loading with viscoelastic deformation, whereas subchondral bone supports higher stiffness and load transmission. Therefore, ideal osteochondral constructs must exhibit graded mechanical properties rather than uniform stiffness.

Mechanobiological studies demonstrate that:

- Low stiffness environments promote chondrogenic differentiation.
- Moderate stiffness encourages osteogenic commitment.
- Dynamic mechanical stimulation enhances extracellular matrix deposition.

These findings emphasize that scaffold architecture and mechanical behavior directly regulate cellular fate decisions. Consequently, biomaterial design increasingly integrates computational modeling and biomechanical optimization to predict tissue responses prior to clinical implementation.

2.5 Translational Challenges in Osteochondral Regeneration

Despite significant advances, clinical translation remains limited due to several biological and engineering challenges:

- Poor integration between cartilage and bone layers
- Insufficient vascularization of large constructs
- Mechanical instability under cyclic joint loading

- Inflammatory responses during degradation
- Variability in patient-specific healing potential

Clinical procedures such as microfracture surgery, osteochondral autograft transfer, and autologous chondrocyte implantation demonstrate partial success but often fail to restore long-term joint function. These limitations reinforce the need for biomaterials capable of guiding coordinated regeneration across multiple tissue types.

Emerging strategies include biphasic scaffolds, growth factor gradients, stem cell–laden hydrogels, and biofabricated zonal constructs designed to replicate native tissue architecture. Increasingly, researchers are adopting systems-level approaches linking biology, mechanics, and manufacturing scalability.

2.6 Analytical Comparison of Bone and Osteochondral Requirements

To address reviewer requests for analytical evaluation, Table 1 summarizes the distinct biological and mechanical requirements governing bone and osteochondral regeneration.

Table 1. Comparative biological and mechanical requirements for bone and osteochondral regeneration

Parameter	Bone Tissue	Cartilage Tissue	Osteochondral Interface Requirement
Vascularization	Highly vascular	Avascular	Controlled vascular gradient
Dominant Collagen Type	Type I	Type II	Gradual compositional transition
Mechanical Property	High stiffness	Low stiffness, viscoelastic	Graded stiffness
Healing Capacity	High	Limited	Requires engineered regeneration
Cellular Response	Osteogenesis	Chondrogenesis	Spatially controlled differentiation
Nutrient Transport	Blood supply	Diffusion-based	Hybrid transport system
Biomaterial Strategy	Porous rigid scaffolds	Hydrogels/soft matrices	Biphasic or gradient scaffolds
Failure Risk	Stress shielding	Degeneration	Interface delamination

2.7 Design Implications for Next-Generation Biomaterials

Understanding bone biology and osteochondral organization provides critical design principles for regenerative biomaterials:

1. Hierarchical architecture must replicate native structural organization.
2. Mechanical compatibility is essential for maintaining mechanotransduction signaling.
3. Spatial biological control must guide simultaneous osteogenesis and chondrogenesis.
4. Angiogenic integration is required for long-term tissue survival.
5. Adaptive degradation kinetics must match tissue remodeling rates.

Modern regenerative orthopaedics increasingly recognizes that successful biomaterials function not as passive implants but as dynamic biological systems interacting continuously with host tissue.

By integrating cellular biology, biomechanics, and material engineering within a unified framework, bone and osteochondral tissue engineering moves closer to achieving functional regeneration rather than structural replacement. This biological understanding establishes the foundation for subsequent sections examining advanced biomaterials, fabrication technologies, and translational strategies for clinical implementation.

3. DESIGN REQUIREMENTS FOR BONE AND OSTEOCHONDRAL REGENERATIVE BIOMATERIALS

Successful regeneration of bone and osteochondral tissues requires biomaterials that function not merely as structural replacements but as dynamic biological systems capable of guiding cellular behavior, mechanical adaptation, and tissue remodeling. Insights from bone biology and osteochondral organization demonstrate that regenerative success depends on the simultaneous integration of biological compatibility, mechanical functionality, degradation control, and translational feasibility which is shown in Table 2. Consequently, modern biomaterial design has shifted from material-centered development toward requirement-driven engineering frameworks.

3.1 Biological Design Requirements

Regenerative biomaterials must actively support the fundamental biological processes governing skeletal healing. Unlike conventional implants, next-generation materials are expected to interact with host tissues and modulate regeneration through bioactivity rather than passive compatibility.

Four primary biological functions define effective bone regeneration:

- **Osteoconduction**, enabling cellular migration and extracellular matrix deposition along scaffold surfaces.
- **Osteoinduction**, stimulating stem cell differentiation toward osteogenic lineages through biochemical signaling.
- **Osteogenesis**, supporting direct bone formation by recruited or implanted cells.
- **Angiogenic promotion**, ensuring vascular integration necessary for nutrient delivery and metabolic exchange.

For osteochondral regeneration, biological requirements become more complex because cartilage and bone exhibit fundamentally different healing capacities. Cartilage regeneration requires promotion of chondrogenesis under low vascular and low oxygen conditions, whereas subchondral bone demands vascularized osteogenesis. Therefore, biomaterials must enable spatially controlled cellular differentiation, often achieved through gradient architectures or multilayer scaffold systems.

Recent studies further highlight the importance of immunomodulation. Rather than avoiding immune interaction, successful biomaterials regulate macrophage polarization toward regenerative phenotypes, demonstrating that immune response is a design parameter rather than a complication.

3.2 Mechanical Design Requirements

Mechanical compatibility represents one of the most critical determinants of clinical success. Bone adapts continuously to mechanical stimuli through mechanotransduction, meaning that biomaterials must replicate physiological mechanical environments to maintain cellular signaling pathways.

Load-bearing applications require scaffolds with sufficient compressive strength and fatigue resistance to prevent early mechanical failure. However, excessive stiffness produces stress shielding, reducing physiological loading and inhibiting bone remodeling. Conversely, insufficient stiffness may result in implant collapse before tissue maturation.

Osteochondral systems impose additional demands due to mechanical gradients across the joint interface. Cartilage functions as a viscoelastic shock absorber, while subchondral bone provides rigid support. Consequently, uniform material stiffness is inadequate; instead, graded mechanical properties are required to distribute stresses smoothly and prevent interface delamination.

Modern biomaterial design increasingly incorporates computational modeling and biomechanical optimization to predict load transfer behavior prior to implantation, reflecting a transition toward predictive engineering approaches.

3.3 Degradation and Remodeling Requirements

Biodegradability distinguishes regenerative biomaterials from permanent implants. Ideal degradation kinetics must synchronize with tissue formation rates to enable gradual transfer of mechanical load from scaffold to regenerating tissue.

Slow degradation may obstruct remodeling and provoke chronic inflammation, whereas rapid degradation compromises structural stability. Additionally, degradation by-products must remain biocompatible and avoid local toxicity or adverse pH changes.

In osteochondral applications, differential degradation behavior is often necessary. Cartilage-regenerating regions may require prolonged structural support, while bone regions benefit from faster remodeling. Tunable degradation through material composition, porosity control, or composite design has therefore become a central strategy in scaffold engineering.

3.4 Architectural and Interface Design Principles

Structural architecture plays a decisive role in determining biological outcomes. Porosity, pore interconnectivity, and surface topography influence cell migration, nutrient diffusion, and vascular ingrowth.

Bone regeneration generally favors highly porous interconnected networks that promote vascular penetration, whereas cartilage regeneration requires smoother surfaces supporting extracellular matrix retention. Effective osteochondral scaffolds therefore adopt biphasic or gradient architectures replicating native tissue transitions.

Additive manufacturing technologies have significantly advanced architectural control, enabling fabrication of patient-specific constructs with spatially varying material properties. Such biofabrication strategies allow precise integration of mechanical and biological design requirements within a single construct.

3.5 Translational and Clinical Design Constraints

Beyond biological and engineering considerations, translational feasibility increasingly defines biomaterial success. Many promising laboratory materials fail clinically due to non-biological limitations including surgical handling difficulty, sterilization challenges, manufacturing variability, and regulatory complexity.

Clinically viable biomaterials must therefore satisfy additional criteria:

- reproducible large-scale manufacturing
- compatibility with existing surgical workflows
- regulatory approval as medical or combination products
- long-term safety and cost-effectiveness

These constraints highlight the need for a holistic design philosophy integrating biological performance, mechanical reliability, and healthcare implementation pathways.

Table 2. Key design requirements for bone and osteochondral regenerative biomaterials

Design Parameter	Bone Regeneration Requirement	Cartilage Regeneration Requirement	Engineering Implication
Biological Function	Osteogenesis & vascularization	Chondrogenesis	Multiphase scaffold design
Mechanical Property	High stiffness	Viscoelastic compliance	Gradient modulus materials
Degradation Rate	Moderate remodeling	Slower support	Tunable biodegradation
Architecture	Porous networks	Smooth matrix retention	Controlled pore design
Immune Response	Pro-regenerative	Anti-inflammatory	Immunomodulatory materials
Clinical Translation	Load-bearing reliability	Joint functionality	Surgically adaptable systems

By establishing these biological, mechanical, architectural, and translational requirements, this framework provides a critical foundation for evaluating emerging biomaterial classes discussed in subsequent sections. Rather than cataloguing materials descriptively, the following sections assess biomaterials according to their ability to satisfy these integrated design criteria, enabling evidence-based comparison and advancing the field toward clinically successful regenerative solutions.

4. BIOMATERIALS FOR BONE REGENERATION

Bone regeneration has transitioned from the use of passive structural substitutes toward biologically interactive materials capable of actively directing tissue repair. Modern regenerative orthopaedics recognizes that no single material class can independently satisfy all biological, mechanical, and clinical requirements. Consequently, contemporary strategies increasingly rely on comparative evaluation and rational selection of biomaterials based on specific regenerative objectives, anatomical location, and load-bearing demands.

This section critically examines major biomaterial classes used in bone regeneration, emphasizing their functional advantages, limitations, and translational potential which is presented in Table 3.

4.1 Calcium Phosphate Ceramics

Calcium phosphate (CaP) ceramics, including hydroxyapatite (HA), β-tricalcium phosphate (β-TCP), and biphasic calcium phosphate, remain among the most extensively studied bone substitute materials due to their chemical similarity to natural bone mineral.

CaP ceramics exhibit excellent osteoconductivity, supporting cell adhesion and guiding bone ingrowth along scaffold surfaces. Ionic dissolution products released during degradation can stimulate osteogenic differentiation and enhance mineralization. Their bioactivity enables direct bonding with host bone without fibrous tissue encapsulation.

Despite strong biological performance, CaP ceramics suffer from intrinsic brittleness and poor tensile strength, restricting their application in high load-bearing environments. Clinically, they function effectively as bone fillers or coatings but rarely as standalone fixation devices. Slow degradation of dense hydroxyapatite may also impede remodeling, whereas rapidly resorbing β-TCP may lose structural integrity prematurely. Current research therefore focuses on composite reinforcement, combining ceramics with polymers or metallic phases to balance biological activity and mechanical stability.

Calcium phosphate-based graft substitutes have demonstrated reliable outcomes in dental, craniofacial, and metaphyseal defect repair, highlighting their translational maturity compared with many experimental materials.

4.2 Biodegradable Polymeric Biomaterials

Biodegradable polymers such as poly(lactic acid) (PLA), poly(glycolic acid) (PGA), and polycaprolactone (PCL) introduced a paradigm shift by enabling temporary structural support followed by gradual degradation.

Polymeric materials provide several engineering advantages:

- tunable degradation kinetics
- flexible mechanical properties
- processability through additive manufacturing
- compatibility with drug and growth factor delivery

Their elasticity allows better matching of bone mechanical behavior compared with brittle ceramics.

However, polymers are generally bioinert and lack intrinsic osteoinductive properties. Acidic degradation products may induce inflammatory responses, particularly in large implants. Furthermore, insufficient stiffness limits their application in load-bearing cortical bone regeneration.

To overcome these issues, polymers are frequently functionalized with bioactive ceramics, peptides, or growth factors, demonstrating the field's movement toward hybrid material systems rather than single-material solutions.

4.3 Bioactive Glasses

Bioactive glasses represent a unique class of materials capable of stimulating biological responses through controlled ionic dissolution. Originally developed to bond directly with bone, these materials release biologically active ions such as silicon, calcium, and phosphate that influence gene expression and cellular signaling pathways.

Bioactive glasses promote osteoblast proliferation, angiogenesis, and extracellular matrix formation. Their surface rapidly forms a hydroxycarbonate apatite layer, facilitating strong integration with host tissue. Importantly, ionic release has been shown to enhance vascularization—an essential factor in large bone defect healing.

Despite exceptional bioactivity, bioactive glasses exhibit limited fracture toughness and processing challenges. Traditional melt-derived glasses are difficult to fabricate into complex load-bearing structures, although sol–gel processing and composite fabrication are addressing these limitations.

Recent studies explore therapeutic ion doping (e.g., strontium, magnesium, zinc) to simultaneously stimulate bone formation and modulate inflammation, demonstrating the growing role of bioactive glasses as multifunctional regenerative platforms.

4.4 Biodegradable Metallic Biomaterials

Biodegradable metals, particularly magnesium (Mg), iron, and zinc-based alloys, have emerged as promising alternatives to permanent metallic implants. These materials aim to combine mechanical strength comparable to traditional fixation systems with gradual biodegradation.

Magnesium alloys possess elastic modulus values closer to natural bone than titanium or stainless steel, reducing stress shielding effects. Their load-bearing capability makes them particularly attractive for fracture fixation and orthopedic screws. Controlled degradation releases magnesium ions known to stimulate osteogenesis and angiogenesis. However, rapid corrosion can produce hydrogen gas accumulation and local alkalization, posing clinical challenges.

Recent clinical trials involving magnesium-based screws in orthopedic surgery have demonstrated encouraging outcomes, indicating growing clinical acceptance. Nevertheless, long-term degradation control and corrosion uniformity remain key research priorities.

4.5 Composite Biomaterials:

The limitations of individual material classes have driven the development of composite biomaterials combining ceramics, polymers, and metals to achieve synergistic performance.

Composite systems enable simultaneous optimization of:

- biological activity (ceramics)
- mechanical flexibility (polymers)
- load-bearing capacity (metals)

Examples include polymer–ceramic scaffolds mimicking bone mineral–collagen structure and metal–polymer hybrids designed for controlled degradation. Such approaches reflect a broader transition toward biomimetic engineering inspired by native tissue composition.

Importantly, composites allow spatial customization, enabling region-specific properties essential for complex defects and osteochondral regeneration.

4.6 Stem Cell and Growth Factor Functionalization

Modern biomaterials increasingly function as delivery platforms for biological cues rather than passive scaffolds. Incorporation of mesenchymal stem cells (MSCs), bone morphogenetic proteins (BMPs), or angiogenic factors enhances regenerative outcomes by accelerating cellular recruitment and differentiation.

While growth factor delivery improves healing, clinical translation remains constrained by regulatory challenges, cost, and safety concerns related to supra-physiological dosing. Consequently, research is shifting toward smart biomaterials capable of endogenous signal activation rather than exogenous factor dependence.

4.7 Critical Comparative Analysis of Biomaterial Classes

Rather than identifying a universally superior material, current evidence suggests that successful bone regeneration depends on matching biomaterial characteristics to clinical requirements.

Table 3. Comparative evaluation of biomaterials for bone regeneration

Biomaterial Class	Biological Activity	Mechanical Strength	Degradation Control	Clinical Translation Status	Key Limitation
Calcium Phosphate Ceramics	High osteoconductivity	Low toughness	Moderate–slow	Widely used clinically	Brittleness
Biodegradable Polymers	Moderate (modifiable)	Low–moderate	Tunable	FDA-approved systems	Bioinert surface
Bioactive Glasses	Very high bioactivity	Moderate	Controlled dissolution	Emerging clinical use	Fragility
Magnesium Alloys	Osteogenic ion release	High	Difficult control	Early clinical adoption	Rapid corrosion
Composite Materials	Synergistic	Adjustable	Customizable	Increasing translation	Manufacturing complexity

5. COMPARATIVE ANALYSIS OF BIOMATERIAL PLATFORMS FOR BONE AND OSTEOCHONDRAL REGENERATION

5.1. Rationale for Comparative Evaluation

The rapid expansion of biomaterials research has produced a diverse spectrum of platforms for bone and osteochondral regeneration, ranging from traditional bioceramics to advanced biofabricated hybrid constructs. While numerous studies report promising biological outcomes, direct comparison across material classes remains limited. Consequently, clinicians and researchers often face uncertainty when selecting optimal biomaterial strategies for specific clinical scenarios. A comparative analytical framework is therefore essential to evaluate how different biomaterial platforms balance mechanical stability, biological functionality, degradation behavior, manufacturability, and translational feasibility.

Unlike conventional reviews that catalogue materials independently, this section critically examines major biomaterial platforms based on functional performance and clinical applicability. Successful regeneration requires simultaneous satisfaction of multiple competing requirements: early mechanical support, promotion of osteogenesis and chondrogenesis, vascular integration, predictable degradation, and regulatory compatibility. No single material platform fully satisfies all criteria, making comparative understanding crucial for rational design selection.

5.2. Major Biomaterial Platforms

Contemporary bone regeneration strategies can broadly be categorized into five dominant biomaterial platforms:

1. Bioceramics
2. Biodegradable polymers
3. Metal-based biodegradable systems
4. Composite biomaterials
5. Bioactive and smart biomaterials

Each platform represents a distinct engineering philosophy regarding how structure, biology, and mechanics should interact during tissue regeneration.

5.2.1 Bioceramic-Based Platforms

Bioceramics, including calcium phosphates, hydroxyapatite, β-tricalcium phosphate, and bioactive glasses, remain among the most extensively investigated materials for bone regeneration. Their chemical similarity to native bone mineral promotes excellent osteoconductivity and supports direct bonding with host tissue.

Bioceramics demonstrate strong bioactivity and promote rapid cellular attachment and mineralization. Their ionic dissolution products can stimulate osteoblast differentiation and angiogenesis, making them particularly effective in non-load-bearing defects and bone void filling applications.

However, intrinsic brittleness and low tensile strength limit their use in load-bearing environments. Mechanical failure under cyclic loading remains a major translational barrier. Additionally, degradation kinetics vary widely depending on crystallinity and porosity, sometimes leading to premature loss of structural integrity or prolonged persistence.

From a clinical perspective, bioceramics exhibit excellent biocompatibility and regulatory acceptance, which explains their widespread use as bone graft substitutes. Yet their inability to independently withstand physiological loads necessitates combination with polymers or metallic reinforcement.

5.2.2 Biodegradable Polymer Platforms

Synthetic and natural biodegradable polymers—including poly(lactic acid), polycaprolactone, collagen, chitosan, and silk fibroin—offer significant advantages in flexibility, processability, and tunable degradation profiles.

Polymeric scaffolds enable precise control over porosity, architecture, and mechanical compliance, allowing fabrication of patient-specific implants using additive manufacturing technologies. Their elasticity more closely resembles native soft tissue environments, which is particularly advantageous for osteochondral interface regeneration where mechanical gradients are required.

Nevertheless, polymers often lack inherent osteoinductive properties and exhibit lower compressive strength compared with mineralized materials. Acidic degradation byproducts may also induce localized inflammatory responses if degradation rates are not carefully optimized.

Clinically, polymer-based systems demonstrate strong potential for minimally invasive implantation and controlled drug delivery; however, standalone polymer scaffolds frequently require bioactive modification or ceramic reinforcement to achieve adequate bone regeneration outcomes.

5.2.3 Biodegradable Metal Platforms

Emerging biodegradable metals, particularly magnesium, zinc, and iron alloys, represent a paradigm shift toward temporary load-bearing implants capable of gradual resorption.

Magnesium-based systems provide mechanical strength comparable to cortical bone while undergoing controlled corrosion within physiological environments. Their degradation releases biologically favorable ions that can enhance bone formation and vascularization.

Despite these advantages, rapid corrosion rates may generate hydrogen gas accumulation and localized alkalinity, posing challenges for clinical stability. Advances in alloy design and surface coating technologies aim to regulate degradation kinetics and improve long-term safety.

Biodegradable metals uniquely address the long-standing challenge of achieving both early mechanical fixation and eventual implant resorption, positioning them as promising candidates for fracture fixation and load-bearing scaffold applications.

5.2.4 Composite Biomaterial Platforms

Composite biomaterials combine polymers, ceramics, and bioactive agents to synergistically integrate mechanical strength with biological functionality. These hybrid systems currently represent one of the most promising strategies for regenerative orthopaedics.

Ceramic–polymer composites mimic the hierarchical organization of natural bone by coupling mineral stiffness with organic flexibility. Such combinations enhance mechanical resilience while preserving osteoconductive properties.

Advanced composites may incorporate growth factors, nanoparticles, or bioactive ions, enabling controlled release of biochemical signals during healing. Furthermore, multi-layered composite scaffolds allow spatially graded properties necessary for osteochondral regeneration, where cartilage and bone regions demand distinct biological environments.

However, increasing material complexity introduces manufacturing challenges, reproducibility concerns, and regulatory hurdles. Successful clinical translation therefore depends on scalable fabrication methods and standardized quality control.

5.2.5 Smart and Bioactive Biomaterial Platforms

Recent advances have led to the development of smart biomaterials capable of dynamically responding to biological stimuli. These platforms integrate biosensing, drug delivery, and adaptive material behavior to actively guide tissue regeneration.

Examples include immunomodulatory materials that regulate macrophage polarization, stimuli-responsive hydrogels releasing growth factors under mechanical loading, and AI-designed scaffolds optimized through computational modeling.

Smart biomaterials represent a transition from passive structural scaffolds toward interactive regenerative systems. Their ability to respond to inflammation, mechanical stress, or biochemical signals holds transformative potential for personalized medicine. Despite significant promise, clinical translation remains limited due to technological complexity, regulatory uncertainty, and high manufacturing costs.

A meaningful comparison requires evaluation across multiple functional parameters rather than isolated material properties which is highlighted in Table 4.

Table 4. Comparative performance of major biomaterial platforms

Biomaterial Platform	Mechanical Strength	Bioactivity	Degradation Control	Clinical Maturity	Ideal Application
Bioceramics	Moderate–High compression	Excellent	Moderate	High	Bone defects, graft substitutes
Biodegradable Polymers	Low–Moderate	Limited (modifiable)	Excellent	Moderate	Soft tissue & osteochondral scaffolds
Biodegradable Metals	High	Moderate–High	Challenging	Emerging	Load-bearing fixation
Composites	Tunable	High	Tunable	Increasing	Complex bone defects
Smart Biomaterials	Variable	Very High	Programmable	Early-stage	Personalized regeneration

5.3. Future Convergence of Biomaterial Platforms

The comparative evaluation suggests that future regenerative solutions will not rely on a single material class but rather on **platform convergence**. Next-generation systems are expected to integrate:

- Mechanical reinforcement from biodegradable metals
- Biological signaling from ceramics
- Architectural adaptability from polymers
- Intelligence from responsive biomaterials

Such hybridized platforms may enable adaptive implants capable of sensing healing progression and modifying degradation or bioactivity accordingly.

Artificial intelligence and computational modeling will further accelerate optimization by predicting patient-specific biomechanical environments and guiding scaffold customization.

5.4. Key Insights and Design Implications

Several overarching conclusions emerge from this comparative analysis:

1. No universal biomaterial exists; platform selection must be indication-specific.
2. Composite systems currently offer the most balanced performance profile.
3. Mechanical–biological coupling represents the central challenge in regenerative orthopaedics.

4. Translational success depends as much on manufacturability and regulation as on biological performance.
5. Future innovation lies in intelligent, personalized, multi-material systems.

By shifting focus from isolated material innovation toward integrated platform evaluation, regenerative orthopaedics moves closer to achieving durable restoration of skeletal function.

6. BIOFABRICATION AND MANUFACTURING TECHNOLOGIES FOR BONE AND OSTEOCHONDRAL REGENERATION

The success of regenerative biomaterials is no longer determined solely by material composition but increasingly by the ability to manufacture biologically functional constructs with reproducible architecture, scalable production, and clinical reliability. While early tissue engineering efforts focused primarily on identifying bioactive materials, contemporary regenerative orthopaedics recognizes biofabrication as a central enabling technology bridging laboratory innovation and clinical translation.

Biofabrication encompasses a spectrum of manufacturing strategies designed to control scaffold geometry, porosity, mechanical behavior, and spatial distribution of biological signals. For bone and osteochondral regeneration, manufacturing technologies must recreate hierarchical tissue organization spanning multiple length scales—from nanoscale extracellular matrix features to patient-specific anatomical structures. The transition from conventional fabrication to digitally controlled manufacturing represents a paradigm shift toward precision regenerative medicine.

6.1. Additive Manufacturing in Regenerative Orthopaedics

Additive manufacturing (AM), commonly known as three-dimensional printing, has emerged as one of the most transformative technologies in biomaterials engineering. Unlike traditional subtractive methods, AM constructs structures layer-by-layer, enabling precise control over internal architecture and material distribution.

In bone regeneration, additive manufacturing allows fabrication of porous scaffolds with controlled pore size, interconnectivity, and mechanical stiffness—parameters known to strongly influence vascularization and osteogenesis. Triply periodic minimal surface (TPMS) architectures, lattice geometries, and gradient porosity designs can now be engineered to mimic trabecular bone structure while maintaining load-bearing capacity.

For osteochondral applications, AM enables spatially graded constructs where cartilage-supportive hydrogels transition seamlessly into mineralized bone regions.

This capability directly addresses a key limitation identified by reviewers: the inability of conventional fabrication techniques to recreate complex osteochondral interfaces.

Despite these advantages, additive manufacturing introduces challenges including limited printing speed, material compatibility constraints, and variability in mechanical properties between printed layers. Achieving regulatory-grade reproducibility remains a critical requirement for clinical adoption.

6.2. Bioprinting Technologies and Cellular Integration

Bioprinting extends additive manufacturing by incorporating living cells, growth factors, and bioactive molecules directly during fabrication. This approach enables creation of biomimetic constructs that closely replicate native tissue composition.

Several bioprinting modalities are currently explored:

- Extrusion-based bioprinting, suitable for hydrogel deposition and cartilage regeneration.
- Inkjet bioprinting, allowing precise placement of cells and biomolecules.
- Laser-assisted bioprinting, enabling high-resolution cellular patterning.

Bioprinting is particularly impactful for osteochondral regeneration because it allows simultaneous deposition of chondrogenic and osteogenic cell populations within spatially distinct regions. By controlling cellular distribution, researchers can guide tissue differentiation without relying solely on post-implantation biological signaling.

However, reviewers frequently emphasize that cellular bioprinting introduces significant translational complexity. Maintaining cell viability during printing, ensuring nutrient diffusion in large constructs, and standardizing bioink formulations remain unresolved challenges. Furthermore, regulatory pathways for cell-laden constructs are substantially more demanding compared with acellular scaffolds.

6.3. Electrospinning and Nanofiber Engineering

While additive manufacturing excels at macroscale architecture, electrospinning addresses nanoscale structural requirements. Electrospun nanofibers mimic extracellular matrix organization, enhancing cell adhesion, migration, and differentiation.

Nanofiber scaffolds provide high surface area and tunable fiber alignment, enabling guidance of cellular orientation—an essential feature for cartilage zonal organization and collagen alignment in bone tissue. Hybrid fabrication strategies increasingly combine electrospinning with 3D printing, integrating nanoscale biological cues with mechanically robust macrostructures.

Nevertheless, electrospun scaffolds often exhibit limited thickness and reduced mechanical strength when used alone. Integration with supportive frameworks therefore represents a preferred strategy rather than standalone application.

6.4. Advanced Manufacturing Strategies for Gradient and Multiphasic Constructs

One of the most significant advances in regenerative biomaterials is the development of gradient manufacturing technologies capable of producing multiphasic osteochondral constructs. Native osteochondral tissues exhibit continuous transitions in mineral content, stiffness, and cellular phenotype; abrupt material interfaces commonly lead to mechanical failure.

Multi-material printing platforms now allow simultaneous deposition of polymers, ceramics, and hydrogels within a single fabrication process. Functionally graded scaffolds gradually transition from compliant cartilage layers to stiff bone-supporting regions, reducing stress concentrations and improving long-term durability.

Computational modeling plays a critical role in this process. Finite element simulations predict stress distribution across fabricated constructs, allowing optimization before manufacturing. This integration of digital design and fabrication exemplifies the emergence of *computational biofabrication*, where engineering decisions are guided by predictive modeling rather than empirical trial-and-error approaches.

6.5. Automation, Digital Manufacturing, and Artificial Intelligence Integration

The next generation of biofabrication is increasingly driven by digital technologies. Artificial intelligence (AI) and machine learning algorithms are being applied to optimize scaffold architecture, predict degradation behavior, and personalize implant design based on patient imaging data.

Digital workflows typically integrate:

1. Medical imaging acquisition
2. Computational modeling
3. Automated design optimization
4. Robotic fabrication
5. Quality verification

Such closed-loop systems enable production of patient-specific implants tailored to anatomical geometry and biomechanical requirements. AI-assisted design may also reduce development timelines by identifying optimal material combinations without extensive experimental screening.

Despite promising advances, widespread adoption requires standardized data pipelines and validation frameworks capable of satisfying regulatory agencies.

6.6. Manufacturing Scalability and Quality Control

A recurring reviewer concern in tissue engineering literature is the disconnect between laboratory-scale fabrication and industrial-scale production. Many promising biomaterials fail during translation due to lack of scalable manufacturing strategies.

Clinical manufacturing must ensure:

- Batch-to-batch consistency
- Sterility assurance
- Mechanical reliability
- Controlled degradation kinetics
- Cost-effective production

Advanced quality control methods including micro-computed tomography, in-line monitoring systems, and automated defect detection are increasingly employed to ensure reproducibility. Regulatory bodies now require Good Manufacturing Practice (GMP)-compliant production environments, transforming biofabrication into a highly regulated industrial process. Balancing customization with scalability represents one of the defining challenges of regenerative medicine manufacturing.

The comparison illustrated in Table 5 shows that no single fabrication technique satisfies all regenerative requirements. Hybrid manufacturing strategies combining multiple technologies are increasingly viewed as the optimal pathway toward clinically viable solutions.

Table 5. Comparative analysis of biofabrication technologies

Technology	Structural Control	Biological Integration	Mechanical Strength	Clinical Scalability	Primary Application
Additive Manufacturing	High	Moderate	High	High	Bone scaffolds
Bioprinting	Very High	Excellent	Moderate	Emerging	Osteochondral repair
Electrospinning	Nanoscale precision	Good	Low	Moderate	ECM mimicry

continued on following page

Table 5. Continued

Technology	Structural Control	Biological Integration	Mechanical Strength	Clinical Scalability	Primary Application
Multi-material Printing	Excellent	Excellent	High	Early stage	Gradient interfaces
Conventional Fabrication	Limited	Low	Moderate	Established	Basic implants

6.7. Translational and Regulatory Considerations

Manufacturing technologies directly influence regulatory classification of regenerative products. Tissue-engineered constructs combining biomaterials, biologics, and devices are often regulated as combination products, requiring extensive validation of safety, efficacy, and reproducibility.

Manufacturing complexity also impacts economic feasibility. Highly sophisticated constructs may demonstrate superior biological performance but remain impractical for routine clinical deployment due to cost and production time. Reviewer comments increasingly emphasize the importance of designing manufacturable biomaterials rather than purely experimental systems.

Collaboration between engineers, clinicians, regulatory experts, and industry partners is therefore essential to translate biofabrication advances into approved therapies.

6.8. Future Outlook: Toward Intelligent Manufacturing Ecosystems

The future of biofabrication lies in the development of intelligent manufacturing ecosystems capable of adaptive, autonomous production. Emerging concepts include:

- Self-optimizing manufacturing platforms
- Real-time process monitoring
- Smart biomaterials responsive to fabrication conditions
- On-demand hospital-based implant printing

Such systems could enable rapid fabrication of personalized regenerative implants directly within clinical environments, fundamentally transforming orthopaedic treatment paradigms.

7. PRECLINICAL AND CLINICAL EVIDENCE FOR BONE AND OSTEOCHONDRAL REGENERATION

7.1. Introduction: Bridging Experimental Innovation and Clinical Reality

Despite rapid advances in biomaterials engineering and biofabrication technologies, successful clinical translation remains the defining challenge of bone and osteochondral tissue engineering. A persistent gap exists between promising laboratory results and routine clinical implementation. Preclinical investigations frequently demonstrate robust osteogenesis, enhanced vascularization, and favorable degradation behavior; however, only a limited subset of biomaterial platforms progress to human clinical application.

The evaluation of regenerative biomaterials therefore requires a translational framework integrating in vitro testing, animal studies, and clinical trials. Preclinical evidence establishes biological feasibility and safety, while clinical studies determine therapeutic effectiveness under complex physiological and patient-specific conditions. Understanding how evidence evolves across these stages is essential for identifying realistic pathways toward clinical adoption.

7.2. In Vitro Evaluation: Biological Validation of Biomaterials

Initial validation of regenerative biomaterials typically begins with controlled in vitro experiments assessing cellular compatibility and functional performance. Key evaluation parameters include cytocompatibility, osteogenic differentiation potential, inflammatory response, and mineralization capacity.

Osteoblast proliferation assays, alkaline phosphatase activity measurements, and calcium deposition analyses are commonly employed to determine osteoinductive behavior. For osteochondral systems, chondrocyte phenotype maintenance and extracellular matrix production serve as critical indicators of cartilage regeneration potential.

Recent studies demonstrate that bioactive ceramics such as calcium phosphates and bioactive glasses enhance osteogenic gene expression through ionic dissolution products that activate signaling pathways involved in bone formation. Similarly, hydrogel-based systems incorporating growth factors or extracellular matrix components promote chondrogenic differentiation.

However, reviewers correctly note that in vitro success does not reliably predict clinical performance. Static culture conditions fail to replicate mechanical loading, immune interactions, and vascular integration present in vivo. Consequently, reliance solely on cell culture outcomes has historically contributed to translational failures.

7.3. Small Animal Models: Early Biological Feasibility

Rodent models represent the first stage of in vivo validation, enabling rapid evaluation of biomaterial degradation, tissue integration, and inflammatory response. Critical-sized defect models in rats and mice are widely used to assess bone regeneration capacity.

Preclinical findings consistently demonstrate that biodegradable scaffolds promote enhanced bone formation compared with empty defects. Composite materials combining polymers and bioactive ceramics show improved outcomes due to simultaneous mechanical support and biological stimulation. In osteochondral research, rabbit knee defect models frequently demonstrate successful cartilage surface restoration and subchondral bone regeneration when gradient scaffolds are employed.

Nevertheless, small animal studies possess inherent limitations. Rodent bone metabolism differs significantly from human physiology, and mechanical loading environments are substantially reduced compared with human weight-bearing conditions. Many biomaterials showing excellent results in rodents later fail under clinically relevant mechanical demands.

7.4. Large Animal Models: Translational Validation

Large animal models—including sheep, goats, pigs, and dogs—represent a critical intermediary step toward clinical application. These models more accurately replicate human bone size, biomechanics, and healing dynamics.

Load-bearing segmental defect models in sheep have demonstrated encouraging outcomes for composite scaffolds and biodegradable fixation systems. Magnesium-based implants, for example, exhibit progressive degradation accompanied by new bone formation without significant systemic toxicity in several large animal studies. Similarly, multiphasic osteochondral scaffolds tested in goat knee models show improved integration compared with monophasic constructs.

Importantly, large animal studies reveal challenges often overlooked in earlier stages:

- Mechanical collapse of highly porous scaffolds
- Delayed vascularization in large constructs
- Variable immune responses
- Interface instability in osteochondral repair

These findings highlight the importance of mechanical engineering strategies alongside biological optimization, reinforcing the interdisciplinary framework emphasized throughout this chapter.

7.5. Clinical Evidence in Bone Regeneration

Clinical translation of biodegradable biomaterials has progressed most successfully in bone regeneration applications. Several material classes have achieved regulatory approval and clinical use.

7.5.1. Calcium Phosphate and Bioactive Ceramic Implants

Calcium phosphate-based bone substitutes are widely used in fracture repair, spinal fusion, and dental reconstruction. Clinical studies demonstrate reliable osteoconduction and gradual remodeling into native bone. However, brittle mechanical behavior limits application in high load-bearing sites.

7.5.2. Biodegradable Polymer Fixation Devices

Polylactic acid and polyglycolic acid fixation systems have been employed in craniofacial surgery and pediatric orthopaedics. Clinical outcomes indicate reduced need for implant removal surgeries compared with metallic systems. Nevertheless, inflammatory reactions related to acidic degradation products have been reported in some patients.

7.5.3. Magnesium-Based Implants

Magnesium alloys represent one of the most promising recent clinical advances. Clinical trials involving biodegradable magnesium screws for orthopedic fixation demonstrate favorable healing outcomes, gradual implant resorption, and reduced long-term complications. Importantly, post-2020 clinical data indicate improved corrosion control and mechanical reliability compared with earlier generations of Mg implants.

These examples illustrate that successful clinical biomaterials typically achieve a balance between mechanical competence, predictable degradation, and surgical practicality.

7.6. Comparative Analysis of Translational Success

A recurring pattern across clinical translation efforts reveals several determinants of success which is shown in Table 6.

Table 6. Factors affecting translational success

Factor	Successful Translation	Common Cause of Failure
Mechanical Stability	Adequate early support	Premature scaffold collapse
Degradation Control	Predictable resorption	Rapid or incomplete degradation
Biological Integration	Vascularized tissue ingrowth	Fibrous encapsulation
Manufacturing Quality	Reproducibility	Batch variability
Regulatory Strategy	Early planning	Late regulatory adaptation

This analysis demonstrates that translational success requires simultaneous optimization of engineering, biological, and regulatory parameters rather than excellence in a single domain.

7.7. Regulatory Evidence and Clinical Approval Pathways

Regulatory approval represents a major barrier identified by reviewers. Agencies evaluate regenerative biomaterials based on safety, efficacy, manufacturing quality, and long-term risk.

Combination products incorporating biomaterials, cells, and biologics undergo particularly stringent evaluation. Requirements typically include:

- Preclinical safety testing
- Biocompatibility assessment
- Mechanical validation
- Controlled clinical trials
- Post-market surveillance

The regulatory landscape increasingly encourages evidence-based design approaches, emphasizing standardized testing and reproducibility.

8. REGULATORY, ETHICAL, AND COMMERCIAL TRANSLATION OF REGENERATIVE BIOMATERIALS

While advances in biomaterials science, biofabrication, and regenerative biology have significantly expanded the possibilities for bone and osteochondral repair, clinical translation remains constrained by regulatory, ethical, and commercial challenges. Many biomaterial systems demonstrating excellent experimental performance fail to progress beyond laboratory research due to difficulties navigating approval pathways, ensuring manufacturing reproducibility, and achieving economic sustainability. Consequently, successful regenerative innovation now depends not only on scientific discovery but also on alignment with regulatory frameworks, ethical responsibility, and viable commercialization strategies.

Regulatory and commercial considerations should therefore be viewed as integral design parameters rather than post-development obstacles. Early integration of translational planning has emerged as a defining feature distinguishing clinically successful biomaterials from purely academic innovations.

8.1. Regulatory Frameworks

Regulatory agencies worldwide classify regenerative biomaterials based on risk profile, biological activity, and intended clinical use. Bone graft substitutes composed solely of inert biomaterials may be regulated as medical devices, whereas constructs incorporating cells, growth factors, or gene therapy components are typically categorized as combination products or advanced therapy medicinal products.

In the United States, the Food and Drug Administration (FDA) evaluates regenerative technologies through pathways including 510(k) clearance, Premarket Approval (PMA), and Human Cells, Tissues, and Cellular and Tissue-Based Products (HCT/P) regulations. Similarly, the European Medicines Agency (EMA) regulates advanced therapies under the Advanced Therapy Medicinal Products (ATMP) framework. These regulatory systems require rigorous demonstration of biocompatibility, mechanical reliability, degradation safety, sterility assurance, and long-term clinical effectiveness.

A major translational challenge arises because regenerative biomaterials often integrate multiple functional components—materials, biologics, and devices—each governed by distinct regulatory standards. This complexity increases development timelines and financial investment requirements, often exceeding the resources available in academic research environments.

8.2. Preclinical Evidence Requirements and Standardization Challenges

Regulatory approval relies heavily on standardized preclinical evidence. Biomaterial developers must demonstrate safety across several domains:

- Cytotoxicity and immunogenicity
- Mechanical performance under physiological loading
- Controlled degradation behavior
- Absence of toxic degradation products
- Long-term systemic safety

However, a critical issue highlighted by reviewers is the lack of standardized testing methodologies across studies. Variability in animal models, evaluation metrics, and reporting practices complicates comparison between technologies and slows regulatory decision-making.

Emerging international standards aim to address this limitation by establishing harmonized protocols for regenerative biomaterial evaluation. Adoption of standardized performance metrics enables more reliable translation and facilitates cross-study comparison—an essential step toward evidence-based regulatory science.

8.3. Ethical Considerations in Regenerative Biomaterial Development

Ethical responsibility plays a central role in regenerative medicine due to the involvement of biologically active materials, living cells, and personalized therapies. Key ethical challenges include patient safety, informed consent, equitable access, and responsible innovation.

Cell-based therapies raise concerns regarding donor tissue sourcing, genetic manipulation, and long-term biological effects that may not be immediately observable during clinical trials. Similarly, personalized implants generated using patient-specific data introduce questions related to data privacy and algorithmic transparency when artificial intelligence is employed in design optimization.

Another ethical issue involves the rapid commercialization of emerging technologies before sufficient long-term evidence is available. Premature clinical adoption may expose patients to unforeseen risks while simultaneously undermining public trust in regenerative medicine. Ethical translation therefore requires balancing innovation speed with rigorous scientific validation.

8.4. Commercial Translation and Market Adoption

Commercial viability represents an equally critical component of translational success. The global orthopaedic biomaterials market continues to expand; however, adoption of regenerative technologies depends on cost-effectiveness, reimbursement policies, and clinical workflow integration.

Several promising biomaterials have failed commercially despite strong scientific evidence due to:

- High manufacturing costs
- Complex surgical procedures
- Limited reimbursement coverage
- Insufficient clinician training

Successful commercialization therefore requires collaboration among engineers, clinicians, industry partners, investors, and healthcare policymakers. Early engagement with industry stakeholders allows biomaterial design to align with real-world clinical and economic constraints.

Health technology assessment frameworks increasingly evaluate regenerative therapies based on long-term economic benefit rather than short-term procedural cost. Biomaterials capable of reducing revision surgeries or accelerating recovery may ultimately demonstrate strong economic value despite higher initial expenses.

8.5. Future Outlook: Toward Integrated Translational Ecosystems

The future of regenerative biomaterial translation lies in creating integrated ecosystems linking academia, clinical practice, regulatory authorities, and industry manufacturing. Emerging trends include:

- Hospital-based manufacturing facilities for personalized implants
- AI-assisted regulatory documentation
- Digital twins for regulatory simulation testing
- Continuous post-market data monitoring

These developments suggest a shift from isolated innovation toward coordinated translational infrastructures capable of accelerating safe clinical adoption.

9. ARTIFICIAL INTELLIGENCE, PERSONALIZATION, AND FUTURE TRANSLATIONAL FRAMEWORK

The evolution of regenerative orthopaedics is entering a new phase characterized by the convergence of artificial intelligence (AI), computational modeling, and personalized medicine. Traditional biomaterial development has largely relied on iterative experimental testing, where materials are synthesized, evaluated, and optimized through trial-and-error approaches. While this methodology has generated significant progress, it remains time-consuming, resource-intensive, and poorly suited to addressing patient-specific variability in tissue regeneration.

Artificial intelligence introduces a transformative paradigm in which biomaterials can be computationally designed, predicted, and optimized before fabrication. Rather than developing universal implant solutions, future regenerative strategies increasingly aim to deliver adaptive and personalized biomaterial systems capable of responding to individual anatomical, biological, and mechanical requirements.

9.1. AI-Driven Biomaterial Design and Optimization

AI algorithms, particularly machine learning and deep learning models, are increasingly applied to analyze large biomaterials datasets linking composition, processing parameters, mechanical properties, and biological outcomes. These systems can identify hidden correlations that are difficult to detect using conventional experimental approaches.

In bone regeneration, AI models are being used to predict optimal scaffold architectures capable of balancing stiffness, porosity, and degradation kinetics. Computational frameworks integrate imaging data with biomechanical simulations to generate implant designs tailored to defect geometry and patient loading conditions. Similarly, predictive modeling enables optimization of composite material ratios to enhance osteoinduction while maintaining mechanical integrity.

For osteochondral regeneration, AI-assisted design supports development of gradient scaffolds that reproduce spatial variations in cartilage and bone properties. By simulating cellular responses to mechanical and biochemical cues, AI platforms reduce reliance on extensive preclinical experimentation and accelerate innovation cycles.

However, reviewers correctly emphasize that AI should not be viewed as a replacement for biological validation. Instead, AI functions as a decision-support system guiding experimental design and improving translational efficiency.

9.2. Personalization in Regenerative Orthopaedics

One of the most significant limitations of conventional implants is their standardized design, which fails to account for interpatient variability in anatomy, bone density, metabolic health, and healing capacity. Personalized regenerative medicine aims to overcome this limitation by integrating patient-specific data into biomaterial development.

Advanced imaging technologies such as CT and MRI enable reconstruction of precise defect geometries, allowing fabrication of customized scaffolds through digital manufacturing workflows. Patient-specific biomechanical models further enable prediction of stress distribution and implant performance under real physiological loading.

Beyond anatomical customization, personalization increasingly incorporates biological profiling. Factors including age, inflammatory status, comorbidities, and genetic background influence regenerative outcomes. Future biomaterials may incorporate adaptive degradation rates or bioactive signaling tailored to individual healing responses.

Such approaches shift regenerative orthopaedics from generalized treatment paradigms toward precision medicine, where therapy selection is guided by predictive analytics rather than empirical decision-making.

9.3. Digital Twins and Predictive Healing Models

A particularly promising emerging concept is the development of digital twins—virtual representations of patients that simulate tissue healing and implant performance over time. Digital twins integrate clinical imaging, biomechanical modeling, physiological data, and AI algorithms to forecast regenerative outcomes before surgical intervention.

In bone repair, digital twin models may predict fracture healing progression, enabling surgeons to select optimal biomaterials and fixation strategies. For osteochondral applications, predictive models can estimate cartilage regeneration quality and long-term joint function.

These systems hold potential to reduce implant failure rates, optimize rehabilitation protocols, and enable adaptive treatment planning throughout the healing process. Importantly, digital twins also provide powerful tools for regulatory evaluation by enabling virtual clinical trials and risk assessment.

9.4. Intelligent and Responsive Biomaterials

Future translational frameworks extend beyond passive implants toward intelligent biomaterials capable of sensing and responding to biological environments. Smart scaffolds incorporating embedded sensors, responsive polymers, or bioelectronic interfaces may monitor mechanical loading, inflammation, or tissue growth in real time.

Adaptive biomaterials could dynamically modify stiffness, release therapeutic agents in response to inflammation, or adjust degradation behavior according to healing progression. Integration of wearable biosensors and remote monitoring platforms further enables continuous post-operative assessment, supporting personalized rehabilitation strategies.

Such systems redefine biomaterials as active therapeutic platforms rather than static structural replacements.

9.5. Translational Framework for Next-Generation Regenerative Medicine

The future translational pathway for regenerative biomaterials can be conceptualized as an integrated framework linking five interconnected domains:

1. Biological Understanding – Cellular and molecular mechanisms of regeneration
2. Engineering Design – Scaffold architecture and mechanical optimization
3. Artificial Intelligence – Predictive modeling and design automation
4. Digital Manufacturing – Personalized fabrication technologies
5. Clinical Translation – Regulatory approval and real-world validation

Successful regenerative solutions will emerge from coordinated interaction among these domains rather than advances in any single field. This integrated framework directly addresses a major research gap identified by reviewers: the absence of a unified model connecting mechanics, biology, and digital innovation.

10. CRITICAL FUTURE RESEARCH DIRECTIONS

Although substantial progress has been achieved in biomaterials engineering for bone and osteochondral regeneration, several fundamental challenges continue to limit widespread clinical adoption. Current research trends increasingly demonstrate that incremental material improvements alone are insufficient to overcome translational barriers. Future advances must instead focus on integrated solutions addressing biological complexity, mechanical functionality, clinical practicality, and regulatory feasibility simultaneously.

The next phase of regenerative orthopaedics therefore requires a shift from material discovery toward system-level innovation, where biomaterials are developed within a comprehensive translational ecosystem.

10.1. Bridging the Mechanics–Biology Interface

One of the most persistent research gaps lies in understanding the dynamic interaction between mechanical loading and biological regeneration. Many biomaterials successfully promote osteogenesis under static laboratory conditions but fail in vivo due to inadequate mechanical performance during early healing stages.

Future research must prioritize mechanobiology-driven design approaches that integrate:

- Load-responsive scaffold architectures
- Adaptive stiffness evolution during healing
- Coupled mechanical–biological modeling frameworks

Developing materials capable of dynamically transferring load as tissue regenerates represents a critical step toward achieving functional rather than purely structural regeneration.

10.2. Vascularization and Immune Modulation

Insufficient vascularization remains a primary cause of failure in large bone defects and osteochondral constructs. Without rapid vascular integration, implanted scaffolds experience nutrient limitations, delayed tissue formation, and fibrotic encapsulation.

Emerging research directions include:

- Angiogenic factor delivery systems
- Prevascularized biofabricated constructs
- Immunomodulatory biomaterials that regulate macrophage polarization

Future biomaterials must be designed not only to support tissue growth but also to actively orchestrate immune responses and vascular development, recognizing regeneration as an immune-mediated process rather than solely a materials engineering challenge.

10.3. Standardization and Quantitative Benchmarking

A major limitation identified by reviewers is the absence of standardized performance metrics across biomaterial studies. Current literature frequently reports qualitative outcomes without enabling meaningful comparison between platforms.

Future investigations should establish quantitative benchmarks including:

- Mechanical strength thresholds for load-bearing applications
- Controlled degradation timelines matched to healing stages
- Long-term clinical success and revision rates
- Functional outcome measures beyond histological evaluation

Standardized reporting frameworks will enable meta-analysis, accelerate regulatory evaluation, and support evidence-based biomaterial selection.

10.4. Integration of Artificial Intelligence and Data-Driven Research

Artificial intelligence offers significant potential to transform biomaterials research, yet its application remains fragmented. Future work must move beyond isolated computational demonstrations toward fully integrated data ecosystems combining experimental, clinical, and manufacturing datasets.

Key research priorities include:

- AI-guided material discovery platforms
- Predictive models linking scaffold design to clinical outcomes
- Digital twin systems for personalized treatment planning
- Real-world data integration for continuous implant optimization

Establishing shared databases and collaborative data infrastructures will be essential for realizing the full potential of AI-driven regenerative medicine.

10.5. Translational Manufacturing and Scalability

Many promising biomaterials fail during commercialization due to manufacturing complexity and cost barriers. Future research should emphasize manufacturability as an early design constraint rather than a late-stage consideration.

Critical directions include:

- Scalable biofabrication workflows
- Automated quality control systems
- Modular manufacturing compatible with personalized implants
- Sustainable and cost-effective production methods

Bridging laboratory innovation with industrial manufacturing capabilities will determine whether regenerative biomaterials achieve global clinical impact.

10.6. Long-Term Clinical Evidence and Real-World Outcomes

A significant gap in the field remains the scarcity of long-term clinical outcome data. Many regenerative technologies are evaluated only through short-term studies, limiting understanding of durability, late complications, and functional recovery.

Future clinical research must prioritize:

- Multicenter clinical trials

- Longitudinal patient monitoring
- Integration of wearable and digital health technologies
- Registry-based outcome tracking

Evidence derived from real-world clinical environments will be essential to validate regenerative therapies and inform next-generation biomaterial design.

10.7. Toward a Unified Translational Framework

The future of bone and osteochondral regeneration lies in convergence across disciplines rather than isolated technological advances. Successful solutions will emerge from coordinated integration of:

- Mechanobiology
- Advanced biomaterials
- Artificial intelligence
- Biofabrication technologies
- Regulatory science and clinical practice

This unified framework directly addresses the central research gap identified throughout this chapter—the lack of cohesive strategies linking engineering innovation with clinical implementation.

11. CONCLUSION

Bone and osteochondral tissue engineering has advanced into a translational field integrating biomaterials, engineering, biology, and clinical medicine. Current evidence shows that no single biomaterial can fully satisfy skeletal regeneration needs. Ceramics, polymers, and biodegradable metals each possess unique advantages and limitations, making composite and biomimetic systems more effective for balancing mechanical strength, degradation, and biological activity. Successful regeneration also depends on scaffold architecture, vascularization, immune response, and patient-specific biomechanics.

Despite progress, challenges in clinical translation, manufacturing scalability, and regulatory approval remain. Emerging integration of artificial intelligence, digital manufacturing, and personalized medicine offers new opportunities for patient-specific regenerative therapies. Continued interdisciplinary collaboration and evidence-based innovation are essential for transforming regenerative orthopaedics from experimental research into reliable clinical practice.

REFERENCES

Bose, S., Roy, M., & Bandyopadhyay, A. (2012). Recent advances in bone tissue engineering scaffolds. *Trends in Biotechnology*, *30*(10), 546–554. DOI: 10.1016/j.tibtech.2012.07.005 PMID: 22939815

Chitra, S.., Rajeshkumar, S., & Mathew, N. K. (2023). Bioceramics: From concept to clinic. Advanced Bioceramics, 199–218.

Daly, A. C., Freeman, F. E., Gonzalez-Fernandez, T., Critchley, S. E., Nulty, J., & Kelly, D. J. (2017). 3D bioprinting for cartilage and osteochondral tissue engineering. *Advanced Healthcare Materials*, *6*(22), 1700298. DOI: 10.1002/adhm.201700298 PMID: 28804984

Fillingham, Y., & Jacobs, J. J. (2016). Bone grafts and their substitutes. *The Bone & Joint Journal*, *98-B*(1, Suppl A), 6–9. DOI: 10.1302/0301-620X.98B.36350 PMID: 26733632

Garimella, A., Ghosh, S. B., & Bandyopadhyay-Ghosh, S. (2025). Biomaterials for bone tissue engineering: Achievements to date and future directions. *Biomedical Materials (Bristol, England)*, *20*(1), 012001. DOI: 10.1088/1748-605X/ad967c PMID: 39577395

Giannoudis, P. V., Dinopoulos, H., & Tsiridis, E. (2005). Bone substitutes: An update. *Injury*, *36*(Suppl. 3), S20–S27. DOI: 10.1016/j.injury.2005.07.029 PMID: 16188545

Hunckler, M. D., & Levine, A. D. (2022). Navigating ethical challenges in the development and translation of biomaterials research. *Frontiers in Bioengineering and Biotechnology*, *10*, 949280. DOI: 10.3389/fbioe.2022.949280 PMID: 36204464

Lee, S. S., Du, X., Kim, I., & Ferguson, S. J. (2022). Scaffolds for bone-tissue engineering. *Matter*, *5*(9), 2722–2759. DOI: 10.1016/j.matt.2022.06.003 PMID: 35673575

Liu, D., Lyu, Y., Lin, Z., Ji, Z., Wang, X., & Shi, X. (2025). 3D printing functional materials with extreme regulation of mechanical performances from hydrogel to engineering plastic. *Chemical Engineering Journal*, *512*, 162310. DOI: 10.1016/j.cej.2025.162310

Makris, E. A., Gomoll, A. H., Malizos, K. N., Hu, J. C., & Athanasiou, K. A. (2015). Repair and tissue engineering techniques for articular cartilage. *Nature Reviews. Rheumatology*, *11*(1), 21–34. DOI: 10.1038/nrrheum.2014.157 PMID: 25247412

Murphy, M. B., Moncivais, K., & Caplan, A. I. (2013). Mesenchymal stem cells: Environmentally responsive therapeutics for regenerative medicine. *Experimental & Molecular Medicine*, *45*(11), e54. DOI: 10.1038/emm.2013.94 PMID: 24232253

Nukavarapu, S. P., & Dorcemus, D. L. (2013). Osteochondral tissue engineering: Current strategies and challenges. *Biotechnology Advances*, *31*(5), 706–721. DOI: 10.1016/j.biotechadv.2012.11.004 PMID: 23174560

Omondi, O. B., Arroyan, Y. N., Onyango, B., Kong, L., Wang, G., & Ye, Z. (2024). Revolutionizing healthcare: Emerging frontiers in 3D bioprinting of tissues and organs. *European Polymer Journal*, *217*, 113210. DOI: 10.1016/j.eurpolymj.2024.113210

Oryan, A., Alidadi, S., Moshiri, A., & Maffulli, N. (2014). Bone regenerative medicine: Classic options, novel strategies, and future directions. *Journal of Orthopaedic Surgery and Research*, *9*(1), 18. DOI: 10.1186/1749-799X-9-18 PMID: 24628910

Pittenger, M. F., Discher, D. E., Péault, B. M., Phinney, D. G., Hare, J. M., & Caplan, A. I. (2019). Mesenchymal stem cell perspective: Cell biology to clinical progress. *NPJ Regenerative Medicine*, *4*(1), 22. DOI: 10.1038/s41536-019-0083-6 PMID: 31815001

Santos-Beato, P., Midha, S., Pitsillides, A. A., Miller, A., Torii, R., & Kalaskar, D. M. (2022). Biofabrication of the osteochondral unit and its applications: Current and future directions for 3D bioprinting. *Journal of Tissue Engineering*, *13*, 20417314221133480. DOI: 10.1177/20417314221133480 PMID: 36386465

Wang, Z., Xu, J., Zhu, J., Fang, H., Lei, W., Qu, X., Cheng, Y. Y., Li, X., Guan, Y., Wang, H., & Song, K. (2025). Osteochondral tissue engineering: Scaffold materials, fabrication techniques and applications. *Biotechnology Journal*, *20*(1), e202400699. DOI: 10.1002/biot.202400699 PMID: 39865414

World Health Organization. (2022). Musculoskeletal conditions. WHO Report.

Zhang, Y. S., & Khademhosseini, A. (2020). Engineering in vitro human tissue models through bio-design and manufacturing. *Bio-Design and Manufacturing*, *3*(3), 155–159. DOI: 10.1007/s42242-020-00080-w

Chapter 12
Bioresorbable and Transient Bioelectronics:
Degradable Conductors, Sensors, and Wireless Transient Implants for Monitoring and Therapy

G. Boopathi Raja
http://orcid.org/0000-0003-3345-4503
Velalar College of Engineering and Technology, India

ABSTRACT

This chapter investigates a new field of biomedical engineering called biocompatible bioelectronics, where bioelectronic devices are designed to serve their therapeutic or diagnostic function and then dissolve naturally in the body without causing any harm. These bioelectronics consist of biodegradable conductors, such as magnesium, zinc, molybdenum, and their alloys, as well as biodegradable substrates, including polylactic acid, polycaprolactone, and silk fibroin, which serve as substrates for silicon nanomembrane semiconductor chips and wireless communications modules that enable precise timing of electronic functionality prior to the complete biological resorption of the device. Because transient implants do not require secondary explanation surgery, they significantly decrease the risk of infection, chronic inflammation and patient morbidity associated with the retention of permanent devices. This chapter will provide an in-depth examination of the materials science, fabrication methods, and clinical pathways for the implementation of biocompatible bioelectronics.

DOI: 10.4018/979-8-2600-0060-1.ch012

1. INTRODUCTION

1.1 Concept and Significance

A major shift in biomedical engineering has occurred due to the advent of bioresorbable or transient electronic devices. Unlike the traditional, permanent implants, which are intended for permanent use within a patient's body and may carry substantial surgical risks for removal, transient electronic devices are designed to function for a predetermined period such as days, weeks, or months. These devices will then dissolve and assimilate into the patient's metabolism without leaving behind foreign matter (Han et al., 2020; Kang et al., 2020; Zhang et al., 2023).

There is a sound clinical justification for using transient bioelectronics as opposed to permanent devices. Permanent implants are associated with a variety of long-term complications, including infections associated with the device, chronic foreign body inflammatory response, fibrotic encapsulation, mechanical fatigue fractures, and complications associated with retrieval surgery. In many diagnostic or therapeutic scenarios, such as post-operative monitoring, temporary neural stimulation for rehabilitation, and drug delivery at wound sites, it is unnecessary to use an implantable device; in fact, an implantable device may present a clinical risk to the patient. Transiently dissolving devices provide solutions for the complexities involved in using permanent devices by matching the duration of the device's usefulness with the desired clinical indications (Ashammakhi et al., 2021; Corsi et al., 2025; Singh et al., 2020).

Transient electronic devices provide a vast range of functionality for biomedical applications. For example, some of the more common uses for transient bioelectronics include temporary cardiac rhythm-mapping devices, intracranial pressure measurement following traumatic brain injury, and temperature/biochemical mapping devices around wound sites. The therapeutic uses of electrical stimulation for peripheral nerve regeneration, enhancement of the healing rate of wounds, healing of broken bones, and controlled release of drugs from electro-responsive potable porous biodegradable matrices are all examples of uses for bioresorbable electronic devices.

In all the above therapies, the electronic device will need to demonstrate reliable electronic performance after the device delivers its therapeutic effect and will need to be complete resorbable with no trace remaining and no adverse biological response after resorption (Cho et al., 2026; Kim et al., 2026; Mirzajani et al., 2025). Figure 1 shows the PRISMA chart for analyzing the available literature survey suitable for the study.

Figure 1. PRISMA flowchart

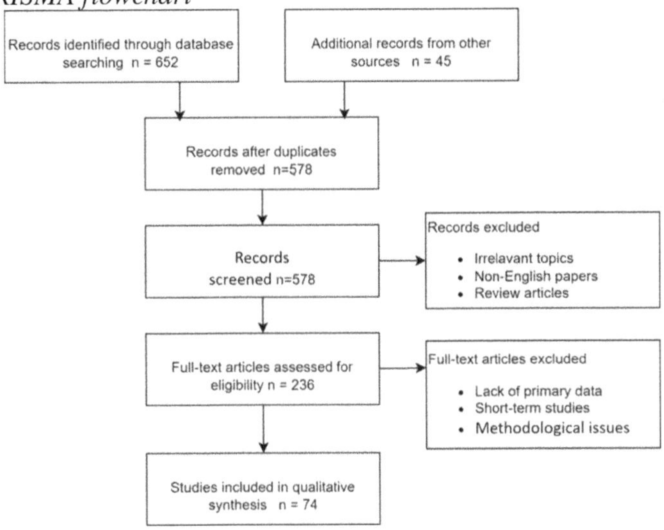

1.2 Historical Evolution

The conceptual underpinnings of the development of bioresorbable electronic devices have stemmed from many years of previous work in the area of biodegradable surgical materials (e.g., absorbable sutures, resorbable orthopedic fixation devices, and drug-eluting biodegradable stents) that have established that it is possible to have functional temporary implants which can be safely integrated into and eliminated from the body. The area of transient electronics has specifically been initialized through the convergence of advances in materials science (i.e., novel biodegradable metals and polymers), micro- and nano-fabrication techniques (i.e., silicon nanomembranes and transfer printing patterns), and the continued miniaturization of wireless communications products (Feiner & Dvir, 2017; Li et al., 2018; Tan et al., 2016).

Figure 2. Evolution of bioresorbable transient bioelectronics

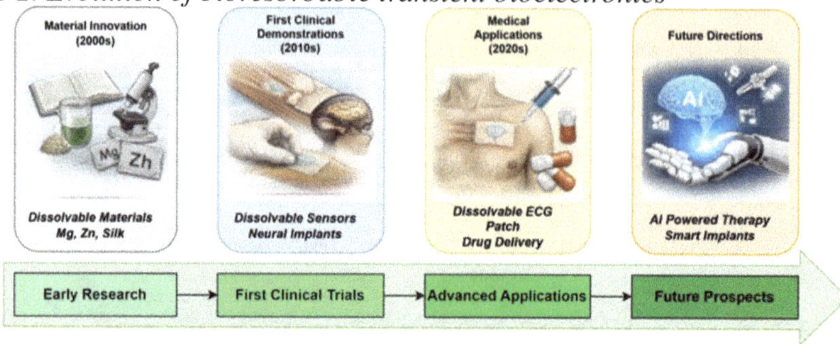

Figure 2 shows how the addition of bioresorbable, transient bioelectronics has changed from the early investigation of new materials to their futuristic vision of A.I.-integrated smart implantable devices over time. The evolution started in the 2000s with the discovery of various dissolvable materials that could be used to create transient bioelectronics, for example; magnesium, zinc and silk fibroin. In the 2010s, the first demonstrations of transient bioelectronics including; transient bioelectric sensors, stimulating devices, and other neural stimulators took place. The 2020s will see continued examples of advanced medical applications of transient bioelectronics including dissolvable ECG patches and drug delivery systems.

The landmark demonstration provided the framework for transient electronics as ultrathin silicon nanomembranes serving as active semiconductors, magnesium thin films used as conductive materials, silicon dioxide as the gate dielectric, and silk fibroin serving as the biodegradable substrate i.e., all 4 components of the transient electronic device can be expected to dissolve predictably in water. This research has triggered a global investigation into materials, device complexity, and successful clinical application of transient bioelectric devices (Dutta & Cheng, 2023; Kang et al., 2020; Zhang et al., 2023).

Numerous research milestones have been achieved including: development of bioresorbable pressure sensors for intracranial pressure monitoring (Tan et al., 2016); biodegradable peripheral nerve stimulators to enhance nerve recovery (Li et al., 2018); and wireless transient implants using near-field technology to support ongoing energy needs (Choi et al., 2020); and recently, mixed biodegradable devices containing clinically valid components such as sensory, stimulatory, and therapeutic drug supply capabilities (Cho et al., 2026; Corsi et al., 2025; Kim et al., 2026). Raja has performed an extensive review of the intersection of IoT and sensor informatics with wearable and implantable health care technologies and has placed transient bioelectrical systems in the larger scope of the digital health support ecosystem (Raja, 2025a).

1.3 Comparison with Permanent Implants

The objective of contrasting permanent implants against transient bioelectrical systems is to identify the clinical environment in which each device is best deployed. Table 1 shows the comparison of permanent implants and transient bioresorbable bioelectronics. Permanent devices (e.g., pacemakers, deep brain stimulators, cochlear implants) are designed for patients with chronic medical problems for which there is a need for long-term electronic intervention. Temporary use devices e.g., acute-care monitoring, post-surgical recovery support, injury rehabilitation, and planned delivery of drugs/biologics meet short-term clinical needs through the delivery or detection of either electrical signals or the injectable form of the respective therapeutics (Fanelli & Ghezzi, 2021; Liu et al., 2023; Singh et al., 2020).

Table 1. Comparison of permanent implants and transient bioresorbable bioelectronics

Permanent Implants	Transient Bioelectronics
Long-term monitoring and therapy	Short-term, targeted interventions
Requires surgical removal	Naturally resorbed — no removal needed
Risk of chronic inflammation and fibrosis	Reduced immune response after resorption
Permanent foreign body burden	Complete biological elimination
Device obsolescence risks	Adaptive, time-matched functionality

1.4 Challenges and Research Directions

Bioresorbable electronics are an area of promise, however, they present a number of significant engineering challenges. The most significant challenge is controlling the rate at which the electronics will degrade; this is difficult since various factors in the body will affect how quickly the materials degrade, such as pH, ion concentration, enzymes and mechanical loading which can be very difficult to model accurately (Alam et al., 2024; Ashammakhi et al., 2021; Liu et al., 2025).

Signal integrity during the degradation of the electronics is another related challenge: as the cross-section of the conductor gets thinner and the mechanical properties of the substrate change, the electrical properties of the device can degrade before the device is completely resorbed. This can lead to inaccurate sensor readings or the device failing to provide enough stimulation amplitude. Maintaining the integrity of wireless communication between the device and the user also adds another layer of complexity as the antenna structures experience degradation and the surrounding tissue is constantly changing (Corsi et al., 2025; Wang et al., 2025). Lastly, the biocompatibility of the degradation by-products, the ability to determine the regulatory approval pathway for the devices as well as the scalability of manufacturing processes are all additional challenges that are facing researchers in developing these devices.

Figure 3. Layered architecture of bioresorbable transient bioelectronics

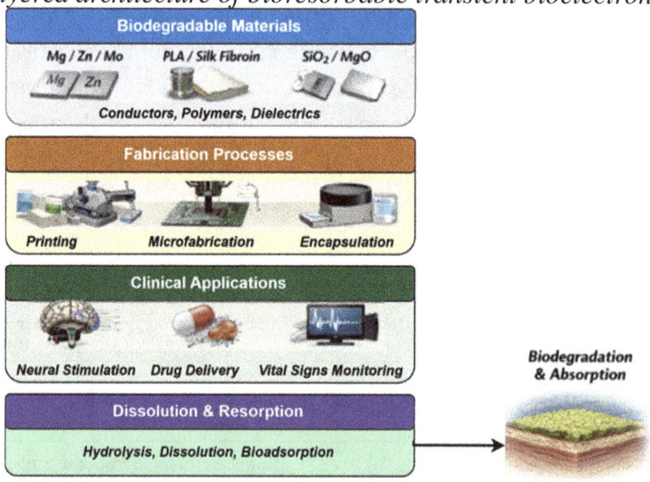

The Hierarchical Structure of Bioresorbable Transient Bioelectronic Devices is shown in Figure 3. The first layer in this hierarchy consists of biodegradable materials used to fabricate transient bioelectronic devices e.g. magnesium, zinc, molybdenum, polylactic acid, silk fibroin and silicon dioxide. The second layer in the hierarchy consists of the fabrication processes used to fabricate transient bioelectronic devices including printing, microfabrication, and encapsulation processes, enabling the precise construction of the devices. The third layer of the hierarchy shows the different clinical applications of the devices demonstrating their functional versatility Neurostimulation, drug delivery, and vital signs monitoring, demonstrating the functional versatility of the bioresorbable transient bioelectronics. The last layer in the hierarchy shows the mechanism for dissolution and bioabsorption, through which the device goes through a controlled process of hydrolysis and bioabsorption that allows for the complete and safe elimination of transient bioelectronic devices from the body.

2. MATERIALS FOR DEGRADABLE CONDUCTORS AND SUBSTRATES

2.1 Metallic Bioresorbable Conductors

The highest electrical conductivity among bioresorbable materials is found in metals, and there have been numerous studies about using metals as conductor layers for transient electronics. The main types of bioresorbable metals are magnesium (Mg) and its alloys, zinc (Zn), iron (Fe) and its alloys, molybdenum (Mo), and tungsten (W). These metals differ from each other in degradation mechanisms, degradation by-products, and potential applications (Fernandes & Taurino, 2022; Kang et al., 2020; Ryu et al., 2021).

2.1.1 Magnesium and Magnesium Alloys

Of the bioresorbable conductors, magnesium is the most commonly used. Magnesium has excellent electrical conductivity (~2.3 x 10^7 S/m), moderate mechanical strength, and a high metabolic safety profile, as magnesium ions are part of more than 300 enzymatic reactions and are removed by the kidneys with high efficiency. The degradation mechanism of magnesium is electrochemical oxidation ($Mg + 2H_2O \rightarrow Mg(OH)_2 + H_2$), and the degradation rate can be varied from days to months by changing the type of alloying element (Zn, Ca, or rare-earth) or the thickness of the magnesium film or by modifying the design of the encapsulation (Cho et al., 2026; Ryu et al., 2021; Zhang et al., 2023).

Thin Mg films (50–300 nm) deposited by physical vapor deposition can be used as transient interconnects and electrode contacts within silicon nanomembrane-based active device architecture. Thicker Mg structures serve as structural conductors within passive array-based sensor circuits. Refinements to alloy engineering have more evenly dispersed the magnesium in the alloy during degradation, leading to a reduction in localized pitting and therefore earlier than designed electrical failure of transient electronics (Bae et al., 2025; Han et al., 2020).

2.1.2 Zinc, Molybdenum, Tungsten, and Iron Alloys

The corrosion rate of zinc (Zn) is slower than that of magnesium (Mg), and it also has a higher ductility making zinc well-suited for flexible transient architectures. In vivo, zinc degrades into zinc ions (Zn^{2+}) and zinc oxide (ZnO); while zinc and zinc oxide are biologically essential at low concentrations, they can be potentially cytotoxic at high concentrations, so they need to be carefully controlled in order to promote proper rates of degradation. Additionally, molybdenum (Mo) and tungsten

(W) dissolve to form water-soluble molybdate and tungstate ions and their thin film coatings are fully compatible with standard microfabrication processes. Iron alloys (Fe–Mn systems) exhibit the best mechanical strength for structural device components, as well as MRI compatibility; however, they may degrade at a slower rate necessitating an optimization of geometry in order to provide adequate prevention of long-term retention of the materials in vivo (Liu et al., 2025; Ryu et al., 2021). Table 2 shows the comparative overview of metallic bioresorbable conductors in transient bioelectronics.

Table 2. Comparative overview of metallic bioresorbable conductors in transient bioelectronics

Material	Conductivity (S/m)	Degradation Rate	Biocompatibility	Key Applications
Magnesium (Mg)	~2.3×10^7	Weeks–months	Excellent	Neural/cardiac electrodes, interconnects
Zinc (Zn)	~1.7×10^7	Months	Good	Flexible sensors, biodegradable interconnects
Iron (Fe) alloys	~1.0×10^7	Months–years	Moderate	Structural reinforcement layers
Molybdenum (Mo)	~2.0×10^7	Weeks–months	Good	Biodegradable gate electrodes
Tungsten (W)	~1.8×10^7	Weeks	Good	Sensor contact pads, interconnects

2.2 Organic and Polymeric Substrates

2.2.1 Polylactic Acid (PLA), PLGA, and Polyglycolide

Polylactic acid (PLA) and the glycolic acid copolymer of PLA (PLGA) are clinically established biodegradable substrate polymers. Both have decades of regulatory history in use for sutures, orthopedic devices, and controlled release formulations. The primary route of degradation for both PLA and PLGA is through hydrolysis of ester bonds; however, both yield lactic acid and glycolic acid metabolites, which are fed into the citric acid cycle. The ratio of lactide to glycolide in PLGA can be effectively used to control the rate of degradation — PLA has a half-life of weeks when it has a high proportion of glycolide and months when it has a high proportion of lactide — thus allowing precise control of biodegradability to meet clinical needs (Ashammakhi et al., 2021; Hu et al., 2024; Lee et al., 2022).

2.2.2 Polycaprolactone (PCL) and Silk Fibroin

Polycaprolactone (PCL) offers exceptional elongation (greater than 300%) and a slow degradation rate (months to years depending on thermal and/or chemical conditions). These attributes make PCL a suitable material for long-term flexible transient architecture in a curved tissue-based anatomical setting (Lee et al., 2022; Wang et al., 2025). Silk fibroin produced by Bombyx mori has a programmable degradation rate, ranging from days (amorphous) to months (highly crystalline), depending on the content of beta-sheets, with excellent biocompatibility and surface chemistry suitable for adhesion of electrodes and integration of biosensors after chemical modification. Yang et al. (2021) demonstrated the use of photocurable silk-based adhesives to resolve long-standing tissue/device delamination challenges(Yang et al., 2021). Table 3 shows the comparison of various polymeric and natural substrates for bioresorbable bioelectronics.

Table 3. Polymeric and natural substrates for bioresorbable bioelectronics

Substrate	Degradation Mechanism	Flexibility	Timeline	Applications
PLA	Hydrolysis of ester bonds	Moderate	Weeks–months	Sensor substrates, packaging
PCL	Enzymatic hydrolysis	High	Months–years	Flexible bioelectronics, chronic monitoring
Silk fibroin	Enzymatic proteolysis	High	Days–months	Neural interfaces, biopotential sensors
PLGA	Hydrolysis (tunable ratio)	Moderate	Weeks–months	Drug delivery, encapsulation
Cellulose derivatives	Enzymatic/hydrolytic	Moderate–high	Weeks–months	Wound monitoring, biosensors

2.3 Hybrid Composites and Inorganic Semiconductors

Modern bioresorbable device platforms utilize metallic conductors, polymeric substrates, inorganic semiconductors, and nanomaterials to create hybrid composites that are rationally designed. Most transient devices incorporate a core conductor/ substrate architecture consisting of a metal films (magnesium or molybdenum) on a silk or poly(lactic-co-glycolic acid) substrate (i.e. bilayer laminate). Nanoparticle-filled polymer composites with conductive metal nanoparticles or polypyrrole/ polyaniline-EDOT dispersed in polylactic acid or silk can provide solution-processable conductive films with conductivities ranging from 10 to 1000 S/m, allowing for biopotential recording (Corsi et al., 2025; Ko et al., 2025). Silicon nanomembranes (Si NMs) are the primary biodegradable semiconductor because they hydrolytically dissolve at a rate that is modulated by both their thickness and doping concentration.

Transistor gate insulators, encapsulants, and other dielectric layers can all be made from biodegradable materials, such as silicon dioxide (SiO_2), silicon nitride, and magnesium oxide. A combination of Mg-conductors added to silk substrates and Si NM materials would create the archetypal fully bioresorbable active electronics platform (Choi et al., 2020; Kang et al., 2020; Zhang et al., 2023). Bioresorbable energy storage batteries with a magnesium anode and supercapacitors also serve as independent power sources for periodic-use implants (Muniraj et al., 2026; Rajaram et al., 2022; Sheng et al., 2021; Xu et al., 2024).

3. FABRICATION TECHNIQUES AND DESIGN STRATEGIES

3.1 Microfabrication and Lithography

The application of semiconductor-based photolithographic microfabrication methods yields the highest level of spatial resolution available for the fabrication of transient electronic devices — specifically, microfabrication methods capable of producing sub-micron scale features necessary for active transistors, as well as active sensor and/or electrode arrays. Typically, there are several process steps involved in the fabrication of a complete bioresorbable active device — the deposition of Si NMs, by peeling them off of silicon-on-insulator (SOI) wafers and onto a biodegradable substrate; patterning the Mg or Mo conductor traces through lift-off lithography; depositing dielectric insulating layers (SiO_2 or MgO) using e-beam evaporation; encapsulating the completed bioresorbable active device in a biodegradable polymer (Han et al., 2020; Kang et al., 2020; Zhang et al., 2023).

The technology of transfer printing is crucial to the successful integration of inorganic semiconductors onto flexible biodegradable substrates. Using an elastomeric stamp to allow selective release of Si NMs or compound semiconductor components from a donor substrate and subsequently depositing them with accurate registration on a target bioresorbable substrate concerns the construction of a device architecture that would not have been otherwise accessible through direct deposition to soft polymer films (Choi et al., 2020; Dutta & Cheng, 2023).

3.2 Printing and Additive Manufacturing

Using inkjet, aerosol jet and screen printed biodegradable conductive inks provides scalable and low-cost fabrication methods for transient electronics without requiring clean room conditions where feature resolution is not required to be below 10μm. Biodegradable conductive inks (i.e. conductive Mg or Zn nanoparticles in polylactic acid (PLA) or polycaprolactone (PCL) binders, or conductive polymers dispersions such as polypyrrole and PEDOT in a biopolymer matrix) enable the printing of printed sensor electrodes, interconnects, and antenna structures directly to flexible biodegradable substrates (Janićijević et al., 2024; Ko et al., 2025).

Three dimensional (3D) and four dimensional (4D) bioprinting gives transient devices additional design space to encompass more complex geometries, such as helical coils for inductive coupling, tubular nerve conduits with stimulation electrodes built-in, and anatomically shaped sensor patches. The incorporation of stimuli-responsive (4D) properties allows printed components to change their shape upon hydration or temperature change following insertion into an animal's body, allowing for the minimally invasive delivery of pre-folded components through small incisions (Corsi et al., 2025; Ding et al., 2026; B. Li et al., 2026).

3.3 Encapsulation and Controlled Degradation Engineering

It can be argued that the encapsulation strategy represents one of most significant design choices when considering the operation lifetime, step change in degradation, and spatial consistency of failed dissolution of a transient electronic device. Biodegradable encapsulants such as thin films of silicon –di-oxide (SiO2), magnesium oxide (MgO), polylactic-co-glycolic acid (PLGA), beeswax, silk and other barrier materials provide the mechanism by which to limit the rate of dissolution of the device as the encapsulant serves as a barrier to premature aqueous ingress into the devices interior during the intended operating window of the device (Ashammakhi et al., 2021; Han et al., 2020; Kuzma et al., 2020).

The design of encapsulation for a given target lifespan requires both numerical modelling of the kinetics of water diffusion, kinetics of hydrolytic degradation, and kinetics of electrical performance (i.e., resistance, mechanical stiffness, structural integrity) of the electronic device. Liu et al. (2025) developed a framework called "transient mechanics" which is both analytical and computational, that can be used to predict the evolution of the electrical resistance, mechanical stiffness, and structural integrity of the device through time based on the dissolution of the encapsulants and conductors within the device(Liu et al., 2025). This framework allows for the rational design of the encapsulation architecture for the specific targeted operational window with predictable safety factors. Park et al. (2025) also reviewed recent advances related to bioresorbable biomedical applications from materials to devices and provided a comprehensive mapping of encapsulant materials to clinical timelines(Park et al., 2025).

Table 4. Comparison of fabrication techniques for transient bioelectronic devices

Technique	Resolution	Scalability	Compatible Materials	Limitations
Photolithography	Sub-micron	High	Si NMs, metals, SiO_2	Rigid substrates; cleanroom needed
Inkjet/aerosol printing	~10–50 µm	High	Conductive inks, polymers	Limited conductivity; ink formulation
Electrospinning	100 nm–10 µm	Moderate	PCL, PLA, silk, PLGA	Fiber alignment control
Transfer printing	~1–10 µm	Moderate	Si NMs, metals on flex	Complex process; yield
3D/4D bioprinting	~100 µm	Moderate	Hydrogels, composites	Low resolution; slow speed

4. TRANSIENT SENSORS FOR PHYSIOLOGICAL MONITORING

Figure 4. IoT-enabled transient bioelectronic systems

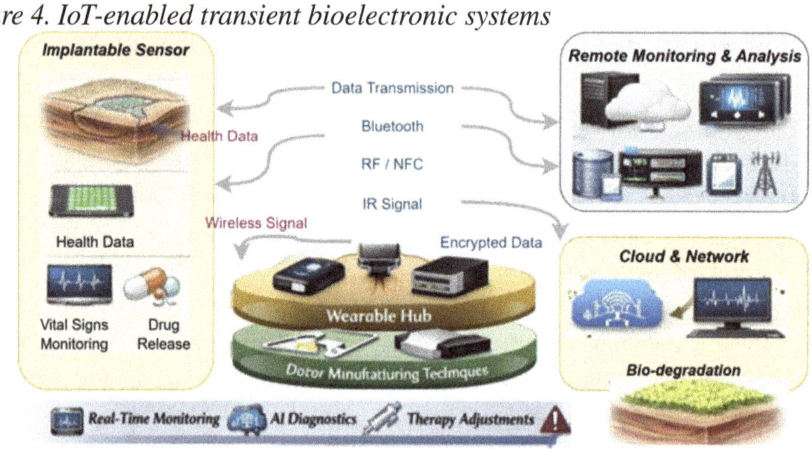

An illustration of how transient bioelectronics used in the body can be incorporated into a complete Internet of Things (IoT) enabled healthcare framework is shown in Figure 4. From the implant site, bioresorbable sensors that are implanted under the skin wirelessly send data related to physiological metrics (or monitoring) from the patient's body - such as vital sign readings, and delivery drug modality measurements (e.g.; time between drug dosages) - to a wearable hub via Bluetooth, radio frequency (RF), or near field communication (NFC). The hub then sends the encrypted data to the cloud, where it will be accessible remotely for IoT healthcare diagnostic assessment supported by Artificial Intelligence analytics (Ajitha et al., 2025; Boopathi, 2026; Raja, 2021b). As IoT-enabled diagnostic assessment is in real time, physicians can use a secure connection to view the analytics provided from the endpoint device during patient evaluation and, therefore, can provide appropriate alerts and modify care based on personal characteristics that impact patient care(Raja, 2021a, 2025b; Raja & Chakraborty, 2023). Additionally, the integration of transient bioelectronics, wireless communication networks and AI provides the basis for an IoT-enabled framework capable of providing continuous monitoring of unique patient data for precision medical care without retaining any permanent device in the body. Table 4 shows comparison of various fabrication techniques for transient bioelectronic devices(Boopathi Raja, 2021; B. Li et al., 2026; Raja, 2025a).

4.1 Overview of Bioresorbable Sensor Architectures

Transient sensors represent the most clinically mature class of wearable bioresorbable electronics. These sensors span a broad spectrum of disciplines, with many device types advancing through in vivo validation in animal models and early-phase clinical feasibility studies. The most clinically relevant sensor categories include temperature, pressure and strain, electrochemical, neural biopotential, and cardiovascular sensors. Each class is designed to address different physiological parameters, resulting in distinct requirements for sensitivity, operational lifetime, spatial form factor, and biocompatibility (Hu et al., 2024; Janićijević et al., 2024; Qi et al., 2025).

All bioresorbable sensor platforms share a common architectural blueprint: a biodegradable substrate (silk, PLGA, PCL) supports patterned metallic conductor electrodes or traces (Mg, Zn, Mo), one or more transduction elements (Si NM piezoresistors, enzyme-modified electrodes, thermoelectric junctions), and an encapsulation layer defining operational lifetime. Wireless readout — via inductive coupling or near-field communication — enables data extraction without percutaneous wiring that would create infection pathways and constrain patient mobility (Abhinav et al., 2025; Bae et al., 2025; Wang et al., 2025).

4.2 Temperature and Pressure Sensors

Temperature sensor that are bio-resorbable use the property of Mg or Si NM thermal resistance or the thermoelectric junction between two dissimilar biodegradable metals to continuously measure temeperature at wound sites or in the core of the body being monitored throughout the post-surgical period. They are constructed from a Mg tracing on a silk substrate and have been found to measure temperature with an accuracy of $\pm 0.1\,^{\circ}C$ over the duration of 1 to 4 weeks during testing in vitro and in rodent models of use (Janićijević et al., 2024; Zhang et al., 2023).

One particularly high-value application of bio-resorbable pressure sensors is in the measurement of intracranial pressure (ICP) in patients after a traumatic brain injury. Current clinical practice employs a permanent or semi-permanent IPC monitoring device that can become infected or cause bleeding during the transcranial implantation of the hardware. Fully bio-resorbable sensors for measuring ICP based on Si NM piezoresistors or capacitive pressure transducers supported on PLGA substrates have been found to measure pressures from 0 to 70 mmHg with a resolution of <1 mmHg in porcine models for a duration of 2 to 8 weeks, which is the amount of time required to monitor the ICP (Chu et al., 2025; Corsi et al., 2025; Mirzajani et al., 2025).

4.3 Electrochemical and Biochemical Sensors

Transitional sensors that are electrochemical in nature will measure various biochemical analytes such as glucose, lactate, pH, dissolved oxygen and electrolytic ions, in either interstitial fluid or wound exudate, giving clinical information on metabolic status necessary for postoperative management of complications including infection, ischemia, and metabolic complications. The device architecture utilizes Zn or Mo working electrodes and the working electrodes are modified with an enzyme-based membrane (glucose oxidase, lactate oxidase) or an ion-selective polymer film. The electrochemical cell also includes a reference electrode made out of Mg or Mo, and a counter electrodes(Hu et al., 2024).

A significant challenge for removable electrochemical sensors is to maintain electrode surface chemistry stability throughout the operational period. Specifically, enzymatic membranes need to remain intact while the underlying metal electrode continues to degrade, while the reference electrode's potential needs to remain stable while the ionic composition at the electrode-to-tissue interface varies. Recent advancements involving the use of conformal silk fibroin coating of enzymatic membranes and molybdenum pseudo-reference electrodes have demonstrated clinically acceptable performance (e.g., <15% drift over 2 weeks of in vitro measurement) of glucose sensing for the purpose of postoperative monitoring of metabolism (Hu et al., 2024; Janićijević et al., 2024).

4.4 Neural and Cardiovascular Monitoring

Bioresorbable neural sensors represent arguably one of the most complex applications of transient bioelectronics, and the simultaneous achievement of low electrode impedance (<100 kΩ at 1 kHz) necessary to record μV-level neural signals, intimate conformability to tissue to reduce damage due to micromotion between the electrode and the tissue, and matched degradation kinetics to the period of time required for monitoring after an injury or surgical procedure (Kim et al., 2023; Lee et al., 2021; Qi et al., 2025).

Biodegradable silicon-based electrode arrays fabricated on silk substrates have been used for recording of single unit and local field potential from the cortex of rodents, with functionality demonstrated over implantation periods of 2-4 weeks (Fanelli & Ghezzi, 2021; Kim et al., 2023; Lee et al., 2021). Upon absorption of moisture, silk dissolves and the array conforms closely to the surface of the cortex, thereby reducing mechanical mismatch that contributes to glial scarring associated with neuroinflammation, a common mode of failure for permanently implanted rigid electrode arrays. Biodegradable peripheral nerve interfaces, in both tubular cuff and flat paddle configurations, have been used for recording of compound nerve action potentials for evaluating the function of repaired peripheral nerves (Fanelli & Ghezzi, 2021; Kim et al., 2023; Lee et al., 2021). Biodegradable electrocardiographic patch electrodes, intra-cardiac pressure sensors, and coronary flow monitors designed for cardiovascular applications address the clinical need for monitoring after heart surgery without permanent hardware implantation. Cardiac devices generally have much more stringent durability requirements than any other type of device, as they must survive the mechanically dynamic cardiac environment (i.e., 1 Hz cyclic deformation; 60-100 mmHg pressure changes) for either one week or four weeks. Multi-layer SiO_2/PLGA encapsulation strategies have shown required durability in ovine cardiac implantation models (Cho et al., 2026; Corsi et al., 2025).

Table 5. Summary of bioresorbable sensor types, materials, and clinical applications

Sensor Type	Target Parameter	Materials Used	Operating Lifetime	Clinical Application
Temperature	Core body/wound temp.	Mg/Zn traces on silk	1–4 weeks	Post-surgical wound monitoring
Pressure/strain	Intracranial/intra-abdominal pressure	Si NMs, PLGA, Mo	2–8 weeks	TBI monitoring, GI diagnostics
Electrochemical	pH, glucose, ions	Zn/Mo electrodes, enzyme membranes	1–6 weeks	Metabolic monitoring post-surgery
Neural (EEG/EMG)	Biopotentials (μV–mV)	Mg/silk/Si NMs	2–12 weeks	Brain–machine interface, neuroprosthetics
Cardiovascular	ECG, blood flow	Mg, PCL, ZnO NWs	1–4 weeks	Post-cardiac surgery monitoring

5. BIORESORBABLE STIMULATORS AND THERAPEUTIC DEVICES

5.1 Electrical Stimulators for Wound Healing

Electric fields that come from inside the body develop by themselves at the wound sites due to moving ions through the skin and breaking the skin's natural electric potential gradient. They also are established signaling mechanisms to encourage galvanotactic movement of cells toward the center of wounds, to support the growth ("proliferation") of keratinocytes, and to speed up the formation of new blood vessels (angiogenesis). Bioresorbable devices that perform external electrical stimulation provide an additional means of stimulating the body's natural electric fields in order to speed the healing of chronic, non-healing wounds (diabetic ulcers, pressure sores) by increasing the effectiveness of the body's natural electrical response (Ashammakhi et al., 2021; Huang et al., 2024; Zhang et al., 2023). Table 5 shows summary of bioresorbable sensor types, materials, and clinical applications.

Temporary stimulators will include arrays of electrodes made from either zinc or magnesium patterned on the surface of PCL or PLA and capable of providing either Direct Current or Pulsed Current electrical fields for a very short but effective time period of 1-100 microamps. The degradation of these electrodes will add to the total amount of bioactive $Zn2+$ ions that Grows a biofilm, as well as have the potential to have anti-microbial effects. In various laboratory and animal studies, the use of bioresorbable electrical stimulation has been shown to reduce wound healing time by 30-50% compared to no stimulation, and complete resorption of the devices will occur within 2-4 weeks after stimulation has stopped (Huang et al., 2024; Lee et al., 2022).

5.2 Nerve Regeneration Stimulators

Repairs of surgical injuries to peripheral nerves are excellent applications for the use of bioresorbable stimulators. Research in humans has demonstrated that if an injured peripheral nerve is electrically stimulated after surgical repair for one to two weeks at a low frequency (20 Hz) from the proximal stump, this can assist with the regeneration of axons and therefore, result in a faster recovery of motor and sensory function to the nerve, and ultimately a better functional outcome. Using bioresorbable nerve stimulators makes it unnecessary to have a second surgical procedure in order to remove the permanent stimulator while applying the established short-course stimulation regime (Fanelli & Ghezzi, 2021; Huang et al., 2024; Kim et al., 2023).

Bioresorbable nerve stimulators have been designed with tubular cuff shapes wrapped around the injured nerve, and flat paddles positioned beside the injured nerve, both delivering charge balanced biphasic electrical stimulation through Mg or Mo electrodes. In rodent models of sciatic nerve transection and repair, bioresorbable stimulation has been shown to promote faster rates of axonal regeneration and improved electrophysiology and functional outcomes as compared to the no-stimulation control (Fanelli & Ghezzi, 2021; Huang et al., 2024; Lee et al., 2021).

5.3 Drug Delivery Systems with Transient Electronics

The combination of programmatically controlled release of drugs and transient electronics provides a new type of therapeutic implant that is capable of delivering on demand pharmacotherapy directly to the targeted tissue for only the amount of time needed before it completely dissolves. Electro-responsive drug delivery systems utilize electrical signals from bioresorbable circuitry to activate drug release from pre-loaded drugs in conducting polymer matrices (polypyrrole, PEDOT in biodegradable composites), in hydrogel reservoirs or from encapsulated nanoparticles (Ashammakhi et al., 2021; Corsi et al., 2025; Huang et al., 2024).

Some clinical applications of electronical drug delivery system technology include providing an anti-inflammatory drug for post-surgical use to reduce the inflammatory response of early implantation for optimal healing; providing localized antibiotic delivery at high risk sites for infection prevention (orthopedic implant beds, closures of abdominal wounds); and providing nerve growth factor for enhancing nerve regeneration due to stimulation by co-integrated electric stimulators. Also, the use of a transient electronic drug delivery system with temporal programmability offers physicians the ability to exercise more control over their patient's drug therapy via externally triggered wireless command or pre-programmed, on-chip electronic circuits that cannot be accomplished with passive drug-eluting biodegradable implants (Ding et al., 2026; Kim et al., 2023).

5.4 Bone and Musculoskeletal Applications

Piezoelectric bioresorbable devices are a very unique design concept that could speed up bone healing following fracture through the use of biodegradable piezoelectric materials (poly-L-lactic acid, barium titanate/PLGA composites, piezoelectric peptide nanofibers) to convert the mechanical energy derived from the physiological loading experienced during walking into local electrical stimulation of the fractured area and thus speed up the activity of osteoblasts, leading to increased deposition of bone matrix, without any external power. The device is physically loaded in the same manner as the injured bone is loaded and will also degrade in a time frame that is concurrent with the completion of the patient's healing process; this represents an ideal alignment of functional performance with time performance (Ashammakhi et al., 2021; Ding et al., 2026; Huang et al., 2024). Table 6 shows the comparative study of modalities, materials, and applications for different bioresorbable stimulators and therapeutic devices.

Table 6. Bioresorbable stimulators and therapeutic devices: modalities, materials, and applications

Device Type	Stimulation Modality	Key Materials	Target Duration	Application
Nerve regeneration stimulator	Electrical	Mg/PLGA/Si NMs	2–6 weeks	Peripheral nerve repair
Wound healing device	Electrical field	Zn/PLA/silk	1–4 weeks	Chronic wound closure, scar reduction
Drug delivery system	Electrophoresis/ thermal	Mo/PCL/drug matrices	Days–months	Targeted pharmacotherapy post-surgery
Bone repair stimulator	Piezoelectric	PVDF composites/Mg	4–12 weeks	Fracture healing acceleration
Anti-infection device	Electrical/chemical	Zn/Ag nanocomposites	1–3 weeks	Implant site infection prevention

6. WIRELESS COMMUNICATION IN TRANSIENT IMPLANTS

Figure 5. Biodegradation pathways of bioresorbable materials in transient bio-electronics

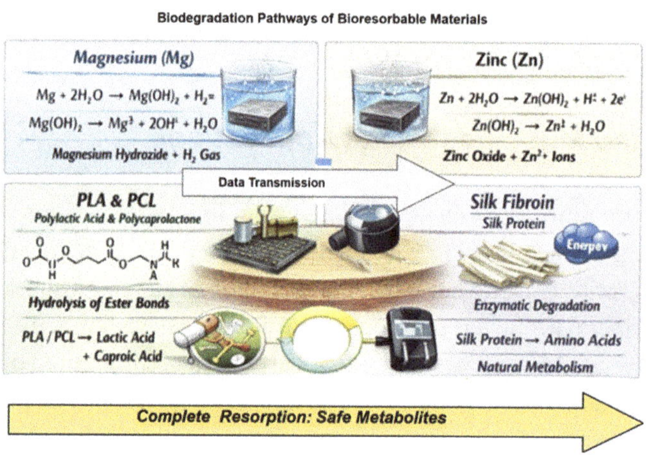

An illustration of the degradation process that occurs with the primary materials used in bioresorbable transient bioelectronics is shown in Figure 5. Electrochemical oxidation occurs with magnesium (Mg), resulting in the formation of magnesium hydroxide and hydrogen gas through the process of electrochemical oxidation. Zinc (Zn) undergoes a controlled process called corrosion to form zinc oxide and Zn2+ ions, while polylactic acid (PLA) and polycaprolactone (PCL) experience hydrolysis and thus cleave the ester linkages that connect the constituent monomers in the polymeric chain, forming lactic acid and caproic acid, which are the metabolites from which the polymers are formed, into citric acid cycle intermediates. Finally, silk fibroin is enzymatically degraded through hydrolysis to produce oligopeptides. The oligopeptides are hydrolyzed to produce L-amino acids which enter the metabolic pathway through naturally occurring biochemical processes. Table 7 shows different wireless communication modalities used for transient bioresorbable implants.

6.1 Biodegradable Antennas and RF Modules

Clinical use of temporary implants requires that they contain a means for wireless communication; without this capability, any data from sensors within the device cannot be accessed, nor can the parameters of an implanted stimulator be modified without making an incision to access the implant. Therefore, biodegradable wireless systems must have this capability using an antenna and circuit components that are biodegradable, which poses a major engineering challenge because radiofrequency performance relies on the geometry of the conductors and on the materials used to make those conductors; and both the geometry and the material properties evolve over time as the device degrades (Cho et al., 2026; Corsi et al., 2025; Wang et al., 2025).

To date, manufacturers have created biodegradable communication antennas that consist of thin-film traces composed of Mg, Zn, or Mo and arranged in planar coil/dipole/patch configurations, using either silk fibroin or PLGA substrates. Antennas operating at a frequency of 13.56 MHz (the NFC frequency) can transmit data passively and harvest power wirelessly at distances on the order of several centimeters, allowing communication with a handheld reader through the skin, without the need for a battery within the device. While the metallic trace of the antenna is dissolving, both the Q-factor of the antenna and it's resonant frequency change in a predictable manner, thereby allowing the antenna to provide a report of its own dissolution in addition to reporting data from the primary sensors within the implant (Dutta & Cheng, 2023; Janićijević et al., 2024; Wei et al., 2021).

6.2 Power Transfer and Energy Harvesting

The delivery of power to bioresorbable implants without the use of percutaneous wiring is achieved through three primary methods: inductive wireless power transfer, harvesting energy from physiological sources, and biodegradable primary batteries. With inductive coupling systems, the power is transmitted through the resonance of magnetic fields from external transmitter and implanted biodegradable receiver coils, providing constant power to enable continuous diagnostic sensing and stimulation. Evidence indicates that implanting biodegradable coils at depths of 1-5 cm into tissue can transfer 30-60% of the power to activate simple sensor and stimulator systems (Sheng et al., 2021; Wei et al., 2021; Xu et al., 2024).

Physiological energy harvesting or energy generated from the body produces energy in the form of mechanical movement (piezoelectric and triboelectric) or heat (thermoelectric) produces electricity that can be utilised to power transitory implants. Biodegradable piezoelectric nanogenerators using PVDF-TrFE and ZnO nanowires have been shown to produce between 10-100μW of electrical output power when physically deformed within a physiological setting. Power generated from biodegradable biofuel cells through the oxidation of glucose or lactate by using glucose oxidase and laccase on molybdenum (Mo) or carbon nanotube electrodes has demonstrated an open circuit voltage of approximately 0.5V, with a total power density of approximately 10μW/cm^2. Biodegradable biofuel cells produce continuous autonomous power without needing an external power supply (Muniraj et al., 2026; Sheng et al., 2021; Xu et al., 2024).

6.3 Signal Integrity during Degradation

To continuously provide the same level of wireless communication performance throughout the entire life cycle of a device despite a decrease in conductor thickness and an increase in the dissolution of the substrate, careful designs must be implemented into devices from the very beginning; ideally with the use of communication protocols that adapt to the expected parameter drift from their respective specifications. The resistance of the conductor increases during the dissolution of magnesium thus decreasing the efficiency of the antenna and creating a shift in the resonant frequency; however, the encapsulation geometry can be built in such a manner to allow for the thickness of the conductor to be maintained until very close to the end point of the expected operational life cycle of the device, at which point the ability to communicate will decrease gradually at the same time the device is no longer functional (Corsi et al., 2025; Liu et al., 2025).

Boopathi Raja (2026) has addressed the issue of security in IoT-connected sensor networks, an issue that is becoming increasingly important as bioresorbable implants become a part of wireless medical device networks. For clinical use of wirelessly networked transient implants to occur, it is paramount that the data transmitted between the devices is encrypted, that the devices can authenticate themselves with one another and that the data cannot be disrupted by a malicious element, especially in critical care monitoring applications where there will be a direct effect on patient therapy (Govindasamy, 2026; Karthika et al., 2026).

Table 7. Wireless communication modalities for transient bioresorbable implants

Technology	Frequency Range	Range	Power Mode	Applications
Near-field (NFC/ RFID)	13.56 MHz	<10 cm	Passive (harvested)	Data readout, passive sensors
Inductive coupling	kHz–MHz	1–10 cm	Wireless power + data	Powering implanted stimulators
UHF RF (sub-GHz)	400–900 MHz	Up to 1 m	Active (battery)	Real-time telemetry, neural recording
Ultrasound	1–10 MHz	1–5 cm tissue	Harvested	Deep implant power, backscatter comm.
Optical (IR)	Near-IR	<1 cm tissue	Harvested/active	Optogenetics, retinal implants

7. BIOCOMPATIBILITY, SAFETY, AND REGULATORY CONSIDERATIONS

7.1 In Vivo Degradation Pathways and Byproduct Safety

The biocompatibility of bioresorbable electronic devices is dependent on both the raw materials inside the device and the suite of by-products that come from the degradation of those materials as the device operates and eventually dissolves in the body. When magnesium degrades, it produces magnesium ions (Mg^{2+}), magnesium hydroxide precipitate, and hydrogen gas; zinc creates zinc ions (Zn^{2+}) and zinc oxide (ZnO); silicon nanomembranes, which are used as substrates in certain bioresorbable electronic devices, when they dissolve, yield orthosilicic acid ($Si(OH)_4$) or, similarly, silica (SiO_2) when it dissolves will have the same chemical composition (i.e., $Si(OH)_4$) while the different types of poly(lactic acid), poly(glycolic acid), polycaprolactone) or other poly(alpha-hydroxy) acid) will yield lactic acid, glycolic acid, and caprolactone acid monomer by-products. It is known from toxicity testing at the cellular (in vitro) and systemic (in vivo) levels that all by-products from the degradation of the various material types will not be produced in enough quantity to be toxic to animal or human subjects.

For example, Mg^{2+} is a naturally occurring electrolytic ion that is cleared from the body via the kidneys while $si(OH)_4$ is the metabolite of silica normally found in everyday diet while lactic and glycolic acids enter into normal biochemical metabolic pathways. However, hydrogen gas produced as a by-product from the degradation of Mg is of the greatest clinical concern because it can create temporary pockets of hydrogen gas in the subcutaneous tissue that can lead to significant clinical complications, which must be addressed with alloy selection and controlling the degradation rate of Mg. Therefore, the most critical design parameter is to ensure that the local peak concentration of all cyto-toxic and systemic-totoxic by-products produced from the degradation process will remain below the established thresholds for cytotoxic and systemic toxicity throughout the duration of the device's dissolution period.

7.2 Immune Response and Tissue Integration

In three phases, there are numerous alterations in the immune response towards bioresorbable devices. In the initial phase (hours to days) of acute inflammation from a foreign body insertion, neutrophils are the most numerous of the inflammatory leukocytes at the site of insertion. In the second phase (days to weeks), macrophage phenotype changes from a pro-inflammatory (M1 phenotype) predominance to a greater proportion of pro-resolving and pro-resorption (M2 phenotype) macrophages due to the favourable characteristics of the bioresorbable device materials. As the bioresorbable devices are resorbed, the continuing stimulus for the foreign body response decreases, thus stopping the continuing development of chronic fibrous encapsulation found in permanent devices, providing a major advantage with respect to biocompatibility with bioresorbable devices (Ashammakhi et al., 2021).

The surface chemistry of the biomaterials used in the construction of a bioresorbable device has a significant impact upon the degree of M1 to M2 macrophage polarization after device insertion. Comparatively, surfaces of silk fibroin and polycaprolactone (PCL) will have an increased tendency to polarize macrophages to M2 when compared to uncoated metallic surfaces. Furthermore, the coating of the surface with anti-inflammatory pharmaceuticals that are biodegraded i.e. dexamethasone, ibuprofen can provide another mechanism for controlling the acute inflammatory response to new bioresorbable devices. Yang et al. (2021) have shown that the introduction of photocurable bioresorbable adhesive interlayers between bioresorbable devices and soft tissue at the time of insertion can reduce the mechanical micromotion that will be encountered at the interface of a bioresorbable device and soft tissue, and thus reduce the chronic inflammation that can result from continued micromotion at the interface (Ashammakhi et al., 2021; Yang et al., 2021).

7.3 Standards and Regulatory Pathways

The rules and guidelines that regulate bioresorbable electronics are currently in a transition phase—more so than other types of medical devices, where the rules are set in stone. There are presently defined regulations on how the devices should be tested, seen as the FDA's Design Controls (21 CFR Part 820) and the ISO 10993 series for biological evaluation of medical devices, to mention a few. But no defined regulations for the characteristics unique to bioresorbable devices, such as time-dependent functionality, intentional dissolution, or novel material combinations; however, the regulatory processes that will provide definitive regulation on these characteristics are currently not validated or exhaustive.

There are several regulatory issues specific to bioresorbable electronics: validating the safety of dissolution products and completeness of dissolution from the start of dissolution to the completion of dissolution, establishing in vitro–in vivo correlations for degradation rate testing, establishing performance based on evolving electrical characteristics for devices, and creating regulatory classifications for novel material combinations, as there is no existing regulatory framework for designs of transient implants. The authors identified the development of regulatory pathways as a bottleneck for clinical translation. Pre-submission interaction with regulatory agencies was strongly encouraged and consideration given to device-specific testing methodologies as early in development as possible (Corsi et al., 2025; Ding et al., 2026; Park et al., 2025).

8. FUTURE DIRECTIONS AND EMERGING TRENDS

8.1 AI-Integrated Transient Systems

Combining AI and ML with bioresorbable bioelectronics is a technology area with enormous opportunity but significant technical challenges. In this vision, transient implants would contain not only passive sensors, but also analog or mixed-signal processor circuits on-chip that would provide feature extraction, anomaly detection, and adaptive stimulus parameter adjustment functions currently performed by external hardware in mainstream neurostimulation and monitoring systems (Corsi et al., 2025; Raja & Salau, 2026). Table 8 shows emerging trends and future directions in bioresorbable bioelectronics.

The key challenge is to implement meaningful computational functionality within the limits of biologically decomposable semiconductor technologies: thin film Si NM transistors would be used as the active device elements, however; the integration density and power efficiencies of these circuits currently fall well short of those of efficient silicon CMOS technology, restricting the complexity of computing that can be performed on-chip prior to transmitting raw data wirelessly for application of the data to external computing systems. An edge-computing approach to the challenge of splitting computation functionality between on-chip analog preprocessing (noise filtering, spike detection) and digital processing on an external system represents an achievable mid-term solution. Fully autonomous on-chip (AI-enabled) transient implants remain a longer-term goal (Boopathi, 2026; Corsi et al., 2025; B. Li et al., 2026; P. Li et al., 2026).

8.2 Multi-Functional Implants for Personalized Therapy

Combining the capabilities of sensing, stimulating, delivering medication through drug delivery, and communicating wirelessly into one bioresorbable device known as a multi-functional transient implant is currently the most complex devices' architecture in the field and has the greatest correlation with precision medicine via having a transient inject-able device implant at the targeted tissue site that autonomously monitors the physiological parameters and provides therapeutic responses such as drug-release and/or stimulation based on sensed abnormal activities; communicates this information to external devices; and dissolves after the completion of the mission (Cho et al., 2026; Corsi et al., 2025; Kim et al., 2026).

Many examples exist of prototype multi-functional bioresorbable devices integrating two of these functions based on evidence from past publications, such as using 2 pressure sensors + 2 NFC communication devices in 2020; using 2 temperature sensors + 1 drug delivery method in 2022; and using 1 neural recording device + 1 electrical stimulation device in 2024; however integrating the four functions incorporating multi-material synchronized degradation on a single fully resorbable platform remains an open area of research and subject for future work (Corsi et al., 2025; Ding et al., 2026; Kim et al., 2026).

Figure 6. Clinical workflow of transient implant deployment and resorption timeline

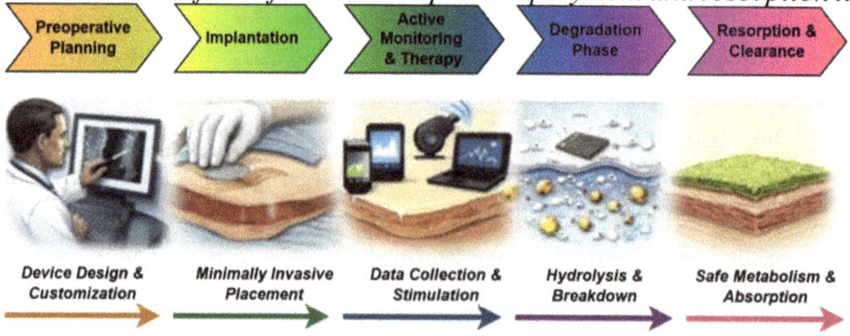

Figure 6 shows the entire clinical workflow for a bioresorbable transient implant begins with pre-operative planning, takes into account all patient-specific factors to determine the implant design and degradation, and continues through implantation and use of the implant followed by the resorptive process or bioresorption. The implant is placed minimally invasively during the implantation phase to allow for short-term monitoring or treatment. During active monitoring the implant captures physiological signals and controls stimulation before transmitting data wirelessly to external receivers. Upon completion of the active monitoring period the device enters into its degradation cycle through hydrolysis and enzymatic degradation while maintaining its biocompatibility throughout. The resorption phase is the conclusion of the cycle, where the metabolically produced by-products of degradation are cleared by the body in a natural way and there is no foreign residue present. The entire clinical workflow represents the clinical advantages of transient bioelectronics through their precise function, reduced risk of infection, and eliminating the need for explant surgery.

8.3 Roadmap toward Clinical Adoption

From demonstrating laboratory results to implementing bioresorbable bioelectronic systems in clinical settings, synced advancement across many areas will be needed:

- ***Material:*** new large libraries of conductor and substrate materials with greater overall in vivo degradation characterization,
- ***Fabrications:*** those processes used to manufacture devices should be scalable, able to reproduce, and be compatible with quality systems found in medical device production,
- ***Device engineering and functionality:*** higher quality encapsulations of devices, wireless functioning, and stable function,
- ***Regulatory models:*** establishing device-specific test requirements, and
- ***Clinical evidence:*** for establishing the level of evidence needed to support treatment of patients using bio-resorbable/biodegradable devices (prospective randomized, controlled studies to demonstrate clinical efficacy versus the present standard of care) (Corsi et al., 2025; Park et al., 2025; Singh et al., 2020).

The bioelectronic devices that are in the near future for clinical use have the most likelihood of successful implementation. This includes devices for intracranial pressure monitoring during persistently elevated pressure following TBI through a new semi-permanent-type monitoring device to replace current fixed bolt-type devices; devices for stimulating the peripheral nerves to increase nerve repair in a patient post-surgery without the need to implant permanent hardware after surgery; and devices for monitoring patients post-surgical cardiac procedures to replace temporary pacing & monitoring wires placed in-situ. Other secular applications of bioresorbable bioelectronic devices include the use of AI-assisted adaptive neuromodulation for treating patients with persistent pain and utilizing multi-modal monitoring methods for creating personalized post-surgical care plans, continuing as the technology and evidence base matures (Corsi et al., 2025; Kim et al., 2026; Mirzajani et al., 2025).

Table 8. Emerging trends and future directions in bioresorbable bioelectronics

Emerging Trend	Current TRL	Key Challenges	Representative Work
AI-integrated transient sensors	TRL 2–3	On-device computation with degradable circuits	Corsi et al., 2025; Raja, 2025
Multi-functional implants	TRL 3–5	Synchronized degradation of multiple functional layers	Kim et al., 2026; Cho et al., 2026
Bioresorbable soft robotics	TRL 1–2	Actuation + degradation compatibility	Li et al., 2026; Ding et al., 2026
Biodegradable energy storage	TRL 3–4	Energy density; safe by-products	Muniraj et al., 2026; Rajaram et al., 2022
Regulatory frameworks	TRL N/A	Novel material classification; standards gap	Corsi et al., 2025; Park et al., 2025

9. CONCLUSION

After decade of growth in the field of bioresorbable and temporary bioelectronics, devices that were once regarded as novelties are now a clinically relevant technology platform. Advances in biodegradable conductible/substrate materials, micro manufacturing methods suitable for soft/transitory systems, miniaturized wireless communication technologies, and increased knowledge of in vivo resorption/degradability and biocompatibility have led to the creation of a new field that can provide solutions to many of today's most pressing clinical challenges, such as carrying out patient monitoring following surgery, providing rehabilitation services, or performing temporary therapeutic interventions. In order to build transient bioelectronics, researchers have developed a collection of suitable materials that span from the use of traditional metal and/or metal alloy conductors (ie. magnesium, zinc, molybdenum, tungsten), to cellulosic polymers (ie. silk fibroin, polylactic acid

(PLA), poly(lactic acid -glycolic acid)(PLGA), and polycaprolactone (PCL)), and silicon nanomembrane semiconductor materials as well as hybrid nanocomposite materials. Having a broad and tunable materials toolkit enables researchers to create a solution that caters to the many varying time and functional requirements of clinical applications as they pertain to one another—ranging from acute/post-operative day-based applications (~days) to chronic/post-operative day-based applications (~months). Fabrication capabilities ranging from microfabrication, through transfer printing, to additive manufacturing, allow for both active device integration at high resolution, as well as scalable sensor array production. Shrinking Challenges such as synchronized multi-material degradation, integrity of wireless communications during dissolution, regulatory pathways development, and compiling clinical evidence have so far retained considerable objective definition while growing rapidly attracting sustained global research interest to their own resolution. The future integration of artificial intelligence; development of fully multi-used transient implants that mix sensing, stimulation, and programmable drug delivery; as well as the expansion of bioresorbable bioelectronics over new clinical territories-from soft robotics to neural prosthetics to personalized cancer therapy-promise to define the next decade of innovation in implantable medical technology. Realization of the resulting potential will require sustained collaborative efforts by experts of an interdisciplinary nature, including material sciences, device engineering, computational modeling, and regulatory expertise-and in this regard, the field is increasingly positioned to address.

REFERENCES

Abhinav, V., Basu, P., Verma, S. S., Verma, J., Das, A., Kumari, S., Yadav, P. R., & Kumar, V. (2025). Advancements in wearable and implantable BioMEMS devices: Transforming healthcare through technology. *Micromachines*, *16*(5), 522. DOI: 10.3390/mi16050522 PMID: 40428648

Ajitha, N., Ashwin, G., Divakar, S., & Dhanushree, M. (2025). Machine Learning-based Panic Syncsight: An IoT-Integrated Panic Attack Detection System for Enhanced Therapeutic Insights. 2025 5th International Conference on Soft Computing for Security Applications (ICSCSA), Alam, F., Ashfaq Ahmed, M., Jalal, A. H., Siddiquee, I., Adury, R. Z., Hossain, G. M., & Pala, N. (2024). Recent progress and challenges of implantable biodegradable biosensors. *Micromachines*, *15*(4), 475. PMID: 40283350

Ashammakhi, N., Hernandez, A. L., Unluturk, B. D., Quintero, S. A., de Barros, N. R., Hoque Apu, E., & Contag, C. (2021). Biodegradable implantable sensors: Materials design, fabrication, and applications. *Advanced Functional Materials*, *31*(49), 2104149. DOI: 10.1002/adfm.202104149

Bae, J.-Y., Choi, M.-K., & Kang, S.-K. (2025). Transient electronics for sustainability: Emerging technologies and future directions. *Beilstein Journal of Nanotechnology*, *16*(1), 1545–1556. DOI: 10.3762/bjnano.16.109 PMID: 40927489

Boopathi, R. G. (2026). Deep learning strategies for advanced wireless communication. In *Deep Learning Approaches in Intelligent Wireless Networking* (pp. 134–169). CRC Press. DOI: 10.1201/9781003751526-6

Boopathi Raja, G. (2021). *An intelligent approach for detecting and preventing heart attacks using wearable technology. Handbook of Research on Lifestyle Sustainability and Management Solutions Using AI Big Data Analytics, and Visualization.* IGI Gobal.

Cho, M., Chae, J. W., Park, Y. J., Yu, K. J., & Won, S. M. (2026). Biodegradable Implantable Electronics with Wireless Technology for Real-Time Clinical Applications. *Advanced Healthcare Materials*, *15*(3), e03424. DOI: 10.1002/adhm.202503424 PMID: 41001961

Choi, Y., Koo, J., & Rogers, J. A. (2020). Inorganic materials for transient electronics in biomedical applications. *MRS Bulletin*, *45*(2), 103–112. DOI: 10.1557/mrs.2020.25

Chu, Z., Zhou, Y., Li, S., Xu, Q., & Pan, L. (2025). Implantable medical electronic devices: Sensing mechanisms, communication methods, and the biodegradable future. *Applied Sciences (Basel, Switzerland)*, *15*(13), 7599. DOI: 10.3390/app15137599

Corsi, M., Bellotti, E., Surdo, S., & Barillaro, G. (2025). Implantable bioresorbable electronic systems for sustainable precision medicine. *Nature Reviews Electrical Engineering*, *2*(8), 572–583. DOI: 10.1038/s44287-025-00190-6

Ding, Z., Qiao, J., Gao, W. C., Wang, C., & Guan, Y. S. (2026). A comprehensive review on biodegradable materials and technologies for transient electronics. *Advanced Materials Technologies*, *11*(6), e01928. DOI: 10.1002/admt.202501928

Dutta, A., & Cheng, H. (2023). Pathway of transient electronics towards connected biomedical applications. *Nanoscale*, *15*(9), 4236–4249. DOI: 10.1039/D2NR06068J PMID: 36688506

Fanelli, A., & Ghezzi, D. (2021). Transient electronics: New opportunities for implantable neurotechnology. *Current Opinion in Biotechnology*, *72*, 22–28. DOI: 10.1016/j.copbio.2021.08.011 PMID: 34464936

Feiner, R., & Dvir, T. (2017). Tissue–electronics interfaces: From implantable devices to engineered tissues. *Nature Reviews. Materials*, *3*(1), 17076. DOI: 10.1038/natrevmats.2017.76

Fernandes, C., & Taurino, I. (2022). Biodegradable molybdenum (mo) and tungsten (w) devices: One step closer towards fully-transient biomedical implants. *Sensors (Basel)*, *22*(8), 3062. DOI: 10.3390/s22083062 PMID: 35459047

Govindasamy, B. R. (2026). Securing IoT and sensor networks with blockchain: Architecture and implementation.

Han, W. B., Lee, J. H., Shin, J. W., & Hwang, S. W. (2020). Advanced materials and systems for biodegradable, transient electronics. *Advanced Materials*, *32*(51), 2002211. DOI: 10.1002/adma.202002211 PMID: 32974973

Hu, C., Wang, L., Liu, S., Sheng, X., & Yin, L. (2024). Recent development of implantable chemical sensors utilizing flexible and biodegradable materials for biomedical applications. *ACS Nano*, *18*(5), 3969–3995. DOI: 10.1021/acsnano.3c11832 PMID: 38271679

Huang, Y., Yao, K., Zhang, Q., Huang, X., Chen, Z., Zhou, Y., & Yu, X. (2024). Bioelectronics for electrical stimulation: Materials, devices and biomedical applications. *Chemical Society Reviews*, *53*(17), 8632–8712. DOI: 10.1039/D4CS00413B PMID: 39132912

Janićijević, Ž., Huang, T., Bojórquez, D. I. S., Tonmoy, T. H., Pané, S., Makarov, D., & Baraban, L. (2024). Design and development of transient sensing devices for healthcare applications. *Advancement of Science*, *11*(20), 2307232. PMID: 38484201

Kang, S.-K., Yin, L., & Bettinger, C. (2020). The emergence of transient electronic devices. *MRS Bulletin*, *45*(2), 87–95. DOI: 10.1557/mrs.2020.19

Karthika, R., Karthikeyan, K., Pavithra, S., Sandeep, T., & Divyaa, N. (2026). IoT-Enabled Accident Prevention System: Real-Time Detection of Sleep Deprivation, Alcohol Influence, and Heart Attack Symptoms. 2026 7th International Conference on Mobile Computing and Sustainable Informatics (ICMCSI), Kim, G., Hong, M., Lee, Y., & Koo, J. (2023). Biodegradable materials and devices for neuroelectronics. *MRS Bulletin*, *48*(5), 518–530.

Kim, J., Seo, Y., & Kim, J. (2026). Bioresorbable Systems in Healthcare for Transient Sensing and Therapeutic Interventions. *Advanced Therapeutics*, *9*(1), e00544. DOI: 10.1002/adtp.202500544

Ko, G. J., Naganaboina, V. R., Goda, E. S., Dutta, A., Cheng, H., & Hwang, S. W. (2025). Towards Polymer Composite-Based Transient Electronic Systems. *Advanced NanoBiomed Research*, *5*(9), 2400126. DOI: 10.1002/anbr.202400126

Kuzma, M., Gerhard, E., Shan, D., & Yang, J. (2020). Advances in bioresorbable electronics and uses in biomedical sensing. In *Interfacing Bioelectronics and Biomedical Sensing* (pp. 29–72). Springer. DOI: 10.1007/978-3-030-34467-2_2

Lee, S., Silva, S. M., Aguilar, L. M. C., Eom, T., Moulton, S. E., & Shim, B. S. (2022). Biodegradable bioelectronics for biomedical applications. *Journal of Materials Chemistry. B*, *10*(42), 8575–8595. DOI: 10.1039/D2TB01475K PMID: 36214325

Lee, W. H., Cha, G. D., & Kim, D.-H. (2021). Flexible and biodegradable electronic implants for diagnosis and treatment of brain diseases. *Current Opinion in Biotechnology*, *72*, 13–21. DOI: 10.1016/j.copbio.2021.07.027 PMID: 34425329

Li, B., Wang, Y., Mei, J., & He, X. (2026). Wearable Bioelectronics: Toward Future Personal Health Monitoring. *BIO Integration*, *7*(1), 993. DOI: 10.15212/bioi-2025-0219

Li, P., Hibino, N., Shi, L. L., & Tian, B. (2026). Minimally invasive bioelectronic implants. *Nature Materials*, •••, 1–17. PMID: 41986453

Li, R., Wang, L., Kong, D., & Yin, L. (2018). Recent progress on biodegradable materials and transient electronics. *Bioactive Materials*, *3*(3), 322–333. DOI: 10.1016/j.bioactmat.2017.12.001 PMID: 29744469

Liu, P., Lee, C., Carusetta, N., & Avila, R. (2025). Transient mechanics: A perspective on bioresorbable electronics. *Journal of Applied Mechanics*, *92*(5), 051007. DOI: 10.1115/1.4067952

Liu, Y., Yu, Q., Yang, L., & Cui, Y. (2023). Materials and biomedical applications of implantable electronic devices. *Advanced Materials Technologies*, *8*(4), 2200853. DOI: 10.1002/admt.202200853

Mirzajani, H., Zolfaghari, P., Akbari Nakhjavani, S., Koca, B. Y., Khodapanahandeh, M., & Urey, H. (2025). Transient implantable electronics for postsurgery preventive medicine. *Advanced Functional Materials*, *35*(3), 2413324. DOI: 10.1002/adfm.202413324

Muniraj, V. K. A., Hammoud, H., & Djenizian, T. (2026). Bioresorbable battery technologies for resorbtronics: A review on energy storage for transient medical devices. *Journal of Materials Research*, *41*(8), 1–14. DOI: 10.1557/s43578-026-01815-8

Park, D. Y., Lee, H. M., Kim, S. H., Sim, Y., Kang, Y., Jang, G., Kim, S., Kwon, K., & Ryu, H. (2025). Recent Advances in Bioresorbable Biomedical Applications: From Materials to Devices. *Advanced Electronic Materials*, *11*(10), 2400997. DOI: 10.1002/aelm.202400997

Qi, L., Wang, Y., & Liang, X. (2025). Emerging Implantable Sensor Technologies at the Intersection of Engineering and Brain Science. *Biosensors (Basel)*, *15*(11), 762. DOI: 10.3390/bios15110762 PMID: 41294772

Raja, G. B. (2021a). Fingerprint-based smart medical emergency first aid kit using IoT. In *Electronic Devices, Circuits, and Systems for Biomedical Applications* (pp. 325-348). Elsevier. DOI: 10.1016/B978-0-323-85172-5.00015-0

Raja, G. B. (2021b). Impact of internet of things, artificial intelligence, and blockchain technology in Industry 4.0. In *Internet of things, artificial intelligence and blockchain technology* (pp. 157–178). Springer. DOI: 10.1007/978-3-030-74150-1_8

Raja, G. B. (2025a). Advancements of IoT and sensor informatics in wearable, implantable, mobile, and remote healthcare. In *Innovations in Biomedical Engineering* (pp. 87–121). Elsevier. DOI: 10.1016/B978-0-443-30146-9.00003-4

Raja, G. B. (2025b). SmartCare—A comprehensive wearable sensor-based remote health monitoring and rehabilitation system. In *Rehabilitation Robotics and Healthcare Devices* (pp. 27–46). Elsevier. DOI: 10.1016/B978-0-443-21505-6.00011-6

Raja, G. B., & Chakraborty, C. (2023). Internet of things based effective wearable healthcare monitoring system for remote areas. In *Implementation of Smart Healthcare Systems using AI, IoT, and Blockchain* (pp. 193-218). Elsevier. DOI: 10.1016/B978-0-323-91916-6.00004-7

Raja, G. B., & Salau, A. O. (2026). Emerging Innovations in Assistive Technologies: Adaptive, Cloud-Based, and IoT-Integrated Brain-Computer Interfaces Platforms for Neurorehabilitation. In *Brain-Computer Interfaces for Neurorehabilitation* (pp. 323-350). IGI Global Scientific Publishing.

Rajaram, K., Yang, S. M., & Hwang, S.-W. (2022). Transient, biodegradable energy systems as a promising power solution for ecofriendly and implantable electronics. *Advanced Energy and Sustainability Research*, *3*(9), 2100223. DOI: 10.1002/aesr.202100223

Ryu, H., Seo, M. H., & Rogers, J. A. (2021). Bioresorbable metals for biomedical applications: From mechanical components to electronic devices. *Advanced Healthcare Materials*, *10*(17), 2002236. DOI: 10.1002/adhm.202002236 PMID: 33586341

Sheng, H., Zhang, X., Liang, J., Shao, M., Xie, E., Yu, C., & Lan, W. (2021). Recent advances of energy solutions for implantable bioelectronics. *Advanced Healthcare Materials*, *10*(17), 2100199. DOI: 10.1002/adhm.202100199 PMID: 33930254

Singh, R., Bathaei, M. J., Istif, E., & Beker, L. (2020). A review of bioresorbable implantable medical devices: Materials, fabrication, and implementation. *Advanced Healthcare Materials*, *9*(18), 2000790. DOI: 10.1002/adhm.202000790 PMID: 32790033

Tan, M. J., Owh, C., Chee, P. L., Kyaw, A. K. K., Kai, D., & Loh, X. J. (2016). Biodegradable electronics: Cornerstone for sustainable electronics and transient applications. *Journal of Materials Chemistry. C, Materials for Optical and Electronic Devices*, *4*(24), 5531–5558. DOI: 10.1039/C6TC00678G

Wang, Y., Wu, C., Zhang, Y., Du, S., & Zhou, X. (2025). Wearable and implantable transient bioelectronics. *Journal of Materials Chemistry. C, Materials for Optical and Electronic Devices*, *13*(29), 14682–14696. DOI: 10.1039/D5TC01911G

Wei, Z., Xue, Z., & Guo, Q. (2021). Recent progress on bioresorbable passive electronic devices and systems. *Micromachines*, *12*(6), 600. DOI: 10.3390/mi12060600 PMID: 34067419

Xu, M., Liu, Y., Yang, K., Li, S., Wang, M., Wang, J., & Castro, F. A. (2024). Minimally invasive power sources for implantable electronics. Exploration, Yang, Q., Wei, T., Yin, R. T., Wu, M., Xu, Y., Koo, J.,...Kandela, I. (2021). Photocurable bioresorbable adhesives as functional interfaces between flexible bioelectronic devices and soft biological tissues. *Nature Materials*, *20*(11), 1559–1570.

Zhang, Y., Lee, G., Li, S., Hu, Z., Zhao, K., & Rogers, J. A. (2023). Advances in bioresorbable materials and electronics. *Chemical Reviews*, *123*(19), 11722–11773. DOI: 10.1021/acs.chemrev.3c00408 PMID: 37729090

Chapter 13
Green Synthesis of Titanium Dioxide Nanoparticles Using Plant Extracts

Aditya Shrivastav
http://orcid.org/0009-0000-8461-8171
Ajeenkya D.Y. Patil University, India

Susanta Das
http://orcid.org/0000-0002-9314-3988
Ajeenkya D.Y. Patil University, India

ABSTRACT

TiO_2 NPs, characterized by their photocatalytic activity, non-toxicity, and high thermal stability, have attracted considerable attention in areas such as environmental remediation, healthcare, energy, and cosmetics. Green synthesis of TiO_2 NPs lessens the use of hazardous chemicals and energy-intensive processes. The process mainly involves harvesting bioactive compounds from the plant materials that help reduce Ti ions to NPs. They exhibit excellent photocatalytic ability to degrade organic pollutants and dyes under UV illumination along with broad-spectrum antibacterial and antifungal abilities. The biocompatibility of green synthesized TiO_2 NPs shows their tremendous opportunities for applications in drug delivery and photothermal therapies in biomedicine. Furthermore, their large surface area promotes the use of these particles in other sectors too. To conclude, the "green" synthesizing of TiO_2 NPs is very promising, sustainable, and innovative by extending their scope of applications while closely following environmental sustainability paradigms.

DOI: 10.4018/979-8-2600-0060-1.ch013

1. INTRODUCTION

Nanotechnology, which functions at the atomic and molecular levels, has transformed fields such as medicine, biotechnology, and energy. It provides tailored drug delivery, which enhances diagnosis and therapy (Wei et al., 2021). Nanoparticles (NPs) produced with precision compositions and consistent sizes, advance nanotechnology rapidly. It is particularly flexible and has wide application in biology and pharmacology as well as much encourage product development. In a bid to eradicate the use of toxic chemicals in the production of different NPs, Traditional (physical & chemical) process involves toxic chemicals, costly, and large instruments to synthesise NPs. The researchers are developing environmentally friendly green synthesis to make the production cost-effective and less hazardous (Santhoshkumar et al., 2014; Nadeem et al., 2018). Figure 1 shows the visualization of the annual trend in publications about Green, Physical and Chemical Synthesis methods listed on the PubMed database. The upward trend for Green Synthesis methods became more pronounced starting from 2018 although both Physical and Chemical methods showed increasing trends in publication numbers from 2012.

TiO_2 NPs have many unique properties which make them viable candidates for numerous applications. They are used in photocatalysis, photovoltaic applications, and for the development of sensors (Irshad et al., 2021). Having good optical properties, thermal stability, and non-toxicity, TiO_2 NPs can be applied to cosmetics, medicine, antibacterial layers and so on. The synthesized TiO_2 NPs exhibit antibacterial and anticancer potentials and are active under UV irradiation by the generation of radical oxygen species (ROS) (Santhoshkumar et al., 2014; Nadeem et al., 2018; Ganesan et al., 2016). TiO_2 NPs, especially the rutile and anatase types, are often used because of their stability and UV light photocatalytic behaviour, although limited by charge separation and surface area. Nonetheless, TiO_2 NPs are utilised in cosmology, electronics, catalysts, and biomedical engineering to enhance functionality with size and a crystalline phase shift (Jassal et al., 2022). The photocatalytic features of TiO_2 NPs assist us to preserve historical monuments as well. They actively safeguard stone surfaces through self-cleaning and depollution processes (Waghmode et al., 2019).

Figure 1. Comparative analysis of PubMed-indexed publications on green, physical, and chemical synthesis

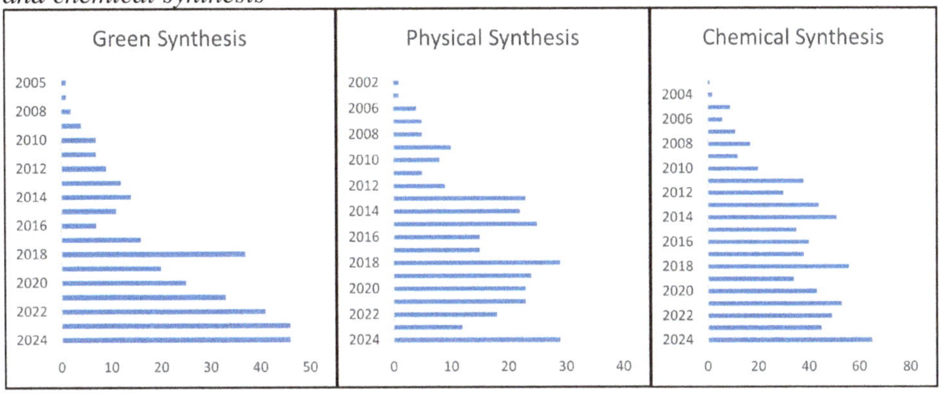

TiO_2 NPs introduced into Ti alloy are used in aerospace industries because they offer the advantage of being lightweight, strong, and highly resistant to corrosion, making them suitable for aerospace applications such as aircraft parts. Possible human health risks associated with TiO_2 exposure are accumulation in tissues and interactions with certain amino acids. This impact of water borne TiO_2 from personal care items on the aquatic life, especially algae, is due to the facts that TiO_2 induces oxidative stress and membrane damage. Points have been raised that burning TiO_2 contaminated sludge is safe for the environment (Waghmode et al., 2019). Synthetic organic dyes are toxic, chemically inert compounds that are non-biodegradable, too. Reactive radicals are generated in advanced oxidation processes with the aim of degrading them. This can be done by applying semiconductor photocatalysis with materials such as TiO_2, zinc oxide (ZnO) and tin oxide (SnO_2). Especially, in TiO_2 anatase, is photocatalytic and relatively economical (Mousa et al., 2022). Further, when we talk about its approaches and uses to have biological process hereby it is noted that they have grown larger as these are more affordable and not deleterious as compared to regular chemical procedures. TiO_2 NPs, with photocatalytic prowess, stability, and safety concerns; they used in cosmetics, medicine, and UV protection. TiO_2 NPs show antibacterial and antioxidant, so they are suitable for water treatment and as an anti-bacterial coating. TiO_2 NPs are typically prepared by different chemical & physical methods (e.g., laser ablation, sol-gel, & hydrothermal). These technologies offer flexibility and the ability to govern NPs size and morphology, yet they are costly, energy injecting, and ecologically unfriendly, thus limiting mass production of NPs (Irshad et al., 2021). Green synthesis methods employ plant extract, micro-organisms, and biological precursor. These are environment-friendly and economic synthetic routes that can be applied in catalysis and degradation. Such green methods are non-toxic, environment friendly & economical, thereby enabling the large-scale synthesis of TiO_2 NPs with different size, morphology, and properties (Irshad et al., 2021).

In this chapter we have discussed the plant-mediated green synthesis of TiO_2 NPs and the steps that we have covered include the synthesis of TiO_2 NPs through leaf extract, the synthesis through herba extract, and synthesis via fruit peel extract. Their characterization as well as use in different ways are also discussed in the chapter.

2. SYNTHESIS OF TIO_2 NANOPARTICLES

2.1. Plant-mediated Biogenic Synthesis of TiO_2 NPs

The process typically involves several steps (Wei et al., 2021; Santhoshkumar et al., 2014; Nadeem et al., 2018; Jassal et al., 2022; Irshad et al., 2021):

A. Extraction: Aqueous extracts are obtained from plant materials such as leaves, roots, or seeds.
B. Reduction: Plant extract is mixed with the titanium precursor (e.g., titanium tetrachloride) to form TiO_2 NPs where phytochemicals reduce titanium ions.
C. Stabilization: Phytochemicals on the same side as the stabiliser facilitate the staking of the nanoparticles, preventing agglomeration and maintaining a uniform size distribution.
D. Characterization: To characterise the synthesised nanoparticles, various techniques like UV-Vis spectroscopy, XRD, TEM, SEM, and FTIR are used. Figure 2 illustrates various benefits.

Figure 2. Various benefits of green synthesis

2.2. *Calendula officinalis* Leaves

The synthesis of titanium nanoparticles (TiNPs) involves mixing 40 ml of the extract with a concentration of 15 g/L with 200 ml of titanium tetrachloride ($TiCl_4$) of concentration 0.15 M. The mixture was refluxed at 120 °C in the dark, and this colour changed to light brown, indicating formation of particles (Wei et al., 2021). UV-Vis spectroscopy confirmed that TiNPs were synthesized with an absorbance peak at 228 nm, Fourier transfer infrared spectroscopy (FT-IR) was found to reveal peaks related to Ti-O bending vibrations and a multitude of functional groups resulting from the plant extract indicating organic compounds for the stabilization of NPs, XRD clearly shows that the synthesized TiNPs were in crystalline form and with an average calculated crystal size ~ 16 nm using the Scherrer equation, SEM images showed spherical morphology of TiNPs with well-spaced growth rates from ~ 16 nm to ~ 41 nm well supported by EDS which confirmed presence of Ti & O in nanoparticles & vine organic compounds from extract (Wei et al., 2021).

2.3. *Psidium Guajava* Leaves

Leaf extract of *Psidium Guajava* (Figure 3(a)) was used to prepare the particles. XRD pattern for synthesised TiO_2 NPs was consisted of both *anatase* and *rutile* phase, which indicated nanocrystalline nature of particles with average size of ~ 33 (Santhoshkumar et al., 2014). FTIR studies identified various functional groups associated with leaf extract. FESEM images depicted spherical surface NPs. EDX analysis further confirmed the formation of TiO_2 NPs. The grown TiO_2 NPs had potent antibacterial properties against *S. aureus* & *E. coli*. DPPH scavenging activity indicated that free radicals were reduced by over 85% by TiO_2 NPs which was considerably higher than that of the guava leaf extract (Santhoshkumar et al., 2014).

2.4. *Altissima* Leaves

Appropriate proportions of *Altissima* leaves and distilled sterile water were mixed and heated (~ 60^0) for some time (~ 10 minutes). The mixture was subsequently cooled and filtered to prepare the extract. The extract was then mixed with $TiO(OH)_4$ solution and kept under light for several hours (Ganesan et al., 2016). The green synthesis of TiO_2 NPs using aqueous extract of *altissima* leaf yielded a colour change from light green to dark brown. From the UV-vis spectroscopy it was observed that the peaks at 332 nm confirmed the existence of TiO_2 NPs. More specifically, FTIR analysis revealed that alcohol, phenolic and other functional groups take part in creating NPs (Ganesan et al., 2016). XRD further supported the anatase crystalline structure of the synthesized TiO_2 NPs. FESEM analysis of the nanoparticles exhibited spherical particles with size of 60 to 100 nm and the EDX analysis confirmed the presence of titanium. The TiO_2 NPs demonstrated high photocatalytic activity, where the Methylene Blue dye was degraded to a rate of up to 86.79% under sunlight (Ganesan et al., 2016).

2.5. *Azadirachta Indica* Leaves

Azadirachta indica leaves (Figure 3(b)) contain antibacterial properties against various pathogens. Thakur et al., (2019) synthesized TiO_2 NPs from the leaf extract of this plant. XRD pattern demonstrated the formation of crystalline nature with average particle size of ~ 42 nm. SEM and TEM analysis determined spherical NPs size of varying sizes (25-87 nm & 15-45 nm) (Thakur et al., 2019). FTIR spectrum indicated the presence of various biomolecules, which were responsible for formation of NPs. These NPs possessed a strong antibacterial activity against various pathogens (*E. coli, S. aureus, B. subtilis, S.typhi* & *K. pneumoniae*) higher than bulk TiO_2.

2.6. *Moringa Oleifera* Leaves

TiO_2 NPs were synthesized from *Moringa oleifera* leaf (shown in Figure 3(c)) extract. Mixture of *Moringa oleifera* (Figure 3(c)) leaf powder and deionized water was heated at moderate temperature for some time to prepare the leaf extract. The extract was then mixed with TiO_2 solution under controlled conditions (Sivaranjani & Philominathan, 2016). The blackish brown colour that emerged after the reaction confirmed the synthesis of TiNPs (Sivaranjani & Philominathan, 2016). The NPs were further processed and purified. UV-Vis spectral analysis (SPR peak at ~ 428 nm) supported the formation of NPs by determining the colour variations in the reaction systems. SEM images confirmed formation of spherical particles with approximate size of 100 nm. Wound healing experiments on rats revealed considerable contraction rates. This study proves that TiO_2 NPs can be beneficial for wound healing, which is more effective than conventional therapies (Sivaranjani & Philominathan, 2016).

2.7. *Trigonella Foenum Graecum* Leaves

Trigonella foenum graecum leaves collected and were used to prepare the plant extract. TiO_2 NPs were synthesised by adding the extract to titanium oxy sulphate ($TiOSO_4$) and adjusting the pH with sodium hydroxide (NaOH). When the filtration and sintering were conducted at 700°C, it led to the formation of well-crystalline nano TiO_2. In this study, FTIR spectra suggested the formation of TiO_2 while XRD presented a clear crystallinity of the anatase phase. UV-Vis measurement showed that there was a high absorbance at 400 nm (Subhapriya & Gomathipriya, 2018). SEM showed that the prepared NPs were spherical with size of 20-90 nm in diameter. The presence of TiO_2 was further confirmed by EDX. The prepared TiO_2 NPs exhibited appreciable bactericidal effects against diverse bacterial and fungal species which suggests that the TiO_2 NPs have practical application in biological system as potential antimicrobial agent (Subhapriya & Gomathipriya, 2018).

2.8. *Aloe vera* Leaves

The materials used were TiO_2, Rhodamine B dye and double distilled water. The leaves of aloe vera (shown in Figure 3(d)) were collected and then its extract was prepared by boiling the inner gel in deionized water for 10 min followed by filtration then refrigerated at 10°C. In synthesis of TiO_2 NPs, 100 mL of the $TiO(OH)_2$ (5 mM) and 20 mL of Aloe vera extract were stirred for 24 hrs to form the light greenish NPs which were then dried at 120°C. The diffused reflectance spectra of TiO_2 NPs had a band gap of 3.1 eV through UV-Vis spectroscopy (Nithya et al., 2013). XRD analysis showed characteristic diffraction pattern at 25.2, 47.9 and 55.0 peaks attributed to the anatase TiO_2 NPs and particle size of the synthesised NPs was 11 nm. Following the fabrication of NPs, absorption peaks of hydroxyl groups at 3396 cm^{-1} and amino groups at 1608 cm^{-1} were observed on the Aloe vera sample using FT-IR. Atomic microscopy showed the particle size to range between 80-90 nm. The photocatalytic activity of the TiO_2 NPs was investigated using Rhodamine B under visible light and proved to be 41% for decolorizing ability compare to TiO_2 (24%) with pseudo first order kinetics in difference absorbance spectra (Nithya et al., 2013).

2.9. *Hypheae Thiebeace* and *Ananos Seneglensisa* Leaves

Leaves of *Hypheae Thiebeace* and *Ananos seneglensisa* were harvested and subsequently dried, ground into powder and subjected to extraction using distilled water. These extracts were employed for synthesizing the TiO_2 NPs. (Hassan et al., 2019). UV-Vis spectroscopy confirmed the synthesis of TiO_2 NPs from *Hypheae Thiebeace* and extracts with showed peaks at 229 nm (*Hypheae Thiebeace*) and 231 nm (*Ananos Seneglensisa*). From the FTIR study it was revealed that biomolecules were actively involved in nanoparticle stabilisation. SEM images indicated formation of spherical particles with different sizes (40 - 50 nm). From the XRD, the material was crystalline TiO_2 with particle size between 42-58 nm. Antibacterial and antifungal investigations suggested their potential applications in medicine. (Hassan et al., 2019).

Figure 3: (a) Psidium Guajava, (b) Azadirachta Indica, (c) Moringa Oleifera and (d) Aloe vera

2.10. *Buddleja Asiatica* Leaves

Buddleja asiatica leaves were first washed to remove any contamination from bacteria, fungi, then shade-dried for 72 hours in a clean environment and finally mixed with distilled water to develop the extract. The end extract was first filtered, then chilled, and used in TiO_2 NPs biosynthesis process. The obtained leaf extract was added to TiO_2 salt solution to prepare NPs. UV-Vis photo spectrometry confirmed the synthesis of TiO_2 NPs (characteristic peak at 397 nm) (shown in Figure 4(a)). SEM images exhibited spherical shapes of different sizes (30 to 111 nm) (shown in

Figure 4(b)). EDX analysis detected presence of Ti & O elements in NPs with ~ 45:55. Solvent extraction qualitative analysis of plant extract by FTIR showed existence of functional groups (such as O-H, N-H, C-H, & carbonyl) (Mustafa *et al.*, 2021).

Figure 4. Results of (a) UV-Vis and (b) SEM images (Mustafa et al., 2021). Creative Commons Attribution 4.0 International (CC BY 4.0)

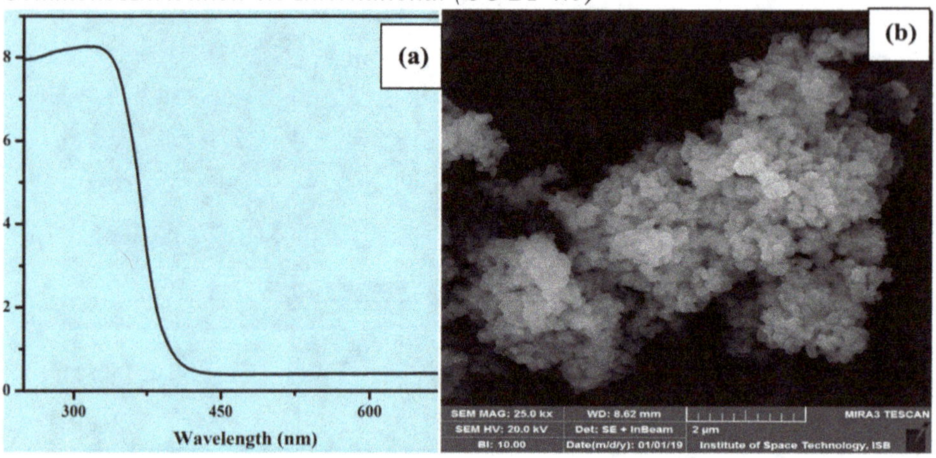

2.11. *Echinacea Purpurea Herba*

Echinacea purpurea herba extract and TiO_2 solution were mixed in appropriate proportions and was stirred for a few hours under controlled conditions. Change of color confirmed the formation of NPs. UV-Vis spectrum indicated SPR peak at ~ 280 nm, SEM images showed formation of spherical particles of average size ~ 120 nm, FTIR peaks indicated presence of stretching/bending of various bonds associate with bio compounds of extract, and XRF exhibited a strong signal at 4.5 kV corresponding to titanium element (Dobrucka, 2017).

2.12. *Prunus domestica, Prunus persica* and *Actinidia deliciosa* Fruit Peel Extracts

The combination of *Prunus domestica* (Plum) and *Prunus persica* (Peach) and *Actinidia deliciosa* (Kiwi) fruit peel extracts allow for a sustainable titanium dioxide (TiO_2) nanoparticle production method that challenges conventional synthesis approaches. The phytochemical extraction begins by washing the peels and exposing them to sun-drying before using an mortar and pestle to make powder that goes into boiling methanol solution (Ajmal et al., 2019). The heated extract receives a slow addition of titanium dioxide powder at 70°C until the formation of a paste material. Heat treatment of the mixture occurs within a ceramic crucible between 400 to 500°C before nanoparticle grinding leads to its light-coloured appearance. XRD analysis verifies the successful production of TiO_2 nanoparticles because it identifies the anatase crystalline structure of TiO_2 through its distinct peaks. The FTIR spectroscopy analysis identifies two functional groups with hydroxyl and carbonyl components that help stabilise and reduce the material. Nanoparticles in SEM images measure 200 nm in size while their shape depends on the fruit peel type used. The absorption spectrum of UV spectroscopy includes an absorption peak area at 200 nm which is distinct for TiO_2. The natural reduction synthesis technique develops promising antibacterial and antioxidant nanoparticles by removing all hazardous elements associated with toxic chemicals (Ajmal et al., 2019).

The summary of the above literature has been provided in Table 1 which outlines different plants employed in the green synthesis of Ti and TiO_2 NPs, specifying the parts of the plants used, methods of approach for synthesis, and corresponding characteristics exhibited by the plant.

Table 1. TiO_2 NPs synthesized from various parts of plants

Plant Name	Precursors Used	Synthesis of TiO_2 NPs	Characteristics Shown by Them	Size of Synthesized TiO_2 NPs	Reference
Calendula officinalis	Plant extract (15 g/L), $TiCl_4$ (0.15 M)	Mixture refluxed at 120 °C in the dark; light brown color change indicated TiO_2 NPs formation. After 12 hours at 40 °C, the precipitate was centrifuged, washed, and dried.	UV-Vis spectroscopy showed absorbance peak at 228 nm; FT-IR revealed Ti-O bending vibrations and plant-derived organic functional groups for stabilization; XRD confirmed crystalline form with peaks for TiO_2; SEM showed spherical morphology; EDS confirmed titanium and oxygen presence.	15.68 nm - 40.74 nm	Wei et al., 2021
Psidium guajava	$TiO(OH)_2$	Leaf extract prepared by boiling, filtrate mixed with $TiO(OH)_2$, light green NPs obtained.	XRD showed anatase and rutile phases; spherical NPs by FESEM; FTIR identified functional groups; antibacterial properties (*Staphylococcus aureus, Escherichia coli*); high antioxidant activity (DPPH scavenging over 85%).	32.58 nm	Santhoshkumar et al., 2014
Ailanthus altissima	$TiO(OH)_4$	Leaf extract prepared, mixed with $TiO(OH)_4$ solution, exposed to light for 12 hours.	UV-Vis spectroscopy confirmed peaks at 332 nm; XRD indicated anatase structure; high photocatalytic activity (86.79% Methylene Blue dye degradation); FESEM showed spherical NPs.	60 nm - 100 nm	Ganesan et al., 2016
Azadirachta indica	TiO_2 (aqueous)	Leaf broth prepared, mixed with TiO_2, centrifuged, and heat-treated.	XRD confirmed anatase and rutile phases; SEM and TEM showed spherical NPs; strong antibacterial activity.	15 nm - 87 nm	Thakur et al., 2019
Moringa oleifera	TiO_2 (5 mM solution)	Aqueous leaf extract mixed with TiO_2, stirred at 50°C, centrifuged, washed, and dried.	UV-Vis spectroscopy confirmed TiNPs; SEM showed polydisperse spherical particles; wound healing experiments on rats showed significant contraction (92.36% by day 12 for TiNPs + gel).	~100 nm	Sivaranjani & Philominathan, 2016

continued on following page

Table 1. Continued

Plant Name	Precursors Used	Synthesis of TiO_2 NPs	Characteristics Shown by Them	Size of Synthesized TiO_2 NPs	Reference
Echinacea purpurea Herba	TiO_2 (1 mM solution)	Herba steeped, filtered, extract mixed with TiO_2 solution, stirred for 4 hours.	UV-Vis spectroscopy confirmed TiNPs; FTIR identified functional groups; SEM showed polydisperse spherical-like particles; EDS confirmed elemental titanium; strong antioxidant activity.	~120 nm	Dobrucka, 2017
Trigonella foenum graecum	$TiOSO_4$	Leaf extract added to titanium oxy sulphate, pH adjusted with NaOH, filtrated, and sintered at 700°C.	FTIR confirmed TiO_2; XRD showed anatase phase; UV-Vis spectroscopy revealed high absorbance at 400 nm; HR-SEM showed spherical particles; strong bactericidal effects against diverse microbial species.	20 nm - 90 nm	Subhapriya & Gomathipriya, 2018
Aloe vera	$TiO(OH)_2$ (5 mM solution)	Inner gel boiled, extract mixed with $TiO(OH)_2$, stirred for 24 hours, dried at 120°C.	UV-Vis spectroscopy showed a band gap of 3.1 eV; XRD confirmed anatase TiO_2 NPs; FTIR revealed functional groups; atomic microscopy showed particle size between 80–90 nm; photocatalytic activity (41% Rhodamine B degradation).	~11 nm	Nithya et al., 2013
Hypheae Thiebeace and *Ananos seneglensisa*	TiO_2 (aqueous)	Leaf extracts prepared, mixed with TiO_2	UV-Vis confirmed TiO_2 NPs; FTIR showed biomolecule involvement in stabilization; SEM showed spherical particles; XRD confirmed anatase crystalline structure; strong antibacterial and antifungal activities (*Bacillus subtilis, Staphylococcus aureus, Escherichia coli, Salmonella typhi*).	40 nm - 58 nm	Hassan et al., 2019

continued on following page

Table 1. Continued

Plant Name	Precursors Used	Synthesis of TiO_2 NPs	Characteristics Shown by Them	Size of Synthesized TiO_2 NPs	Reference
Buddleja asiatica	TiO_2 salt solution	Leaf extract mixed with TiO_2 salt solution, precipitated by centrifugation, and washed.	UV-Vis confirmed TiO_2 NPs (peak at 397 nm); SEM showed spherical particles; EDX detected Ti and O elements; FTIR confirmed functional groups; agronomic and physiological improvements in wheat seedlings under salt stress.	30 nm - 111 nm	Mustafa et al., 2021
Prunus domestica (Plum)	Plum peel extract, Methanol, Titanium dioxide powder	A mixture of dried peel powder extracted in methanol and TiO_2 heated to paste at 70°C underwent calcination at 400-500°C.	XRD analysis verified the sample contained the anatase phase while FTIR detected hydroxyl and carbonyl groups through its spectrums and SEM imaging revealed a uniform surface with UV absorption at nearly 200 nm wavelength	~200 nm	Ajmal et al., 2019
Prunus persica (Peach)	Peach peel extract, Methanol, Titanium dioxide powder	A procedural approach like Plum's involved mixing peel extract with TiO_2 while heat processing followed by drying and calcination to generate NPs.	XRD data revealed crystalline anatase structure and FTIR showed organic functional groups together with moderate size variations based on SEM analysis along with UV-active properties.	~210 nm	Ajmal et al., 2019
Actinidia deliciosa (Kiwi)	Kiwi peel extract, Methanol, Titanium dioxide powder	The production process involved extracting phytochemicals from peel with methanol followed by TiO_2 reaction under heat and final calcination to synthesise nanoparticles.	The spectroscopic analysis of FTIR demonstrated stabilising groups while XRD confirmed the structure remained in the anatase form and SEM imaging showed irregular particles.	~190 nm	Ajmal et al., 2019

3. APPLICATIONS OF TiO_2 NPS

TiO_2 has recently received much attention in numerous commercial applications because of its desirable characteristics such as photocatalytic activity, non-toxicity, and thermal stability. The graph shown in Figure 5 shows how the number of PubMed-indexed publications about TiO_2 NPs applications evolved from 2000 through 2024. The research interest toward Titanium Dioxide Nanoparticles has shown a steep growth after 2008 which led to recurring peaks before maintaining elevated levels until 2024. Research activities involving TiO_2 NPs continue to rise across multiple sectors of photocatalysis, biomedicine, environmental remediation, and energy applications.

Figure 5. Annual number of PubMed publications on the applications of TiO_2 NPs

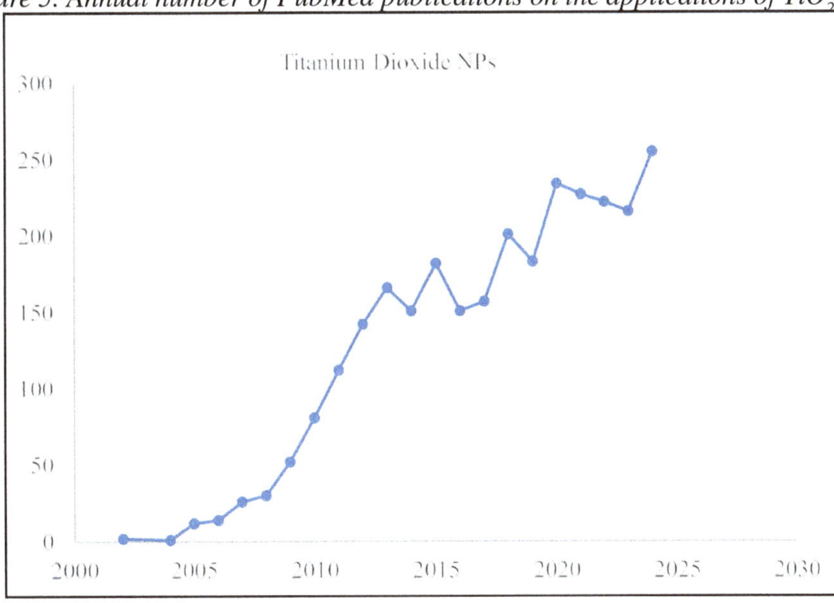

Also, bibliometric analysis (shown in Figure 6) is conducted using PubMed data demonstrates the research diversity regarding TiO_2 NPs between catalytic, biomedical, and environmental applications. The central map terminology consists of "titanium," "titanium dioxide" and "nanoparticles" which showcase their essential role in the analysis. The red cluster demonstrates research in catalysis and photochemistry as well as materials science applications but the green cluster centres on biologically relevant studies about cytotoxicity and oxidative stress while the blue cluster prioritises environmental pollution remediation that includes water and soil pollutant treatment. Analysis of the network reinforces interdisciplinary collaborations through research on functional applications that focuses on biosafety and toxicity as central research areas.

TiO_2 are also employed as the precursor material to prepare titanium metal by direct electrolytic reduction in a molten Calcium Chloride ($CaCl_2$) bath. This process is performed at 950°C in the presence of air with the sintered TiO_2 pellets as the cathode, which eliminates the requirement for inert gas. Pure titanium is obtained in the form of a powder or consolidated product, dependent upon the preparation conditions of the TiO_2 pellets and conditions under which the reaction takes place. The titanium obtained is of very high purity (>99%), as confirmed by XRD, SEM, and Electron Probe Microanalysis (EPMA), showing the ability of TiO_2 to make high-purity titanium for a vast array of applications. In the past, the preparation of these compounds was tedious, and used poisonous precipitating agents and high energy in most cases. On the other hand, green synthesis is an environmentally friendly process that uses natural extracts to synthesise nanoparticles. Biological syntheses of TiO_2 NPs are directed to employ plant extracts as reductive and capping agents (Aravind et al., 2021; Yitagesu et al., 2023; Saka et al., 2022).

TiO_2 NPs are widely used in photocatalysis for environmental remediation. They can degrade organic pollutants in wastewater when exposed to UV light. The mechanism involves the generation of electron-hole pairs that create ROS, leading to the breakdown of contaminants. They have high anti-oxidant, anti-bacterial, anti-fungal, photocatalytic and dye-degradation properties, to name a few, and because of that they can be used in a variety of fields (Yitagesu et al., 2023; Aravind et al., 2021; Saka et al., 2022; Srinivasan et al., 2019).

Figure 6. Bibliometric network of PubMed keywords related to applications of TiO_2 NPs.

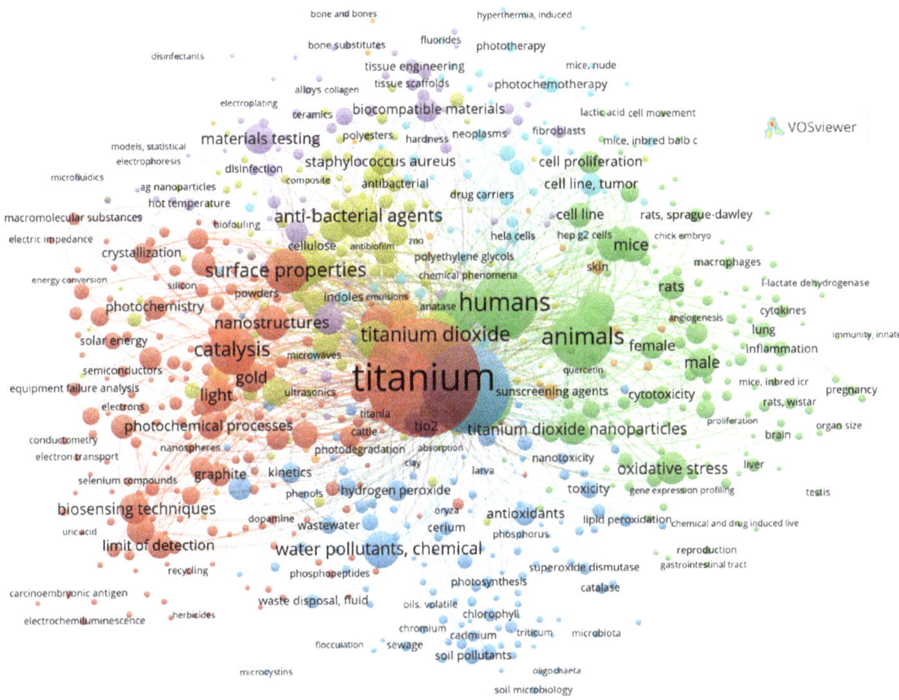

4. GREEN SYNTHESIS OF TIO_2 NPS: INDUSTRIAL IMPLICATIONS AND COMMERCIALIZATION PATHWAYS

The use of green synthesis of TiO_2 NPs is a paradigm of making nanomaterials sustainable by employing the use of biological systems in terms of plant extracts, microorganisms and biopolymers in the production of reducing and capping agents. The green synthesis methods are more reliable than other conventional, chemical/physical approaches because they are low-toxicity, energy-efficient and economical processes when compared to other chemical/physical methods that uses hazardous reagents and is energy-intensive (Shrivastav et al., 2025). Not only does this decrease the load on the environment but also increases the application of nanoparticles through photocatalysis, antimicrobial coatings, and wastewater remediation since the natural compounds used result in surface functionalization to boost performance (Verma et al., 2022; Shakeel et al., 2025; Wu et al., 2019). The benchmarking of the available expertise shows that regular methods are scalable and offer production-management capabilities, but they lack sustainability scores (Shakeel et al., 2025). The comparison of the three primary strategies of synthesis may be done directly on the major industrial parameter, as indicated in the table 2 below:

Table 2. Comparative analysis of green, chemical, and physical synthesis routes for TiO_2 nanoparticles

Parameter	Green Synthesis	Chemical Synthesis	Physical Synthesis	References
Environmental Impact	Low (uses agricultural waste, no toxic byproducts)	High (hazardous precursors, toxic waste)	High (energy-intensive, gas emissions)	Shakeel et al., 2025; Wu et al., 2019; Verma et al., 2022
Scalability	Moderate (batch-dependent, limited standardization)	High (established protocols)	High (continuous processes)	Shakeel et al., 2025; Wu et al., 2019
Cost-Effectiveness	Low (reduced reagent costs, waste valorization)	Moderate (precursor expenses)	High (energy/ equipment costs)	Shakeel et al., 2025; Wu et al., 2019
Energy Consumption	Low (ambient temperatures)	Moderate (moderate heating)	Very high (plasma/ evaporation)	Shakeel et al., 2025; Verma et al., 2022; Wu et al., 2019
Production Efficiency	Variable (dependent on biomass quality)	High (controlled yields)	High (rapid production)	Shakeel et al., 2025; Wu et al., 2019

Green synthesis can cause 30-50% fewer life cycle environmental impacts than chemical methods due to a lack of toxic solvents and processing at high temperatures (Wu et al., 2019; Verma et al., 2022). In the plant mediated pathway, there is the control of particle morphology and stability. The use of phytochemicals in plant extracts like polyphenols, flavonoids, terpenoids, and alkaloids results in the chelation of Ti^{4+} ions and regulates the nucleation rates subsequently forming homogenous anatase-phase TiO_2 NPs with sizes varying between 9-200 nm. These molecules are then natural stabilizers, which help lessen aggregation and get better colloidal stability. By way of an example, a 67.6% crystalline anatase TiO_2 NPs, 33.28 nm in size, was produced through a reaction that used Lactobacillus bulgaricus mediates the synthesis of TiO_2 NPs on using food-grade biomass (Shakeel et al., 2025). Further, phytochemicals provide reactive functional groups (e.g. -OH, -COOH) on nanoparticle surface, which increases their ROS generation facilitating photocatalytic degradation (e.g. 81.12% of phenol removal) and intrinsic antibacterial and antioxidant potential when compared to conventional synthesis of TiO_2 NPs (Al-Shabib et al., 2020; Verma et al., 2022).

Nonetheless, even though it is beneficial, industrial-scale green synthesis has obstacles related to engineering and operations. Biological fluctuations in biomass result in inconsistency between batches (different batches) of nanoparticles in terms of their size and purity of their crystalline phase. Quality control is challenging because no specific process parameters (pH, temperature, extract concentration) are established, whereas polyphenol-rich extract might favor anatase phase, whereas amine-rich ones might promote rutile, therefore, this requires particular tuning (Shakeel et al., 2025; Verma et al., 2022). Moreover, the requirement of energy is low (25-80 °C) but the design of scalable continuous-flow reactors is a challenge. Biomass waste also makes downstream purification harder which adds ~15% to the costs of operation (Wu et al., 2019).

Commercially speaking, economic and regulatory environment are the determining factors. The waste-derived materials can lower the total cost of production, in general, by 20-40% in green synthesis. The TiO_2 NPs market and especially the uses of the nano particles in wastewater reclamation and antimicrobial surfaces will reach a mark of more than 3.5 billion by 2030 (Shakeel et al., 2025). However, the application of the industry is hindered due to the absence of particular recommendations of regulatory authorities such as FDA and EMA to the biogenic nanoparticles. Plant-mediated nanoparticle-related patenting activity was ~60% of the total in 2020-2024 indicating commercial interest, yet microbial pathways have intellectual property clashes because of strain sharing (Al-Shabib et al., 2020; Verma et al., 2022). The plant-based synthesis has advanced to technology readiness levels (TRL) 6 (pilot scale), but the microbial methods are at TRL 4 (lab scale) because of the complexity and cost of their operation (Wu et al., 2019; Al-Shabib et al., 2020). There are also obstacles like the competition with chemically synthesized TiO_2 NPs (price ~$50/kg compared to ~$70/kg of the green counterparts), the distrust of nano-bio products by the mass market, and the inability to supply the biomass chain logistics on a commercial scale, contributing to the inability to commercialise this technology on an industrial scale (Shakeel et al., 2025).

Disadvantages of the green synthesis are that one will have difficulties in attaining high phase purity (i.e. mixed anatase rutile combo), poor morphology control compared to the sol-gel method, and post-synthesis temperatures (in excess of 500 °C) will be necessary to enhance crystallinity (Shakeel et al., 2025; Verma et al., 2022). To address these, the research directions relate to AI- based optimization of extraction protocol, use of a hybrid solar-assisted green synthesis to lessen dependency on calcination and genetic modification of hyperaccumulator plants to increase titanium uptake (Verma et al., 2022; Al-Shabib et al., 2020). Other technological processes have been found to be promising, including microfluidic reactors in continuous flow, and bio-waste templating approaches based on cellulose scaffolds to create orderly mesoporous TiO_2 NPs (Verma et al., 2022).

Mechanism of green synthesis of TiO_2 NPs usually follows three steps: reduction of Ti^{4+} ions (precursors include $TiCl_4$ or $TiO(OH)_2$) by phytochemicals or microbial enzymes; nucleation and growth where shape is directed by biomolecules; and capping/ stabilization where functional groups such as -OH or -COOH groups are attached to the surface of the particles to prohibit aggregation and introduce activity (Shakeel et al., 2025; Verma et al., 2022). A hypothetical reaction scheme comprises in Figure 7.

Figure 7. Mechanism of green synthesis of TiO_2 NPs.

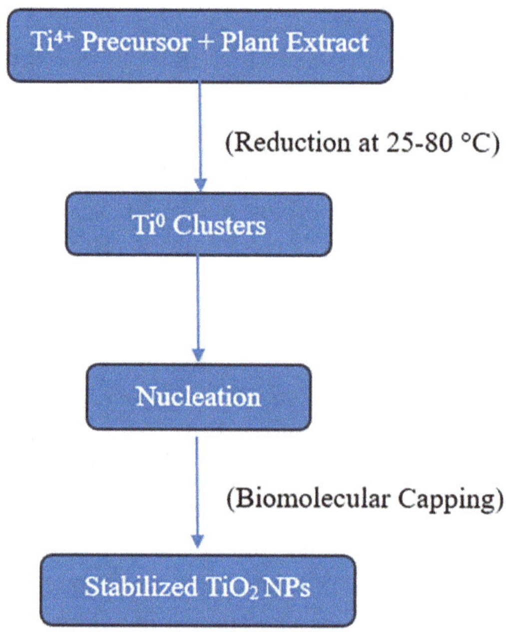

To summarize, green synthesis of TiO_2 NPs is an alternative to conventional methodology that is both sustainable and functionally superior, but commercialization is limited by reproducibility, standardization, and market competitiveness. Nevertheless, green TiO_2 synthesis is ready to be moved out of the lab and into the green nanotechnology movement with the incorporation of process control enabled by AI methodologies, hybrid synthesis practices, bioengineering advancements, as well as enhancement of regulatory pathways (Shakeel et al., 2025; Verma et al., 2022).

5. ENVIRONMENTAL, HEALTH, AND SAFETY (EHS) ASPECTS OF GREEN-SYNTHESIZED TIO_2 NPS

TiO_2 NPs can be synthesized using a green method, in which plant extracts, microorganisms or other biocompatible reducing and capping agents can be utilized, providing an eco-friendly alternative to traditional chemical and physical synthesis procedures (Purkait et al., 2020; Shakeel et al., 2025). The process limits the duty of dangerous chemicals, energy use and liability to the environment. Research also affirms that green-synthesized TiO_2 NPs have their main chemical characteristics preserved but with better biocompatibility and low toxicity compared to chemically generated NPs (Shakeel et al., 2025). Ecology risk assessment also focuses on life-cycle implications of TiO_2 NPs where the persistence, transformation, and release in the environment is considered. The green-synthesized ones tend to be safer due to the lack of toxic products of the synthetic process and improvement of the stability level reducing the chances of undirected release (Shakeel et al., 2025; Shrivastav et al., 2025). However, the build-up in the water and land ecosystems is commendable. Assesments of acute toxicity in zebrafish have shown that green-synthesized TiO_2 NPs has a relatively lower acute toxicity level when expressed (e.g. 300 mg/L at 48 h) (Jafari et al., 2022).

According to the toxicological findings, TiO_2 NPs have the ability to cross the cell membranes into the body, distribute in biological tissues, and may cause oxidative stress, inflammation, or genotoxicity (Fiordaliso et al., 2022). Nevertheless, rather good compatibility with mammalian cells and significantly reduced cytotoxicity can be explained by their unique surface chemistry and lower residual contaminants in green-synthesized NPs (Shakeel et al., 2025). They are of high biocompatibility thus capable of acting as a biomedical product in terms of drug delivery, wound dressing as well as antimicrobial coatings, among others, with the help of capping agents skewered by plants that helps in decreasing immunogenicity and increasing cellular logistics (Shakeel et al., 2025).

The green TiO_2 NPs have tunable surface properties and low immunotoxicity and are ideal in targeted drug delivery and photodynamic treatment in the biomedical field (Shakeel et al., 2025). They also exhibit antibacterial and antioxidative functions, which confirm the use of these nanoparticles in wound dressing and therapy, as well as implant coating (Shakeel et al., 2025). The nanoparticles have also been studied as a type of nano-fertilizer and pest control agent with reduced phytotoxicity and persistence beginning in agricultural fields with the nanoparticles improving the nutrient bioavailability and resulting plant growth (Shakeel et al., 2025). Environmentally, these have effective photocatalytic activities that allow easy destruction of organic and microbial contaminants in drinking water treatment, and air filtering systems and there are minimal chances of secondary pollution since they are synthesized in a benign approach (Purkait et al., 2020; Shakeel et al., 2025).

Nevertheless, such benefits do not diminish the fact that regulatory measures to assess the hazard potential of TiO_2 NPs are outdated and tend to be insensitive to the complex risk aspects of nanoscale objects since it is still based on bulk material data (Jovanović, 2014). Emerging requirements involve the need to renounce the current framework to involve the properties of particle but rather particle-specific properties, including size, surface area and reactivity (Jovanović, 2014). A successful process of risk assessment should incorporate full profile of hazard identification, exposure assessment, and risk characterization (Bergamaschi et al., 2015). Green synthesis is getting more embraced in regulatory systems because of it being recognized as a safer and more sustainable route (Shakeel et al., 2025). Examples of risk mitigation strategies are engineering controls which involve closed-loop manufacturing process, personal protective equipment, and process optimization which limit the potential risk of exposure and release into the environment. Furthermore, the concepts of safety-by-design, which consists in incorporating hazard mitigation at the early phases of materials creation, are becoming popular (Bergamaschi et al., 2015). Green synthesis is also supported by sustainability analyses, showing decreased input energy, absence of harmful chemicals and conversion to renewable materials during the life cycle of the nanoparticle (Shakeel et al., 2025).

When combined with synthesis processes, artificial intelligence (AI) brings considerable precision and scalability improvement to the synthesis workflows. Optimization algorithms by machine learning refine synthesis conditions through parameters including pH, temperature, and precursor concentrations, to produce reproducible nanoparticle sizes, shapes, and crystalline domains, and reduce overall waste. The prospective application of machine learning to safety-by-design is that ML models can forecast the effects of particular conditions of synthesis on toxicity and aggregation properties (Bergamaschi et al., 2015). Moreover, machine learning models of toxicity based on empirical data can be used to predict biological and ecological effect of newly developed TiO_2 NPs at an early stage, and thus inform the choice of safe and efficient synthesis pathways.

In practice, AI was used to develop TiO_2 NPs that favor smart agriculture applications where controlled nutrient or pesticide release is obtained with limited non-target hazard and low persistence of the materials deposited in the environment. The use of AI-based simulations has enabled the biomedical field to develop the design of surface-functionalized NP tailored to specific cellular uptake and release of medicine to reduce off-target effects and systemic toxicity (Shakeel et al., 2025). The latest research proves the effectiveness of AI-enabled green synthesis, which allows achieving more improved photocatalytic and antimicrobial activity and reduced toxicity. Case studies also show that by integrating the nanoparticle development pipeline with AI, it is possible to make real-time changes in the synthesis conditions, thus avoiding the production of hazardous byproducts, and regulatory compliance (Shakeel et al., 2025).

On the whole, there is a distinct combination of decreased toxicity, improved biocompatibility, and increased environmental friendliness in green-synthesized TiO_2 NPs (Shakeel et al., 2025; Jafari et al., 2022). AI-based approaches also accelerate the work of creating safer and more specific forms of nanomaterial design supporting the creation of novel forms of medicine, agriculture, as well as environmental engineering. Nevertheless, the regulatory standards should remember some concerns about the long-term environmental concentrations, possibility of bioaccumulation, and chronic toxicity, which should remain monitored and improved (Jafari et al., 2022). Already it is apparent that the current state of technology development in nanotechnology is already being developed considerably faster than regulatory frameworks, thus highlighting the importance of the development of harmonized, flexible risk assessment measures (Jovanović, 2014; Bergamaschi et al., 2015).

6. CONCLUSION AND FUTURE TRENDS

Plant mediated green synthesis of biogenic TiO_2 NPs using different plants has been discussed. This technique is cos-effective, less hazardous, and environmentally friendly. Various spectroscopic analytical techniques such as XRD, UV-Vis, TEM, SEM, FTIR confirmed the formation of crystalline nanoparticles particles of various sizes and shapes, and the important roles played by numerous biochemicals associated with leaf extract in the formation and stabilization of these NPs. These NPs exhibit strong anti-bacterial, anti-fungal, and antioxidant properties. Various results further confirmed that they can be used as promising agents in many fields (water purification, drug delivery, dye degradation, wound healing, to name a few).

The green synthesis of TiO_2 NPs coupled to artificial intelligence AI will transform material development, as it allows a detailed manipulation of the synthesis conditions and the predictive simulation of the nanoparticle's qualities. Machine learning, optimization techniques, and other AI-based methods can be applied to accelerate the iterative process of finding optimal plant extracts, conditions to use in reactions, and the scaling up approaches and reduce the trial-and-error experimental process. Predictive models will help correlate phytochemical profiles and nanoparticle size, morphology and functional performance with the objective of enhancing reproducibility and effectiveness. Also, monitoring could help with the real-time quality control and scaling together with AI support, which acts as one of the biggest bottlenecks of sustainable nanomanufacturing. Such merging of green chemistry and AI fits within the wider agenda of sustainable nanotechnology and Industry 5.0, to create eco-sensitive but high-performance materials to suit a variety of activities.

REFERENCES

Ajmal, N., Saraswat, K., Bakht, M. A., Riadi, Y., Ahsan, M. J., & Noushad, M. (2019). Cost-effective and eco-friendly synthesis of titanium dioxide (TiO_2) nanoparticles using fruit's peel agro-waste extracts: Characterization, in vitro antibacterial, antioxidant activities. *Green Chemistry Letters and Reviews*, *12*(3), 244–254. DOI: 10.1080/17518253.2019.1629641

Al-Shabib, N. A., Husain, F. M., Qais, F. A., Ahmad, N., Khan, A., Alyousef, A. A., Arshad, M., Noor, S., Khan, J. M., Alam, P., Albalawi, T. H., & Shahzad, S. A. (2020). Phyto-Mediated Synthesis of Porous Titanium Dioxide Nanoparticles From Withania somnifera Root Extract: Broad-Spectrum Attenuation of Biofilm and Cytotoxic Properties Against HepG2 Cell Lines. *Frontiers in Microbiology*, *11*, 1680. Advance online publication. DOI: 10.3389/fmicb.2020.01680 PMID: 32849352

Aravind, M., Amalanathan, M., & Mary, M. S. M. (2021). Synthesis of TiO_2 nanoparticles by chemical and green synthesis methods and their multifaceted properties. *SN Applied Sciences*, *3*(4), 409. Advance online publication. DOI: 10.1007/s42452-021-04281-5

Bergamaschi, E., Murphy, F., Poland, C. A., Mullins, M., Costa, A. L., McAlea, E., Tran, L., & Tofail, S. M. (2015). Impact and effectiveness of risk mitigation strategies on the insurability of nanomaterial production: Evidences from industrial case studies. *Wiley Interdisciplinary Reviews. Nanomedicine and Nanobiotechnology*, *7*(6), 839–855. DOI: 10.1002/wnan.1340 PMID: 25808636

Dobrucka, R. (2017). Synthesis of Titanium Dioxide Nanoparticles Using *Echinacea purpurea* Herba. *Iranian Journal of Pharmaceutical Research : IJPR*, *16*(2), 756–762. PMID: 28979329

Fiordaliso, F., Bigini, P., Salmona, M., & Diomede, L. (2022). Toxicological impact of titanium dioxide nanoparticles and food-grade titanium dioxide (E171) on human and environmental health. *Environmental Science. Nano*, *9*(4), 1199–1211. DOI: 10.1039/D1EN00833A

Ganesan, S., Babu, I. G., Mahendran, D., Arulselvi, P. I., Elangovan, N., Geetha, N., & Venkatachalam, P. (2016). Green engineering of titanium dioxide nanoparticles using Ageratina altissima (L.) King & H.E. Robines. medicinal plant aqueous leaf extracts for enhanced photocatalytic activity. *Annals of Phytomedicine*, *5*(2), 69–75. DOI: 10.21276/ap.2016.5.2.8

Hassan, H., Omoniyi, K., Okibe, F., Nuhu, A., & Echioba, E. (2019). Evaluation of Antibacterial Potential of Biosynthesized Plant Leave Extract Mediated Titanium Oxide Nanoparticles using Hypheae Thiebeace and Anannos Seneglensis. *Journal of Applied Science & Environmental Management, 23*(10), 1795–1804. DOI: 10.4314/jasem.v23i10.5

Irshad, M. A., Nawaz, R., Rehman, M. Z. U., Adrees, M., Rizwan, M., Ali, S., Ahmad, S., & Tasleem, S. (2021). Synthesis, characterization, and advanced sustainable applications of titanium dioxide nanoparticles: A review. *Ecotoxicology and Environmental Safety, 212*, 111978. DOI: 10.1016/j.ecoenv.2021.111978 PMID: 33561774

Jafari, A., Rashidipour, M., Kamarehi, B., Alipour, S., & Ghaderpoori, M. (2022). Toxicity of green synthesized TiO_2 nanoparticles (TiO_2 NPs) on zebra fish. *Environmental Research, 212*, 113542. DOI: 10.1016/j.envres.2022.113542 PMID: 35660565

Jassal, P. S., Kaur, D., Prasad, R., & Singh, J. (2022). Green synthesis of titanium dioxide nanoparticles: Development and applications. *Journal of Agriculture and Food Research, 10*, 100361. DOI: 10.1016/j.jafr.2022.100361

Jovanović, B. (2014). Critical review of public health regulations of titanium dioxide, a human food additive. *Integrated Environmental Assessment and Management, 11*(1), 10–20. DOI: 10.1002/ieam.1571 PMID: 25091211

Mousa, S. A., Shalan, A. E., Hassan, H., Ebnawaled, A., & Khairy, S. (2022). Enhanced the photocatalytic degradation of titanium dioxide nanoparticles synthesized by different plant extracts for wastewater treatment. *Journal of Molecular Structure, 1250*, 131912. DOI: 10.1016/j.molstruc.2021.131912

Mustafa, N., Raja, N. I., Ilyas, N., Ikram, M., Mashwani, Z., & Ehsan, M. (2021). Foliar applications of plant-based titanium dioxide nanoparticles to improve agronomic and physiological attributes of wheat (Triticum aestivum L.) plants under salinity stress. *Green Processing and Synthesis, 10*(1), 246–257. (Creative Commons Attribution 4.0 International (CC BY 4.0)) (https://creativecommons.org/licenses/by/4.0/deed.en)DOI: 10.1515/gps-2021-0025

Nadeem, M., Tungmunnithum, D., Hano, C., Abbasi, B. H., Hashmi, S. S., Ahmad, W., & Zahir, A. (2018). The current trends in the green syntheses of titanium oxide nanoparticles and their applications. *Green Chemistry Letters and Reviews, 11*(4), 492–502. DOI: 10.1080/17518253.2018.1538430

Nithya, A., Rokesh, K., & Jothivenkatachalam, K. (2013). Biosynthesis, characterization and application of titanium dioxide nanoparticles. *Nano Vis, 3*(3), 169–174.

Purkait, P., Bhattacharyya, A., Roy, S., Maitra, S., Das, G., & Chaudhuri, M. G. (2020). Green synthesis of TiO_2 nanoparticle : its characterization and potential application in Zoxamide photodegradation. *DOAJ (DOAJ: Directory of Open Access Journals)*. https://doi.org/DOI: 10.22090/jwent.2020.03.001

Saka, A., Shifera, Y., Jule, L. T., Badassa, B., Nagaprasad, N., Shanmugam, R., Dwarampudi, L. P., Seenivasan, V., & Ramaswamy, K. (2022). Biosynthesis of TiO_2 nanoparticles by Caricaceae (Papaya) shell extracts for antifungal application. *Scientific Reports, 12*(1), 15960. Advance online publication. DOI: 10.1038/s41598-022-19440-w PMID: 36153393

Santhoshkumar, T., Rahuman, A. A., Jayaseelan, C., Rajakumar, G., Marimuthu, S., Kirthi, A. V., Velayutham, K., Thomas, J., Venkatesan, J., & Kim, S. K. (2014). Green synthesis of titanium dioxide nanoparticles using Psidium guajava extract and its antibacterial and antioxidant properties. *Asian Pacific Journal of Tropical Medicine, 7*(12), 968–976. DOI: 10.1016/S1995-7645(14)60171-1 PMID: 25479626

Shakeel, N., Piwoński, I., Iqbal, P., & Kisielewska, A. (2025). Green Synthesis of Titanium dioxide Nanoparticles: Physicochemical Characterization and Applications: A review. *International Journal of Molecular Sciences, 26*(12), 5454. DOI: 10.3390/ijms26125454 PMID: 40564917

Shrivastav, A., Khapekar, Y., Soni, D., Sankathala, S., & Das, S. (2025). Green synthesis of palladium nanoparticles using medicinally important plant extracts. *Transactions of the Indian Institute of Metals, 78*(4), 95. Advance online publication. DOI: 10.1007/s12666-025-03572-4

Sivaranjani, V., & Philominathan, P. (2016). Synthesize of Titanium dioxide nanoparticles using Moringa oleifera leaves and evaluation of wound healing activity. *Wound Medicine, 12*, 1–5. DOI: 10.1016/j.wndm.2015.11.002

Srinivasan, M., Venkatesan, M., Arumugam, V., Natesan, G., Saravanan, N., Murugesan, S., Ramachandran, S., Ayyasamy, R., & Pugazhendhi, A. (2019). Green synthesis and characterization of titanium dioxide nanoparticles (TiO_2 NPs) using Sesbania grandiflora and evaluation of toxicity in zebrafish embryos. *Process Biochemistry (Barking, London, England), 80*, 197–202. DOI: 10.1016/j.procbio.2019.02.010

Subhapriya, S., & Gomathipriya, P. (2018). Green synthesis of titanium dioxide (TiO_2) nanoparticles by Trigonella foenum-graecum extract and its antimicrobial properties. *Microbial Pathogenesis, 116*, 215–220. DOI: 10.1016/j.micpath.2018.01.027 PMID: 29366863

Thakur, B., Kumar, A., & Kumar, D. (2019). Green synthesis of titanium dioxide nanoparticles using Azadirachta indica leaf extract and evaluation of their antibacterial activity. *South African Journal of Botany*, *124*, 223–227. DOI: 10.1016/j.sajb.2019.05.024

Verma, V., Al-Dossari, M., Singh, J., Rawat, M., Kordy, M. G. M., & Shaban, M. (2022). A review on green synthesis of TIO2 NPs: Photocatalysis and antimicrobial applications. *Polymers*, *14*(7), 1444. DOI: 10.3390/polym14071444 PMID: 35406317

Waghmode, M. S., Gunjal, A. B., Mulla, J. A., Patil, N. N., & Nawani, N. N. (2019). Studies on the titanium dioxide nanoparticles: Biosynthesis, applications, and remediation. *SN Applied Sciences*, *1*(4), 310. Advance online publication. DOI: 10.1007/s42452-019-0337-3

Wei, X., Liu, Y., El-Kott, A., Ahmed, A. E., & Khames, A. (2021). Calendula officinalis-based green synthesis of titanium nanoparticle: Fabrication, characterization, and evaluation of human colorectal carcinoma. *Journal of Saudi Chemical Society*, *25*(11), 101343. DOI: 10.1016/j.jscs.2021.101343

Wu, F., Zhou, Z., & Hicks, A. L. (2019). Life Cycle Impact of Titanium Dioxide Nanoparticle Synthesis through Physical, Chemical, and Biological Routes. *Environmental Science & Technology*, *53*(8), 4078–4087. DOI: 10.1021/acs.est.8b06800 PMID: 30912646

Yitagesu, G. B., Leku, D. T., & Workneh, G. A. (2023). Green Synthesis of TiO_2 Using Impatiens rothii Hook.f. Leaf Extract for Efficient Removal of Methylene Blue Dye. *ACS Omega*, *8*(46), 43999–44012. DOI: 10.1021/acsomega.3c06142 PMID: 38027313

ADDITIONAL READINGS

Al Malki, J. S., Hussien, N. A., Akkad, L. M., Al Thurmani, S. O., & Al Motiri, A. E. (2023). Green Synthesis of Silver and Titanium Oxide Nanoparticles Using Tea and Eggshell Wastes, Their Characterization, and Biocompatibility Evaluation. *Sustainability (Basel)*, *15*(15), 11858. DOI: 10.3390/su151511858

Khan, F., Shariq, M., Asif, M., Siddiqui, M. A., Malan, P., & Ahmad, F. (2022). Green Nanotechnology: Plant-Mediated Nanoparticle Synthesis and Application. *Nanomaterials (Basel, Switzerland)*, *12*(4), 673. DOI: 10.3390/nano12040673 PMID: 35215000

Mohanty, J., Rath, P. C., Subbaiah, T., Mishra, K. G., & Paramguru, R. K. (2009). Preparation of titanium metal by electrochemical reduction of titanium dioxide. *Transactions of the Indian Institute of Metals*, *62*(3), 249–253. DOI: 10.1007/s12666-009-0028-7

KEY TERMS AND DEFINITIONS

Biogenic Synthesis: The synthesis in which biological systems which include plants together with bacteria and fungi and algae are employed for the production of nanoparticles is called biogenic synthesis. The sub-field of green synthesis features natural reducing agents for delivering an eco-friendly along with scalable alternative to traditional chemical production methods.

Green Synthesis: It is the environmentally friendly synthesis of materials and nanoparticles specifically uses biological agents including plant extracts along with enzymes and microbial entities. The methodology adopts sustainable practises and safe reagents and efficient energy consumption as core components which implement the principles of green chemistry.

Nanomaterials: Materials with dimensions between 1-100 nanometers fall into the category of nanomaterials. Their diminutive size allows nanoparticles to show distinctive physical properties together with chemical traits along with biological characteristics dissimilar to bulk materials.

Nanotechnology: It is the field dedicated to controlling matter at atomic and molecular levels through supramolecular manipulations to generate novel structures and systems featuring enhanced operational characteristics.

Phytochemicals: Plants contain different bioactive compounds known as phytochemicals which include flavonoids and terpenoids as well as tannins and saponins and alkaloids. The compounds function as key elements which contribute to reducing and capping nanoparticles throughout the green synthesis methodology.

Reducing Agents: The conversion of metal ions into metallic nanoparticles through electron donation processes occurs with the use of reducing agents. The use of plant extracts during green synthesis provides reducing agents in place of toxic chemical reducers.

Stabilizing Agents: This agents stop nanoparticles from forming clumps by constructing steric or electrostatic barriers between them. Plant-derived phytochemicals function as natural stabilising additives in green synthesis methods.

Chapter 14
Trio Insect Bio Clusters: An AI-Enabled Entrepreneurial Blueprint for a Circular Climate-Resilient Rural Economy

Priti Saxena
DAV Degree College, Lucknow, India

Mohammad Irfan
http://orcid.org/0000-0002-4956-1170
Chandigarh University, Lucknow, India

Alka Singh
Deen Dayal Upadhyaya Gorakhpur University, Gorakhpur, India

ABSTRACT

The growing challenges of food insecurity, climate change, biodiversity decline, and waste mismanagement require practical and regenerative solutions. This paper proposes an integrated bio-cluster model built around three insect systems: Black Soldier Fly (Hermetia illucens), silkworm (Bombyx mori), and bees (Apis species and other pollinators). Each contributes unique ecological and economic value. Black Soldier Fly supports organic waste conversion into protein and compost; silkworms generate high-value silk and biomaterials; bees enhance pollination and agricultural productivity. When combined within a localized bio-cluster, these systems create a circular framework that transforms waste into resources, improves soil fertility, and diversifies rural income. The study presents a holistic approach linking biology, ecology, and economics to demonstrate how insect-based clusters can reduce emissions, lower landfill use, support carbon balance, and strengthen climate-resilient rural development through integrated, multi-stream livelihoods.

DOI: 10.4018/979-8-2600-0060-1.ch014

INTRODUCTION

The 21st century is considered as deep rooted interconnected, multi-disciplinary economic, social and environmental crisis that challenges old and conventional development paradigms. As we all know that world population is growing rapidly and projected to reach up to 9.7 billion by 2050, is actually putting a lot of pressure on the food system, fresh water availability, energy infrastructure (United Nations 2019) and land resources. And at the same time Climate change is increasing the frequency of unusual weather events, altering precipitation patterns, and disrupting agricultural productivity (IPCC, 2022). Biodiversity loss has speed up in many folds to unprecedented rates, with approximately dangering 1 million species at risk of extinction due to anthropogenic pressure (IPBES, 2019). Such types of crises are compounded by increasing organic waste generation, which is directly linked to urbanisation and Changes in food intake habits. At the global level, approximately 1/3 of all food products that are consumed by human are lost or wasted, contributing significantly to Greenhouse gas emissions (FAO 2011). Organic waste composition in land fields produces Methane Gas, a green house gas with a global warming potential is increasingly higher than carbon dioxide in last 20 years time period (UNEP 2021). Combining all these factors, which exposed the limitations of the linear ecosystem based on extraction production, consumption and disposition.

To answer this problem, one very powerful bio economic Framework has been immersed as a strategic tool for an alternative that emphasises sustainable utilisation of renewable biological resources for the production of material, chemical food and energy (European Commission, 2018). The bio economy integrates biotechnology, agriculture management, commerce life sciences, forestry, agriculture, fishery and industrial processing into a value generating system that minimises the over-reliance on Fossil resources (OECD, 2009).

This study is central to the Paradigm shift towards the circular economy principle, which will promote resource efficiency, regenerative design and reduce waste (Allen Macarthur Foundation 2013). This circular bio economy objective is to close material loops by changing organic waste streams into profitable inputs for new production cycles (Kirchherr et al. 2017). These types of systems are closely aligned with the Sustainable Development Goals which are related to climate action, responsible production and zero hunger (United Nations 2015).

Within this changing landscape, the bio cluster gives geographically concentrated ecosystems of three interconnected yet separate Enterprises, farmers, research institutions processing units specialising in bio based production (Porter, 1998; Delgado et al., 2010). Clustering enhances involves innovation, reduced transportation costs, increases competitiveness and improve knowledge spillovers (Ketels, 2013). In the context of bio economy development ecosystem enables interconnected integrated value change from where biological inputs are processed into high value multiple outputs which will maximise efficiency of resources (Asheim et al., 2011). Geographical regional bio clusters have shown extremely close potential in sectors such as biofuels, agri biotechnology and bioplastics (McCormick & Kautto, 2013). However maximum existing models remain Technology Complex and capital intensive, which limits accessibility of peri Urban and rural communities.

Figure 1. Global interlinked bio economy and addressing crisis framework

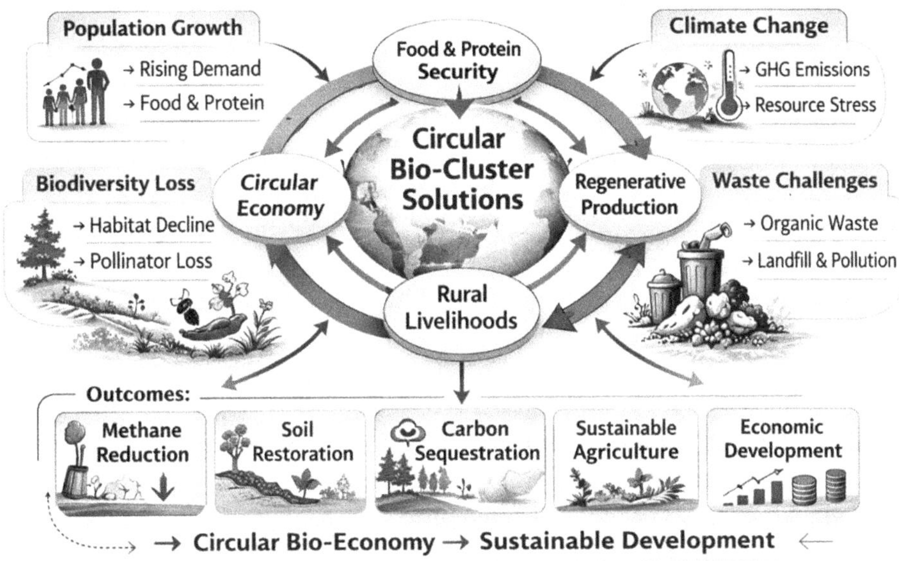

(Conceptual diagram linking climate change, population growth, biodiversity loss and waste to circular bio cluster solutions)

Beyond their ecological sustainability benefits, integrated insect based bio clusters also create important business and economic opportunities. The conversion of organic waste into commercially valuable products like insect protein for aquaculture and poultry feed, Organic fertilizers, Silk based bio materials, honey, Bees wax and other Apiculture products enables the development of diversified value chains. These bio clusters can support the emergence of micro, small and medium enterprises (MSMEs), Agri Biotech startups and rural Processing Units while attracting green investment and accelerating innovation in sustainable production systems. By lowering entry barriers and utilising locally available biological resources, insect based by bio cluster can stimulate rural entrepreneurship generate employment and strengthen circular bio economy markets thereby transforming sustainability solutions into a scalable business ecosystem.

Insects present a transformative opportunity to operationalise decentralized and inclusive bio cluster. Insects are among the most preferred biological converters known, they are capable of transforming low value organic matter into high value protein fibres bioactive and lipid compounds (Van Huis et al., 2015). When we compare in sex with conventional livestock, insect production requires substantially less land feed and water inputs and also they emit very little greenhouse gases (Oonincx et al., 2010; Halloran et al., 2016). Insects have very short life cycle very low feed requirement very high rapue reproductive rate and high conversion efficiency making them particularly suited to scale bio economic system (Makkar et al., 2022). The organisation like the Food and Agriculture Organisation highlight edible and industrial insects as promising contributed toward sustainable food systems, waste management Strategies and business opportunities (Van Huis et al., 2015).

This research paper explores the integration of three very high impact insect systems that are black soldier fly (Hermetia Illu cerns), Bees (Apis spp. And related pollinators) and silkworms (Bombyx mori) within a unified bioclastor framework. Among all three the black soldier fly BSF has gained very high attention across the globe for its capability to convert organic waste into protein rich larval Biomass suitable for animal feed, and same time it produce nutrient rich frass fertiliser (Lalander et al., 2019). Bees function as keystone pollinators essential for biodiversity maintenance crop productivity and multi product Apiculture like Honey wax Royal Jelly and propolis (Potts et al., 2016). Silkworms represent one of the oldest Legacy of generating silk, protein rich pupae and sericin with its application extending from textile to biomedical engineering (Kundu, 2014). All these three systems work independently to contribute ecological and economic value, and when they are integrated with each other, they create synergetic value loops characteristics of circular bio economies.

Collectively, these three insects offer decentralised, modular and regenerative model adaptable to rural and very urban contexts. Buy a lining agricultural productivity waste management and biometrical production within the clustered ecosystem integrated insect bio clusters may address multiple sustainability challenges simultaneously. Such system shows a shift from extractive growth model toward generative development that works with biological processes rather than against them (Raworth, 2017).

CASE STUDIES AS MOTIVATORS FOR ENTREPRENEURSHIP

An entrepreneur can begin with the integration of the insect base business. Here, advantage of leverage will complement each other. Different species like silkworm, black soldier fly and bees are different and perform distinct same time they are supportive of each other's growth. Above the environmental sustainability converting towards circular bio economy is very crucial and important for strengthening the national economy by implementing innovation lead growth yet sustainable and helping to increase bio based ecosystem. countries that actively help increasing bio based businesses and entrepreneurship they shows tremendous opportunities in contributing toward development of greener environment, Such countries also shows trend of high level of productivity, which helps in reduction of unemployment from the society and increases the possibility of export diversification. Entrepreneurial activities boost the commercialisation help in quality research, support the micro, small and medium enterprises, attracts foreign funding and create positive impact in global scenario. For example countries like Netherland and Germany have always taken advantage of bioeconomics strategies by leveraging the circular economy to create waste to energy sectors and robust agri tech sectors. Such countries are contributing toward GDP growth significantly and same time rural development is significant. By providing sufficient resources it helps in reducing dependency on other countries and also dependency on fossil based products which are not economically viable. The trio bio sector is a circular economy that is attracting markets.

SUCCESS STORIES RELATED TO SUSTAINABLE ENTREPRENEURSHIP

When we talk about Indian context many successful Case studies demonstrate entrepreneurial innovations in this inclusive growth bio economy sector. Ecos and company has developed cold storage which is powered by solar energy, helping in reducing post harvest losses and generating more income for the farmers. One more success story like phool.co Has shown the perfect example of temple floral waste converted into biodegradable incense product and creating the livelihood for under privileged community. One more example of success story is related to chekr innovation. 4 Kimi has given a device that collects Emissions from diesel generator and convert into usable ink. This is an example of waste to wealth innovation. Search type of Ventures are curving the challenges related to environment along with the Employment generation. Rural India is heart of the country and Entrepreneurship developing in rural areas will lead to the progress of country so it also helps in contributing towards India's vision of sustainability. It also helps in self sustaining, self Reliance economy. Such examples give the idea of entrepreneurial ecosystem within the circular bio economy which can accelerate economic growth, ensure long term financial stability, enhance competitiveness and environmental sustainability. When we see the business at a broader and higher level, were entrepreneurial activities contribute hugely toward the national economic growth, and it provides scalable opportunity to the businessman or those who want to start the business.

It is a long term solution to the Employment generation problem where not only providing the employment but also increasing living standard of the employee. This type of environmentally friendly business it promotes innovation in the field of agriculture and supports the rural economy and also increase the productivity through multi level ecosystem. Lot of integrating activities are involved in bioeconomics models like insect are the main seed for this type of farming it promotes efficiency in the utilisation of resource it also convert all the waste product into valuable outputs and same time it reduces cost of living and improve sustainability by creating new economic value chain. It leads to the development of all the associated industries which are aligned with this bio eco friendly model like production of organic farming fertiliser, animal feed production, Food processing and textile manufacturing. Textile manufacturing is supported by silk food processing is supported by Honey Bee and animal feed is supported by Black Soldier fly. As these factors spent in circular way they help in stimulating development of rural areas by increasing the cluster and supporting in increasing farmers' income in many folds. It reduces the dependence on different products that are imported from outside the country and help in strengthening the production system locally. Nowadays, sustainable Enterprises contribute to the reduction of carbon emissions and support climate change. This is key factor that will help to grow economic stability in long term and also answer the environmental uncertainties.

The projected growth of Trio Insect Bio Clusters from 2026 to 2030 indicates a steady increase in bio clusters, technology-supported units, rural entrepreneurs, and village participation. The data reflects the expanding role of sustainable rural enterprises and modern technological practices in strengthening local economies. The forecast also shows promising opportunities for employment creation, entrepreneurship promotion, and environmentally sustainable rural development.

Table 1. Forecasted cluster expansion (2026–2030)

Year	Bio Clusters Established	AI-enabled Units	Rural Entrepreneurs Supported	Villages Connected
2026	25	12	180	40
2027	32	20	260	58
2028	41	31	375	82
2029	53	45	520	110
2030	68	60	740	148

Source: Compiled by the authors using projected estimates based on rural bioeconomy development policies, agricultural innovation initiatives, circular economy practices, and entrepreneurship development reports published by NITI Aayog, Ministry of Rural Development, Food and Agriculture Organization, and United Nations Development Programme for the period 2026–2030.

ROLE OF ARTIFICIAL INTELLIGENCE IN AN INSECT CLUSTER

In the current scenario, where the use of artificial intelligence is increasing, the proposed bio based circular economy like insect based bio cluster model has a very important sector to be enhanced through decision making by using data, Technology automation has played a vital role in the development of rural economy. Use of market algorithm for the prediction of the market has helped farmers to know the customer and also locate area wise usage of the product. The use of AI Technology enables the monetary rear time biological system like the growth of larvae of black soldier fly, Help of silkworm and behaviour of we call me by using sensors. Devices like internet of Things and computer vision help in collecting the data for future needs. Machine learning algorithms help in the prediction of market can be optimised for knowing the feeding cycle. It can also be used to know the condition of humidity and temperature which is highly important for reducing the mortality rate of the insects and producing the high quality of output. AI powered segregation of waste will ensure the organic input which will increase deficiency of bio conversion process and reduce the risk of contamination. Lot of digital platforms backed by AI can be workable for the rural producer and it will help in connecting the urban market and different types of financial services which are available for farmers same time many policies given by the government are available to the farmer at real time, which helps in trending start up eco system. Further climate modelling can help in adapting different AI strategies for crop productivity, pollination and making these model for Resilient toward environmental changes. Thus, inclusion of AI in the bio insect economy model not only helps in increasing operation efficiency but also accelerates inclusivity scalability and sustainable eco friendly economic growth in the emerging bio industrial landscape.

The integrated insect based circular bio classes help in many ways for the businesses. First it creates sustainability and eco friendly environment, second it help in optimisation of resources, third cost efficiency and profitability. By converting low organic waste into increased value output like Silk Honey protein feed and bio fertilieer, with his businesses can diversify revenue streams while reducing low raw material cost. This circular approach is excellent in minimising the dependency on traditional way of getting inputs and fossil based scarce resources. It in hell the long term resilience and profitability. Additionally AI and cutting edge technologies enable quality control predicting the maintenance and optimising the process overall operational efficiency and competitiveness in emerging green economy is boosted.

For entrepreneurs who are interested in such type of business in rural and semi urban areas this eco friendly model opens up low barrier entry opportunity and inclusive economic growth. Insect base model requires relatively very limited land and shorter production cycle and with very less capital investment such type of startups and micro entrepreneur ideal. Entrepreneur scanning gauge in multiple value chains such as waste collection product marketing insect wearing processing which will diversify the income sources for many associated entrepreneurs. The model also encourages innovation developed local Employment generation and skill which is associated. From societal point of view, when bio clusters are integrated then contribution towards environmental sustainability economic inclusiveness and food security are enhanced. they help in reducing greenhouse gas effect, carbon emission and over all improving soil health through organic fertiliser production. Increase pollination services through bees help in supporting food system farmer income and Agriculture productivity. Moreover such model helps in reduction of urban Migration of Manpower and marginal community stayback with good financial support for livelihood. overall it represents Holistic solution for economic growth and social well being through environmental conservation.

The employment and social impact forecast indicates continuous growth in rural livelihood opportunities between 2026 and 2030. Direct and indirect employment are expected to rise steadily with the expansion of bio-cluster activities. Increased participation of women and youth highlights the inclusive nature of the model. The forecast also reflects significant improvement in skill development and entrepreneurial engagement, supporting sustainable rural development, income generation, and community-based economic transformation through environmentally responsible practices.

Table 2. Employment and social impact forecast

Impact Area	**2026**	**2027**	**2028**	**2029**	**2030**
Direct Employment	420	610	860	1,180	1,620
Indirect Employment	750	1,050	1,420	1,920	2,550
Women Employment Share (%)	32	36	40	44	48
Youth Entrepreneurship Participation (%)	24	30	37	45	53
Skill Development Beneficiaries	500	820	1,250	1,760	2,450

Source: Developed by the authors using estimated trends from rural employment programmes, entrepreneurship promotion initiatives, women workforce participation reports, and skill enhancement frameworks published by NITI Aayog, Ministry of Rural Development, National Skill Development Corporation, Food and Agriculture Organization, and United Nations Development Programme for forecasting the 2026–2030 period.

BIO CLUSTERS: CONCEPT AND CONTEMPORARY RELEVANCE

Bio clusters are localised ecosystems that integrate biological production processing research and market distribution within defined geographical areas. They combine primary producers, academic institutions, technology producers and industries into synergic networks that enhance competition and innovation (Porter, 1998; Delgado et al., 2010). In bio based sectors, coordinated supply change, clustering support share infrastructure and collaborative research accelerate commercialisation of biological innovations (Asheim al., 2011).

The contemporary importance of the bio cluster is rooted in 4 major Global drivers. First and foremost is the increasing Global demand for protein, which is putting unprecedented strain on conventional livestock systems. Livestock production occupies roughly 77% of the world's agricultural land while contributing significantly towards greenhouse gas emissions and fresh water depletion (Poore & Nemecek, 2018). There was a need for an alternative protein system with a lower ecological footprint (Springmann et al, 2018). Insect based protein demonstrated a favourable feed conversion ratio and reduced emissions compared to ruminant livestock (Oonincx et al., 2010).

Second, waste management challenges are escalating. urban organic waste streams exceed Municipal processing capacities, leading to landfill expansion and methane emissions (UNEP,2021). Biological waste conversion by insect larvae offers Rapid decomposition, nutrient recycling and pathogen suppression (Diener et al., 2011; Lalander et al., 2019).

Third, climate and biodiversity pressures necessitate generative agriculture models. Intensive monoculture systems reduce pollinator habitats and soil biodiversity (IPBES, 2019). Pollinated dependent crops account for a substantial portion of global agricultural output (Klein et al., 2009). Integrating pollinated supportive landscapes within bio clusters enhances ecosystem services and resilience (Potts et al., 2016).

Fourth, rural economic stagnation underscores the need for decentralised low capital Industries capable of generating exclusive livelihoods. cluster based development strategies have historically in the hands of regional competitiveness and employment (Ketels, 2013). Insect bio clusters, requiring comparatively less infrastructure, may produce accessible entrepreneurship opportunities for women and youth (Halloran et al., 2016).

The development of the bio cluster also intersects closely with the growth of micro, small, and medium enterprises (MSMEs). Such a decentralised biological production system enables small scale entrepreneurs to participate in value change through relatively low entry barriers and modular production units. Access to targeted MSME financing including micro credit impact investment and cluster based credit facilitation can significantly enhance the scalability of these enterprises by enabling investment in processing infrastructure quality assurance systems and waste conversion Technologies. Consequently, a bio cluster can function not only as an ecological production hub but also as an entrepreneurial ecosystems that stimulate rural enterprise development.

Figure 2. Circular material flow model in an integrated insect bio cluster

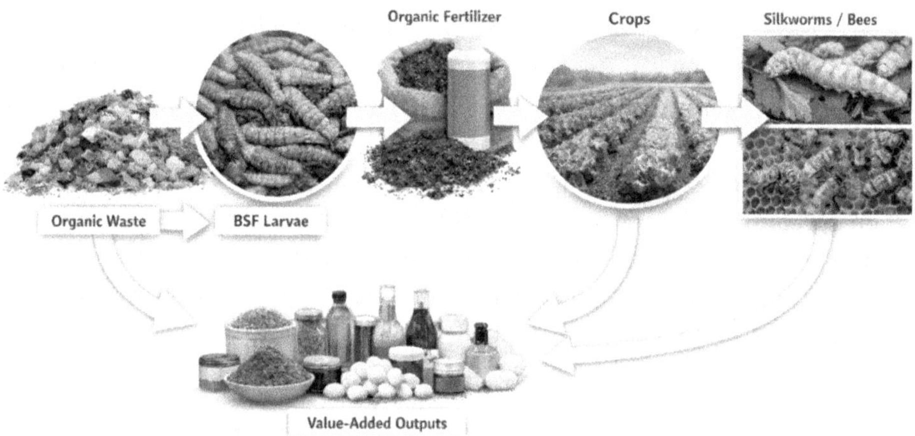

(Flowchart: waste BSF fertiliser Crops Silkworms/ bees value added output)

By converting biological inputs into multiple values streams, which includes fertilizer feed Honey fibre and bioactive compounds, bio cluster embodies circular economy principles (Elle n mAC Arthur Foundation, 2013). Waste from one subsystem becomes input for another, reducing environmental externalities while increasing economic resilience. in this sense, integrated insect bio cluster exemplifies regenerative production system capable of addressing food security, climate mitigation, rural development and biodiversity conservation simultaneously.

BLACK SOLDIER FLY (HERMETIA ILLUCENS): WAYS TO PROTEIN BIO CONVERSION

The black soldier fly has emerged as one of the most promising insect species for circular bio economy applications due to its exceptional capacity for Organic waste bioconversion. native to America but now globally distributed in tropical and temperate regions, BSF is a non pest, non vector insect whose adults do not feed on or transmit disease, making it particularly suitable for controllable industrial production (Diener et al., 2011). The larval stage constitutes the primary bio conversion phase, during which larva rapidly consume decomposing organic substrates and convert them into high value Biomass. Under suitable conditions the life cycle is completed within 30 to 40 days, which enables multiple production cycles and valley and ensures scalability for commercial operations (Makkar et al., 2022).

When we talk about nutritional values of BSF larvae it is extraordinary and model for other insects. Those larvae die, Their Biomass contains approximately 40- 45% Raw protein and substantial lipid fractions with lysine is sufficient proportion found in total fatty acid (Makkar et al., 2022). One more antimicrobial property like lauric acid is found in it. This increases the functional value of BSF oil in speciality appliances and animal feed. When BSF protein or insect protein is compared with the traditional source of proteins like soybean, fish meal etc it has been shown more benefits in terms of competitive amino acid composition and digestibility (Van Huis et al., 2015).

Traditional livestock have lower conversion efficiency than insects. BSF system has the capability to process diverse organic waste streams which makes it unique. Larvae of the BSF stage help in metabolising the agricultural residuals market discards, household food waste and slaughterhouse by products, This helps in reducing waste mass up to 80% (Diener et al., 2011). When larvae eat waste, It suppresses foul odours and reduces pathogenic loads through its microbial activity and it increases the sanitisation as outcome in decentralized waste management systems Lalander et al., 2019). BSF system is one of the sustainable alternatives to conventional composting and landfill disposal.

One of the most compelling attributes of the BSF system is its capacity to process diverse organic wave streams. Larva can efficiently metabolise household food waste, slaughterhouse by products, agricultural residues and market discards, achieving waste mass reduction of up to 80% in control environments (Diener et al., 2011). During larval feeding, microbial activity is altered in ways that suppress foul ordours and reduce pathogenic load, enhancing sanitation outcomes in decentralised waste management systems (Lalander et al., 2019). Search capabilities position BSF technology as a sustainable alternative to landfill disposal and conventional composting.

Beyond protein and oil, BSF system generates multiple co products: The residuals substrate, known as Frass, contains nitrogen, phosphorus potassium and beneficial microbial community that improves soil fertility and structure (Lalander et al., 2019). Frass application has been shown to enhance plant growth and contribute to the restoration of soil organic matter. Additionally, chitin extracted from larval exoskeleton serves as a precursor for chitosan, biopolymer with applications in water treatment biomedical and Agriculture. The multi product nature of BFS farming is an example of the circular economy principle by maximising resource efficiency and minimising waste.

Figure 3. Life cycle and waste conversion pathway of black soldier fly

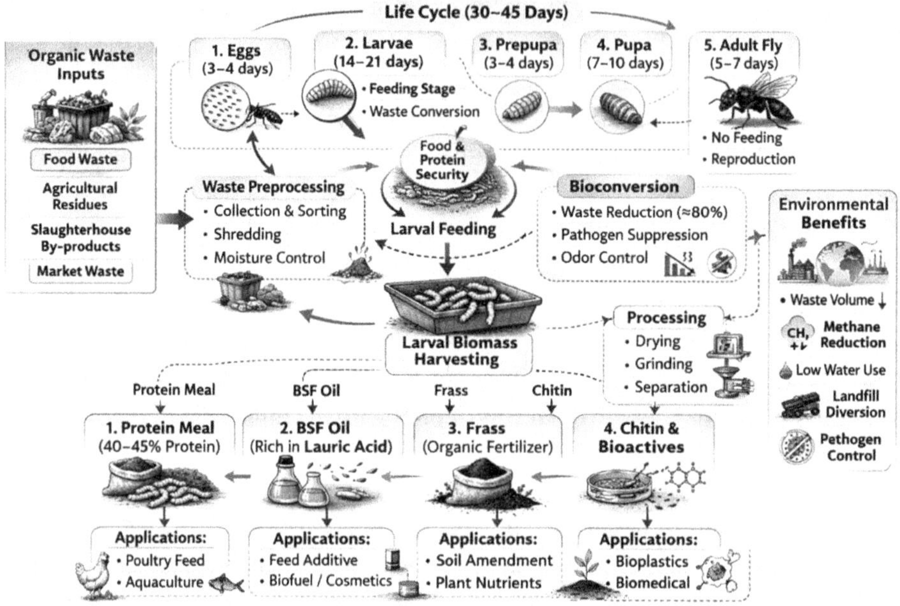

From the entrepreneurial perspective, Bio Classic created an opportunity for MSMEs to expand across multiple nodes of the value chain, including insect cultivation waste collection value added product development and processing. However the realisation of these opportunities of an depends on an accessible financing mechanism. MSMEs oriented credit program Cooperative financing models and blended finance instruments can support early stage and scalable in overcoming capital constraints while facilitating Innovation and market entry within the cluster ecosystem. Importantly, the distributed nature of bio cluster activities Aligns well with MSME-led entrepreneurial development. Small scale Enterprises can operate specialised components of the production cycle while benefiting from shared infrastructure and collective market access. With an appropriate MSME financing framework and institutional support such cluster can stimulate local Anderson while strengthening the economic viability of circular bio based systems.

SILKWORMS (BOMBYX MORI): CULTURAL LEGACY AND BIO MATERIAL PRODUCTION

Sericulture is one of the most ancient industrial activities, with records insect base industry. The traces of its original origin are found long back i.e. 5,000 years legacy. It started from China and spread across the world. Sericulture has over the years attracted the development of global trade routes which includes very famous Silk Route and it is also considered as cultural heritage of most of the country in Asia and Europe. In the sericulture business, silkworms exclusively eat the leaves of mulberry plants which are called Morus spp., Silkworm convert C Biomass into Silk protein efficiently.

In the study of the life cycle of silkworm, it has 4 stages eggs, larvae, pupae and adult moths. In these stages, only the larva produces continuous silk fibre, which can be 1500 m long in length. The Silk fibre is made of protein named as fibroin, which provides strength and the other protein sericin provides holding of the fibre together (Altman et al., 2003). The mechanical properties that are inbuilt in silk are elasticity, which also provides high tensile strength and this is easily biodegradable. Because of these features, it is considered the most versatile natural eco friendly polymers.

Figure 4. Circular integration of BSF, Silkworms and Bees in regenerative systems

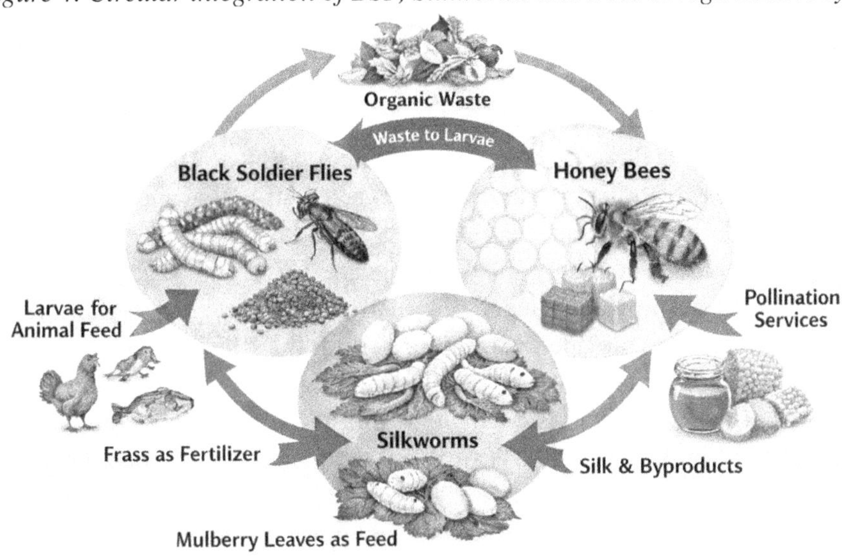

Beyond it traditional role in luxury textiles, Silk has gained prominence in advanced biomedical applications. Silk fibrone exhibits exceptional biocompatibility and control biodegradability, enabling its use in wound dressing tissue, engineering scaffolds, surgical sutures and drug delivery systems. (Kundu, 2014). It's structural stability and ability to be processed into hydrogels, films, nanofibres, etc. have expanded its relevance to biosensor technologies and regenerative medicines.

Modern sericulture has transitioned from a single mono output textile industry to the multi product bio resource system. Sericin, once discarded during silk degumming, has now been recognised for its moisturising, antioxidant and antimicrobial properties, leading to application in cosmetics and Pharmaceuticals (Kundu, 2014). Silkworm pupae HAVE around 50 to 60% Protein, which is commonly utilised for animal feed and in nutraceutical markets (Van Huis et al., 2015). Businesses like mulberry cultivation, provides agroecological advantages by improving the soil quality, erosion can be reduced and reducing atmospheric carbon by perennial Biomass accumulation.

In the context of socio economical environment, Sericulture is dominated by women and predominantly Cottage based industry. When talking about the labour requirement here it is more related to silkworm rearing and Silk processing. These two work provide employment opportunities to the rural people, also continue the traditional knowledge system and give gender equity too. Bringing into the bio cluster model, sericulture provides rural resilience by diversifying earnings and collaborating with agriculture with value added bio material in businesses. From the sericulture business, people may diversify bio based products for extending the available opportunities for the rural starters and the government also support through MSMEs which leads to the value creation. With the support of government initiative toward rural development, small scale Industry can come out as different stages of sericulture value chain system It includes mulberry cultivation, production of cocoon, Silk reeling, development of sericin or pupae based products and biomaterial processing. First type of production and distribution ancillaries can be well aligned with the micro and small

Enterprises with very nominal Capital requirement. This enhances the employment of the local people and provides value addition and cluster development. Micro, small and medium enterprises provide financial support for the schemes like microcredit schemes, Cooperative financing, impact investment fund, seed money requirement facility and cluster based facility. Such a type of arrangement helps entrepreneurs to adopt better technology to establish small or micro processing units and in return it will provide higher value returns. In this way, sericulture cluster can contribute to the entrepreneurial ecosystem old traditional knowledge system and livelihood with modern bio based Innovation and market expansions.

BEES: ECOSYSTEM ENGINEERS AND KEYSTONE SPECIES

Bees are considered as the main insect pollinators. They are very important insects for the ecological balance and their significance is more than biomass. Around 75% of world's crops depend on bee pollination, these bees serve as an important pollinating group (Klein et al., 2009). Through pollination, crop yield and fruit quality increases and generative diversity can be enhanced, which stabilises ecosystem productivity.

With the passage of time pollinated population has declined due to several reasons like pesticide exposure, Habitat loss and climate change. This has Trigger the Global concern (Potts et al., 2016). Taking this sector as an entrepreneurial activity, bee friendly habitats will provide integration within bio cluster and support biodiversity conservation with increased agricultural productivity.

In beekeeping enterprise mainly honey Bee, stingless bees and wild solitary pollinators, together contribute to flexibility in pollination networks. In the business of beekeeping, it really generates very high products which are used by the people at their home like Royal jelly honey Bee Venom be wax and propolis type of product have antioxidant and anti microbial properties which make it very important it has a medicinal value to so far was it will company also use it by and large (Pasupuleti et al. 2017). Many products made by bees like beeswax also help in cosmetic industry which is very popularly used in Cosmetic manufacturing and also it is used in medicinal industry. When we talk about it is explored that it has biomedical properties and it is commonly used in treatment by medical Research. The multiple product lines are given by the Apiculture that is why now we can proudly say that Industries or businesses related to be keeping our first of all, nursing the sustainable environment It requires very little investment, and it gives lot many products that are commonly used by household as well as the cosmetic industry medical industry and it makes a very viable cluster. Even when we talk about sustainability so it is very rightly said that be reduces dependency or many medicines and reduces the burden of costly medicines because it has a property which naturally improve human health as well as crop productivity. Honey bees are very important for the pollination, and in the presence of pollination they are integral part to provide market opportunity and make sustainable product through this particular industry (Potts at el. 2016). Honey bees also contribute toward environmental Change, especially in the agriculture products because they stabilise the crop under different climate conditions. When incorporated with different bio cluster honey bees not only functions as a marketable good but it also help in increasing ecological ecosystem and supporting productivity of Mulberry plants same time it also increases productivity of different flowering crops and maintains the healthy rich green environment. there role is very important in completing the regenerative cycle by connecting economic production with biodiversity conservations.

Table 3. Comparative biological and production characteristics of insect systems

Parameter	Silkworm Bombyx Mori)	Black Soldier Fly (Hermetia Illucens)	Honey Bee (Apis spp.)
Basic function	Silk biometrical production	protein production and waste bio conversion	agriculture and pollination
Feed source	Mulberry leaves	organic waste streams	Pollen and nectar
Duration of life cycle	45 to 60 days	30 to 45 days	365 days
Feet conversion ability	moderate	very	ecosystem mediated
Main outputs	Sericin, pupae protein, silk fibroin	Oil, chitin, frass, protein meal	Propolis, wax, royal jelly, honey
Content of protein %	50-60% (pupae)	40-45% (Dry Larvae)	1% (Honey trace protein)
Role of environment	Integration of agroforestry	Methane mitigation, waste reduction	Biodiversity, Crop pollination
Requirement of capital	Moderate	Low moderate	Low
Scalability	Medium	High	High

Figure 5. Comparison between protein content, lifecycle duration and scalability index

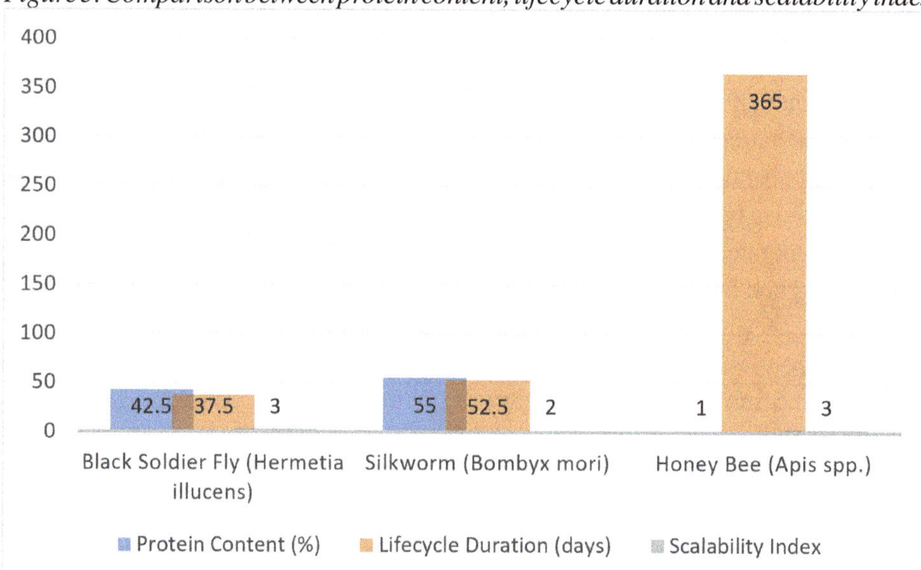

BIO CLUSTER MODEL INTEGRATING ALL THREE INSECTS

Traditionally we have seen that production changes are always linear where one produces a research flow from extraction to disposal but in this trial system we are considering integrating bio cluster system which will think the size waste management, bio material production, regenerating agriculture and pollination services in 2a single unified circular model. unlike linear production chain, here the integrated trio insect bio clusters operate as a chain of interconnected biological networks where one insect's output works as input for another insect and another insect's output functions as input for third insect and third insect's output again comes back to input for 1st insect.

This type of circular ecosystem complements to each other enhancing economic resilience ecological efficiency and social inclusion. The conceptual basic foundation of this type of integration is deeply rooted in circular economy theory, where pressure is placed on resources to be circulated and regenerated, minimising waste and increasing the profitability (Geissdoerfer et al., 2017; Kirchherr et al., 2017). These types of business should be promoted in all rural areas with two or three different types of collaboration where entrepreneurs will have to invest only minimum and sustainability and maximum output will be the result and this way the economic status is improved. In a circular manner, when silkworm bees and soldier flies work together, geographically proximate bio cluster synergistic value creation is built across the multiple biological Pathways. Each insect plays a distinct role and same time complements functional niche inside the cluster. black soldier fly larvae Access main decomposer and very soon it is converted into heterogeneous organic waste which has nutrient Frass and protein rich biomasses.

MAIN ASPECT TO BE ADOPTED BY BUSINESSMAN

An entrepreneur can begin with integrating these three insects and leveraging the concept of complementary system. Different species like silkworm BSF and bees perform distinct mutually collaborative role. The process should be started with a small pilot model, where a small set up with the Black Spider fly unit will be kept. This insect will create a organic waste this waste will help in growing healthy a plant of Mulberry which will be used by the silkworm rearing. Third the waste of silk wearing will help in making a colony of bees for pollination and production of honey. Here for the beginner who is starting the Enterprise the experience and risk will be low and resources will also be economical. Once the idea has come then proper infrastructure can be developed which will include control rearing atmosphere and availability of organic waste resources. same time entrepreneur has to take technical knowledge through the training program offered by many Institutes for example Krishi Vigyan Kendra aur different Apiculture departments and Sericulture department organised training. Lack of proper technical knowledge and behaviour of insect may lead to the business not working as planned and finally it can be wound up. So businesses should start with very good training and knowledge about the insect's behaviour before going for the large scale.

Challenges for the entrepreneur

Once you have decided to start with cluster there will be several problems will hamper the establishment and operation of such type of system. it includes non availability of funds, Right weather, government policies, technical know how, available types of insects, Initial high investment cost, market uncertainty, precaution from the diseases and above all maintaining the will power of the entrepreneur. These all situations generally hampered the entrepreneurial mindset. There is another very critical issue that is environmental sensitivity as all the three insects requires specific temperature, ecological conditions and humidity to grow. Additionally Mulberry leaves floral resources and organic waste should be supplied sufficient and timely quantities to nurture the insects. in the absence of any above said will lead to the closing of the business.

Overcoming from the challenges

To reduce the challenges and minimise the risk of entrepreneur why should focus on professional development and try to have required skills before starting the business. Second limitation one should always overcome through starting with a small business and slowly gradually as experience comes one should think of expanding the business. here I would like to mention that entrepreneurs sometimes without realising take huge risks in deploying finance without having Knowledge of environment and the behaviour of the insect which leads to death of many insect. Entrepreneurs should be sensitive enough to understand the nature of each insects before starting the business. no coming to the building a partnership with the local farmers who are basically going to provide resources as well as their also going to help in growing the entrepreneur activity by supporting with many type of work in the Enterprise. nearby local restaurants industries and markets should be exploded and taken care as after the production we should know where the product to be sold. A study of market who are the consumer of product should be very clear before starting the Enterprise. Who will be the raw material supplier from where all the support for the business will come and to whom product has to be sold should be analysed before going for the entrepreneurial activity. One More very important aspect is associated such type of bio cluster that is climate control because such type of insects grow in a certain temperature and they need to be monitor continuously for the better and improve productivity and minimise losses. room temperature humidity and cert certain other factors to be carefully examined so that all the challenges should be taken care of. Finally integrated trio insect bio cluster itself actor risk management strategy as the failure of one product will be the compensated for the other insect and scalability of the business will be stable.

Government initiative

The government is known for supporting and increasing entrepreneurial activities. The government of India is giving many types of support to encourage such sustainable ventures and innovative activities there are schemes like be keeping and honey mission offers training for Apiculture with lot many subsidies to start the business. The government is providing the seed money and supporting throughout the business so that anyone with less knowledge can also easily start the business with the support of government. Another initiative of the government is promoting the sericulture department by conducting seminars and supporting the cultivation of Mulberry so that high quality of silkworm wearing can be done. A lot of financial support is provided by the government under such type of business. Institutes like NABARD and SIDBI are actively providing financial assistance and also providing very low interest rate on loans to promote the rural entrepreneurship and ecosystem so that rural cluster can be developed efficiently. Lots of initiatives, Like Start-up India and waste to wealth program, Provide boost to the economy and encourage a circular economy. Together these supports can significantly reduce operational and financial burden on businessmen, making this integrated insect based business Sustainable and viable. India's most important Enterprises, MSME micro, small and medium enterprises play a very crucial and vital role in development of economy. The government has introduced so many diverse schemes to support the innovation growth and sustainability. Few very evident initiatives are Prime Minister Employment Generation Program which is very successful program providing financial assistance to the new budding entrepreneurs for setting up the business through subsidies and also helping in getting the credit availability. One more important enterprise that is Credit Guarantee Fund Trust for micro and small Enterprises provides collateral free loans with easy access to loan. Additionally there is one more scheme known as MSME champions which focuses on the upgradation of technology development of skills and checking the competitiveness of the market. in reality these schemes collectively help in promoting entrepreneurial activities in the Rural environment and helping in emerging startups in these areas related to bio economy agri tech and encouraging industrial growth. It is evident that these steps of MSME help in improving the Employment generation and collective economic development of the country.

CONCLUSION

The convergence of Artificial Intelligence (AI), entrepreneurship, and MSME-driven development with integrated insect-based bio-clusters marks a transformative pathway toward sustainable and inclusive economic growth. The synergistic integration of Black Soldier Fly (Hermetia illucens), silkworms (Bombyx mori), and bees within a circular bio-economic framework demonstrates how biological efficiency can be amplified through technological intelligence and decentralised enterprise models. AI-enabled monitoring systems, predictive analytics, and smart resource optimisation enhance the productivity and scalability of these bio-clusters, ensuring precision in waste bioconversion, biomaterial production, and pollination management. This fusion not only improves operational efficiency but also reduces risks, enhances quality control, and enables real-time decision-making, thereby making such systems viable for large-scale adoption. From an entrepreneurial and MSME perspective, this integrated model offers a highly adaptable and low-barrier opportunity structure, particularly suited to rural and semi-urban economies. Government-backed MSME schemes, access to digital platforms, and AI-powered market intelligence collectively empower small enterprises to participate in high-value bio-economy value chains. Entrepreneurs can diversify income streams across insect protein production, silk-based biomaterials, apiculture products, and organic fertilisers, reducing dependency on single markets and enhancing financial resilience. The cluster-based approach further promotes shared infrastructure, skill development, and knowledge spillovers, fostering innovation ecosystems that drive competitiveness and local economic development. In this context, MSMEs act as critical vehicles for translating sustainability into scalable business models.

At a societal level, the integration of AI with bio-cluster systems contributes to a regenerative development paradigm that aligns economic growth with environmental stewardship and social inclusion. These systems address critical global challenges by reducing organic waste, lowering greenhouse gas emissions, restoring soil health, and enhancing agricultural productivity through natural pollination. Simultaneously, they generate employment, empower marginalised communities, and reduce rural-urban migration by creating localised livelihood opportunities. By internalising ecological processes into economic systems, integrated insect bio-clusters redefine development as a balance between profitability, sustainability, and resilience. As nations strive to achieve climate goals and sustainable development objectives, the convergence of AI, entrepreneurship, and MSME-led bio-economy models offers a scalable, future-ready solution for building resilient economies and sustainable societies.

At societal level when fusion of such bio cluster system is adopted it helps in GDP of that country creates environmentally friendly business And it help including all the sector of country to develop the rural economy. This system critical Global challenges by reducing the organic waste and bringing down Greenhouse gas emission increasing the soil health and making profitable agriculture productivity through natural green business model. Simultaneously empowering marginalised communities helping in generating employment upliftment of rural urban economy and only providing localised lively hold opportunities. all these benefits are very good for the ecological support system. As Nation needs to achieve climate goals and sustainable development goals by converging the entrepreneurship green bio clusters MSME lead financing schemes together help in building resilient and Sustainable Societies.

REFERENCES

Altman, G. H., Diaz, F., Jakuba, C., Calabro, T., Horan, R. L., Chen, J., ... & Kaplan, D. L. (2003). Silk-based biomaterials. *biomaterials, 24*(3), 401-416.

Asheim, B., Boschma, R., & Cooke, P. (2011). Constructing regional advantage. *Regional Studies, 45*(7), 893–904. DOI: 10.1080/00343404.2010.543126

Delgado, M., Porter, M. E., & Stern, S. (2010). Clusters and entrepreneurship. *Journal of Economic Geography, 10*(4), 495–518. DOI: 10.1093/jeg/lbq010

Diener, S., Zurbrügg, C., & Tockner, K. (2011). Conversion of organic material by BSF larvae. *Waste Management & Research, 29*(11), 1177–1184. PMID: 20837558

Ellen MacArthur Foundation. (2013). *Towards the circular economy.*

European Commission. (2018). *A sustainable bioeconomy for Europe.*

FAO. (2011). *Global food losses and food waste.*

Geissdoerfer, M., Savaget, P., Bocken, N. M., & Hultink, E. J. (2017). The Circular Economy–A new sustainability paradigm? *Journal of Cleaner Production, 143*, 757–768. DOI: 10.1016/j.jclepro.2016.12.048

Halloran, A., Roos, N., Eilenberg, J., Cerutti, A., & Bruun, S. (2016). Life cycle assessment of edible insects for food protein: a review: Life cycle assessment of edible insects for food protein. *Agronomy for Sustainable Development, 36*(4), 57. DOI: 10.1007/s13593-016-0392-8 PMID: 32010238

IPBES. (2019). *Global assessment report on biodiversity.*

IPCC. (2022). [Impacts, adaptation.]. *Climatic Change*, •••, 2022.

Ketels, C. (2013). Recent research on competitiveness and clusters. *Competitiveness Review, 23*(4), 265–284.

Kirchherr, J., Reike, D., & Hekkert, M. (2017). Conceptualizing the circular economy: An analysis of 114 definitions. *Resources, Conservation and Recycling, 127*, 221–232. DOI: 10.1016/j.resconrec.2017.09.005

Klein, A. M., Müller, C., Hoehn, P., & Kremen, C. (2009). Understanding the role of species richness for crop pollination services. *Biodiversity, ecosystem function and human wellbeing. New York*, 195-208.

Kundu, S. C. (Ed.). (2014). *Silk biomaterials for tissue engineering and regenerative medicine*. Elsevier.

Lalander, C. D. S. Z., Diener, S., Zurbrügg, C., & Vinnerås, B. (2019). Effects of feedstock on larval development and process efficiency in waste treatment with black soldier fly (Hermetia illucens). *Journal of Cleaner Production, 208*, 211–219. DOI: 10.1016/j.jclepro.2018.10.017

Makkar, H., Heuzé, V., & Tran, G. (2022). Locusts and grasshoppers: nutritional value, harvesting and rearing for animal feed, and other applications. *CABI Reviews*, (2022).

McCormick, K., & Kautto, N. (2013). Bioeconomy in Europe. *Sustainability (Basel)*, *5*(6), 2589–2608. DOI: 10.3390/su5062589

OECD. (2009). *The bioeconomy to 2030.*

Oonincx, D. G., Van Itterbeeck, J., Heetkamp, M. J., Van Den Brand, H., Van Loon, J. J., & Van Huis, A. (2010). An exploration on greenhouse gas and ammonia production by insect species suitable for animal or human consumption. *PLoS One*, *5*(12), e14445. DOI: 10.1371/journal.pone.0014445 PMID: 21206900

Pasupuleti, V. R., Sammugam, L., Ramesh, N., & Gan, S. H. (2017). Honey, propolis, and royal jelly: A comprehensive review of their biological actions and health benefits. *Oxidative Medicine and Cellular Longevity*, *2017*(1), 1259510. DOI: 10.1155/2017/1259510 PMID: 28814983

Poore, J., & Nemecek, T. (2018). Reducing food's environmental impacts. *Science*, *360*(6392), 987–992. DOI: 10.1126/science.aaq0216 PMID: 29853680

Porter, M. E. (1998). Clusters and the new economics of competition. *Harvard Business Review*. PMID: 10187248

Potts, S. G., Imperatriz-Fonseca, V., Ngo, H. T., Aizen, M. A., Biesmeijer, J. C., Breeze, T. D., Dicks, L. V., Garibaldi, L. A., Hill, R., Settele, J., & Vanbergen, A. J. (2016). Safeguarding pollinators and their values to human well-being. *Nature*, *540*(7632), 220–229. DOI: 10.1038/nature20588 PMID: 27894123

Raworth, K. (2017). *Doughnut economics*. Chelsea Green.

Springmann, M., Wiebe, K., Mason-D'Croz, D., Sulser, T. B., Rayner, M., & Scarborough, P. (2018). Health and nutritional aspects of sustainable diet strategies and their association with environmental impacts: A global modelling analysis with country-level detail. *The Lancet. Planetary Health*, *2*(10), e451–e461. DOI: 10.1016/S2542-5196(18)30206-7 PMID: 30318102

UNEP. (2021). *Food waste index report.*

United Nations. (2015). *Transforming our world: The 2030 agenda.*

United Nations. (2019). *World population prospects.*

Van Huis, A., Dicke, M., & van Loon, J. J. (2015). Insects to feed the world. *Journal of Insects as Food and Feed, 1*(1), 3–6. DOI: 10.3920/JIFF2015.x002

Chapter 15

Core-Shell Nanostructure Design in Polymer Bio-Nanocomposites

Elif Esra Altuner
http://orcid.org/0000-0001-7663-6898
Kocaeli Health and Tecnology University, Turkey

Fatih Sen
Dumlupinar University, Turkey

ABSTRACT

In this chapter, the design, formation, and importance of core-shell nanoparticles are emphasized. At the same time, the importance of the production of core-shell nanoparticles from bio-composite polymer materials was also mentioned. Today, nanotechnological studies continue to develop considerably. Nanoparticle research from nanotechnological studies attracts considerable attention. Catalytic effects and performances of nanoparticles have been reported in the studies. However, the controlled release of nanoparticles, especially in drugs, is very important for human health. Direct nanoparticles reaching cells, tissues, or organs can threaten human health. Therefore, it is extremely important to cover the core with a harmless shell, and a harmless controlled drug release is extremely important. While forming core-shell nanoparticles, obtaining from bio nanocomposite polymers will further increase the advantage.

DOI: 10.4018/979-8-2600-0060-1.ch015

INTRODUCTION

Nanocrystals are a subject of study used in many applications such as biomarking, solar fuel cells and LED screens. Likewise, it is one of the important materials in industry and science because of its stable chemical structure in polymers, and similar properties like nanocrystals (Eryılmaz, 2016). There are numerous studies in the literature made with nanocomposite materials (Altuner, 2025; Cetinkaya et al., 2025; Esra Altuner et al., 2025; Lounissa et al., 2025). *Sen Research Group* has reported many studies in the literature such as oxidation, sensor, catalytic activity values with nanocomposite / nanoparticles (Eris et al., 2018; Erken et al., 2015, 2016; Karatepe et al., 2016; Nesrin et al., 2020; Ozturk et al., 2012; B. Sen, Demirkan, Şavk, et al., 2018; B. Sen, Demirkan, Şimşek, et al., 2018; B. Şen et al., 2018; F. Sen et al., 2014; F. Şen & Gökağaç, 2007, 2008; S. Şen et al., 2011; Sert et al., 2016, 2017). Layer-by-layer assembly is an increasingly advanced technology. This technology is used in applications such as drugs, vitamins, nano delivery systems (encapsulations). Core-shell nanoparticles can be created with layer-by-layer techniques. This technique is the process of collecting cationic and anionic bio layers around the desired nanoscale particles. The thickness of the shell, its compatibility with the environment, its charge ability, and permeability are important parameters for this process (S. Chen et al., 2019). Thus, with these parameters, the molar mass, conductivity, structure and electrical characteristic structure of the nanoscale polymer structure are determined with the cationic, and anionic bio layers gathered around the nanoparticles (S. Chen et al., 2019; Correa et al., 2016). Multiple uses of these systems in related applications create a greater synergistic effect than single-use. Therefore, multiple uses are more advantageous. Especially in cancer or chronic patients, the application of multiple drugs and these systems instead of single drugs increases the synergistic effect much more and results are obtained much faster (L. Ma et al., 2013; Muhsin et al., 2025; Taner et al., 2025). Also, drug encapsulation through the single delivery system and the use of this process is difficult due to the release of drugs. Because polarization of substances and differences in molecular properties are among the factors affecting this situation. However, in the use of encapsulated drugs in the use of multiple systems with a very strong synergistic effect, the substances have different polarities and can be located in various regions around the core shell (Gong et al., 2018; Lee et al., 2012). This promotes the layer-by-layer feature of core-shell nanostructured materials. Apolar particles can settle inside the hydrophobic core. However, particles that are polar settle in the hydrophilic shell (S. Chen et al., 2019; Haidar et al., 2008; L. Ma et al., 2013). To investigate this situation, researchers have conducted studies on tumor-infected mice, and these studies have taken place in the literature. According to these studies, layer-by-layer technology was applied to increase the effectiveness of anticancer agents in tumor-contaminated mice. Thus, the nanoparticles are encapsulated in the liposome using the layer-by-layer method (S. Chen et al., 2019; Oh et al., 2013). Biopolymers are examined in 3 categories according to their monomeric structure (Desai et al., 2023; Fazal et al., 2023). These are polysaccharides, proteins/polypeptides, and polynucleotides. This scheme is given in figure 1. Biopolymers are examined in 3 categories according to their monomeric structure. These are polysaccharides, proteins/polypeptides, and polynucleotides. This scheme is given in figure 1. Examples of polysaccharides are chitin and chitosan, alginate, and cellulose. Examples of proteins or polypeptides are silk (spider), collagen, and keratin (Xiong et al., 2018). These polymers are found in insect shells (chitin), hair and nails (collagen and keratin), plants (cellulose), and animal products. Examples of polynucleotides are deoxyribonucleic acid (DNA) and ribonucleic acid (RNA). Deoxyribonucleic acid (DNA) is responsible for transmitting genetic codes in living organisms and viruses (Kondratyeva et al., 2022; Xiong et al., 2018). It is ribonucleic acid (RNA), on the other hand, deoxyribonucleic acid (DNA) enables the transformation of the transported genetic information into proteins.

Figure 1. The structures and sources of biopolymer for nanocomposites

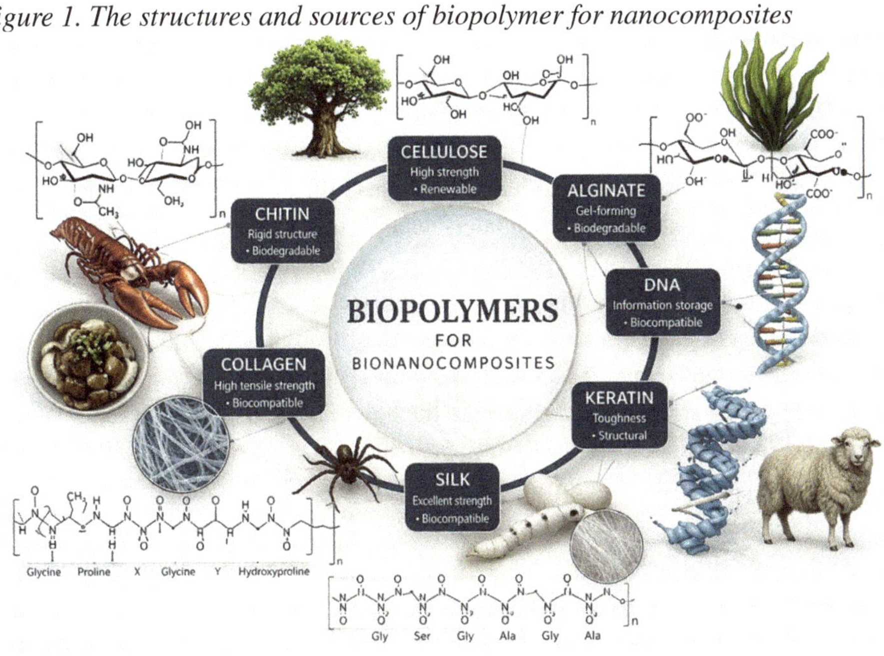

"Illustrated by the authors based on literature data"

In this chapter, detailed explanations have been made with the studies on the core-shell design of nanocomposite materials produced from biopolymers.

1. NANO-STRUCTURED BIOMATERIALS

Biomaterials have a structure that imitates natural tissue properties and they have properties such as biocompatibility and biodegradability. Biomaterials include natural and synthetic polymers, metals, composites, and particles with these parameters. Functional biomaterials that can mimic the properties of natural tissue development are the most important limiting component of biomaterial studies (Lalegül Ülker, 2020). Nanobiomaterials, on the other hand, have structures that easily adapt and mimic the size of extracellular matrix components of tissues due to their nano size (Kelleher & Vacanti, 2010). Due to their nanostructure, they easily enter and adapt to target drugs in the environment, tissues, and cells. The word nano in Latin and Greek means "*dwarf*" (Latu, 2018). A human hair is on average 80000 nm. In other words, 1 nm is 80000 times smaller than a human hair (Natarova et al., 2019) and 1 nm is one-billionth of a meter (Lalegül Ülker, 2020; Tan et al., 2016). Studies on nanocomposites and nanoparticles in nanotechnological studies have been reported in numerous studies in the literature (Hiremath et al., 2021). In the studies on nano-composites in nanotechnology, the particles of the materials were investigated and the physical and chemical properties of the materials were investigated, and studies to further elaborate these properties were presented (Lalegül Ülker, 2020; Boulaiz et al., 2011). Nanostructured materials are highly preferred in optronics and biomedical fields. Researchers are aiming to develop biomaterials and biopolymers with much more practical, functional, and durable properties than macro-implementations for the development of innovative products and three-dimensional architecture (Lalegül Ülker, 2020; Kim et al., 2013). Biomaterials used in medicine and healthcare must be biocompatible. Also, it is of particular importance that it does not show an important immunogenic effect and reaction (Baker et al., 2010). It is important that cells and tissues in the nanoscale area these enter through adhesion, migration, and differentiation significantly direct their behavior, functions, and functions (Lalegül Ülker, 2020). In pharmaceutical studies in the health field, nanostructured materials easily enter the target cell due to their nanostructure (Goldberg et al., 2007). Thus, they direct cells with their biocompatibility. Scientific and technological advances have made nanotechnology more consistent (X. Li et al., 2012). These processes are followed as (a) new and advanced synthesis methods that allow intervention and size control of nanoscale building blocks (b) new characterization tools developed to work at nano size (c) a better understanding of the relationship between nanostructure and properties and how they can be designed. The performance of the materials depends on their properties. These properties depend on the atomic structure, composition, microstructure, and interface, which are in turn controlled by the kinetics and thermodynamics of the synthesis process (A.N. & E.A., 2018; Bououdina et al., 2006). Today, the process of synthesis of advanced materials from the atomic or molecular scale to macroscopic scale emphasizes the orderly formation of atoms and molecules (L. Chen et al., 2025). Nano-structured materials are generally characterized by physical size, surface, and the interface between 1-100 nm; known materials (Altuner, Bozbeyoğlu, et al., 1 C.E.; Altuner et al., 2023; Erduran et al., 2023). It attracts a lot of attention due to its unique properties that have been proven or hoped compared to. Nanostructured materials with the "top-down" approach, from the macro scale to the nanoscale, with the abrasion of coarse-grained materials or its complete opposite. It can be formed by the combination of atoms or particles with the "bottom-up" approach (Altuner, Bozbeyoglu, et al., 1 C.E.). The control of the arrangement of atoms from the nanoscale to the macroscale is directly proportional to the chemical character of the material (Teo & Sun, 2006). It is not surprising that interest in chemical synthesis and processing is increasing day by day. Chemical reactions for material synthesis can be carried out in solid, liquid, and gaseous forms (Teo & Sun, 2006). In the best known solid-state synthesis; the solid starting material (precursor) is ground and mixed to increase the surface contact area, and then this mixture is mixed at high temperature (Teo & Sun, 2006). The heat treatments are applied to facilitate the diffusion of atoms or ions. The diffusion of the atoms depends on the reaction temperature and grain boundary contact (Kreider & Lipiński, 2018; Kumar et al., 2022; Suzuki & Mishin, 2005). The transition between grain boundaries is affected by impurities and errors in the environment. The mixing and grinding steps are repeated throughout the heating cycle, and it often takes a lot of effort in mixing nanoscale materials as well as preparing new surfaces for further reactions (F. Chen et al., 2022; Kishore et al., 2022; Tulinski & Jurczyk, 2017). Generally, solids with large particle sizes are obtained in systems where an inhibitor to stop grain growth is not used (Ersoz et al. 2018). As nanostructured materials develop, many names and descriptions are used in this field. Some of these definitions are as follows:

Nanocluster: It is formed by the combination of more than fifty units (atoms or reactive molecules). Surrounded by a ligand that allows for the isolation of the beam-forming compounds or molecular species (stable, soluble, isolable) [32].

Colloids: It is the general name for solid-liquid mixtures where particles suspended in the liquid phase can maintain their position with stability (Atiemo-Obeng et al., 2004). A colloidal particle has a size between 1-1000 nm (Ersoz et al. 2018)

Nanoparticles: It is the nano-sized form of non-crystalline, single crystal or crystalline aggregates in the range of 1-1000 nm (Ersoz et al. 2018).

Nanocrystals: Nanometer-sized single crystalline solid particles (Ersoz et al. 2018).

Quantum dots: At least one-dimensional particles that have the effect of determining the size (Ersoz et al. 2018).

1.1. Polymer Bio-Nanocomposites

Biocomposites are nanostructured materials made up of biopolymers or synthetic polymers at nano-size. Nanocomposite materials obtained from synthetic polymers act as a supporter of some parameters of products obtained from natural sources. These tasks are features such as increasing the thermal stability, functionalizing the barrier properties (Bitinis et al., 2011). Today, due to the damage caused by materials such as petroleum and plastic to the environment and human health, there is a need for nanocomposites obtained from natural biopolymers and studies with these bio-nano composite materials. According to studies reported in the literature, the biodegradability of natural and pure biopolymer is accelerated by the addition of organo-clay due to the role of the catalytic activity of the nano filter (Bitinis et al., 2011). Also, bio nano-composite polymer materials are used in biomedical and electrochemical applications due to their non-toxicity and easy adaptability. In the literature, potentiometric sensor measurements of hydride metals attached to Na + - montmorillonite / chitosan were taken (Bitinis et al., 2011; Darder et al., 2005, 2006, 2007; Ruiz-Hitzky et al., 2005). Many studies on the investigation of the properties of polymer bio nanocomposites from nature and biomass have been reported in the literature (El Achaby et al., 2014; Raemdonck et al., 2013). Generally, polymer bio nanocomposites are called green polymer composites and polymer nanocomposites. Generally, polymer bio nanocomposites are called green polymer composites and polymer nanocomposites. As we mentioned earlier, polymer bio nanocomposites can easily be compost and dispose due to their environmentally friendly, non-toxic, sustainable and dergradable properties (Dong et al., 2015). Eco-friendly polymer bio nanocomposites are obtained from green fillers, cellulose, polylactic acid, gelatin, lactic acid, collagen from a mixture of polymers, keratin, and various biomass and naturally occurring polymers.

2. CORE-SHELL NANOSTRUCTURE DESIGN

It consists of a multi-layered inorganic core and dynamic organic surfactant shell in colloidal nano-structured materials (Qamar et al., 2024; Srivastava et al., 2026). Therefore, these nanostructured materials are nanostructured materials that have an inorganic-organic component in terms of a core-shell relationship (Chiozzi & Rossi, 2020; Shafiee et al., 2022; Tiwari et al., 2021). Compact polymer fibers with sizes below 10 nm and 10 nm are used in various fields such as protective suits, different filter applications, fiber reinforcement, and biomedical studies. In tissue engineering and drug delivery systems, nanostructured particles loaded with catalysts are important for optical applications, particles having a nanoscale diameter leads to positive properties in changes in the release rate of drugs (Sun et al., 2003; J. Zhu et al., 2012). In the core-shell properties of nanoparticles, studies with core-accepting fibers made from conductive materials made of metal, semiconductor and conductive polymers have been reported in the literature (Jones et al., 2024). In these reports, studies based on gold (Abed et al., 2019), silver (Banik et al., 2018), nickel (Y. Zhang et al., 2019), silicon-carbon (Cui et al., 2009), cadmium-selenium (Bhanoth et al., 2016), and polypyrrole (Qiao et al., 2016) applications are reported. When the structure of the spider silk is examined in nanoscale, it shows a very durable feature compared to other fibers (Mehrotra et al., 2019). It shows us a well-known example of ± shell fibers in combination (Sun et al., 2003). According to the information reported in the literature, nano cable monomers are obtained by laser ablation method on a template. Likewise, nano and meso cables from Polymer ± metals and polymer ± polymer were obtained by the TUFT technique (Sun et al., 2003; Caruso et al., 2001). Co-extrusion assistance is used to transform core ± coated polymer or glass optical fibers. The liquid flowing from the core and an nular nozzles is a mixture of so-called polymer solution and molten materials. Co-electrospinning is a method used to create polymer nanotubes (Sun et al., 2003). Core and shell fibers of 2% and 3% by weight of loan polyethylene oxide have been reported in the literature and the core-shell relationship of the optical images of these two polymers is shown in figure 2. At the same time, TEM (transmission electron microscopy) images of these two identical polymers are given in Figure 3 to better display the core and shell relationship. According to Figure 2, 2% and 3% polyethylene oxides were compared with bromophenol. As a result, the bead diameter in the two nanofibers is approximately 3000 nm according to figure 3. In Figure 3, it was observed that 3% polyethylene oxide and 8% polysulfone combined nanofibers had a shell diameter of 60 nm and a core diameter of 40 nm. It has also been observed that the shell interface of this combined nanofiber is not smooth (Sun et al., 2003).

Figure 2. The illustration image of the poly ethylene oxides (2% and 3%) optical images for the core and shell properties

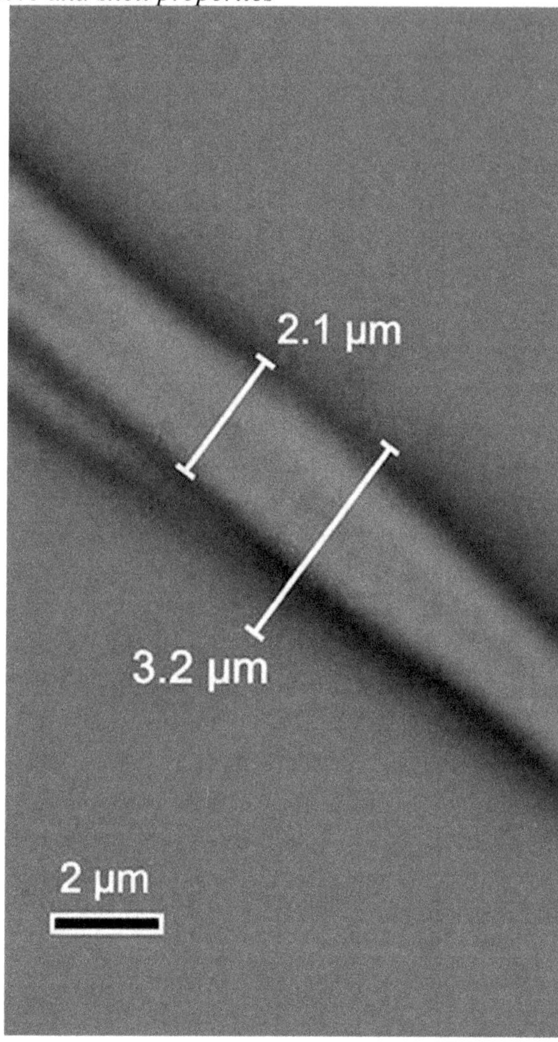

"Illustrated by the authors based on literature data."

Figure 3. TEM (transmission electron microscopy) images of nanofibers (poly ethylene oxide and polysulfone ssolutions).

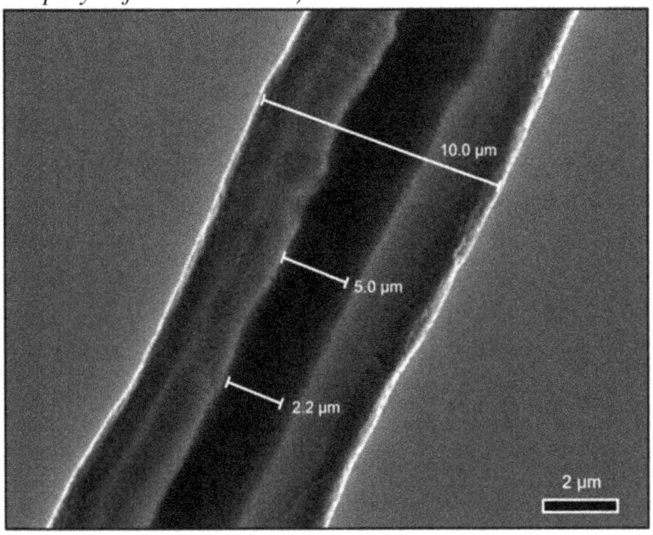

"Illustrated by the authors based on literature data."

Chemistry is based on the arrangement of atoms and molecules and indeed has a long history in the synthesis of materials including nanostructures (Kondratyeva et al., 2022). Colloids and catalysts are some examples (Quinson et al., 2018). The current popularity of nanoscience has not only led to a revival of many old chemical methods used but also has led to the continuous development of new and modified chemical methods for the synthesis of nanostructured materials (Bayda et al., 2019; Cai et al., 2018; Malik et al., 2023). The scope of chemical synthesis and processes of nanostructured materials is very wide, electronic, magnetic, biological, catalytic, and biomedical materials, etc (Alhalili, 2023; K. Zhu et al., 2018). Particle synthesis in solution occurs through a chemical reaction that develops with stable core formation and subsequent particle growth (Ghosh Chaudhuri & Paria, 2011). The term precipitation is commonly used to describe this series of events. Precipitation of solids in solutions is a highly studied topic (Wang et al., 2005). Elemental or multi-component particles can be precipitated. When a multi-component material is desired, special attention is required in controlling co-precipitation conditions to achieve the chemical homogeneity of the product (Bhat et al., 2024; Z. Li et al., 2020). The reason for this; different ions are usually precipitated under different pH and temperature conditions and the products have different solubility constants (Bhattachar et al., 2006). Upon the addition of reagents such as reducing or oxidizing agents to solutions containing the reactants, chemical reactions occur and the solution becomes too saturated with the product (Fanning, 2000). This state of saturation leads the chemical system to deviate from the minimum free energy order. The thermodynamic equilibrium state of the system is restored by the condensation of the product's core. Meanwhile, two types of nucleation may occur. The homogeneous nucleation process does not contain any foreign species as a nucleation aid (Xu et al., 2021). On the other hand, heterogeneous nucleation has a mechanism that develops by nucleation on foreign species. During the growth process, kinetic factors, as well as the thermodynamics of the system, come to the fore. In particle growth, kinetic factors such as reaction rate, rate of passage of reactants, proper removal of the substance from a solution, and redistribution of the substance compete with thermodynamic effects (Vehkamäki & Riipinen, 2012). Reaction and rate of transition; It is affected by the concentration of the reactants, temperature, pH, regulation of the reagents added to the solution, and mixing (Bourne, 2003). The structure and crystallinity of particles can be affected by reaction rate and impurities. Factors such as saturation, nucleation and growth rate, colloidal stability, recrystallization, and dwell time affect the morphology of particles (Thanh et al., 2014). Generally, extreme saturation has a large effect on the morphology of the precipitates. The low saturation particles are small, dense, well-formed, and the particle shape has a crystalline structure, and depends on the surface energy (Jiang et al., 2025). Large and branched particles are formed at high saturation levels (Liu et al., 2015; Müller et al., 2014). At higher saturation, smaller but compacted, agglomerated particles are formed (Bai et al., 2017). While the particles are small, growth in solution is interface controlled; becomes diffusion-controlled after reaching critical size (Peukert et al., 2003). Electrostatic repulsion occurs between two particles carrying the same charge (Matsusaka & Masuda, 2003). Van der Waals gravitational forces are proportional to the inverse of the distance (Orlov, 2017; Raicheff & Zahariev, 2001; L. Zhang, 2013). The net pull or repulsion forces of the particles in the suspension are the sums of the electrostatic repulsion and Van der Waals attraction forces (Tadros, 2011). An electrical couple around particles in a suitable solvent layer is formed (Sainis et al., 2007). The sufficient distance that allows the particles to remain stably separated is not only due to the charges but also to other particles in the diffuse zone of the bilayer (Agudo-Canalejo & Lipowsky, 2015). It also depends on the concentration of the ions. Charge repulsion will be neutralized when there are sufficient ions or highly charged ions in the diffuse layer (Attard et al., 2000; Trefalt et al., 2015). Narrowing the double layer leads to coalescence and agglomeration of particles (Özer et al., 2017). As a result of van der Waals gravitational forces and the tendency to reduce the total surface or interfacial energy of the system, nanostructured particles have a large surface area and often form clusters (Luo et al., 2015; Ouyang et al., 2009). Coagulation leads to the formation of dense and robust clusters, and flocculation causes the formation of a loose particle network (Goudeli et al., 2015). Aggregation of particles can be carried out in any of the synthesis, drying, pouring, and processing steps (Ayyala et al., 1993; S. Li et al., 2025). In many processes and applications where dispersed particles or stable dispersions are required, unwanted aggregation must be avoided in each step of the synthesis and processing (Anastasiadis et al., 2022; Moinard-Checot et al., 2006; Peukert et al., 2005). To produce non-agglomerated particles or surfactants that control the dispersion during chemical synthesis can be used to disperse the produced agglomerate fine particles (Mueller et al., 2004). Nanomaterials create a bridge between the molecular phase and the condensed phase (Yun et al., 2023). Therefore, several properties such as optical properties, magnetic properties, melting points, and crystal morphologies can vary widely. New applications are at a very high level (Berman et al., 2001). To understand the size of nanocrystalline structures, it is possible to compare them with other little things found on earth. Thousands of substances that are solid at normal temperatures and pressures; can be subdivided into metals, ceramics, semiconductors, composites, and polymers. A classification into biomaterials, catalytic materials, coatings, glasses, magnetic and electronic materials is also possible. Quercetin succinylated chitosan-alginate core-shell-corona nanoparticle for diabetes treatment was reported in the literature (Mukhopadhyay et al., 2018), (Loganathan & Karthikeyan, 2013). Researchers synthesized trimetallic nanoparticles as gold core and palladium / platinum as well as gold/palladium/platinum in the literature and looked at the core relationship with the shell by MW irradiation (Figure 4).

Figure 4. The MW irradiation schema for a method for the synthesis of trimetallic gold/palladium/platinum nanoparticles

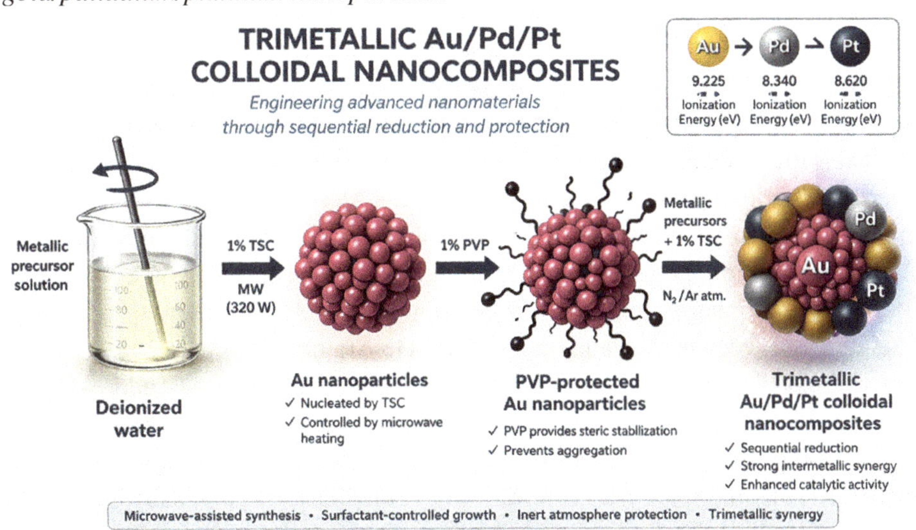

"Illustrated by the authors based on literature data."

In particular, core-shell nanoparticles from those described in this chapter are coated with other materials. So it has another core. Thus, their properties become better than other nanocomposite materials. These properties are parameters such as better dispersion, and optimization of cytocompatibility (Law et al., 2008). Especially in the pharmaceutical field, the method of encapsulation makes core/shell nanoparticle studies more important. Because some toxic properties of nanoparticles can directly damage human cells, tissues and organs. Thus, nanoparticles are coated with a more benign substance and encapsulation studies are carried out in this direction. In other words, core / shell nanoparticles are much more advantageous than simple nanoparticles (Chatterjee et al., 2014). It is especially important for biomedical applications that core/shell nanoparticles increase the binding parameter with relevant ligands and receptors (Gilmore et al., 2008). The opposition of core/shell nanoparticles to each other is usually due to the core. Because the shell is responsible for biocompatibility and conjugation properties along with the presence of reagents (Chatterjee et al., 2014), (Pinho et al., 2010). We can see the capsulation mechanism, drug delivery by core-shell nanoparticles in figure 5. Several recent reviews, covering the years 2021 to 2025, have noted that core–shell drug delivery systems based on polymeric structures are among the most advanced and promising for achieving controlled, targeted delivery of stimuli-responsive therapeutics. In such systems, the core is used for encapsulating hydrophobic or hydrophilic drugs, while the shell is useful for providing the desired level of colloidal stability, enhanced biocompatibility, prolonged circulation in vivo, cellular interactions, and controlled delivery (Ramadhan & Ali, 2025; Shi et al., 2024). Biodegradable polymeric nanocarrier systems using natural and synthetic polymers are of considerable interest in that they help stabilize highly labile therapeutic agents and also enhance their bioavailability and reduce toxicity following administration to patients. Stimuli-responsive shell structures consisting of pH-responsive, redox-responsive, enzyme-responsive, temperature-responsive, and hypoxia-responsive polymers are particularly interesting because they enable localized or microenvironment-triggered delivery of encapsulated drugs. Nevertheless, although such core–shell polymer systems have shown considerable promise in preclinical trials, their clinical application presents challenges associated with reproducible large-scale production, batch-to-batch uniformity, long-term safety issues, regulatory approval, in vivo fate, interpatient variability, and poor correlations between animal and human data (De Araujo Silva et al., 2025). Future investigations of such core–shell polymeric bio-nanocomposites must address these problems at the design stage. Recent breakthroughs in the area of polymeric core-shell bio-nanocomposites have redirected research interest from the design of traditional descriptive nanomaterial systems toward more functional and clinically applicable nanomaterial platforms. In particular, biodegradable and stimuli-responsive core-shell nanodelivery systems have been gaining significant attention in recent times due to their potential for controlled and targeted drug delivery, enhancing therapeutic efficacy while minimizing systemic toxicity. Biocompatible and stimuli-sensitive shell materials based on natural polymers, such as chitosan, alginate, gelatin, and polyethylene glycol, as well as synthetic polymer based on pH sensitivity, have been successfully incorporated into core-shell systems with improved colloid stability and responsiveness towards environmental stimuli. Such systems have shown great promise in cancer therapies, bioimaging, biosensing, tissue engineering, antimicrobial applications, and theranostics. Despite the promising progress that has been achieved in this field, certain limitations need to be addressed before large-scale clinical application can be considered. Some of these limitations include reproducibility in synthesis, toxicity, scale-up procedures, regulatory approvals, and lack of translational research (Shi et al., 2024).

Figure 5. **Illustration** *image of core-shell nanoparticles nanoparticle for drug delivery and therapy*

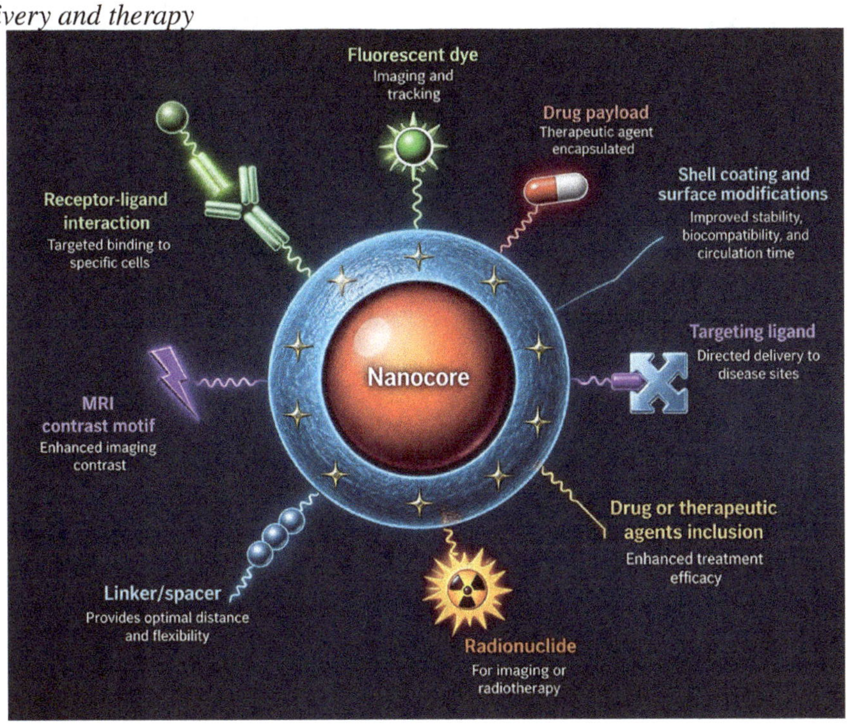

"Illustrated by the authors based on literature data."

Researchers have reported that the core / SiO_2 core/shell nanoparticles of multiple SPIO treated with imidazolinium have approximately seven times more effective contrasts than simple nanoparticles (Tanaka et al., 2010). The contrast of SPIO nanostructured particles embedded in polyvinyl alcohol (PVA) and poly (DL-layer) (PLA) inner layer is given in Figure 6.

Figure 6. The core-shell system of designed SPIO-inclusion nanoparticles encapsulated microbubble.

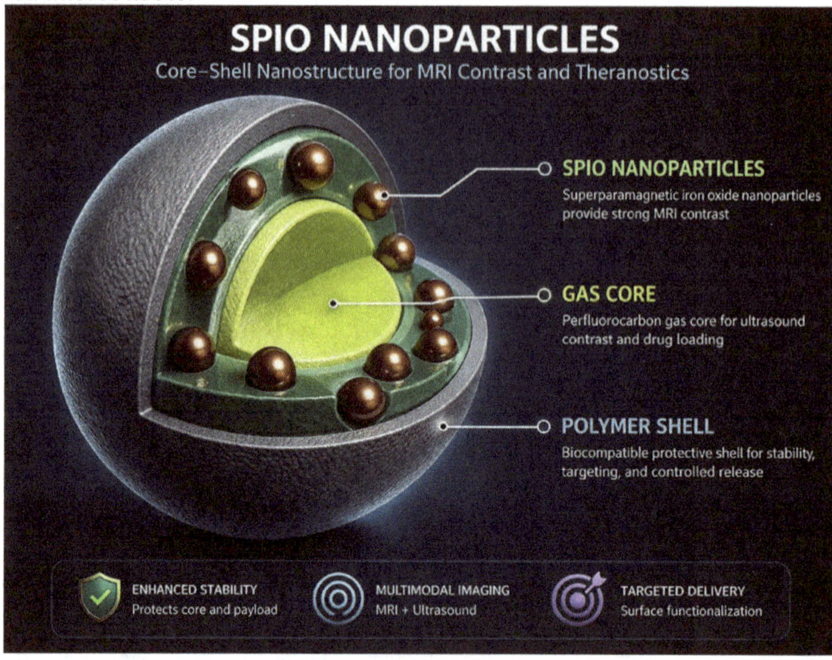

"Illustrated by the authors based on literature data."

There are also core-shell studies made with quantum dots in academic studies. The researchers combined a quantum dot core with a deposited ultrathin gold shell to create a core-shell nanoparticle that reveals fluorescent and plasmonic perforations (Figure 7) (Jin & Gao, 2009). pH-responsible nanogels have a high drug loading capacity (Anastasiadis et al., 2022; Nirmal et al., 2023; Yu et al., 2013). These nanogels, on the other hand, perform pH-controlled drug release (B. Ma et al., 2024). The schematic diagram has been given in Figure 8.

Figure 7. Gold-shell-encapsulated quantum dot (Chatterjee et al., 2014)

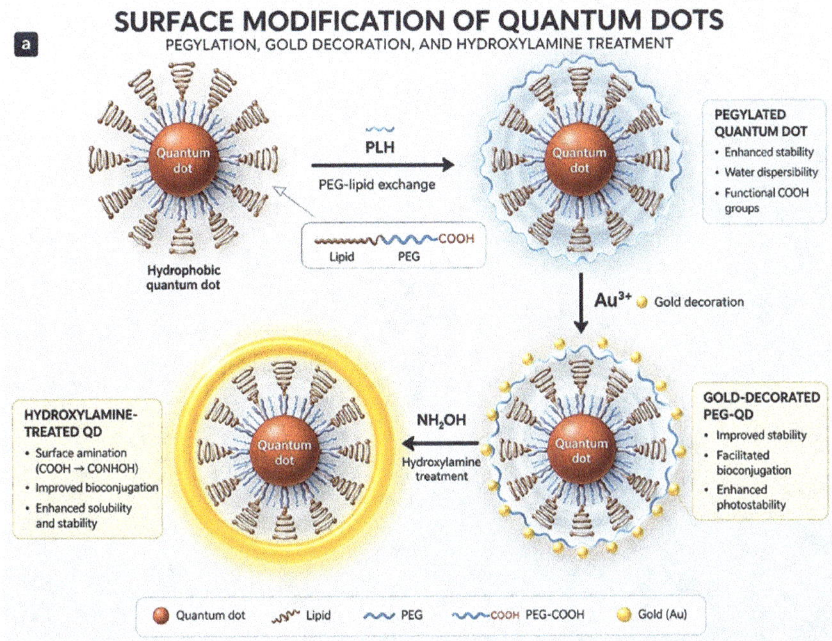

In core-shell nanogels, the core is metallic and the shell is hydrophilic (Chatterjee et al., 2014). Ag-Au bimetallic nanogels have been reported in the literature (Wu et al., 2010). Figure 9 shows the diagram of carbon quantum dots with fluorescent properties in cancer biology.

Figure 8. The Functional roles of core-shell hybrid nanogel (Chatterjee et al., 2014)

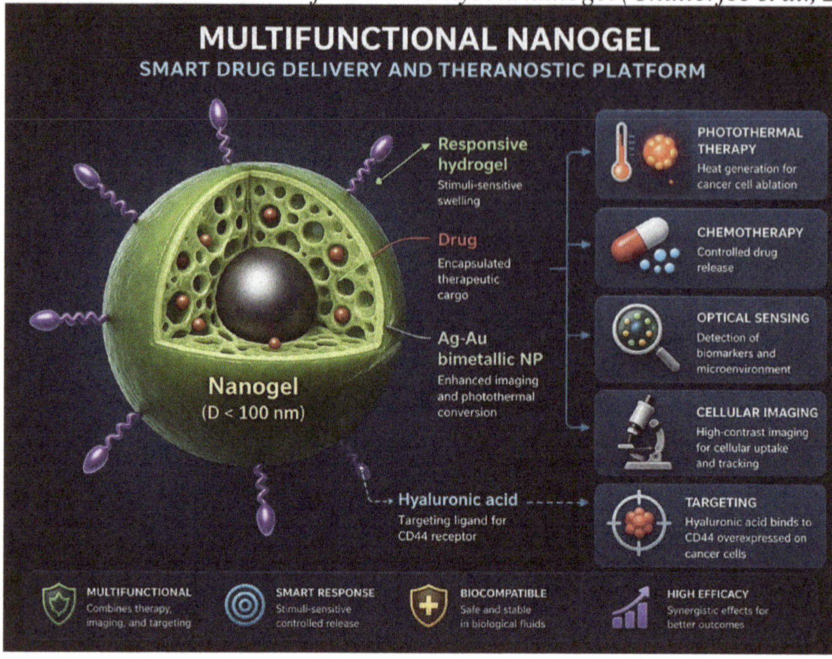

Figure 9. Carbon dot core/shell fluorescent silica nanoparticle (A) the SEM images of C dot core/shell fluorescent silica nanoparticle (30 nm –diameter) (scale bar 50 nm) (B) .

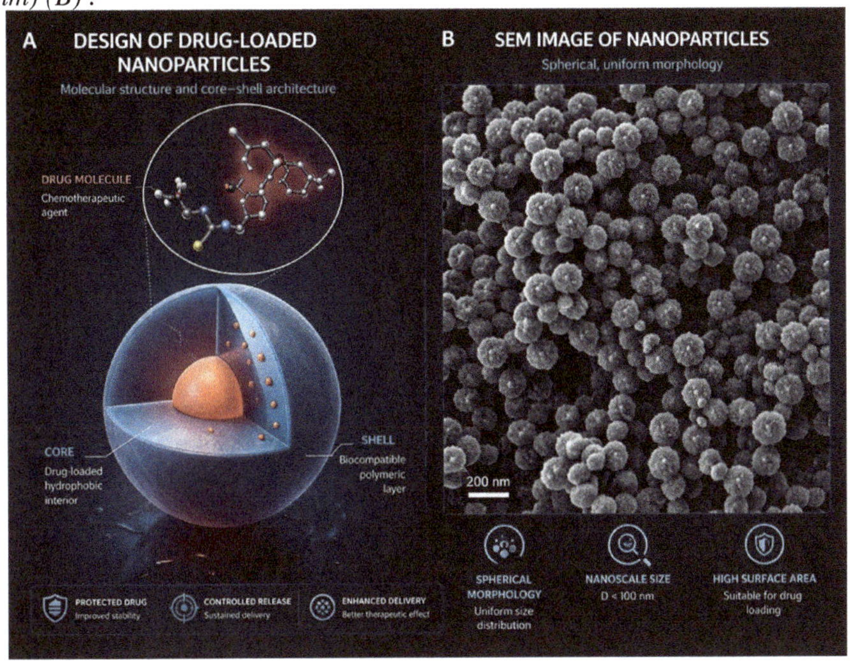

"Illustrated by the authors based on literature data."

3. CONCLUSION

Studies in nanotechnology are very popular. One of the most striking and concentrated studies among nanotechnological studies is the studies on nanoparticles. Nanoparticles are used in many areas including sensors, electrochemistry, industry, medicine, military, and biomedical fields. There are many derivatives such as nanoparticles, graphene, nanotubes, nanocomposite materials, nanomaterials, mesoporous nanoparticles. There are also bimetallic and trimetallic studies of nanoparticles. Nanoparticles, especially when used in medicines, easily enter the target cell with its nanostructure. However, in health, nanoparticles can show toxic effects when they enter the target cell, tissue, or organs in the human body. Therefore, core-shell nanoparticle studies are important to minimize this disadvantage and increase the efficiency of the nanoparticle. Because in the studies of encapsulated drugs, a shell is formed by covering the core with another harmless material. Thus, a controlled and slow release of the kernel with the harmless shell is initiated. This is especially important in the release of drugs encapsulated in cancer cells. In this chapter, information about the design and properties of core-shell nanoparticles is given. At the same time, the production of core-shell nanoparticles and bio nanocomposite materials creates a distinct advantage. Because minimizing the harm of these nanoparticles to human health as much as possible is important for human health.

REFERENCES

Abed, Z., Beik, J., Laurent, S., Eslahi, N., Khani, T., Davani, E. S., Ghaznavi, H., & Shakeri-Zadeh, A. (2019). Iron oxide–gold core–shell nano-theranostic for magnetically targeted photothermal therapy under magnetic resonance imaging guidance. *Journal of Cancer Research and Clinical Oncology*, *145*(5), 1213–1219. DOI: 10.1007/s00432-019-02870-x PMID: 30847551

Agudo-Canalejo, J., & Lipowsky, R. (2015). Critical Particle Sizes for the Engulfment of Nanoparticles by Membranes and Vesicles with Bilayer Asymmetry. *ACS Nano*, *9*(4), 3704–3720. DOI: 10.1021/acsnano.5b01285 PMID: 25840649

Alhalili, Z. (2023). Metal Oxides Nanoparticles: General Structural Description, Chemical, Physical, and Biological Synthesis Methods, Role in Pesticides and Heavy Metal Removal through Wastewater Treatment. *Molecules 2023, Vol. 28, 28*(7). DOI: 10.3390/molecules28073086

Altuner, E. E. (2025). Synthesis, characterisation and photocatalytic activation of zinc nanoparticles via. biogenic methods. *International Journal of Environmental Analytical Chemistry*, *105*(19), 1–11. DOI: 10.1080/03067319.2025.2451356

Altuner, E. E., Bozbeyoglu, P., Özkan, V., & Ozkan, A. (1 C.E.). Catalytic Nanocomposites in Oxidative on Adsorptive Delsulfurization Process: A Comprehensive Review. Https://Services.Igi-Global.Com/Resolvedoi/Resolve.Aspx?Doi=10.4018/979-8-3373-5966-3.Ch001, 1–34. DOI: 10.4018/979-8-3373-5966-3.ch001

Altuner, E. E., Bozbeyoğlu, P., Tekeli, Y., & Sen, F. (1 C.E.). Nanotechnological Green Synthesis and Its Impact on Plant Science. Https://Services.Igi-Global.Com/Resolvedoi/Resolve.Aspx?Doi=10.4018/979-8-3693-9869-2.Ch015, 433–482. DOI: 10.4018/979-8-3693-9869-2.ch015

Altuner, E. E., Erduran, V., & Sen, F. (2023). Green synthesized nanomaterials for bioimaging. In *Synthesis of Bionanomaterials for Biomedical Applications Micro and Nano Technologies* (pp. 265–286). Elsevier., DOI: 10.1016/B978-0-323-91195-5.00023-4

Anastasiadis, S. H., Chrissopoulou, K., Stratakis, E., Kavatzikidou, P., Kaklamani, G., & Ranella, A. (2022). How the Physicochemical Properties of Manufactured Nanomaterials Affect Their Performance in Dispersion and Their Applications in Biomedicine: A Review. *Nanomaterials 2022, Vol. 12, 12*(3). DOI: 10.3390/nano12030552

Atiemo-Obeng, V. A., Penney, W. R., & Armenante, P. (2004). Solid–Liquid Mixing. *Handbook of Industrial Mixing: Science and Practice*, 543–584. https://doi.org/DOI: 10.1002/0471451452.CH10;JOURNAL:JOURNAL:BOOKS;WGROUP:STRING:-PUBLICATION

Attard, P., Antelmi, D., & Larson, I. (2000). Comparison of the Zeta Potential with the Diffuse Layer Potential from Charge Titration. *Langmuir, 16*(4), 1542–1552. DOI: 10.1021/la990487t

Ayyala, S., Pugh, R. J., & Forssberg, E. (1993). Aggregate Characteristics in Coagulation and Flocculation. *Mineral Processing and Extractive Metallurgy Review, 12*(2–4), 165–184. DOI: 10.1080/08827509508935257

Bai, Y., Wagner, G., & Williams, C. B. (2017). Effect of Particle Size Distribution on Powder Packing and Sintering in Binder Jetting Additive Manufacturing of Metals. *Journal of Manufacturing Science and Engineering, 139*(8), 081019. Advance online publication. DOI: 10.1115/1.4036640

Baker, M. P., Reynolds, H. M., Lumicisi, B., & Bryson, C. J. (2010). Immunogenicity of protein therapeutics: The key causes, consequences and challenges. *Self Nonself: Immune Recognition and Signaling, 1*(4), 314–322. DOI: 10.4161/self.1.4.13904 PMID: 21487506

Banik, M., Patra, M., Dutta, D., Mukherjee, R., & Basu, T. (2018). A simple robust method of synthesis of copper-silver core-shell nano-particle: Evaluation of its structural and chemical properties with anticancer potency. *Nanotechnology, 29*(32), 325102. DOI: 10.1088/1361-6528/aac372 PMID: 29742069

Bayda, S., Adeel, M., Tuccinardi, T., Cordani, M., & Rizzolio, F. (2019). The History of Nanoscience and Nanotechnology: From Chemical–Physical Applications to Nanomedicine. *Molecules 2020, Vol. 25, 25*(1). DOI: 10.3390/molecules25010112

Berman, F., Chien, A., Cooper, K., Dongarra, J., Foster, I., Gannon, D., Johnsson, L., Kennedy, K., Kesselman, C., Mellor-Crumme, J., Reed, D., Torczon, L., & Wolski, R. (2001). The GrADS project: Software support for high-level grid application development. *International Journal of High Performance Computing Applications, 15*(4), 327–344. DOI: 10.1177/109434200101500401

Bhanoth, S., Kshirsagar, A. S., Khanna, P. K., Tyagi, A., & Verma, A. K. (2016). Biotoxicity of CdS/CdSe Core-Shell Nano-Structures. *Advances in Nanoparticles, 05*(01), 1–8. DOI: 10.4236/anp.2016.51001

Bhat, M., Luo, S., Zhang, J., Zhang, C., Zhou, B., & Deng, S. (2024). Multi-component precursor droplet evaporation in spray synthesis of cathode materials. *Chemical Engineering Journal*, *479*, 147417. DOI: 10.1016/j.cej.2023.147417

Bhattachar, S. N., Deschenes, L. A., & Wesley, J. A. (2006). Solubility: It's not just for physical chemists. *Drug Discovery Today*, *11*(21–22), 1012–1018. DOI: 10.1016/j.drudis.2006.09.002 PMID: 17055411

Bitinis, N., Hernandez, M., Verdejo, R., Kenny, J. M., & Lopez-Manchado, M. A. (2011). Recent advances in clay/polymer nanocomposites. *Advanced Materials*, *23*(44), 5229–5236. DOI: 10.1002/adma.201101948 PMID: 22299134

Boulaiz, H., Alvarez, P. J., Ramirez, A., Marchal, J. A., Prados, J., Rodríguez-Serrano, F., Perán, M., Melguizo, C., & Aranega, A. (2011). Nanomedicine: Application Areas and Development Prospects. *International Journal of Molecular Sciences*, *12*(5), 3303–3321. DOI: 10.3390/ijms12053303 PMID: 21686186

Bououdina, M., Grant, D., & Walker, G. (2006). Review on hydrogen absorbing materials—Structure, microstructure, and thermodynamic properties. *International Journal of Hydrogen Energy*, *31*(2), 177–182. DOI: 10.1016/j.ijhydene.2005.04.049

Bourne, J. R. (2003). Mixing and the Selectivity of Chemical Reactions. *Organic Process Research & Development*, *7*(4), 471–508. DOI: 10.1021/op020074q

Cai, Z., Dwivedi, A. D., Lee, W. N., Zhao, X., Liu, W., Sillanpää, M., Zhao, D., Huang, C. H., & Fu, J. (2018). Application of nanotechnologies for removing pharmaceutically active compounds from water: Development and future trends. *Environmental Science. Nano*, *5*(1), 27–47. DOI: 10.1039/C7EN00644F

Caruso, R. A., Schattka, J. H., & Greiner, A. (2001). Titanium Dioxide Tubes from Sol–Gel Coating of Electrospun Polymer Fibers. *Advanced Materials*, *13*(20), 1577. DOI: 10.1002/1521-4095(200110)13:20<1577::AID-ADMA1577>3.0.CO;2-S

Cetinkaya, O., Altuner, E. E., Yaylali, F. V., Tatli, H. N., & Najah Saud, S. (2025). *An Investigation into the Analyte-Dependent Fluorescence Quenching of Batio3 Nanoparticles for the Fluorimetric Determination of Glucose and Paracetamol.* DOI: 10.2139/ssrn.5336356

Chatterjee, K., Sarkar, S., Jagajjanani Rao, K., & Paria, S. (2014). Core/shell nanoparticles in biomedical applications. In *Advances in Colloid and Interface Science* (Vol. 209, pp. 8–39). Elsevier B.V., DOI: 10.1016/j.cis.2013.12.008

Chen, F., Yan, T. H., Bashir, S., & Liu, J. L. (2022). *Synthesis of nanomaterials using top-down methods* (2nd ed.). Advanced Nanomaterials and Their Applications in Renewable Energy., DOI: 10.1016/B978-0-323-99877-2.00007-2

Chen, L., Ting, V. P., Zhang, Y., Coventry, J., Rahbari, A., Yin, Z., Wang, F., Tian, M., Rochat, S., Zhang, Z., Deng, S., Krebsz, M., Bhomick, P., & Wang, X. (2025). Hydrate-based H 2 storage with porous materials as heterogeneous promoters: State of the art and challenges. *Journal of Materials Chemistry. A, Materials for Energy and Sustainability*, *13*(33), 27001–27049. DOI: 10.1039/D5TA04503G

Chen, S., McClements, D. J., Jian, L., Han, Y., Dai, L., Mao, L., & Gao, Y. (2019). Core-Shell Biopolymer Nanoparticles for Co-Delivery of Curcumin and Piperine: Sequential Electrostatic Deposition of Hyaluronic Acid and Chitosan Shells on the Zein Core. *ACS Applied Materials & Interfaces*, *11*(41), 38103–38115. DOI: 10.1021/acsami.9b11782 PMID: 31509373

Chiozzi, V., & Rossi, F. (2020). Inorganic–organic core/shell nanoparticles: Progress and applications. *Nanoscale Advances*, *2*(11), 5090–5105. DOI: 10.1039/D0NA00411A PMID: 36132014

Correa, S., Dreaden, E. C., Gu, L., & Hammond, P. T. (2016). Engineering nanolayered particles for modular drug delivery. *Journal of Controlled Release*, *240*, 364–386. DOI: 10.1016/j.jconrel.2016.01.040 PMID: 26809005

Cui, L. F., Yang, Y., Hsu, C. M., & Yi, C. (2009). Carbon-silicon Core-shell nanowires as high capacity electrode for lithium lon batteries. *Nano Letters*, *9*(9), 3370–3374. DOI: 10.1021/nl901670t PMID: 19655765

Darder, M., Aranda, P., & Ruiz-Hitzky, E. (2007). Bionanocomposites: A New Concept of Ecological, Bioinspired, and Functional Hybrid Materials. *Advanced Materials*, *19*(10), 1309–1319. DOI: 10.1002/adma.200602328

Darder, M., Colilla, M., & Ruiz-Hitzky, E. (2005). Chitosan-clay nanocomposites: Application as electrochemical sensors. *Applied Clay Science, 28*(1-4 SPEC. ISS.), 199–208. DOI: 10.1016/j.clay.2004.02.009

Darder, M., López-Blanco, M., Aranda, P., Aznar, A. J., Bravo, J., & Ruiz-Hitzky, E. (2006). Microfibrous chitosan - Sepiolite nanocomposites. *Chemistry of Materials*, *18*(6), 1602–1610. DOI: 10.1021/cm0523642

De Araujo Silva, R., Arruda, I. E. S., Chaves, L. L., de La Roca Soares, M. F., & Sobrinho, J. L. S. (2025). Exploring the potential of polymers: Advancements in oral nanocarrier technology. *Beilstein Journal of Nanotechnology*, *16*, 1751–1793. DOI: 10.3762/bjnano.16.122 PMID: 41103493

Desai, N., Rana, D., Salave, S., Gupta, R., Patel, P., Karunakaran, B., Sharma, A., Giri, J., Benival, D., & Kommineni, N. (2023). Chitosan: A Potential Biopolymer in Drug Delivery and Biomedical Applications. *Pharmaceutics 2023, Vol. 15, 15*(4). DOI: 10.3390/pharmaceutics15041313

Dong, P., Prasanth, R., Xu, F., Wang, X., Li, B., & Shankar, R. (2015). Eco-friendly polymer nanocomposite—Properties and processing. *Advanced Structured Materials, 75*, 1–15. DOI: 10.1007/978-81-322-2470-9_1

El Achaby, M., Essamlali, Y., El Miri, N., Snik, A., Abdelouahdi, K., Fihri, A., Zahouily, M., & Solhy, A. (2014). Graphene oxide reinforced chitosan/polyvinylpyrrolidone polymer bio-nanocomposites. *Journal of Applied Polymer Science, 131*(22), app.41042. Advance online publication. DOI: 10.1002/app.41042

Erduran, V., Altuner, E. E., & Şen, F. (2023). Stability and validation of bionanomaterials. *Synthesis of Bionanomaterials for Biomedical Applications*, 251–263. DOI: 10.1016/B978-0-323-91195-5.00021-0

Eris, S., Daşdelen, Z., Yıldız, Y., & Sen, F. (2018). Nanostructured Polyaniline-rGO decorated platinum catalyst with enhanced activity and durability for Methanol oxidation. *International Journal of Hydrogen Energy, 43*(3), 1337–1343. DOI: 10.1016/j.ijhydene.2017.11.051

Erken, E., Esirden, I., Kaya, M., & Sen, F. (2015). A rapid and novel method for the synthesis of 5-substituted 1H-tetrazole catalyzed by exceptional reusable monodisperse Pt NPs@AC under the microwave irradiation. *RSC Advances, 5*(84), 68558–68564. DOI: 10.1039/C5RA11426H

Erken, E., Yildiz, Y., Kilbaş, B., & Şen, F. (2016). Synthesis and characterization of nearly monodisperse Pt nanoparticles for C1 to C3 alcohol oxidation and dehydrogenation of dimethylamine-borane (DMAB). *Journal of Nanoscience and Nanotechnology, 16*(6), 5944–5950. DOI: 10.1166/jnn.2016.11683 PMID: 27427655

Eryılmaz, Z. (2016). *CdSe/CdS Nano Çubuk Sentezi ve Polimer Ligantlar ile Modifikasyonu*. Osmangazi Üniversitesi Fen Bilimleri Enstitüsü.

Esra Altuner, E., Yousif, M., & Al Janabi, A. (2025). Synthesis and Characterization of Iron-Cobalt@Activated Carbon Nanoparticles for Electrochemical Sensing of Dopamine Neurotransmitters. *Chemistry Africa, 2025*(3), 1–9. DOI: 10.1007/s42250-025-01197-0

Fanning, J. C. (2000). The chemical reduction of nitrate in aqueous solution. *Coordination Chemistry Reviews, 199*(1), 159–179. DOI: 10.1016/S0010-8545(99)00143-5

Fazal, T., Murtaza, B. N., Shah, M., Iqbal, S., Rehman, M. U., Jaber, F., Dera, A. A., Awwad, N. S., & Ibrahium, H. A. (2023). Recent developments in natural biopolymer based drug delivery systems. *RSC Advances, 13*(33), 23087–23121. DOI: 10.1039/D3RA03369D PMID: 37529365

Ghosh Chaudhuri, R., & Paria, S. (2011). Core/Shell Nanoparticles: Classes, Properties, Synthesis Mechanisms, Characterization, and Applications. *Chemical Reviews, 112*(4), 2373–2433. DOI: 10.1021/cr100449n PMID: 22204603

Gilmore, J. L., Yi, X., Quan, L., & Kabanov, A. V. (2008). Novel nanomaterials for clinical neuroscience. *Journal of Neuroimmune Pharmacology, 3*(2), 83–94. DOI: 10.1007/s11481-007-9099-6 PMID: 18210200

Goldberg, M., Langer, R., & Jia, X. (2007). Nanostructured materials for applications in drug delivery and tissue engineering. *Journal of Biomaterials Science. Polymer Edition, 18*(3), 241–268. DOI: 10.1163/156856207779996931 PMID: 17471764

Gong, T., Liu, T., Zhang, L., Ye, W., Guo, X., Wang, L., Quan, L., & Pan, C. (2018). Design Redox-Sensitive Drug-Loaded Nanofibers for Bone Reconstruction. *ACS Biomaterials Science & Engineering, 4*(1), 240–247. DOI: 10.1021/acsbiomaterials.7b00827 PMID: 33418691

Goudeli, E., Eggersdorfer, M. L., & Pratsinis, S. E. (2015). Coagulation–Agglomeration of Fractal-like Particles: Structure and Self-Preserving Size Distribution. *Langmuir, 31*(4), 1320–1327. DOI: 10.1021/la504296z PMID: 25560979

Haidar, Z. S., Hamdy, R. C., & Tabrizian, M. (2008). Protein release kinetics for core-shell hybrid nanoparticles based on the layer-by-layer assembly of alginate and chitosan on liposomes. *Biomaterials, 29*(9), 1207–1215. DOI: 10.1016/j.biomaterials.2007.11.012 PMID: 18076987

Hiremath, A., Murthy, A. A., Thipperudrappa, S., & Bharath, K. N. (2021). Nanoparticles Filled Polymer Nanocomposites: A Technological Review. *Cogent Engineering, 8*(1), 1991229. Advance online publication. DOI: 10.1080/23311916.2021.1991229

Jiang, Q., Luo, X., Li, L., Li, B., & Zhang, X. (2025). Interfacial freeze-thaw stability of Pickering emulsions stabilized by sodium Caseinate-modified solid lipid particles: Influence of lipid crystalline structure and polymorphic transitions. *Food Chemistry, 492*, 145481. DOI: 10.1016/j.foodchem.2025.145481 PMID: 40645057

Jin, Y., & Gao, X. (2009). Plasmonic fluorescent quantum dots. *Nature Nanotechnology, 4*(9), 571–576. DOI: 10.1038/nnano.2009.193 PMID: 19734929

Jones, C. F., Resina, L., Ferreira, F. C., Sanjuan-Alberte, P., & Esteves, T. (2024). Conductive Core–Shell Nanoparticles: Synthesis and Applications. *The Journal of Physical Chemistry. C, Nanomaterials and Interfaces*, *128*(27), 11083–11100. DOI: 10.1021/acs.jpcc.4c02012

Karatepe, Ö., Yildiz, Y., Pamuk, H., Eris, S., Dasdelen, Z., & Sen, F. (2016). Enhanced electrocatalytic activity and durability of highly monodisperse Pt@PPy -PANI nanocomposites as a novel catalyst for the electro-oxidation of methanol. *RSC Advances*, *6*(56), 50851–50857. DOI: 10.1039/C6RA06210E

Kelleher, C. M., & Vacanti, J. P. (2010). Engineering extracellular matrix through nanotechnology. *Journal of the Royal Society, Interface*, *7*(suppl_6), S717–S729. DOI: 10.1098/rsif.2010.0345.focus PMID: 20861039

Kim, N. J., Lee, S. J., & Atala, A. (2013). Biomedical nanomaterials in tissue engineering. In *Nanomaterials in Tissue Engineering: Fabrication and Applications* (pp. 1–25e). Elsevier Ltd. DOI: 10.1533/9780857097231.1

Kishore, K., Sinha, M. K., & Singh, A. Archana, Gupta, M. K., & Korkmaz, M. E. (2022). A comprehensive review on the grinding process: Advancements, applications and challenges. *Proceedings of the Institution of Mechanical Engineers, Part C: Journal of Mechanical Engineering Science, 236*(22), 10923–10952. https://doi .org/DOI: 10.1177/09544062221110782;CTYPE:STRING:JOURNAL

Kondratyeva, L. G., Dyachkova, M. S., & Galchenko, A. V. (2022). The Origin of Genetic Code and Translation in the Framework of Current Concepts on the Origin of Life. *Biochemistry (Moscow) 2022 87:2, 87*(2), 150–169. DOI: 10.1134/ S0006297922020079

Kreider, P., & Lipiński, W. (2018). High-Temperature Gas–Solid Reactions in Industrial Processes. *Reviews in Mineralogy and Geochemistry*, *84*(1), 499–514. DOI: 10.2138/rmg.2018.84.13

Kumar, A., Dutta, S., Kim, S., Kwon, T., Patil, S. S., Kumari, N., Jeevanandham, S., & Lee, I. S. (2022). Solid-State Reaction Synthesis of Nanoscale Materials: Strategies and Applications. *Chemical Reviews*, *122*(15), 12748–12863. DOI: 10.1021/ acs.chemrev.1c00637 PMID: 35715344

Lalegül Ülker, Ö. (2020). *Elektromanyetik Nanokompozitlerin Biyoteknolojik Uygulamalarına Yönelik Olarak Üretimi ve Geliştirilmesi*. Ankara üniversitesi, Fen Bilimleri Enstitüsü.

Latu, M. (2018). Nanoworld and Megaworld in the Mirror of Language. *Балканистичен Форум*, (1), 276–284.

Law, W. C., Yong, K. T., Roy, I., Xu, G. X., Ding, H., Bergey, E. J., Zeng, H., & Prasad, P. N. (2008). Optically and magnetically doped organically modified silica nanoparticles as efficient magnetically guided biomarkers for two-photon imaging of live cancer cells. *The Journal of Physical Chemistry. C, Nanomaterials and Interfaces*, *112*(21), 7972–7977. DOI: 10.1021/jp712090y

Lee, E. J., Khan, S. A., Park, J. K., & Lim, K. H. (2012). Studies on the characteristics of drug-loaded gelatin nanoparticles prepared by nanoprecipitation. *Bioprocess and Biosystems Engineering*, *35*(1–2), 297–307. DOI: 10.1007/s00449-011-0591-2 PMID: 21909678

Li, S., Liu, Y., Wang, Z., Dou, C., Zhao, W., & Shu, H. (2025). Characteristic analysis of floc size distribution and image texture evolution in chemical coagulation process. *Process Safety and Environmental Protection*, *199*, 107298. DOI: 10.1016/j.psep.2025.107298

Li, X., Wang, L., Fan, Y., Feng, Q., & Cui, F. Z. (2012). Biocompatibility and Toxicity of Nanoparticles and Nanotubes. *Journal of Nanomaterials*, *2012*(1), 548389. DOI: 10.1155/2012/548389

Li, Z., Chai, F., Yang, L., Luo, X., & Yang, C. (2020). Mechanical properties and nanoparticles precipitation behavior of multi-component ultra high strength steel. *Materials & Design*, *191*, 108637. DOI: 10.1016/j.matdes.2020.108637

Liu, Y., Zhang, L., Willis, B. G., & Mustain, W. E. (2015). Importance of Particle Size and Distribution in Achieving High-Activity, High-Stability Oxygen Reduction Catalysts. *ACS Catalysis*, *5*(3), 1560–1567. DOI: 10.1021/cs501556j

Loganathan, B., & Karthikeyan, B. (2013). Au core Pd/Pt shell in trimetallic Au/Pd/Pt colloidal nanocomposites - Physicochemical characterization study. *Colloids and Surfaces. A, Physicochemical and Engineering Aspects*, *436*, 944–952. DOI: 10.1016/j.colsurfa.2013.08.035

Lounissa, M. I., Tiri, R. N. E., Aygun, A., Halvaci, E., Altuner, E. E., Ucar, M., & Sen, F. (2025). Synthesis, characterization, photocatalytic and biochemical applications of new generation biogenic trimetallic nanomaterials. *Next Research*, *2*(2), 100326. DOI: 10.1016/j.nexres.2025.100326

Luo, D., Yan, C., & Wang, T. (2015). Interparticle Forces Underlying Nanoparticle Self-Assemblies. *Small*, *11*(45), 5984–6008. DOI: 10.1002/smll.201501783 PMID: 26436692

Ma, B., Li, Q., Mi, Y., Zhang, J., Tan, W., & Guo, Z. (2024). pH-responsive nanogels with enhanced antioxidant and antitumor activities on drug delivery and smart drug release. *International Journal of Biological Macromolecules*, *257*, 128590. DOI: 10.1016/j.ijbiomac.2023.128590 PMID: 38056756

Ma, L., Kohli, M., & Smith, A. (2013). Nanoparticles for combination drug therapy. In *ACS Nano* (Vol. 7, Number 11, pp. 9518–9525). American Chemical Society. DOI: 10.1021/nn405674m

Malik, S., Muhammad, K., & Waheed, Y. (2023). Nanotechnology: A Revolution in Modern Industry. *Molecules 2023, Vol. 28, 28*(2). DOI: 10.3390/molecules28020661

Matsusaka, S., & Masuda, H. (2003). Electrostatics of particles. *Advanced Powder Technology*, *14*(2), 143–166. DOI: 10.1163/156855203763593958

Mehrotra, S., Chouhan, D., Konwarh, R., Kumar, M., Jadi, P. K., & Mandal, B. B. (2019). Comprehensive Review on Silk at Nanoscale for Regenerative Medicine and Allied Applications. *ACS Biomaterials Science & Engineering*, *5*(5), 2054–2078. DOI: 10.1021/acsbiomaterials.8b01560 PMID: 33405710

Moinard-Checot, D., Chevalier, Y., Briançon, S., Fessi, H., & Guinebretière, S. (2006). Nanoparticles for Drug Delivery: Review of the Formulation and Process Difficulties Illustrated by the Emulsion-Diffusion Process. *Journal of Nanoscience and Nanotechnology*, *6*(9–10), 2664–2681. DOI: 10.1166/jnn.2006.479 PMID: 17048474

Mueller, R., Kammler, H. K., Pratsinis, S. E., Vital, A., Beaucage, G., & Burtscher, P. (2004). Non-agglomerated dry silica nanoparticles. *Powder Technology*, *140*(1–2), 40–48. DOI: 10.1016/j.powtec.2004.01.004

Muhsin, A. H., Torlak, Y., Altuner, E. E., Aldemir, O., Khan, R., Issa, G., Zafar, S., Bibi, A., Khan, M. R., Huma, H., & Ali, G. (2025). Synthesis of β-Lactam Conjugates of Thiadiazine Thiones as a New Class of Anti-Nociceptive Agents. *Iranian Journal of Chemistry and Chemical Engineering*, 775–791. DOI: 10.30492/IJCCE.2025.2069202.7257

Mukhopadhyay, P., Maity, S., Mandal, S., Chakraborti, A. S., Prajapati, A. K., & Kundu, P. P. (2018). Preparation, characterization and in vivo evaluation of pH sensitive, safe quercetin-succinylated chitosan-alginate core-shell-corona nanoparticle for diabetes treatment. *Carbohydrate Polymers*, *182*, 42–51. DOI: 10.1016/j.carbpol.2017.10.098 PMID: 29279124

Müller, K. H., Motskin, M., Philpott, A. J., Routh, A. F., Shanahan, C. M., Duer, M. J., & Skepper, J. N. (2014). The effect of particle agglomeration on the formation of a surface-connected compartment induced by hydroxyapatite nanoparticles in human monocyte-derived macrophages. *Biomaterials, 35*(3), 1074–1088. DOI: 10.1016/j.biomaterials.2013.10.041 PMID: 24183166

Natarova, A. O., Kokodii, N. G., Pogorelov, S. V., & Priz, I. A. (2019). Physical Methods of Analysis of the Human Hair Properties. *Proceedings of the International Conference on Advanced Optoelectronics and Lasers, CAOL, 2019-September*, 300–305. DOI: 10.1109/CAOL46282.2019.9019565

Nesrin, K., Yusuf, C., Ahmet, K., Ali, S. B., Muhammad, N. A., Suna, S., & Fatih, Ş. (2020). Biogenic silver nanoparticles synthesized from Rhododendron ponticum and their antibacterial, antibiofilm and cytotoxic activities. *Journal of Pharmaceutical and Biomedical Analysis, 179*, 112993. Advance online publication. DOI: 10.1016/j.jpba.2019.112993 PMID: 31780283

Nirmal, G. R., Liao, C. C., Lin, Z. C., Alshetaili, A., Hwang, E., Yang, S. C., & Fang, J. Y. (2023). Topically applied pH-responsive nanogels for alkyl radical-based therapy against psoriasiform hyperplasia. *Drug Delivery, 30*(1), 2245169. Advance online publication. DOI: 10.1080/10717544.2023.2245169 PMID: 37585684

Oh, K. S., Lee, H., Kim, J. Y., Koo, E. J., Lee, E. H., Park, J. H., Kim, S. Y., Kim, K., Kwon, I. C., & Yuk, S. H. (2013). The multilayer nanoparticles formed by layer by layer approach for cancer-targeting therapy. *Journal of Controlled Release, 165*(1), 9–15. DOI: 10.1016/j.jconrel.2012.10.013 PMID: 23103984

Orlov, S. (2017). Gravitational Properties of Atom. *Journal of Physics & Astronomy, 5*(3). https://www.tsijournals.com/articles/gravitational-properties-of-atom-13432.html

Ouyang, G., Wang, C. X., & Yang, G. W. (2009). Surface Energy of Nanostructural Materials with Negative Curvature and Related Size Effects. *Chemical Reviews, 109*(9), 4221–4247. DOI: 10.1021/cr900055f PMID: 19670888

Özer, M., Basha, O. M., & Morsi, B. (2017). Coal-Agglomeration Processes: A Review. *International Journal of Coal Preparation and Utilization, 37*(3), 131–167. DOI: 10.1080/19392699.2016.1142443

Ozturk, Z., Sen, F., Sen, S., & Gokagac, G. (2012). The preparation and characterization of nano-sized Pt-Pd/C catalysts and comparison of their superior catalytic activities for methanol and ethanol oxidation. *Journal of Materials Science, 47*(23), 8134–8144. DOI: 10.1007/s10853-012-6709-3

Peukert, W., Schwarzer, H. C., Götzinger, M., Günther, L., & Stenger, F. (2003). Control of particle interfaces — The critical issue in nanoparticle technology. *Advanced Powder Technology*, *14*(4), 411–426. DOI: 10.1163/156855203769710645

Peukert, W., Schwarzer, H. C., & Stenger, F. (2005). Control of aggregation in production and handling of nanoparticles. *Chemical Engineering and Processing*, *44*(2), 245–252. DOI: 10.1016/j.cep.2004.02.018

Pinho, S. L. C., Pereira, G. A., Voisin, P., Kassem, J., Bouchaud, V., Etienne, L., Peters, J. A., Carlos, L., Mornet, S., Geraldes, C. F. G. C., Rocha, J., & Delville, M. H. (2010). Fine tuning of the relaxometry of γ-Fe2O 3@SiO2 nanoparticles by tweaking the silica coating thickness. *ACS Nano*, *4*(9), 5339–5349. DOI: 10.1021/nn101129r PMID: 20795638

Qamar, S. A., Junaid, M., Riasat, A., Jahangeer, M., Bilal, M., & Mu, B. Z. (2024). Carrageenan-Based Hybrids with Biopolymers and Nano-Structured Materials for Biomimetic Applications. *Starch/Staerke, 76*(1–2), 2200018. https://doi.org/DOI: 10.1002/STAR.202200018;CSUBTYPE:STRING:SPECIAL;PAGE:STRING:ARTICLE/CHAPTER

Qiao, M., Lei, X., Ma, Y., Tian, L., Su, K., & Zhang, Q. (2016). Well-Defined Core-Shell Fe3O4@Polypyrrole Composite Microspheres with Tunable Shell Thickness: Synthesis and Their Superior Microwave Absorption Performance in the Ku Band. *Industrial & Engineering Chemistry Research*, *55*(22), 6263–6275. DOI: 10.1021/acs.iecr.5b04814

Quinson, J., Neumann, S., Wannmacher, T., Kacenauskaite, L., Inaba, M., Bucher, J., Bizzotto, F., Simonsen, S. B., Theil Kuhn, L., Bujak, D., Zana, A., Arenz, M., & Kunz, S. (2018). Colloids for Catalysts: A Concept for the Preparation of Superior Catalysts of Industrial Relevance. *Angewandte Chemie*, *130*(38), 12518–12521. DOI: 10.1002/ange.201807450 PMID: 30051948

Raemdonck, K., Martens, T. F., Braeckmans, K., Demeester, J., & De Smedt, S. C. (2013). Polysaccharide-based nucleic acid nanoformulations. In *Advanced Drug Delivery Reviews* (Vol. 65, Number 9, pp. 1123–1147). Elsevier B.V. DOI: 10.1016/j.addr.2013.05.002

Raicheff, R. G., & Zahariev, H. Z. (2001). *Electrophoretic Deposition*. 2733–2738. DOI: 10.1016/B0-08-043152-6/00487-3

Ramadhan, S. A., & Ali, D. S. (2025). Innovations in Core-Shell Nanoparticles: Advancing Drug Delivery Solutions and Precision Medicine. *OMICS: A Journal of Integrative Biology*, *29*(3), 73–86. DOI: 10.1089/omi.2024.0182 PMID: 39981668

Ruiz-Hitzky, E., Darder, M., & Aranda, P. (2005). Functional biopolymer nanocomposites based on layered solids. *Journal of Materials Chemistry, 15*(35–36), 3650–3662. DOI: 10.1039/b505640n

Sainis, S. K., Germain, V., Mejean, C. O., & Dufresne, E. R. (2007). Electrostatic Interactions of Colloidal Particles in Nonpolar Solvents: Role of Surface Chemistry and Charge Control Agents†. *Langmuir, 24*(4), 1160–1164. DOI: 10.1021/la702432u PMID: 18062711

Şen, B., Akdere, E. H., Şavk, A., Gültekin, E., Paralı, Ö., Göksu, H., & Şen, F. (2018). A novel thiocarbamide functionalized graphene oxide supported bimetallic monodisperse Rh-Pt nanoparticles (RhPt/TC@GO NPs) for Knoevenagel condensation of aryl aldehydes together with malononitrile. *Applied Catalysis B: Environmental, 225*, 148–153. DOI: 10.1016/j.apcatb.2017.11.067

Sen, B., Demirkan, B., Şavk, A., Karahan Gülbay, S., & Sen, F. (2018). Trimetallic PdRuNi nanocomposites decorated on graphene oxide: A superior catalyst for the hydrogen evolution reaction. *International Journal of Hydrogen Energy, 43*(38), 17984–17992. DOI: 10.1016/j.ijhydene.2018.07.122

Sen, B., Demirkan, B., Şimşek, B., Savk, A., & Sen, F. (2018). Monodisperse palladium nanocatalysts for dehydrocoupling of dimethylamineborane. *Nano-Structures and Nano-Objects, 16*, 209–214. DOI: 10.1016/j.nanoso.2018.07.008

Şen, F., & Gökağaç, G. (2007). Different sized platinum nanoparticles supported on carbon: An XPS study on these methanol oxidation catalysts. *The Journal of Physical Chemistry. C, Nanomaterials and Interfaces, 111*(15), 5715–5720. DOI: 10.1021/jp068381b

Şen, F., & Gökağaç, G. (2008). Improving catalytic efficiency in the methanol oxidation reaction by inserting Ru in face-centered cubic Pt nanoparticles prepared by a new surfactant, tert-octanethiol. *Energy & Fuels, 22*(3), 1858–1864. DOI: 10.1021/ef700575t

Sen, F., Karatas, Y., Gulcan, M., & Zahmakiran, M. (2014). Amylamine stabilized platinum(0) nanoparticles: Active and reusable nanocatalyst in the room temperature dehydrogenation of dimethylamine-borane. *RSC Advances, 4*(4), 1526–1531. DOI: 10.1039/C3RA43701A

Şen, S., Şen, F., & Gökağaç, G. (2011). Preparation and characterization of nanosized Pt-Ru/C catalysts and their superior catalytic activities for methanol and ethanol oxidation. *Physical Chemistry Chemical Physics, 13*(15), 6784–6792. DOI: 10.1039/c1cp20064j PMID: 21403959

Sert, H., Yıldız, Y., Okyay, T. O., Gezer, B., Dasdelen, Z., Sen, B., & Sen, F. (2016). Monodisperse Mw-Pt NPs@VC as Highly Efficient and Reusable Adsorbents for Methylene Blue Removal. *Journal of Cluster Science, 27*(6), 1953–1962. DOI: 10.1007/s10876-016-1054-3

Sert, H., Yildiz, Y., Okyay, T. O., Sen, B., Gezer, B., Bozkurt, S., Başkaya, G., & Sen, F. (2017). Activated carbon furnished monodisperse Pt nanocomposites as a superior adsorbent for methylene blue removal from aqueous solutions. *Journal of Nanoscience and Nanotechnology, 17*(7), 4799–4804. DOI: 10.1166/jnn.2017.13776

Shafiee, A., Rabiee, N., Ahmadi, S., Baneshi, M., Khatami, M., Iravani, S., & Varma, R. S. (2022). Core–Shell Nanophotocatalysts: Review of Materials and Applications. *ACS Applied Nano Materials, 5*(1), 55–86. DOI: 10.1021/acsanm.1c03714

Shi, Y., Zhang, Y., Zhu, L., Miao, Y., Zhu, Y., & Yue, B. (2024). Tailored Drug Delivery Platforms: Stimulus-Responsive Core–Shell Structured Nanocarriers. *Advanced Healthcare Materials, 13*(1), 2301726. DOI: 10.1002/adhm.202301726 PMID: 37670419

Srivastava, N., Chandra Singh, A., & Chattopadhyay, J. (2026). Multi-shelled nanospheres in biosensing: Classification, synthesis and applications. *International Journal of Materials Research, 117*(4), 137–161. Advance online publication. DOI: 10.1515/ijmr-2024-0230

Sun, Z., Zussman, E., Yarin, A. L., Wendorff, J. H., & Greiner, A. (2003). Compound Core–Shell Polymer Nanofibers by Co-Electrospinning. *Advanced Materials, 15*(22), 1929–1932. DOI: 10.1002/adma.200305136

Suzuki, A., & Mishin, Y. (2005). Atomic mechanisms of grain boundary diffusion: Low versus high temperatures. *Journal of Materials Science 2005 40:12, 40*(12), 3155–3161. DOI: 10.1007/s10853-005-2678-0

Tadros, T. (2011). Interparticle interactions in concentrated suspensions and their bulk (Rheological) properties. *Advances in Colloid and Interface Science, 168*(1–2), 263–277. DOI: 10.1016/j.cis.2011.05.003 PMID: 21632031

Tan, A., Chawla, R., Natasha, G., Mahdibeiraghdar, S., Jeyaraj, R., Rajadas, J., Hamblin, M. R., & Seifalian, A. M. (2016). Nanotechnology and regenerative therapeutics in plastic surgery: The next frontier. In *Journal of Plastic, Reconstructive and Aesthetic Surgery* (Vol. 69, Number 1, pp. 1–13). Churchill Livingstone. DOI: 10.1016/j.bjps.2015.08.028

Tanaka, K., Narita, A., Kitamura, N., Uchiyama, W., Morita, M., Inubushi, T., & Chujo, Y. (2010). Preparation for highly sensitive MRI contrast agents using core/shell type nanoparticles consisting of multiple SPIO cores with thin silica coating. *Langmuir*, *26*(14), 11759–11762. DOI: 10.1021/la1015077 PMID: 20527913

Taner, B., Mahmood Mahmood, A. H., Karakurt, S., Altuner, E. E., & Dar, U. A. (2025). Synthesis, characterization and biological evaluation of new Coumarin-Ferrocene conjugates as antitumor agents. *Inorganica Chimica Acta*, *578*, 122534. DOI: 10.1016/j.ica.2024.122534

Teo, B. K., & Sun, X. H. (2006). From top-down to bottom-up to hybrid nanotechnologies: Road to nanodevices. *Journal of Cluster Science*, *17*(4), 529–540. DOI: 10.1007/s10876-006-0086-5

Thanh, N. T. K., Maclean, N., & Mahiddine, S. (2014). Mechanisms of Nucleation and Growth of Nanoparticles in Solution. *Chemical Reviews*, *114*(15), 7610–7630. DOI: 10.1021/cr400544s PMID: 25003956

Tiwari, A., Tripathi, A. K., & Khare, P. (2021). A comprehensive study of synthesis and applications of core/shell nanoparticles. *International Journal of Engineering Science and Technology*, *13*(1), 153–157. DOI: 10.4314/ijest.v13i1.23S

Trefalt, G., Behrens, S. H., & Borkovec, M. (2015). Charge Regulation in the Electrical Double Layer: Ion Adsorption and Surface Interactions. *Langmuir*, *32*(2), 380–400. DOI: 10.1021/acs.langmuir.5b03611 PMID: 26599980

Tulinski, M., & Jurczyk, M. (2017). Nanomaterials Synthesis Methods. *Metrology and Standardization for Nanotechnology: Protocols and Industrial Innovations*, 75–98. DOI: 10.1002/9783527800308.ch4

Vehkamäki, H., & Riipinen, I. (2012). Thermodynamics and kinetics of atmospheric aerosol particle formation and growth. *Chemical Society Reviews*, *41*(15), 5160–5173. DOI: 10.1039/c2cs00002d PMID: 22710807

Wang, L. K., Vaccari, D. A., Li, Y., & Shammas, N. K. (2005). *Chemical Precipitation*. 141–197. DOI: 10.1385/1-59259-820-x:141

Wu, W., Shen, J., Banerjee, P., & Zhou, S. (2010). Core-shell hybrid nanogels for integration of optical temperature-sensing, targeted tumor cell imaging, and combined chemo-photothermal treatment. *Biomaterials*, *31*(29), 7555–7566. DOI: 10.1016/j.biomaterials.2010.06.030 PMID: 20643481

Xiong, R., Grant, A. M., Ma, R., Zhang, S., & Tsukruk, V. V. (2018). Naturally-derived biopolymer nanocomposites: Interfacial design, properties and emerging applications. In *Materials Science and Engineering R: Reports* (Vol. 125, pp. 1–41). Elsevier Ltd. DOI: 10.1016/j.mser.2018.01.002

Xu, S., Cao, D., Liu, Y., & Wang, Y. (2021). Role of Additives in Crystal Nucleation from Solutions: A Review. *Crystal Growth & Design*, *22*(3), 2001–2022. DOI: 10.1021/acs.cgd.1c00776

Yu, S., Wu, G., Gu, X., Wang, J., Wang, Y., Gao, H., & Ma, J. (2013). Magnetic and pH-sensitive nanoparticles for antitumor drug delivery. *Colloids and Surfaces. B, Biointerfaces*, *103*, 15–22. DOI: 10.1016/j.colsurfb.2012.10.041 PMID: 23201714

Yun, Q., Ge, Y., Shi, Z., Liu, J., Wang, X., Zhang, A., Huang, B., Yao, Y., Luo, Q., Zhai, L., Ge, J., Peng, Y., Gong, C., Zhao, M., Qin, Y., Ma, C., Wang, G., Wa, Q., Zhou, X., & Zhang, H. (2023). Recent Progress on Phase Engineering of Nanomaterials. *Chemical Reviews*, *123*(23), 13489–13692. DOI: 10.1021/acs.chemrev.3c00459 PMID: 37962496

Zhang, L. (2013). *The van der Waals force and gravitational force in matter*. https://arxiv.org/pdf/1303.3579

Zhang, Y., Zhang, Y., Rong, J., Wu, J., Dong, P., Xu, M., Feng, J., & Gu, C. (2019). Design and controllable synthesis of core-shell nanostructured Ni-P particles with an ionothermal strategy. *Journal of Alloys and Compounds*, *795*, 177–186. DOI: 10.1016/j.jallcom.2019.04.279

Zhu, J., Wei, S., Gu, H., Rapole, S. B., Wang, Q., Luo, Z., Haldolaarachchige, N., Young, D. P., & Guo, Z. (2012). One-pot synthesis of magnetic graphene nanocomposites decorated with core@double-shell nanoparticles for fast chromium removal. *Environmental Science & Technology*, *46*(2), 977–985. DOI: 10.1021/es2014133 PMID: 22126606

Zhu, K., Ju, Y., Xu, J., Yang, Z., Gao, S., & Hou, Y. (2018). Magnetic Nanomaterials: Chemical Design, Synthesis, and Potential Applications. *Accounts of Chemical Research*, *51*(2), 404–413. DOI: 10.1021/acs.accounts.7b00407 PMID: 29412634

Chapter 16
Understanding the Challenges Through Climate Microbiota Mental Health Trilogy: A Systematic Review

Hajira Gul
University of Lahore, Pakistan

Ibrahim Shah
http://orcid.org/0009-0007-6212-4411
University of Lahore, Pakistan

Muhammad Younus
http://orcid.org/0000-0001-9654-1546
Universitas Muhammadiyah Yogyakarta, Pakistan

Halimah Abdul Manaf
Universiti Utara Malaysia, Malaysia

Bhimo Widyo Andoko
Institut Seni Indonesia, Indonesia

Andi Adawiah
Universitas Lamappapoleonro, Indonesia

Arif Zainudin
Universitas Pancasakti Tegal, Indonesia

Eliza Meiyani
Universitas Muhammadiyah Yogyakarta, Indonesia

Abdul Rehman
PAF College, Sargodha, Pakistan

Rashid Minhas
Superior University, Pakistan

ABSTRACT

Climate change is an occurrence that has occurred globally and affects the natural world, having overarching effects on human health and well-being, like mental illness. These types of interactions are significant at a time when environmental

DOI: 10.4018/979-8-2600-0060-1.ch016

adjustments are an ugly reality, as is mental illness. Here, it seeks to find studies that document the interaction between gut microbiome and climate change and its impact on mental health. A systematic review of the literature was sought in a number of databases (PubMed, SCOPUS and ProQuest), between 2000 and 2023. Out of 677 studies in the first instance, 10 were finally chosen carefully against pre-decided criteria of inclusion and exclusion. The significant results of this review discovered that there exists an interdependent relationship of such factors that we refer to as brain-gut-climate change vicious cycle, the modification in one compound affects the other, this might be triggered because of climate change. Several mechanisms of modulation of gut microbiota in mental health were also investigated.

1 INTRODUCTION

Climate change is increasingly recognized not only as an environmental challenge but also as a multidimensional public health issue that influences human physiological and psychological well-being. Emerging evidence suggests that climate-related stressors, including rising temperatures, air pollution, food insecurity, biodiversity loss, and extreme weather events, can alter the composition and diversity of the gut microbiota. These microbial changes may disrupt the gut–brain axis, a bidirectional communication network linking the gastrointestinal system and the central nervous system through neural, endocrine, immune, and metabolic pathways. Consequently, dysbiosis of the gut microbiota has been associated with a higher risk of mental and neurological disorders such as anxiety, depression, cognitive decline, and neurodegenerative diseases. Rather than viewing climate change, microbiome health, and brain disorders as separate phenomena, this chapter conceptualizes them as interconnected components of a complex socio-ecological system. By synthesizing evidence from environmental science, microbiology, neuroscience, and public health, the chapter demonstrates how climate-induced environmental changes may indirectly affect mental health through microbiota-mediated mechanisms and discusses the implications of these relationships for achieving the health and environmental targets of the Sustainable Development Goals (SDGs).

Climate change is a global phenomenon that has been impacting the world in several different ways. One of the largest issues that has to be solved is how climate change is able to impact people's mental health (Lerner et al., 2017). Around the world, the physical and psychological well-being of people is already being affected by global warming (Vera-urbina et al., 2022). According to the World Health Organization (WHO), environmental factors lead to 12.6 million deaths annually that could be prevented, and climate change is adding to the problem (World Health Organization, 2021). The effects on human health due to climate change can either be direct or indirect. Some of the direct effects include weather conditions, global warming, and changed ecosystems. People are going to be even more stressed, nervous, and depressed due to the changes, so it will be easier for mental disorders to occur or worsen already existing mental disorders. On the other hand, alteration of dietary habits might affect the intestinal microbiota to cause or worsen mental status, and indirectly, climate change affects modifications in the intestinal microbiota to affect gut-brain axis function. However, the exact mechanisms by which climate change causes changes in mental status are not yet completely understood. Here, following this fact, this article will attempt to comprehend the effect that climate change has on the function and structure of the gut microbiota and their influence on mental health.

1.1 Background

The relationship between climate change, gut microbiota, and mental health can be understood through a cascading pathway of environmental and biological interactions. Climate change alters environmental conditions through rising temperatures, increased air pollution, extreme weather events, biodiversity loss, and disruptions to food and water systems. These environmental changes influence human exposure to microorganisms, dietary patterns, stress levels, and environmental toxins, all of which are important determinants of gut microbiota composition and diversity. Alterations in the gut microbiota, commonly referred to as dysbiosis, can disrupt the gut–brain axis, a complex communication network that connects the gastrointestinal tract and the central nervous system through neural, immune, endocrine, and metabolic mechanisms. Such disruptions may increase systemic inflammation, impair neurotransmitter production, and affect stress-response regulation, thereby contributing to mental health conditions such as anxiety, depression, cognitive impairment, and other neurological disorders. Consequently, climate change may indirectly influence mental health by reshaping the environmental factors that regulate microbial ecosystems within the human body. This interconnected framework highlights the need for an integrated approach that considers environmental sustainability, microbiome health, and mental well-being as mutually reinforcing dimensions of human health and sustainable development.

1.2 Sustainable Development Goals, Climate Change, and Health

Climate change and its effects on health are one of the challenges of today's era, which is a serious threat to health and likely to become even more dangerous in the coming decades. By doing so, the scientific community is concerned about the potential that climate change may affect civilizations (Ojha, S. et al., 2023). Government agencies take an interest as well. The United Nations formulated the 17 SDGs when it came to an awareness of the need for action on this issue. SDG 13 is quite specific and clear in identifying the significance of climate change, and seeks to encourage immediate action to combat climate change and its effects (Health et al., 2018). It is linked with all of them, but today, we are going to point out how it is related to SDG 3 (good health and well-being) and SDG 12 (responsible consumption and production). Climate change is a multidimensional threat to food systems through land use, CO2 emissions, global warming, and water ecosystems.

This climate change impacts not only agricultural practices and supply chain vulnerabilities but also the nutritional quality of the food that we consume. Molding food systems successfully to adapt to climate change is necessary, but a huge challenge. It requires an integrated approach that coordinates climate- and nutrition-sensitive practices, prioritizing sustainable agricultural practices, reduction of food losses, and fair access to nutritious food across the whole food system. It involves more than avoiding environmental degradation to protect food quality and availability in a changing climate (Zelenski et al., 2023). Climate change impacts food systems, and the impact is thus not just on CO2 emissions, land use, global warming, and water ecosystems. But on the food that we consume as well. Change in food systems is required, but it's a big ask. Fertilizer application is paramount in contemporary farming, but microorganisms of the soil microbiome are at the heart of nutrient cycling in agroecology.

Biofertilizers, obtained from favorable organisms, provide eco-friendly alternatives to chemical fertilizers. Probiotic microbes such as Lactobacillus are employed in the use of functional food for the purpose of developing sustainable agriculture and human health. Plant protection practices are a part of integrated pest management (IPM) practices. Plant-based products serve as a secure and efficient alternative to chemical insecticides. Biopesticides and biofungicides obtained from soil microbiome are eco-friendly alternatives to plant protection in agriculture (Logan et al., 2015). Shifting our dietary and food pattern can contribute significantly to our health and the environment; shifting our dietary pattern and food choice towards healthier foods can become our health benefit as well as environmental benefit; consumption of more qualitative foods, like whole grain cereals, fruits and vegetables, legumes, nuts, olive oil, and fish, is linked with better health and decreased environmental impacts (Deans, E., 2017).

1.3 Food and impact on the Gut-brain connection

The new consumption pattern of food is a major deviation in the consumption trend, with a propensity towards rapid and processed food, spurred by the exigencies of contemporary times and limited time. Such a change raises a real need to challenge the suitability of such a tendency, followed by a strong sense of consciousness that such dominant patterns are certainly not going to be in harmony with the requirements of optimum well-being. This awakening is supplemented by an exponential increase in health literacy, with individuals becoming increasingly aware of the part that eating habits contribute to good health (Bested, A. et al., 2013). Due to this increased awareness, there is a highly discernible interest in increasing healthier food intake. A new trend perceives the virtue of pro- and prebiotics in promoting well-being, valuing their beneficial effects on the gut. The same foods are now recognized as constituting part of a healthy diet. The invitations to consume pro- and prebiotics are a response to enhanced understanding of the relationship between food choice and overall health (Bachner, H. A., 2016). Recognition that the gut microbiota is implicated in the gut-brain axis has given rise to fresh insights into disease pathophysiology, and this marks a radical change in the discipline of neuroscience. The finding that the gut microbiota plays a role in controlling the gut-brain axis has provided new insight into the understanding of the pathophysiology of disease, a paradigm shift in neuroscience research (Rajkumar, R. P., 2021).

Most importantly, the recognition that our diets profoundly impact not only body health but also mind health emphasizes the holistic basis of the relationship between nutrition and health. The recognition emphasizes the necessity of a holistic approach to ways of eating, not just an analysis of their impact on the body but also on the mind. As people become more aware of such interconnectedness, there is an increasing pressure towards healthier diets that also care for the body and mind dimensions of well-being (Liang, S. et al., 2022). And on the other hand, though, the more we are aware of the not-so-good habits of ours, which we must change in a hurry, the greater the rise in health literacy. There is a worry here about the consumption of healthier foods. It is now understood that the consumption of pro- and prebiotic foods is a way to promote well-being.

We are sure that what we consume impacts not just our body wellness but also our brain. The gut microbiota (GM), or "second brain," harbors ~100 trillion bacteria, i.e., upgraded fungi and viruses, and 10 times more cells than in the human body. Besides, the total genome of GM, i.e., the gut microbiome, comprises ~3 million genes, 150 times that of a human genome (Logan, 2015). Gut microbiota (GM) has been reported to be implicated in gut function, metabolism, and the energy cycle. GM is involved in neurological outcomes through various mechanisms. Gut microbiota is also thought to play a crucial role in mediating communication involved in brain function and development. Microbiota-gut communication tends to be positive and bidirectional within a healthy host. The effect of the environment on the determination of the structure and function of microbiota in the variability of their number and species could make the causation of certain diseases simpler, a state of dysbiosis.

1.4 Brain-Gut-Climate Vicious cycle

Having performed a comprehensive literature review, we developed a new conceptual model, the "brain-gut-climate change vicious cycle." The model depicts how inextricably intertwined are the relationships between human mind health, diet, and resultant environmental impacts, with particular focus on climate change. Stress and other mental illnesses can lead to unhealthy nutrition habits, like the intake of comfort foods that are usually rich in carbohydrates, bad fats, and processed foods (Paprocka, J. et al., 2020). These unhealthy nutritional intakes will not only condition physical illnesses (e.g., obesity, diabetes, chronic diseases, cancer), but also provide space for the improvement or worsening of mental pathologies. Consequently, unhealthy eating is linked to environmentally unsustainable food processing with considerable environmental loads (e.g., a high-meat diet), leading to climate change by amplifying greenhouse gases emitted, land use, and water scarcity. Climate change effects, including climate-related hazards, food insecurity, and ecosystem degradation, can amplify stress and mental illness in persons and communities and thus widen the cycle. Accordingly, it is important to examine all facets and promote literacy on the topic. The role of education is critical to increase knowledge and to provide people with skills in order to allow them to make sustainable and informed choices.

Instructing and educating citizens as to the kind of impact climate/climate change will have on physical and psychological well-being is going to make more sustainable behavior work. For instance, global warming is not going to raise the temperature on this planet and make us physically and psychologically uncomfortable, but it's also going to make it impossible to produce good food, and then that will ensure physical and psychological well-being. This cause-and-effect cycle has to be reversed. Embrace healthy diets, with health production practices (minimum use of fertilizers, prevent soil loss, recycle resources, utilize water, prevent food losses, among other practices). Healthy people are more self-assured and vital in the face of a changing world, and are eager to learn and to assist in designing healthier and greener societies. By identifying and stopping the vicious cycle between gut, brain, and climate change, we can ensure our future is healthier and sustainable for humanity and the world (Figure 1).

Figure 1. Schematic representation of the interconnection between climate change, gut microbiome, and mental health

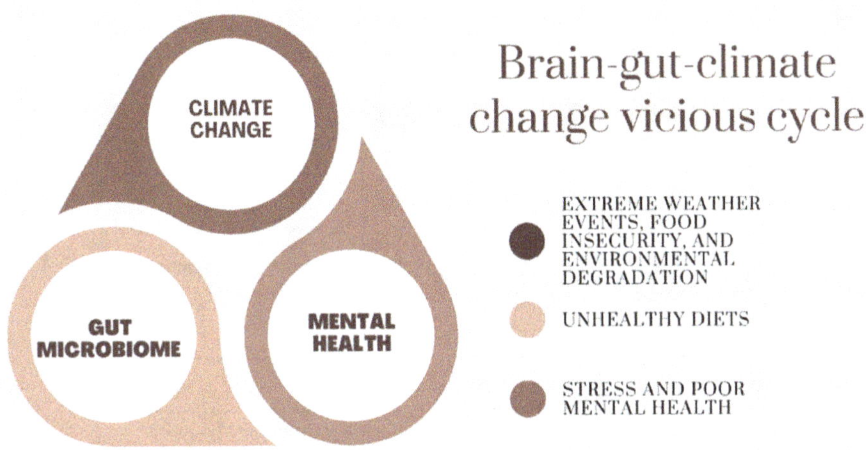

2 METHODOLOGY

During this literature review, the electronic databases ProQuest, Scopus, and PubMed were searched between July 1st, 2023, and September 15th, 2023. We carefully reviewed 200 articles rigorously screened for relevance and significance to our study. The descriptors in this study were: ("climate change" OR "environmental temperature") AND "gut microbiota". Upon analyzing the inclusion criteria, we performed a scoping review on this subject with 11 articles. Exclusion and inclusion criteria for studies are laid out in Table 1.

Table 1. Criteria for inclusion and exclusion from the study

Inclusion Criteria	Exclusion Criteria
- studies with methodology based on "human population survey", - articles published until May 2025; - articles written in English	- interviews and other data collected methodologies; - articles published after May 2025; - articles written in other languages

The methodology of carrying out the review was guided by PRISMA guidelines (Preferred Reporting Items for Systematic Reviews and Meta-Analyses) (Figure 2). The following data were retrieved from included manuscripts: author, year, study title, general area of research, area of application, implications in gut-microbiota, implication mechanisms, contributing to climate change, and mental health implications.

Figure 2. Flowchart applying the inclusion and exclusion criteria

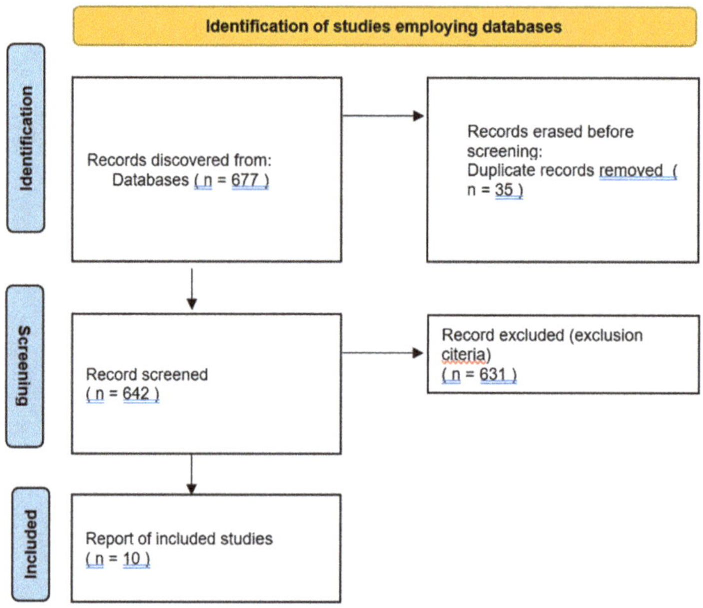

As a complement to this literature review, we employed an artificial intelligence program, namely the VOSviewer package, to build bibliometric maps. This is a computer program that builds and visualizes bibliometric networks.

2.1 Bibliometric Analysis Procedure

To complement the systematic literature review, a bibliometric analysis was conducted to identify the intellectual structure and emerging research themes related to climate change, gut microbiota, and mental health. Bibliographic data were retrieved from the Scopus, PubMed, and ProQuest databases using the search string:

("climate change" OR "environmental temperature") AND ("gut microbiota" OR "gut microbiome") AND ("mental health" OR "psychological health").

The search was performed between July 2023 and May 2025 and limited to English-language publications. After screening and removing duplicate records, the bibliographic information, including titles, abstracts, author keywords, and citation data, was exported in CSV format for analysis. The bibliometric mapping was performed using VOSviewer version 1.6.20. Co-occurrence analysis of author keywords was employed to identify major research themes and relationships among concepts. A minimum threshold of two keyword occurrences was established to improve network visualization and reduce noise. The resulting network maps were used to examine thematic clusters, research trends, and the conceptual relationships among climate change, gut microbiota, and mental health research.

2.2 Conceptual Framework

Table 2.

Climate Change
↓
Environmental Stressors
↓
Gut Microbiota Dysbiosis
↓
Gut–Brain Axis Dysfunction
↓
Mental Health Outcomes
↓
Sustainable Development Challenges

The framework proposes that climate change generates environmental stressors such as rising temperatures, pollution, biodiversity loss, and food insecurity. These stressors can alter the composition and diversity of the gut microbiota, resulting in dysbiosis. Changes in the gut microbiota may disrupt the gut–brain axis through immune, endocrine, and neural pathways, thereby increasing susceptibility to mental health and neurological disorders. Consequently, these health impacts pose challenges to achieving the Sustainable Development Goals, particularly those related to health, climate action, and environmental sustainability.

3 RESULTS AND DISCUSSION

3.1 Bibliometric Analysis

Table 3. Annual publication trend of included studies

Year	Number of Publications
2013	1
2015	3
2016	1
2017	1
2018	1
2019	1
2021	1
2022	4
2023	4

Table 3 presents the annual publication trend of studies included in this review. The findings indicate a growing scholarly interest in the relationship between climate change, gut microbiota, and mental health over the last decade. Although early contributions appeared between 2013 and 2019, research activity remained relatively limited. A notable increase in publications was observed during 2022–2023, suggesting heightened academic attention to the interconnected effects of environmental change on human health and well-being. This growth may reflect increasing concerns regarding climate-related health risks and the expanding recognition of the gut–brain axis as a critical mechanism influencing mental health outcomes.

Table 4. Country distribution of authors' institutional affiliations

Country	Frequency
United States	5
Australia	4
Canada	3
United Kingdom	2
Saudi Arabia	1
Croatia	1
Malaysia	1

Table 4 summarizes the geographical distribution of authors' institutional affiliations within the reviewed literature. The results demonstrate that research on climate change, gut microbiota, and mental health is predominantly concentrated in developed countries, particularly the United States, Australia, Canada, and the United Kingdom. These countries possess strong research infrastructures and have invested substantially in environmental health and microbiome studies. The limited representation of developing nations highlights an important research gap, particularly considering that many low- and middle-income countries face disproportionate exposure to climate-related health challenges. Future studies should encourage broader international participation to improve the global applicability of findings.

Table 5. Major publishers represented in the reviewed literature

Publisher	Number of Articles
Frontiers Media	3
Springer Nature	3
MDPI	2
PLOS	1
Oxford/Other Academic Publishers	2

Table 5 illustrates the distribution of publishers contributing to the literature included in this review. Frontiers Media and Springer Nature emerged as the most prominent publishers, reflecting their strong engagement with interdisciplinary topics related to environmental health, neuroscience, and microbiology. The presence of publications from MDPI, PLOS, and other major academic publishers further demonstrates the multidisciplinary nature of this research field. The diversity of publication outlets indicates growing recognition of the complex interactions among climate change, gut microbiota, and mental health across multiple scientific disciplines.

Table 6. Research domains represented in the included studies

Research Domain	Frequency
Environmental Science	8
Public Health	7
Mental Health/Psychiatry	6
Microbiology	5
Nutrition	3
Neuroscience	2

Table 6 presents the major research domains represented within the reviewed studies. Environmental science and public health were the most frequently occurring disciplines, followed closely by mental health and microbiology. This distribution confirms that research on climate change, gut microbiota, and mental health is inherently interdisciplinary. The integration of environmental, biological, and psychological perspectives provides a more comprehensive understanding of how climate-related factors influence human health through microbial and neurological pathways. Such interdisciplinary collaboration is essential for developing effective interventions and evidence-based policy responses to emerging environmental health challenges.

Table 7. Most frequently occurring keywords in the bibliometric dataset

Keyword	Frequency
Climate Change	18
Mental Health	16
Gut Microbiota	15
Gut-Brain Axis	12
Microbiome	10
Depression	9
Anxiety	8
Environmental Stress	7
Probiotics	6
Sustainability	5

Table 7 presents the most frequently occurring keywords identified through the bibliometric analysis. “Climate Change,” “Mental Health,” and “Gut Microbiota” emerged as the dominant terms, confirming the central focus of the reviewed literature. The frequent appearance of “Gut-Brain Axis,” “Depression,” and “Anxiety” indicates that researchers are increasingly investigating the biological mechanisms through which environmental changes influence psychological outcomes. Furthermore, the occurrence of keywords such as “Probiotics” and “Sustainability” suggests a growing interest in preventive and intervention-oriented approaches that simultaneously address human health and environmental challenges. Overall, the keyword distribution highlights the multidisciplinary nature of the field, integrating perspectives from environmental science, microbiology, neuroscience, nutrition, and public health.

3.2 Key Insights

In this review, we identified 11 papers that were included in the inclusion criteria. In order to make understanding easy and study analysis simpler, a summary table was constructed (Table 7). The table had several relevant items, including: author, year, study title, general area, area of relevance, gut-microbiota implications, mechanisms of implication, input towards climate change, and mental health implications.

Table 8. Summary of information from 11 relevant articles met our inclusion criteria

Author/Year	Title	General Domain	Gut Microbiota Implications	Mechanism of Implication	Climate Change and Mental Health Implications	Type of Study	Instrument/ questionnaire used	Number of participants
John Zelenski, Sara Warber, Jake M. Robinson, Alan C. Logan, Susan Prescott / Mar 2023	*Nature Connection: Providing a Pathway from Personal to Planetary Health*	- Environmental Science, - Health, - Psychology	Discusses the importance of a connection to nature for personal and planetary health.	Highlights the role of a declining connection to nature in the modern health crisis and the erosion of planetary health.	Suggests that restoring the human connection to nature can promote physical and spiritual well-being, community cohesion, and environmental responsibility, addressing climate change and mental health issues.	Review or conceptual paper.	Nature Relatedness (NR) scale and other measures related to nature connection.	Not applicable, as it does not report the results of a specific empirical study with a participant sample.

continued on following page

Table 8. Continued

Author/Year	Title	General Domain	Gut Microbiota Implications	Mechanism of Implication	Climate Change and Mental Health Implications	Type of Study	Instrument/ questionnaire used	Number of participants
Silva Vujanovic Josko Vujanovic And Vladimir Vujanovic/ Nov 2022	*Microbiome-Driven Proline Biogenesis in Plants under Stress: Perspectives for Balanced Diet to Minimize Depression Disorders in Humans*	Microbiology, Nutrition, Mental Health	Explores the role of proline consumption and gut microbiota in depression.	Discusses the potential links between gut microbiota, proline levels, and depression severity.	Proposes that understanding the microbiome-plant-food axis can help design proline-balanced diets, potentially impacting mental health and environmental sustainability under climate change.	does not explicitly mention the specific type of study or the number of participants. However, based on the content, it appears to be a literature review or a review article that synthesizes existing research and case reports related to the risks and effects of consuming contaminated cannabis.	Regarding the use of questionnaires as instruments, there is no indication in the text that primary data collection, such as surveys or questionnaires, was conducted as part of a research study. Instead, the information is derived from existing case reports, meta-analyses, and findings from different studies.	Without specific details on the number of participants, the nature of the study, or the use of structured instruments like questionnaires for primary data collection, it is challenging to classify it as qualitative, quantitative, or exploratory in a traditional research sense. It seems more aligned with a comprehensive review and synthesis of existing literature on the topic.

continued on following page

Table 8. Continued

Author/Year	Title	General Domain	Gut Microbiota Implications	Mechanism of Implication	Climate Change and Mental Health Implications	Type of Study	Instrument/ questionnaire used	Number of participants
Nicolas J. C. Stapelberg, Grace Branjerdporn, Sam Adhikary, Susannah Johnson, Kevin Ashton, and John Headrick / Oct 2022	*Environmental Stressors and the PINE Network: Can Physical Environmental Stressors Drive Long-Term Physical and Mental Health Risks?*	Environmental Science, Health, Psychiatry	Investigates the impact of physical environmental stressors on the PINE network and mental health.	Suggests that physical environmental stressors disrupt the PINE network, possibly leading to long-term mental health impacts.	Emphasizes the need for more empirical evidence to establish links between physical stressors, mental health, and climate change, discussing the concept of syndemics.	The study conducted in this article is exploratory and qualitative in nature. It involves discussions, presentations, and workshops to explore the relationship between sustainable, healthy, and organic diets and mental health and social well-being (MHSW).	The primary instruments used in the study are questionnaires. These questionnaires are employed to assess MHSW outcomes related to diets. The article discusses both the strengths and weaknesses of using questionnaires in this context. Strengths include the ability to combine several aspects in one tool, applicability to diverse interest groups, and the potential for validated questionnaires for adults to be applied to adolescents. Weaknesses include comparability issues between different cultural settings, potential translation challenges, and the impact of distribution methods on outcomes.	The exact number of participants in the workshop is not explicitly mentioned in the provided text. Typically, workshops involve a smaller group of experts or stakeholders. The emphasis is on in-depth discussions and exchanges of ideas, and the quality of insights takes precedence over a large sample size.

continued on following page

Table 8. Continued

Author/Year	Title	General Domain	Gut Microbiota Implications	Mechanism of Implication	Climate Change and Mental Health Implications	Type of Study	Instrument/ questionnaire used	Number of participants
Emmanuel A. Ayeni Ahmad M. Aldossary Daniel A. Ayejoto, Lanre A. Gbadegesin Abdullah A. Alshehri Haya A. Alfassam Henok K. Afewerky Fahad A. Almughem Saidu M. Bello And Essam A. Tawfik / Sept 2022	*Neurodegenerative Diseases: Implications of Environmental and Climatic Influences on Neurotransmitters and Neuronal Hormone Activities*	Neuroscience, Environmental Science, Health	Explores the environmental and climatic influences on neurodegenerative diseases and neurotransmitters.	Investigates the implications of environmental changes on neurotransmitters and neuronal hormones in neurodegenerative diseases.	Addresses the role of environmental and climatic factors in neurodegenerative diseases and their potential impact on mental health, but specifics are not provided.	The provided text appears to be part of a scientific review article. It does not represent a specific primary research study but rather synthesizes existing knowledge on the association between climate, environmental factors, and neurodegenerative diseases.	The text does not indicate the use of questionnaires or specific instruments, as it is a review article discussing various studies and findings. The information is presented in a narrative style, discussing concepts, studies, and potential associations without detailing specific instruments or questionnaires.	As this is a review article and not a primary research study, there is no specific number of participants mentioned. The content summarizes information from various studies and literature, providing a broad overview of the topic without reporting on a specific participant group.

continued on following page

Table 8. Continued

Author/Year	Title	General Domain	Gut Microbiota Implications	Mechanism of Implication	Climate Change and Mental Health Implications	Type of Study	Instrument/ questionnaire used	Number of participants
David H. Nelson, Alan C Logan, Susan L Prescott, Jeffrey Bland/ Feb 2019	*Clinical Ecology—Transforming 21st-Century Medicine with Planetary Health in Mind*	Medicine, Environmental Science, Health	Discusses clinical ecology and its role in understanding interconnected health and environmental challenges.	Emphasizes the concept of clinical ecology, viewing health from an ecological perspective.	Advocates for viewing medicine as clinical ecology to address non-communicable diseases, socioeconomic inequalities, and climate change, but specific mechanisms are not detailed.	The content provided does not describe a specific study but instead continues the discussion and narrative from the previous sections. Therefore, it does not fit into a specific type of study like qualitative, quantitative, or exploratory research.	No specific instruments or questionnaires are mentioned in this section. The content focuses on presenting a perspective and discussing the implications of the microbiome and omics revolution in the context of clinical ecology.	As this is not a study with a defined participant group, there is no specific number of participants mentioned in this section. The content discusses ideas, perspectives, and potential practical implications related to clinical ecology, the microbiome, and omics.

continued on following page

Table 8. Continued

Author/Year	Title	General Domain	Gut Microbiota Implications	Mechanism of Implication	Climate Change and Mental Health Implications	Type of Study	Instrument/ questionnaire used	Number of participants
Alison C Bested, Alan C Logan & Eva M Selhub / Mar 2013	*Intestinal microbiota, probiotics and mental health: from Metchnikoff to modern advances: Part I - autointoxication revisited*	Microbiology, Mental Health	Reviews the history of probiotics and the gut-brain-microbiota connection in mental health.	Provides a historical perspective on the relationship between gut microbiota and mental health.	Discusses the historical context of probiotics and gut-brain interactions but does not provide specific mechanisms or implications for climate change.	It covers a range of experimental and clinical research studies related to the influence of probiotics on mental health. These studies involve both animal models and human participants. suggesting a mixed-methods approach with both qualitative and quantitative elements	The questionnaires and instruments mentioned include: Quality of life scores, Beck Anxiety Inventory, validated anxiety, stress, and depression scales.	The first formal investigation involved 132 otherwise healthy adults. A separate placebo-controlled pilot study with 39 chronic fatigue syndrome patients. An evaluation of a Lactobacillus helveticus and Bifidobacterium longum combination probiotic in a placebo-controlled study involving otherwise healthy adults. A study involving 44 patients with irritable bowel syndrome

In the drastically affected world of climate change, environmental stresses, food choices, human-nature relationships, gut microbiome, and personal and planetary well-being, these 11 interrelated articles travel varied ground to find intricate relationships between health at the individual and global levels. They highlight the necessity of reoccupying our health connection to nature, suggesting that this return can begin with body and soul well-being and promote eco-awareness in the face of climate change and mental disease. Moreover, the investigation of the possible contribution of gut microbiota to depression and proline-balanced diets presents promising avenues for tackling mental illness through food choices in the context of climate change. Investigation of environmental stressors and their possible long-term implications within the field of mental health brings with it the concept of syndemics, highlighting the description of complex interrelated health issues within the dynamic climate. Its emphasis on Eco-Regions and sustainable diets further helps to highlight the comprehensive approach towards dietary impacts on mental health and their wider environmental sustainability issues. Overall, these studies highlight the necessity of taking into account the complex interplay between gut microbiota, ecological factors, and human health in addressing the complex issues of climate change, leading to more integrative research and forethought-based measures in maintaining individual and planetary welfare. This review article provides valuable insight into the specific mechanisms by which climate change affects the intestinal microbiota composition and function. Understanding how climate change affects the intestinal microbiome will allow us to identify potential biomarkers or microbial signatures for mental health outcomes.

This will allow us to implement early detection and intervention to assist those vulnerable to climate change mental health impacts. Apart from this, while determining the interaction of climate change-induced changes in gut microbiota and mental health outcomes, we may identify important pathways and biological processes through which the development or aggravation of mental disorders occurs. We may then identify some treatment modalities, such as probiotic supplementation or dietary control, to manage the gut microbiota so that climate change-induced mental health symptomology is reversed. Second, examining the potential for the use of the gut-brain axis to connect the influence of climate change on mental health can also give useful information regarding the mechanisms involved. This can be employed to guide intervention deployment that will confront the gut-brain interaction directly, such as psychobiotics or neuroregulatory drugs, to restore balance between the brain and the gut and to foster mental health during climate change. By identifying potential pathways by which climate change affects the gut-brain axis and psychological well-being, we can then establish an understanding of these pathways. This information will dictate that we can create novel interventions, such as lifestyle modification, therapeutic treatment, and environmental modification, that will work to reverse the adverse impacts of climate change on mental health.

Incorporation of the gut-brain axis into policy for adaptation and mitigation of climate change can ultimately increase our resilience to climate change's psychological impacts. With an appreciation of interaction between climate change, the gut microbiota, the gut-brain axis, and mental health, evidence-informed policy and intervention can drive forward mental well-being in a changing climate, with increased global public health relevance to populations and individuals. In our study, in order to enhance understanding of this theme, we applied machine learning methods in a bid to evaluate interaction amongst studies. Artificial intelligence allows you to connect the fields of climate change, mental health, and microbiota easily. This is superior to recent bibliometric studies as well as literature reviews. We conducted references. Figure 3 illustrates the main themes of the interactions.

3.3 Discussion

Figure 3. Keyword co-occurrence networks in artificial intelligence research in the fields of climate change, mental health, and microbiota

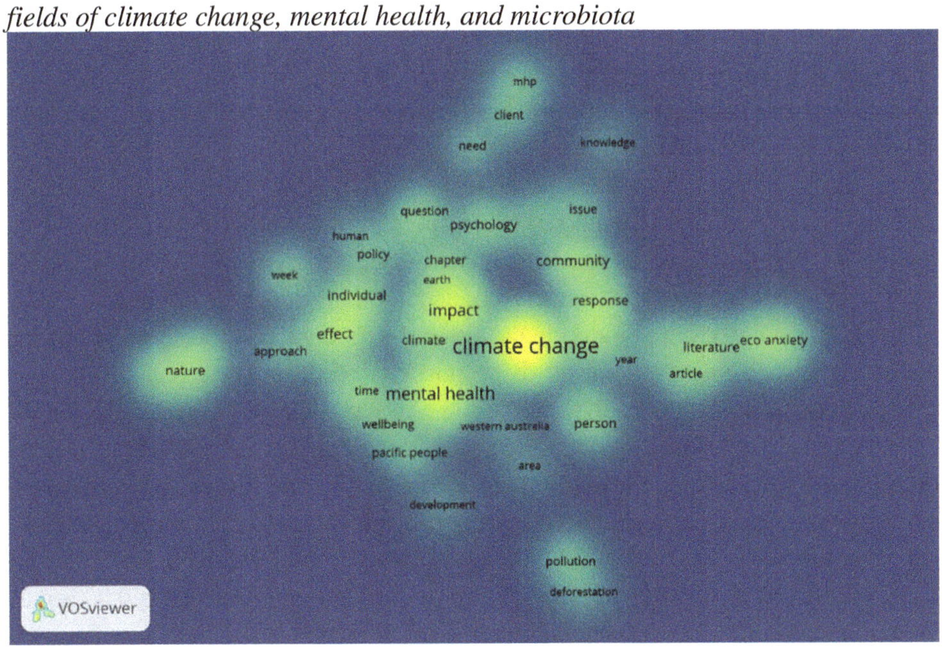

Perhaps the most important consequence of these discoveries is that public health and environmental policy will need to start incorporating microbiota health into mental health risk models—particularly in climates vulnerable to climate disruption. Mental health clinicians can be required to ask about nutrition, environmental exposure, and gut health when taking patient histories. Apart from this, activation of resilient microbiota by diet, probiotic therapy, and lifestyle change may perhaps be a preventive measure for climate-susceptible mental disorders. Also observed in the review were serious methodological problems with existing studies. Longitudinal or cohort study designs to track long-term climate-microbiota-mental health paths were employed in a few studies. Most were based on self-reported mental health questionnaires without physiological validation. Microbiota studies were largely restricted to fecal samples, omitting oral, skin, or nasal microbiomes. Also, climate variables tended to be operationalized with coarse proxies (e.g., urban-rural, mean temperature), lacking exposure precision modeling. These gaps mean that interdisciplinary collaboration, and more specifically between climate researchers, microbiologists, psychiatrists, and public health scientists, is warranted.

3.4 Recommendations

Recommendations that arise from this systematic review are: First, subsequent research will utilize combined frameworks and interdisciplinary teams in considering causal relationships between climate, microbiota, and mental health on sound multi-omic data, climate models, and psychometric evaluations. Second, governments and institutions need to invest in early warning systems for mental health hazards in the context of climate catastrophes, including microbial health surveillance and classical mental health interventions. Third, policymakers must take microbiota resilience into account in environmental health policy, in spending on food security, clean water access, and sanitation as foundations of not only physical but also psychological health. Fourth, high-priority status must be accorded to vulnerable populations such as children, the elderly, and individuals with pre-existing gastrointestinal or mental disease, who will inevitably suffer disproportionately from the effects of climate change that are mediated by the microbiota. Another central suggestion is public education and awareness. Activist groups working on climate and mental health must incorporate messages regarding the gut-brain axis and its environmental sensitivity. Promoting fiber-rich, sustainable eating habits, including fermented foods and micronutrients, could benefit planetary and microbial wellbeing at the same time. Medical and psychology schools must start integrating microbiota-mental interaction training in the context of ecological disruption.

3.5 Limitations

This chapter is subject to several limitations. First, the discussion is primarily based on existing literature and bibliometric evidence, which may not fully capture the complex causal relationships among climate change, gut microbiota, and mental health. Second, the interdisciplinary nature of the topic means that findings originate from diverse fields with varying methodologies, making direct comparisons challenging. Third, the rapidly evolving nature of microbiome and climate-health research may result in the omission of newly emerging evidence. Consequently, the conclusions presented should be interpreted as a synthesis of current knowledge rather than definitive proof of causality.

3.6 Implications

The findings of this chapter have important theoretical, practical, and policy implications. Theoretically, they contribute to a more holistic understanding of health by linking environmental change, microbial ecosystems, and neurological outcomes within a single conceptual framework. Practically, the chapter highlights the need for preventive health strategies that address environmental determinants of health alongside biological and psychological factors. From a policy perspective, the identified relationships support the integration of climate action, environmental protection, and mental health initiatives to advance sustainable development objectives, particularly those related to health, climate resilience, and ecosystem preservation.

4 CONCLUSION

This article contributes to our understanding of the complex interactions among the gut–brain axis, climate change, and mental health. The gut–brain axis is where researchers are focused due to the fact that it is the center around which overall physical and mental well-being occurs. There are three large channels of neuro, immune, and hormonal conduits wherein two-way brain–gut communication occurs. Any disruption of gut microbiota can affect the gut–brain axis or, in other words, the physical as well as mental state of the host. The difference in the proportion of Bacteroidetes and Firmicutes implies a significant role played in terms of sustaining mental status, and the same is true even with corresponding genus-level data. Such methodological direction determines the role of the function of microbiota. Therefore, it needs to be explored further in the human population. Since the association between obesity and mood disorders already exists, gut–adipose–brain axis communication is the ticket to solving this increasingly common issue.

4.1 Future Research

Future research should move beyond descriptive analyses and focus on empirically testing the pathways linking climate change, gut microbiota alterations, and mental health outcomes. Longitudinal studies, cross-national comparisons, and multidisciplinary investigations are needed to establish causal relationships and identify key mediating mechanisms within the gut–brain axis. Researchers should also explore how specific climate-related stressors, such as air pollution, heat exposure, and food insecurity, influence microbial diversity and mental health across different populations. Additionally, future studies could examine the effectiveness of microbiome-based interventions and climate adaptation strategies in mitigating adverse mental health outcomes associated with environmental change.

REFERENCES

Bachner, H. A. (2016)... *Integrative Therapies in Anxiety Treatment.*, *89*, 397–422.

Bested, A. C., Logan, A. C., & Selhub, E. M. (2013). Intestinal microbiota, probiotics and mental health : from Metchnikoff to modern advances : Part I – autointoxication revisited. *Gut Pathogens*, *5*(1), 1. DOI: 10.1186/1757-4749-5-5 PMID: 23369162

Deans, E. (2017). Microbiome and mental health in the modern environment. *Journal of Physiological Anthropology*, *1–5*(1), 1. Advance online publication. DOI: 10.1186/s40101-016-0101-y PMID: 27405349

Health, O., Foods, F., & Microbiota, G. (2018). *One Health*. Fermented Foods, and Gut Microbiota., DOI: 10.3390/foods7120195

Lerner, A., Matthias, T., & Aminov, R. (2017). Potential effects of horizontal gene exchange in the human gut. *Frontiers in Immunology*, *8*, 1630. DOI: 10.3389/fimmu.2017.01630 PMID: 29230215

Liang, S., Jin, F., & Jia, C. (2022). Editorial : The effect of gut microbiota on the brain structure and function. 2020–2024.

Logan, A. C. (2015). Dysbiotic drift : mental health, environmental grey space, and microbiota. ???, 1–17. DOI: 10.1186/s40101-015-0061-7

Logan, A. C., Katzman, M. A., & Balanzá-martínez, V. (2015). Natural environments, ancestral diets, and microbial ecology : is there a modern " paleo-deficit disorder " ? Part II. 1–22. DOI: 10.1186/s40101-014-0040-4

Obradovich, N., Migliorini, R., Paulus, M. P., & Rahwan, I. (2018). Empirical evidence of mental health risks posed by climate change. 115(43). DOI: 10.1073/pnas.1801528115

Oren, H. (2021). European Journal of Oncology Nursing. The effect of virtual reality on pain, fear, and anxiety during access of a port with huber needle in pediatric hematology-oncology patients : Randomized controlled trial. 50(December 2020), 1–7.

Paprocka, J., & Emich-Widera, E. (2020). Brain Sciences: Neuroimaging of Basal Ganglia in Neurometabolic Diseases in Children. 1–17.

Pasha, O., Esamai, F., Patel, A., Garces, A., Wright, L. L., Koso-, M., Moore, J., Althabe, F., & Kodkany, B. S. (2015). NIH Public Access. 29(8), 649–656. https://doi.org/DOI: 10.1055/s-0032-1314885.Neonatal

Potter, T., & Potter, T. M. (2023). Mental Health Impacts of Climate Change for Birthing People and the Provider's Role. *Journal of Midwifery & Women's Health, 10–13*(3), 320–323. Advance online publication. DOI: 10.1111/jmwh.13492 PMID: 37042448

Rajkumar, R. P. (2021). Gut Hormones as Potential Therapeutic Targets or Biomarkers of Response in Depression : The Case of Motilin.

Vera-urbina, F., Dos, F., & Godoy-vitorino, F. (2022). The Gut Microbiome May Help Address Mental Health Disparities in Hispanics : A Narrative Review. 1–17. Ojha, S., Patil, N., Jain, M., & Kole, C. (2023). Probiotics for Neurodegenerative Diseases. *Systematic Reviews.*

Wong, Y. S., & Osborne, N. J. (2022). Biodiversity Effects on Human Mental Health via Microbiota Alterations.

Zelenski, J., Warber, S., Robinson, J. M., Logan, A. C., & Prescott, S. L. (2023). Nature Connection. *Providing a Pathway from Personal to Planetary Health Anthropocene : Figure, 1*, 1–24.

KEY TERMS AND DEFINITIONS

Brain Disorders: Brain disorders are medical conditions that affect the structure, function, or biochemical processes of the brain, leading to impairments in cognition, emotion, behavior, movement, or neurological functioning. These disorders encompass a broad range of conditions, including neurodegenerative diseases such as Alzheimer's Disease and Parkinson's Disease, psychiatric disorders such as Depression and Schizophrenia, as well as developmental and neurological conditions. Brain disorders can result from genetic, environmental, infectious, traumatic, or lifestyle-related factors and often require multidisciplinary approaches for diagnosis, treatment, and management.

Climate Change: Climate change refers to long-term alterations in global or regional climate patterns, particularly changes in temperature, precipitation, wind systems, and the frequency of extreme weather events. Although climate variations can occur naturally, contemporary climate change is primarily driven by human activities such as the burning of fossil fuels, deforestation, and industrial processes that increase the concentration of greenhouse gases in the atmosphere. These changes have significant consequences for ecosystems, biodiversity, human health, food security, water resources, and economic development worldwide.

Environment: The environment encompasses all living and non-living components that surround and influence organisms. It includes natural elements such as air, water, soil, plants, animals, and ecosystems, as well as human-made features and social conditions. The environment provides essential resources and services that sustain life, including clean air, fresh water, fertile land, and climate regulation. Environmental quality is closely linked to human well-being, and environmental degradation resulting from pollution, resource depletion, and habitat destruction can adversely affect ecological balance and public health.

Gut Microbiota: Gut microbiota refers to the diverse community of microorganisms, including bacteria, fungi, viruses, and archaea, that reside in the gastrointestinal tract of humans and other animals. These microorganisms play crucial roles in digestion, nutrient absorption, immune system regulation, metabolism, and the production of bioactive compounds. A balanced gut microbiota contributes to overall health, while disruptions in microbial composition, known as dysbiosis, have been associated with various conditions such as obesity, inflammatory bowel disease, diabetes, and neurological disorders.

Sustainable Development Goals (SDGs): The Sustainable Development Goals (SDGs) are a set of 17 interconnected global objectives adopted by the United Nations in 2015 as part of the 2030 Agenda for Sustainable Development. The SDGs aim to address major global challenges, including poverty, inequality, climate change, environmental degradation, health, education, gender equality, and economic growth. Designed to promote sustainable and inclusive development, the goals encourage governments, businesses, civil society organizations, and individuals to work collaboratively toward creating a more equitable, prosperous, and environmentally sustainable future for all by the year 2030.

About the Contributors

Safaa Al-Humairi is a postdoctoral researcher and PhD graduate in Mechanical/ Biomaterials Engineering from the University of Teknologi Malaysia. His research focuses on the design and development of novel biomaterials and their related applications. He is an associate professor and program coordinator at Management and Science University in Shah Alam, Malaysia. Dr. Safaa is a registered professional engineer in a number of engineering associations and societies and has more than 15 years of experience in biosensor/materials design and implementation. He is the author of more than 150 research articles published in well-known, highly cited journals. He has also been awarded numerous special gold, silver, and bronze medals for innovations at national and international engineering and technology exhibitions. He is also an editor and reviewer for several well-known journals and publishers

Abdollah Bahador is a leading expert in materials engineering, specializing in advanced materials such as titanium and shape memory alloys. He holds a Ph.D. in Mechanical Engineering from Universiti Teknologi Malaysia, with research focusing on laser welding and powder metallurgy. He serves as Principal Researcher at the Materials Processing Institute, UK, following academic roles at Universiti Teknologi Malaysia and Osaka University, Japan. Dr. Bahador's work spans additive manufacturing, tribology, corrosion, and thermal properties of materials, with applications in biomedical and industrial fields. With over 40 high-impact publications and numerous awards, he is a recognized innovator in advancing materials science and engineering

Fida Majdoub is an Assistant Professor at the Department of Material and Mechanical Structures, ECAM La Salle, University of Lyon, France. She specializes in mechanical engineering with expertise spanning advanced vibrations, tribology, general mechanics, and materials science. Her research focuses on friction law modeling, biomedical materials, and energy efficiency in transmission systems. Dr. Fida holds a PhD in Mechanical Engineering from Ecole Centrale de Lyon and has earned multiple distinctions, including the Silver Poster Award (STLE 2012, USA) and Best Poster Award (JIFT 2011, France). Her academic journey includes postdoctoral work at École Arts et Métiers ParisTech and research roles in France, with a focus on mechanical modeling and numerical simulations. She is a prolific author with numerous publications in leading journals, especially on friction, lubrication, and innovative materials for biomedical and energy applications.

Nattapon Chantarapanich, Ph.D., is the Vice Dean for Research and Innovation at the Faculty of Engineering at Sriracha, Kasetsart University in Thailand. He is also affiliated with the Digital Industrial Design and Manufacturing Research Unit at the same university and serves as a Senior Researcher at the Research Unit in Dental Biomechanics, Thammasat University. His expertise lies in clinical biomechanics and orthopaedic/dental biomechanics, specifically in areas such as fracture fixation design, dental implant design, surgical guide design, and personalized medical devices. Assoc. Prof. Chantarapanich's academic career includes positions as an Associate Professor (2020-Present), Assistant Professor (2015-2020), and Lecturer (2013-2015) at the Faculty of Engineering at Sriracha, Kasetsart University Sriracha Campus. He has also held administrative roles such as Head of the Digital Industrial Design and Manufacturing Research Unit and Head of the Mechanical Engineering Department at Kasetsart University Sriracha Campus. He is involved in various capacities with government agencies and associations, including roles with the Thailand Productivity Institute, the Food and Drug Administration (Thai Ministry of Public Health), the Thai Subcontracting Promotion Association, and the Thai Tool and Die Industry Association (TDIA). Additionally, Assoc. Prof. Chantarapanich has served in private/corporate sector roles, including Chief Technical Officer (CTO) at Digital Orthopaedic Solutions Co., Ltd., and Research & Development Manager at Customized Technology Co., Ltd. His educational background includes a Ph.D. in Biomedical Engineering from Prince of Songkla University, an M.Sc. in Engineering from Thammasat University, and a B.Eng. in Aerospace Engineering from Kasetsart University

***

Adinarayana Andy is a licensed pharmacist and research scientist with over sixteen years of experience bridging clinical pharmacy practice, pharmaceutical innovation, and operations management. A multidisciplinary expert, Dr. Andy holds a unique combination of credentials: a PhD in Biotechnology, an MBA in Project Management, and active licensure as a pharmacist. Served as Pharmacist-in-Charge at Weatherwax Family Pharmacy in Michigan, Dr. Andy directs clinical operations with a focus on medication therapy management (MTM), immunization initiatives, and public health outreach. His leadership is grounded in extensive industry experience with global pharmaceutical organizations, including BMS-Biocon, Cadila Pharmaceuticals, and Dr. Reddy's Laboratories. In these roles, he contributed significantly to drug development, regulatory affairs, and quality assurance frameworks. Andy earned his PhD in Biotechnology from Acharya Nagarjuna University and an MBA from the University of the Cumberlands, USA. He also holds a Master of Pharmacy in Pharmaceutical Analysis & Quality Assurance from Biju Patnaik University of Technology and a Bachelor of Pharmacy from Andhra University.

Garima Awasthi is a plant and environmental biologist specializing in plant stress physiology, heavy metal and metalloid toxicity, and sustainable remediation strategies. Her research focuses on understanding molecular, biochemical, and proteomic mechanisms governing plant responses to toxic elements, particularly arsenic, with emphasis on sulfur-mediated detoxification pathways, antioxidant defense systems, thiol metabolism, and metal transporter regulation. She completed her Ph.D. at CSIR–National Botanical Research Institute, Lucknow, under AcSIR, where her work provided important insights into reducing arsenic accumulation and improving stress tolerance in rice. She was awarded the SERB–DST National Postdoctoral Fellowship and subsequently expanded her research to nano-bioremediation, biochar-based adsorption, phytoremediation, and industrial wastewater treatment. Her interdisciplinary approach integrates plant molecular biology with environmental toxicology and nanotechnology to develop eco-friendly solutions for soil and water pollution. Prof. Awasthi has published over eighty peer-reviewed articles in leading international journals, contributed book chapters.

Kumud Kant Awasthi is currently serving as Professor at KR Mangalam University, Gurugram India. As the founding Head of the Departments of Life Sciences and Forensic Science, at Vivekananda Global University (VGU), Jaipur, India, I also served as the Administrative Head for the Departments of Chemistry, Physics, and Mathematics at VGU Jaipur. Previously, I worked at the National Institute of Animal Welfare (NIAW), Faridabad, India, an institute under the Ministry of Environment, Forest and Climate Change (MoEF&CC). I have also served as the course coordinator for Animal Welfare programs offered by Jawaharlal Nehru University (JNU), New Delhi, and was invited to the Oxford Centre for Animal Ethics in the UK. My research centers on toxicity assessment of various nanomaterials using both in vivo and in vitro models, with a focus on their safe application in drug development. I am also actively engaged in bioremediation research and latent fingerprint development. I have authored and edited books with esteemed publishers, including Springer Nature, Taylor & Francis, Wiley, and CRC Press. Additionally, I hold copyrights and patents in Germany and India for my innovative contributions.

Muralidhar L B is a professionally qualified dental surgeon. He holds Ph.D. from Canara Bank School of Management Studies, Bangalore University after his MBA from Christ University. He has 22 years of total teaching experience spanning across dental education, healthcare management and general management and 4 years of corporate experience. He has worked as research associate in Institute for Social and Economic Change, Bangalore in a project entitled as "Management By and For Common Man: An Exploration of the Concept and Direction for Future"post Ph.D. He had secured 28th rank 10th Standard examination for Karnataka State in 1993. He has been the co-facilitator of a MOOC course on "Social Entrepreneurship and Sustainability",offered through SWAYAM portal, jointly sponsored by GoI and IIMB. His 32 patents have been published and 2 design patents have been granted. He has 41 research publications to his credit out of which 36 are Scopus-indexed and 5 are ABDC indexed. He has co-authored 13 books. He is currently working as Assistant Professor (Grade-3) in Department of Business Management Studies, School of Commerce, Jain Deemed to be University, Bangalore.

Debika Chaudhuri obtained her B. Tech and M. Tech from West Bengal University of Technology, West Bengal. She is pursuing her Ph.D in VLSI Technology. She has more than 12 years of teaching experience. She has published 19 papers and 2 patents. She has also 1 UK design patent grant and 2 Indian design patent grant in her credit. Apart from her academic duties as Assistant Professor in the Electronics Department, School of Engineering & Technology, IFTM University she is also engaged in administrative work like coordinator, Departmental IQAC committee at IFTM University.

Susanta Das is a seasoned educator with a proven record of teaching and mentoring diverse students, working at multiple institutions, and providing service to the various initiatives of organizations. He received his Ph.D. and M.A. degrees from Western Michigan University (WMU), USA, and M.Sc. degree from Banaras Hindu University (BHU), India all in Physics. He continued his research as a Marie-Curie post-doctoral fellow at Stockholm University, Stockholm, Sweden on beam diagnostics for the DESIREE (Double ElectroStatic Ion Ring ExpEriment) in the project DITANET (Diagnostic Techniques for particle Accelerators – a Marie-Curie initial training NETwork), at the Indian Institute of Science Education and Research-Kolkata, India on high-pressure physics as a Project Scientist-B, and at the University of Electro-Communications, Tokyo, Japan on ion-surface interactions as a Post-doctoral fellow. Throughout his career, Dr. Das worked and collaborated with many researchers, Ph.D., master, and visiting students from different countries. He further visited many countries to discuss research and an international conference participant (UK, Italy, Germany, Greece, Belgium, Bulgaria, Romania, Brazil etc.). He worked at several institutions before assuming his current role at Ajeenkya D Y Patil University. He received the Marie-Curie post-doctoral fellowship in Sweden, Science Academies' Summer Research Fellowship in India, Gowen Frostic Doctoral Fellowship, Department Graduate Research and Creative Scholar Award by WMU, and Leo R. Parpart Doctoral Fellowship by Dept. of Physics, WMU, among many others, throughout his academic journey. He co-authored several research articles in WoS/Scopus indexed international journals, conference proceedings, and scholarly book chapters and co-edited several books published by renowned international publishing houses (IGI-Global, Springer Nature, Nova Science Publishers, Wiley, & Scrivener-Willey, Notion Press). At present, Dr. Das continues to delve into cutting-edge fields, with a keen interest in nanotechnology, UN SDGs, and data science. His multifaceted contributions, spanning teaching, research, and administrative leadership, showcase a dedicated professional who is instrumental in advancing the vision and mission of the institutions he serves.

Sara El Aimani holds a Master's degree in Human Pathology Biology, specialized in Cancer Sciences, from Mohammed V University, Morocco. Her research interests focus on natural products and their antioxidant, anti-inflammatory, and anticancer activities, with an emphasis on molecular mechanisms involved in tumor development.

Mohammad Irfan is a Professor at the School of Business and Commerce, Chandigarh University, Lucknow (Unnao), India, with over eighteen years of teaching and research experience. He has published more than 112 Scopus-indexed research papers, accumulating 860+ citations with an H-index of 15. He has authored 14 books with leading international publishers, including Springer and IGI Global, and is recognized among the World's Top 2% Scientists. His research expertise covers Security Analysis and Portfolio Management, Artificial Intelligence, Machine Learning, Blockchain, Cryptocurrency, Financial Engineering, FinTech, Green Finance, and Alternative Finance. He earned his Ph.D. from the Central University of Haryana and holds an MBA in Finance, an M.Com in Accounts and Law, and an MA in Economics. A UGC-JRF/SRF/NET qualifier in Management and Commerce, he also holds professional certifications from NSE (NCFM) and BSE.

Mangaiyarkkarasi J. is currently working as an Associate Professor of Physics in Nadar Mahajana Sangam S. Vellaichamy Nadar College, Nagamalai, Madurai-625 019, Tamil Nadu, India. She has 25 years of teaching experience. Her major research interest includes synthesis of ceramic materials, powder X-ray analysis and electron density studies. She has published 11 research papers in the refereed International journals and 17 book chapters. She has prepared study materials (UG & PG Physics) for Madurai Kamaraj University Distance Education Course.

Shanthalakshmi Revathy J holds a Bachelor of Engineering (B.E.) in Information Technology from Mohamed Sathak Engineering College, Kilakkarai, a Master of Engineering (M.E.) in Computer Science and Engineering from PSG College of Technology, Coimbatore, and a Ph.D. in Information and Communication Engineering from Anna University, Chennai. I am currently working as an Assistant Professor at University College of Engineering, Ramanathapuram, with 17 years of teaching experience in higher education. My research interests focus on the dynamic field of Data Analytics. I have contributed extensively to academia by publishing more than 10 research papers in reputed international journals and authoring 25 book chapters, reflecting my strong domain expertise. I also hold two patents, demonstrating my innovative research contributions. My academic journey represents my continuous commitment to teaching, research, and advancing knowledge in data analytics.

Sheeba K P., M. Sc, PhD (pursuing). She is having a academic Experience of 5 years and 8 years in health care industries like hospitals, health insurance organizations and TPA companies both in abroad and in India. She is a post graduate in Hospital Administration from Rajiv Gandhi University of Health Sciences, currently pursuing her PhD in management. She is a member of associate ship in general insurance from Insurance Institute of India (III). She has attended many workshops and Faculty Development Programs, seminars, and orientation programs. . She has presented more than 3 research articles at National and International conferences. She has published few research articles in UGC care listed journals. She is currently working as Assistant Professor in Department of Business Management Studies, School of Commerce, Jain Deemed to be University, Bengaluru, India.

Shamim Ahmad Khan completed their Ph.D. in Chemistry at Jamia Millia Islamia, New Delhi, where I focused my research on advanced topics in chemical sciences. I am currently working as an Assistant Professor of Chemistry at Shibli National College, Azamgarh, India. My academic journey reflects my passion for both teaching and research, and I strive to guide students while contributing to the broader field of chemistry through scholarly work.

Kasturi Laturkar is currently working as a Validation Engineer for Validation Associates LLC. She graduated with a M.S. in Chemical Engineering from Syracuse University and a B.Tech in Chemical Engineering from Guru Gobind Singh

Indraprastha University, Delhi, India. She has more than 7 years of experience working in commissioning, qualification and validation of upstream and downstream bioprocessing equipment and critical utilities.

Kaustubh Laturkar is currently working as a Process Engineering Manager at Procon Global LLC after completing his M.S in chemical engineering from University of Florida and B.E. in chemical engineering from Panjab University, Chandigarh, India. He has more than 14 years of experience working in the field of process engineering, refinery operations, utility systems design and operation with a special focus on design and commissioning of engineering systems.

Atanu Nag, obtained his B. Sc., M.Sc. and Ph.D. degree from University of Kalyani, West Bengal and is presently working as Associate Professor and Head in the Department of Physics, School of Sciences, IFTM University, Moradabad, Uttar Pradesh. He is a recognized innovation ambassador by the Ministry of Education Innovation Cell and is currently working as the Convener of the Institution's Innovation Council at IFTM University with a mission to develop a robust start-up ecosystem there. He has been teaching since last 17 years. He has published more than 50 Research papers in high impact Journals including SCI, Scopus, WoS indexing and over 40 proceeding papers in conferences. One research scholar had completed her Ph.D. and five scholars are perusing their Ph.D. under his supervision. He is the author of more than 10 books written for engineering and science students and working as editorial and reviewer capacity for regular and special issue of reputed journals and conferences. He has 1 UK design patent grant and 2 Indian design patent grant in his credit. He had organized two national conferences and many workshops, faculty development programmes as convener or coordinator.

Praveen Javali R is an Assistant Professor in the Department of Pharmaceutics at Sri Raghavendra College of Pharmacy in Chitradurga, Karnataka. With over four years of teaching experience across various institutions, he specializes in drug formulation, pharmacokinetics, and advanced pharmaceutical technologies, delivered through interactive lectures, hands-on labs, and dynamic discussions. His research focuses on innovative drug delivery systems and cutting-edge formulation techniques. Passionate about fostering critical thinking and real-world applications, he aims to guide the next generation of pharmaceutical scientists through collaborative projects and knowledge-sharing.

Khawla Rachmoune is a junior researcher in biology and materials engineering, specializing in the study of biomaterials and their applications in health and environmental fields. She is currently a PhD candidate in biology at Cadi Ayyad

University (Marrakech), where her research lies at the interface of biotechnology, biochemistry, and radiobiology.

Sandhiya S is a postgraduate student in Pharmacology with a strong academic background in pharmaceutical sciences. She completed her Bachelor of Pharmacy with distinction and has demonstrated consistent academic excellence, including recognition for her outstanding performance. Her research interests focus primarily on endocrine and metabolic disorders, particularly polycystic ovary syndrome (PCOS), with an emphasis on understanding its complex pathophysiology and identifying potential therapeutic strategies. She is also interested in neuropharmacology, especially in exploring the neuroprotective effects of natural compounds in experimental models of neurodegenerative diseases. Currently, her research work involves evaluating the neuroprotective potential of herbal extracts in rotenone-induced Parkinson's disease models, aiming to develop cost-effective and safer treatment approaches. Her work reflects a growing interest in translational research that connects experimental findings with clinical applications. She has experience in scientific writing, literature review, and data interpretation, and is committed to contributing to evidence-based research. Her long-term goal is to advance knowledge in pharmacology through innovative research and meaningful scientific contributions.

Janarthanan S is a postgraduate student pursuing Master of Pharmacy (M.Pharm) in Pharmacology at Dr. M.G.R. Educational and Research Institute. His academic training focuses on experimental pharmacology, toxicology, and drug discovery, with a strong emphasis on translational research. His primary research interests include neuropharmacology, particularly the evaluation of neuroprotective agents against chemically induced neurotoxicity using preclinical models such as zebrafish. He is especially interested in studying the therapeutic potential of natural compounds, including polyphenols like ellagic acid, in mitigating oxidative stress, neuroinflammation, and neuronal damage. He is also actively engaged in pharmacovigilance and adverse drug reaction (ADR) monitoring, with an interest in improving drug safety and clinical outcomes. His broader interests extend to novel drug delivery systems, experimental design in pharmacological research, and the integration of emerging technologies such as artificial intelligence in drug discovery. Janarthanan is committed to contributing to high-quality scientific research and aims to publish in peer-reviewed journals while adhering to ethical research practices. He continuously seeks to expand his knowledge through academic collaborations, conferences, and advanced training in pharmacology and related biomedical sciences.

Rojaa SA is an undergraduate Biomedical Engineering student with a strong focus on biomedical data science, biomechanics, and intelligent healthcare technologies. Her research centers on integrating artificial intelligence with wearable biosensing and medical data analytics for personalized and preventive healthcare solutions. She has worked on projects involving cuffless blood pressure estimation, rehabilitation engineering, and advanced biomaterials for fracture fixation. Actively engaged in interdisciplinary innovation and academic initiatives, she demonstrates a commitment to translational healthcare research. She aims to advance AI-driven biomedical systems through future research.

Sameer Saharan is Assistant Professor and Head of the Department (Forensic Science) at Mody University of Science and Technology, Rajasthan. He holds a Ph.D. in Forensic Science from Amity University, Haryana, focused on identifying forged fingerprints using polyvinyl acetate. His core areas of expertise include fingerprint analysis, cyber and digital forensics, DNA fingerprinting, and drone forensics. Dr. Saharan has developed a patented system for differentiating real and forged fingerprints. He has been recognized for mentoring students to national awards, delivering keynotes, and organizing numerous academic programs and workshops. With robust experience in both academic and laboratory forensic environments, he is proficient with a range of platforms including Linux and Windows and advanced forensic software such as FTK Imager, Cellebrite UFED, and Wireshark. Known for his patience, learning attitude, and friendly leadership, Dr. Saharan continues to be a key contributor to research and practical innovation in forensic science.

Aditya Shrivastav is an aspiring technologist currently pursuing a Bachelor of Technology (B. Tech.) in Computer Science and Engineering at Ajeenkya D. Y. Patil University. He holds solid programming expertise in C, C++, and Python, alongside strong command over data and business tools such as Microsoft Excel and Tally software. Aditya is a member of Team Vac-Vectoor Hyperloop and participated with the team in the Global Hyperloop Competition organized by IIT Madras, contributing to cutting-edge transportation solutions. He actively engages in hackathons and various technical events to expand his skills and contribute to innovative problem-solving. He has authored and co-authored book chapters in the domains of solid waste management, bioplastics, nanotechnology applications, and artificial intelligence, showcasing his multidisciplinary approach and research acumen. Aditya is also working on an innovative project on predictive e-waste management. The project is targeted toward IT departments, schools, and colleges, aiming to support sustainable practices and resource optimization. In parallel with his academic journey, Aditya is serving as a faculty intern at Disha Computer Institute, where he teaches C, C++, basic Python, basic Java, Excel formulas, and

Tally software. His dedication and teaching proficiency earned him a Letter of Appreciation and the opportunity to conduct a workshop on Advanced Excel at G. H. Raisoni College of Arts, Commerce and Science. With a strong foundation in technology, a passion for sustainability, and hands-on experience in both teaching and project development, Aditya Shrivastav is committed to driving impact through research, education, and innovation.

Gayathri T. is currently working as Assistant Professor in the Department of Biomedical Engineering, KPR Institute of Engineering and Technology, Coimbatore, India. She has completed her Bachelor's degree in Biomedical Engineering, Master's degree in Nanotechnology and Ph.D. in Nanotechnology. Her research interests includes biomaterials and nanomaterials for healthcare applications. She has published her work in various international journals and conferences and continues to contribute to advancements in biomedical research.

Aman Suresh T is an Assistant Professor in Pharmacy Practice, currently serving at the Faculty of Pharmacy, Dr. M.G.R. Educational and Research Institute, Chennai. With a Pharm.D and ongoing PhD in Pharmacy Practice from SIMATS, he brings academic depth and clinical insight to his teaching and research. His teaching repertoire includes subjects such as pharmacotherapeutics, biostatistics, and pharmacoepidemiology for undergraduate and doctoral pharmacy programs. Dr. Aman is an active contributor to scholarly literature with multiple publications in Scopus and Web of Science-indexed journals, including research on diabetes, nephrology, and pharmacovigilance. His work has earned recognition such as subject-wise ranks at RGUHS and a prestigious ASCPT LMIC Travel Grant. Beyond academia, Dr. Aman has reviewed over 200 manuscripts and serves on editorial boards of reputed journals like Acta Scientific Pharmaceutical Sciences, PLOS ONE, and Cureus. He is also a frequent workshop resource person and guest lecturer on clinical pharmacy and regulatory topics. With strong skills in academic mentorship, interdisciplinary collaboration, and public health education, he actively bridges classroom instruction with real-world clinical practice.

Muhammad Younus graduated in 2019 from the Karachi Institute of Economics and Technology University in Karachi, Pakistan, and got a Bachelor's degree in Computer Science. The four-year undergraduate program in Computer Science formed the basis of Younus's perseverance towards achieving an in-depth knowledge in the field. After graduation, Younus worked for the past few years as a business IT analyst in the product and research development of a logistics startup in Pakistan called TPL Logistics. They aim to Digitize Logistics Operations in Pakistan from the First Mile to the Last Mile. After this, Younus graduated in 2024 from the Universitas Muhammadiyah Yogyakarta in Yogyakarta, Indonesia, and got a Master's Degree in Government Science. Younus's research interest is transforming governance and the public sector through technology. This experience has given Younus the confidence and also kindled a desire to do research. The interdisciplinary nature has convinced Younus in taking this decision as it is the ideal choice for an exciting research career. At the same time, I am confident of contributing originally to the ongoing work. By working under the guidance of distinguished Team, Younus shall certainly be able to exploit my potential to the fullest.

Index

Symbols

A

B

C

T

www.ingramcontent.com/pod-product-compliance
Lightning Source LLC
LaVergne TN
LVHW081318300926
849161LV00008B/1796